Marine Ecology

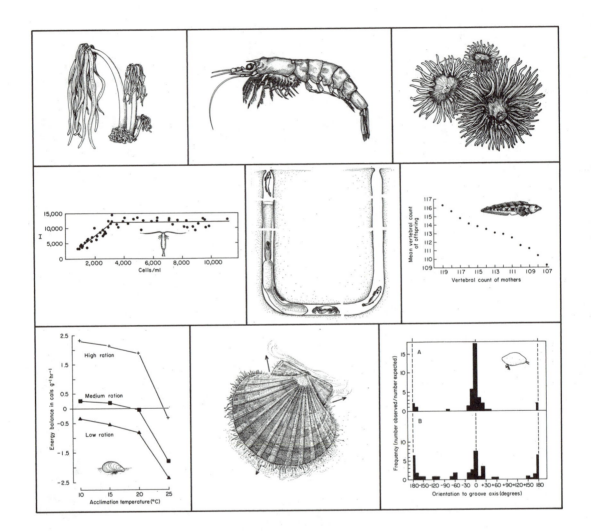

Marine Ecology

Jeffrey S. Levinton

State University of New York
at Stony Brook

PRENTICE-HALL INC., Englewood Cliffs, New Jersey 07632

Library of Congress Cataloging in Publication Data

LEVINTON, JEFFREY S.
  Marine ecology.

  Bibliography: p.
  Includes index.
  1.–Marine ecology. I.–Title.
QH541.5.S3L48   574.5′2636   81-12064
ISBN 0-13-556852-8

Printed in the United States of America

10   9   8   7   6   5

Editorial/production supervision by Ellen W. Caughey
Interior design by Ellen W. Caughey and S. Specque
Drawings by Marie Gladwish and Mitzi Eisel
Cover design by Edsal Enterprises
Manufacturing buyer: John Hall

ISBN 0-13-556852-8

Prentice-Hall International, Inc., *London*
Prentice-Hall of Australia Pty. Limited, *Sydney*
Prentice-Hall of Canada, Ltd., *Toronto*
Prentice-Hall of India Private Limited, *New Delhi*
Prentice-Hall of Japan, Inc., *Tokyo*
Prentice-Hall of Southeast Asia Pte. Ltd., *Singapore*
Whitehall Books Limited, *Wellington, New Zealand*

For Joan

contents

preface

Level and Scope

Marine Ecology aims to be a text or background reading for junior–senior undergraduate and first-year graduate courses in marine ecology and biological oceanography. The book reflects my experience in teaching a course, Marine Ecology, to 40 to 120 students over the last nine years. I complete this course in one semester. If your college or university works under the quarter system, two quarters would probably be needed.

The text combines the following: (1) Introductory material covering oceanography and adaptations of marine organisms; (2) the framework and principles needed to understand the ecological structure and dynamics of marine communities; and (3) accounts of the ecology of major marine habitats. I have attempted to combine aspects of both the plankton and benthos in one text. I feel that students will benefit from an overview of the perspectives and approaches (which are often quite different) that have emerged from research in these two areas of study. This book is unique in presenting the student with a variety of approaches.

I have also tried to give the student a strong sense of current research in marine ecology. To this end, the text is filled liberally with literature citations. In some cases oversimplication has been necessary because of the size and scope of the text. This broad scope will expose, perhaps to an embarrassing degree, some of my ignorance. I welcome edification from all who discover errors, or problems of interpretation or emphasis.

Organization

Marine Ecology is subdivided into six sections with 21 total chapters.

Part I, The Ocean and the Effects of Its Properties on Marine Organisms, introduces basic aspects of the oceans, both physical and chemical. The effects of physical (e.g., temperature) and chemical (e.g., oxygen) variables on marine organisms, and the adaptations of those organisms are then discussed. I find Part I to be an effective introduction and review. An environmental framework is provided within which community interactions can be later placed.

Part II, Some Models and Principles in Marine Ecology, introduces the principles needed to understand the ecological significance of the interactions within marine communities. The purpose is to provide a unifying framework for a diverse array of natural history and ecological processes. Demography, interspecies interactions, resources, biogeography, and some evolutionary aspects are emphasized.

Part III, Reproduction, Dispersal, and Larval Ecology, discusses constraints on reproduction, life histories and sexuality, migration, and dispersal. The question of how and why larvae disperse is also covered.

Part IV, Plankton and Productivity in the Oceans, introduces the planktonic biota and then discusses the parameters that determine the dynamics of planktonic communities. Water column stability, light, nutrients, and grazing by the zooplankton are especially emphasized. Food webs in the plankton are examined, and the techniques used to estimate primary productivity are introduced. Part IV also covers geographic patterns of primary production and shows how planktonic systems can be modeled mathematically.

Part V, Substrata and Life Habits of Benthic Organisms, introduces the aspects of sediments important to benthic organisms. Life habits, adaptations to substratum, and feeding are discussed at length. The purpose of Part V is to present the natural history of sediments and organisms so essential to an understanding of interactions of marine species within communities.

Part VI, Nearshore and Benthic Habitats, discusses in a series of chapters a variety of nearshore and benthic habitats (intertidal, estuaries, subtidal, coral reefs). The important habitat features and determinants of community structure are discussed and those problems upon which a given habitat sheds light are emphasized. No attempt is made to be encyclopedic; rather, some habitats are discussed in great detail. For example, I feel that coral reefs are the sites of future exciting research in marine ecology and have devoted two chapters to them.

General Emphasis

This text emphasizes those aspects of ecology related to the distribution and abundance of species, and to interactions of species within communities. The constraints imposed by adaptation of organisms to their environment and how these constraints influence interspecies interactions are emphasized. I try to portray the plurality of approaches taken by marine ecologists to solve problems. Theory is crucial to our understanding of marine ecology. An understanding of biological complexity, however, is equally central to a workable model of how marine communities operate. Those lovely and grotesque, graceful and clumsy, obvious and obscure marine organisms we wish to understand have their little features that often have great significance in community interactions. Do not overlook those spines, peculiar physiological features, oddball behaviors, and so on. These details may be instrumental to the knowledge of a system.

Nutrient-cycling, mass balances of elemental cycles, energy flow in ecosystems, and systems ecology are not emphasized. I have no prejudice concerning the value of these subjects. Rather, the subjects I do cover constitute a body of theory and data relevant to the distribution and abundance of organisms. I have also de-emphasized marine vertebrates. Most courses in ichthyology adequately cover these organisms.

Acknowledgments

I am very grateful to more people than I can mention. I feel especially grateful to Donald C. Rhoads and Howard L. Sanders, whose encouragement and inspiration crystallized my love for marine organisms. My outlook was perhaps even more influenced by my fellow graduate students in geology and biology at Yale University. Similarly, I have benefited greatly over the years from pleasant associations with many colleagues at the Woods Hole Oceanographic Institution, Marine Biology Laboratory of Woods Hole, Friday Harbor Laboratories, Duke University Marine Laboratory, the University of Århus (Denmark), and, of course, at the State University of New York at Stony Brook.

I thank the following individuals who have patiently reviewed all or parts of the text manuscript: P. G. Meiers, R. Wolcott, D. C. Rhoads, R. R. L. Guillard, T. H. DeWitt, and P. W. Sammarco. These individuals are not liable for my errors, but they surely must be given credit for whatever quality this text might have. My wife, Joan Miyazaki, kept me going on this project and alternately provided the necessary love, prodding, and tolerance I needed. I am grateful to Mitzi Eisel and Marie Gladwish, who did the artwork for this text. Faye Schillaci, Christine Helquist, Shirley Felicetti, and (mostly) Wanda Mocarski typed the manuscript in various stages. Finally, I am grateful to my friends and colleagues who provided information and photographs so necessary for this text. To all I am deeply grateful.

JEFFREY S. LEVINTON

I do not know what I may appear to the world; but to myself I seem to have been only like a boy playing on the seashore, and diverting myself in now and then finding a smoother pebble, or a prettier shell than ordinary, whilst the great ocean of truth lay all undiscovered before me.

Isaac Newton

Clear cascades!
Into the waves scatter
Blue pine needles.

Matsuo Basho

1 The Ocean and the Effects of its Properties on Marine Organisms

In Part I we learn of the basic configuration of the ocean basins, properties of seawater important in understanding the dynamics of oceanic climate and chemistry, and the role of climate, rotation, and seawater features in oceanic circulation. We explore processes important at the coastline. We next examine how marine organisms adapt to the physical and chemical variation in the ocean.

1 geography, physics, chemistry, and movement in the oceans

1-1 INTRODUCTION

The earth is truly a planet dominated by its oceans. Oceans cover approximately 71% of the earth's surface. About 80% of the surface of the Southern Hemisphere is covered by ocean whereas 61% of the Northern Hemisphere is oceanic. Most of the world's oceans are deep and 84% of the sea bottom lies at depths greater than 2000 m. The average depth of the oceans is 4000 m and some ocean trenches are as deep as 10,000 m. Inasmuch as our knowledge of marine organisms is, for all practical purposes, largely confined to those living at depths less than 100 m, our ignorance of most of the oceans is obvious.

The Pleistocene ice ages had dramatic effects on the marine biota. The periodic advance of continental glaciers during cold periods lowered sea level because of the transfer of some of the earth's water into ice. Fluctuations of sea level on the order of 100 m occurred (we are now in a period of relatively high sea level). Thus during the last glacial maximum 11×10^3 years ago the continental shelves of the world were mostly exposed to air. Semienclosed bays, such as Long Island Sound, were freshwater lakes or dry land. Retreats and advances of the ice also greatly changed the distribution of water masses and climatic belts in the ocean. Consequently, the recent evolutionary history of the marine biota has been in a framework of fluctuating planetary climatic change. This pattern is not peculiar to the Pleistocene but extends back through the history of the oceans and marine biotas. Glacial periods are known for the Ordovician and Permian; other climatic fluctuations have been documented as well. These patterns must have influenced the evolution of the marine biota.

In this chapter we introduce the ocean. We discuss its geography, topography, and water movements. Properties of seawater are described, with some consideration of the factors controlling these properties. We explain the effects of physical and chemical properties of seawater on marine organisms in Chapter 2.

1-2 GEOGRAPHY, TOPOGRAPHY, AND STRUCTURE

The major ocean areas are the Southern (Antarctic) Ocean, the Atlantic Ocean, the Pacific Ocean, the Indian Ocean, and the Arctic Ocean (Fig. 1-1). Smaller bodies of water known as seas are nearly enclosed by land or island chains and have local distinct oceanographic characteristics. Examples are the Gulf of Mexico, the Mediterranean Sea, the Baltic Sea, and the Japan Sea.

The Southern Ocean (Antarctic Ocean) is unique in having its major boundary with other oceans and in having a continuous water connection along lines of latitude around the earth. The northern surface boundary of the ocean is the subtropical convergence, where colder and more saline water descends northward below the surface. A general pattern of west winds in the 40 to 60° range of south latitude generates a surface current circling Antarctica (Fig. 1-1).

The Pacific is the largest ocean and is relatively little affected by the landmasses surrounding it. Island chains are most numerous in the Pacific and volcanic activity around the margins is pronounced. In contrast, the Atlantic is relatively narrow and is bordered by large marginal seas (the Gulf of Mexico, Mediterranean Sea, Baltic Sea, and North Sea). Its average depth is less than the Pacific Ocean. Furthermore, many of the great rivers of the world (the Mississippi River, Amazon River, Nile River, and Congo River) drain into the Atlantic system. The surface area of the Atlantic is only 1.6 times the surface of the land areas drained by the rivers flowing into it. The Indian Ocean is intermediate between the Atlantic and Pacific in depth and drainage.

Marginal seas often have unique oceanographic characteristics because of restricted circulation. Reduced mixing with the ocean permits local river drainage or the precipitation–evaporation balance to affect seawater properties. A shallow-water barrier, or sill, for instance, restricts circulation between the Atlantic and Mediterranean at the Strait of Gibraltar. Because of an excess of evaporation over precipitation and heat input, the Mediterranean is more saline and warmer than the adjacent Atlantic Ocean. The Baltic Sea, in contrast, has an excess of precipitation and river runoff over evaporation and averages only 100 m in depth. The salinity is low as a result. In other places (e.g., Black Sea, many Norwegian fjords), restricted circulation and consumption of oxygen by organisms result in deep waters devoid of oxygen.

The oceans share three main topographic features: (a) *continental margins,* composed of a shallow *continental shelf* and deepening *slope* complex, (b) *ocean basin floors,* and (c) *oceanic ridge systems.* The continental margin (Fig. 1-2) consists of the *shelf,* a low-sloping (1:500—about 0.1°) platform adjoining the continent and extending from a few miles to over a hundred miles from the shoreline. At the *shelf–slope break* (usually a depth of 100 to 200 m) the change in depth with distance from shore increases to 1:20 (about 2.9°), forming the *slope.* The slope is usually dissected by submarine canyons that act as channels for downslope transport of bottom material. Rapidly moving turbidity

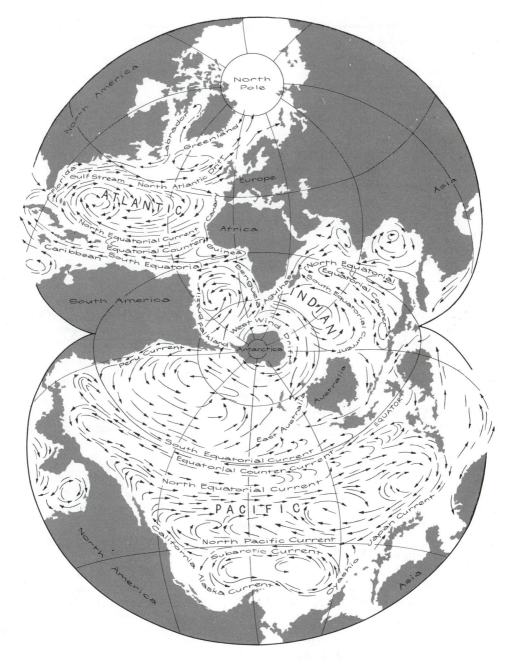

Figure 1-1 Surface currents of the world's oceans. (From "The Circulation of the Oceans" by Walter Munk, Copyright © 1955 by Scientific American, Inc. All rights reserved.)

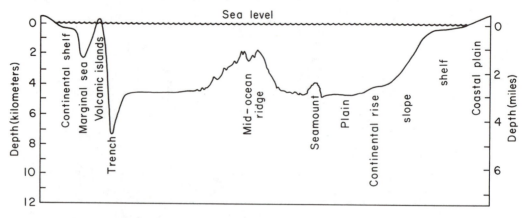

Figure 1-2   Two types of oceanic margins. From the right: shelf-slope rise. From the left: shelf-marginal sea-volcanic islands-slope-trench.

currents carry dense slurries of material. The slope descends to the deep-sea floor, but the topographic pattern at depth varies regionally. Two major sequences (Fig. 1-2) exist: (a) shelf–slope–rise and (b) shelf–slope–(marginal sea–volcanic islands)–trench. In the first case (e.g., the eastern North Atlantic), the foot of the slope merges with a deeper and gentler depositional feature known as a *rise,* which descends to the almost level *abyssal plain,* at an average depth of 4000 m. In the second case (e.g., the Aleutian Islands, Peruvian coast), the slope descends to a very deep (10,000 m) and narrow *trench,* parallel to shore. In the western Pacific the slope typically descends to a basin of about 2000-m depth, seaward of which is a chain of volcanic islands—island arcs—bordered by a still farther seaward trench. Seaward of these trenches is the abyssal plain, or ocean basin floor.

The *oceanic ridge systems* (Fig. 1-2) are a series of topographically high, linear ridges rising 2000 to 4000 m above the ocean basin floor. Usually they are submarine but often have emergent islands (e.g., Iceland, on the mid-Atlantic ridge). Ridges are volcanic in origin and are cut by transverse faults. Rift valleys, at the ridge system center, are parallel to the map trend of the ridge.

The deep ocean-basin floor consists of soft sediments derived from deposition of the following:

1. Mineralized skeletons of planktonic (freely floating) organisms, such as foraminifera, coccolithophorids, radiolarians, and diatoms derived principally from the upper 200 m of the water column.

2. Clay and other minerals that settle out of the water column to great depths.

3. Volcanic products consisting of basaltic rock and volcanic glass in various states, derived from oceanic volcanic islands.

4. Certain deposits actively forming through precipitation on the seabed, such as deposits of nodules of hydrous manganese and iron oxides. Shelf sediments derive

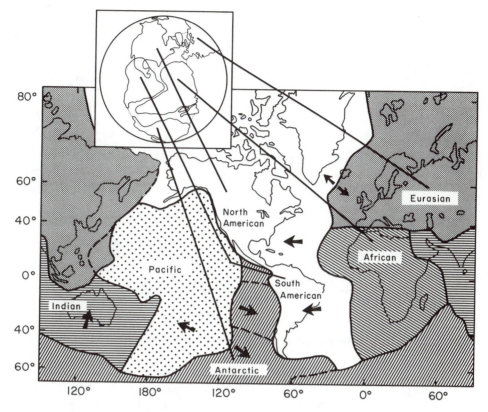

Figure 1-3 Sea-floor structure, showing plates, ridge systems, and directions of crustal movement. Inset shows arrangement of continents 225 million years before present.

from land erosion and from shelf sediments eroded subaerially during lower stands of sea level.

Beneath the soft-sediment mantle of the ocean basin floor is a crust made of (a) sediments in varying degrees of consolidation into sedimentary rock and (b) an underlying dense layer of volcanic rocks beneath the sedimentary cover. Several lines of evidence demonstrate that the ocean basin floors are moving horizontally (Fig. 1-3) a few centimeters per year. This phenomenon is known as sea-floor spreading. New crust is formed through volcanic activity at the oceanic ridges, carried as on a conveyor belt away from the ridges and consumed at trenches usually near ocean margins (Fig. 1-4). At the sites of crust destruction, crustal material is dragged downward and melted into the upper mantle below. The great depth of the trenches is a topographic reflection of the downward dragging of crust. The mechanism behind sea-floor spreading is probably a convective process involving a deep layer of the earth, the mantle.

The earth's crust consists of several major plates whose boundaries are trenches,

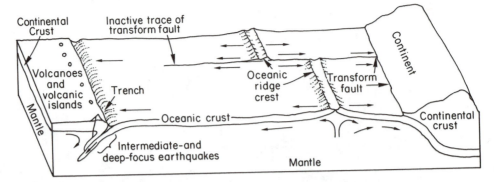

Figure 1-4   Schematic diagram of the oceanic crust, showing the formation of crust at ridges and downward transport-destruction at trenches.

ridges, or major breaks in the crust. Continents are enmeshed in plates and are conveyed along, via sea-floor spreading, in the plate in which they reside. So continents have drifted large distances over geologic time (Fig. 1-3).

Sea-floor spreading explains several well-known phenomena.

1. Volcanic rocks increase in age with distance to either side of a ridge system.
2. Oceanic sediments older than Cretaceous are relatively rare because they have been consumed.
3. Continents exhibit a jigsaw puzzle fit as if one large block had been fragmented (e.g., Africa and South America). This fit includes rock outcrop patterns and fossil distributions, as well as a general coastline match (see inset of Fig. 1-3).

The biological significance of these movements is based on these facts: (a) changes in the arrangement of the continents reorient current systems and water temperature regimes and (b) continental movement and changes in volcanic island systems probably influence biogeographic processes, such as dispersal and isolation of populations to specific ocean basins.

1-3 PHYSICAL AND CHEMICAL PROPERTIES OF SEAWATER

Water

Water molecules are asymmetric in charge distribution. Hydrogen bonds form between molecules, allowing a liquid state at atmospheric pressure and earth surface temperatures. Charge asymmetry allows water molecules to combine with other charged ions. Thus water is a good solvent for salts and biologically important molecules in cells.

Water has a high *specific heat;* that is, it takes 1 calorie to raise 1 gram of pure water by 1°C (at 15°C). Seawater of 35‰ at 17.5°C has a specific heat of 0.9. Thus it takes a great deal of heat to change the temperature of the ocean and the ocean can store and transfer a great deal of heat.

Salinity

Salinity (S) is the number of grams of dissolved salts in 1000 g of seawater (after all bromine has been replaced by chlorine, all carbonate converted to oxide, and all organic matter destroyed). It is expressed in *parts per thousand* (‰ or ppt) and ranges from 33 to 38 ‰ in the open ocean.

Because of the constancy of the *ratios* (see later) of major components, *chloride* has been used as an index of salinity. *Chlorinity* (Cl) is the number of grams of chloride ions in 1000 g of seawater. Chlorinity is measured by titration with silver nitrate (Knudsen method) and related to salinity through the following experimentally determined relationship:

$$S\ (‰) = 1.80655 \times Cl\ (‰)$$

Seawater is an electrolyte and can therefore conduct an electric current. Seawater conductivity can thus be used as an index of salinity. At a given temperature electrical conductivity increases with salinity. Conductivity techniques of salinity estimation are far more accurate ($\pm 0.003\%$ for 32 to 38 ‰) than chlorinity measurements ($\pm 0.02\%$).

The refractive index of water is directly related to salinity. Modern refractometers are temperature compensated and provide a convenient alternative method for measuring salinity.

In the open ocean salinity is controlled by evaporation and sea-ice formation, which increase salinity, and by dilution processes, such as rainfall and river runoff. Latitudinal variation in precipitation and evaporation leads to two salinity maxima (Fig. 1-5) (ca. 35.5 ‰) at latitudes 30 N and 30 S. Slight excesses of precipitation relative to evaporation result in salinity minima at the equator (ca. 34.5 ‰) and at latitudes higher than 40 (ca. 34.0 to 34.5 ‰). Along coastlines more dramatic variation in salinity is conceivable because of the influence of river input.

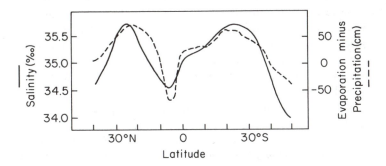

Figure 1-5 Latitudinal variation in surface salinity of the open oceans. Balance of evaporation and precipitation also shown. (After Sverdrup et al., 1942)

TABLE 1-1 MAJOR SEAWATER CONSTITUENTS

Constituent	Concentration, g/kg of 35 ‰ seawater	Constituent	Concentration, g/kg of 35 ‰ seawater
Cl^-	19.353	K^+	0.387
Na^+	10.76	HCO_3^-	0.142
SO_4^{2-}	2.712	Br^-	0.067
Mg^{-2}	1.294	Sr^{+2}	0.008
Ca^{+2}	0.413		

Salinity can increase in bodies of water when restricted circulation is accompanied by an excess of evaporation. Such conditions exist, for example, on the Bahama Bank west of Andros Island in the Bahamas, where the salinity is usually 40 ‰. Strong salinity changes also occur in tidal pools, where rainfall or evaporation can change seawater from 0 ‰ to concentrations at which various salts precipitate from solution.

Seawater is a complex solution containing nearly all known elements. The typical concentration of the main components of seawater is shown in Table 1-1. Chloride and sodium ions dominate. The relative abundances of all the components listed in Table 1-1 are nearly constant over almost all the world oceans despite overall changes in total concentration of solids. The close to constant ratios of concentrations of major ions is known as *Marcet's principle* or the *Forchhammer principle*. Such constancy is related to the residence time of an element in seawater or the mean length of time that a unit mass of a given constituent remains in the ocean. The respective residence times of most of the major ions in seawater are far greater than the time required to mix them evenly throughout the ocean.

Temperature

Temperature is a major factor regulating the distribution and abundance of marine organisms. The latitudinal thermal gradient is accompanied by major biogeographic changes in pelagic and bottom assemblages of organisms. At the lower extreme the freezing of seawater results in the formation of ice crystals that disrupt cells and terminate metabolic activity. At lethally high temperature physiological integration is impaired and enzymes are inactivated. Cytoplasm properties are altered and behavior is severely affected. Most marine organisms do not regulate their own body temperature (are poikilothermic). Temperature variation within lethal extremes thus has great effects on biochemical reactions and metabolism (discussed later).

Seawater has a much narrower range of temperature than air. Whereas air temperatures can range as low as $-68.5°C$ (Siberia) to as high as $+58°C$ (Libya), seawater ranges between $-1.9°C$ (freezing point) to $40°C$. The upper limit is relatively unusual and only obtained in water of a meter or less, as in sand flats of tropical Jamaica (Jackson, 1973). Open ocean seawater ranges from $-1.9°C$ to about $27°C$ (Fig. 1-6).

The most significant factor affecting ocean water temperature is the latitudinal gradient of insolation or influx of solar energy. At low latitudes there is a net capture of

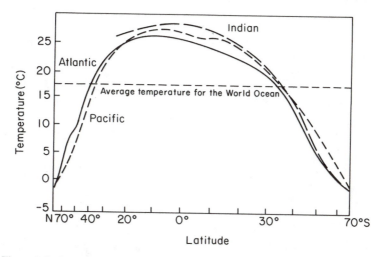

Figure 1-6   Latitudinal variation in sea-surface temperatures in the Atlantic, Pacific, and Indian oceans. (After Anikouchine and Sternberg, 1973)

heat from solar energy, but at high latitudes the earth loses heat. Other less important sources include (a) geothermal heating, (b) transformation of kinetic energy into heat (from water mass interactions), and (c) atmospheric processes, such as water vapor condensation and convection of heat from the atmosphere. Heat is lost to the atmosphere by (a) back radiation from the sea surface, (b) convection of heat to the atmosphere, and (c) evaporation. Most of the solar heat energy intercepted by the ocean is absorbed in very shallow water. So deep water is only about 2 to 4°C, even in the tropics. Latitudinal gradients in temperature are thus pronounced only in surface waters.

Surface water temperatures at the equator are seasonally stable as well as warm (Fig. 1-7). Seasonal fluctuations in the open ocean reach a maximum at intermediate latitudes (ca. 40 N and 40 S). At higher latitudes the temperature regime is again stable,

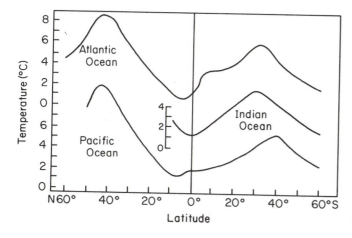

Figure 1-7   Annual range of sea-surface temperatures as a function of latitude. (After Sverdrup et al., 1942)

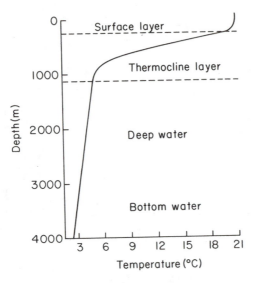

Figure 1-8   Temperature profile with depth in the open tropical ocean.

although cold. Seasonal temperature stability varies regionally as well; coasts whose climate is dominated by weather systems emanating from the continental interior (e.g., coast of New York) are less stable than coasts dominated by oceanic weather systems (e.g., West Coast of United States).

Because of solar heating, a stratified profile of temperature usually develops as in Fig. 1-8. The *surface layer* is warm. Due to wind-induced mixing, temperature of the surface layer is uniform with depth (isothermal). Another virtually isothermal *deep layer* is cooler. An intermediate *thermocline* is the depth range where temperature decreases from the surface layer to the deep layer. This overall vertical structure is generated by small eddy currents that transfer heat from the surface to deep water. The scale of this general structure varies with location. In the open tropical ocean the surface mixed layer is about 100 m deep, the thermocline extending to 1500 m. But this general vertical structure exists in small and relatively shallow bodies of water, such as Long Island Sound, New York (less than 50 m in depth), in spring and summer. Such vertical thermal structure induces a density gradient with cool dense water below and warm, less dense water above.

Density

Although density is usually expressed in grams per cubic centimeter (values ranging from 1.02400 to 1.03000 g cm^{-3} in the open ocean), the following convention is more convenient for oceanographically meaningful values:

$$\sigma_{S,T,P} = (\text{density} - 1) \times 10^3$$

Thus water with a density of 1.02400 would have a $\sigma_{S,T,P}$ of 24.0. This parameter, expressing density at a given salinity (S), temperature (T), and pressure (P) is usually abbreviated to σ_T ("sigma-tee"). Unless otherwise specified, σ_T refers to the density of water of a given salinity and temperature at atmospheric pressure (pressure increases density significantly).

Density increases with increasing salinity and decreasing temperature. At higher temperatures changes in temperature have more pronounced effects on density than at lower temperatures. The combined latitudinal variation of salinity and temperature results in a minimum surface density near the equator, with an increase northward and southward.

The depth zones of rapid temperature (thermocline) and salinity (halocline) change must also be regions of density change *(pycnocline)*. When water of low density resides above water of greater density, the water column is stable. In an estuary water of low salinity will float on top of denser, higher salinity water. In the presence of a thermocline, warmer and less dense water lies above colder and denser water. The stability imparted by this arrangement decreases the likelihood that wind can vertically mix the water column. Cold and saline surface waters, however, can be produced during sea-ice formation, as in the Arctic Sea and Antarctic Ocean. In this case, dense cold and saline water may sit on top of less dense water of lower salinity. This situation will be unstable, with sinking of the denser water.

Ice Formation

The formation of ice requires an orderly internal structure of water molecules. Ice crystals, however, are less dense (0.92 g/cm^3) than liquid distilled water (1.0 g/cm^3). Thus the formation of ice is accompanied by its flotation on liquid water. Although ice structure is loose, salts fit poorly in the crystalline matrix and are generally excluded during ice formation. Consequently, the formation of sea ice may increase the local seawater salinity. This situation occurs in the Weddell Sea in the Antarctic Ocean.

The presence of salts alters ice formation in two ways. First, the freezing point of seawater is lowered by increasing salinity. Secondly, water of salinity greater than about 15 ‰ will steadily increase in density as the temperature decreases toward the freezing temperature ($-1.9°C$). This situation is in contrast to freshwater, which has a density maximum at 4°C, with decreasing density as the temperature decreases toward the freezing point (0°C). Therefore seawater near the freezing point does not ascend in the water column due to density decrease, as occurs in freshwater lakes in winter.

pH of the Ocean

The acidity or pH of water is defined as the negative $\log_{10}$ of the activity of hydrogen ions. Because the activity of hydrogen ions in distilled water is 10^{-7}, the pH is 7. More acid waters have a lower pH whereas more basic waters have a pH greater than 7. Ocean water usually has a pH of about 8, making it a mildly basic solution. The pH of the ocean can be explained through chemical reactions involving dissolved carbon dioxide, carbonate and a few other weakly ionized chemical species, and crystalline calcium carbonate. Carbon dioxide is a source of carbon in terrestrial photosynthetic production of carbohydrates and is also the product of respiration. Table 1-2 shows the reactions in

TABLE 1-2 CARBONATE EQUILIBRIA

CO_2(dissolved) +	H_2O	$\rightleftharpoons$	H_2CO_3	
	water		carbonic acid	
	H_2CO_3	$\rightleftharpoons$	H^+	+ HCO_3^-
			hydrogen ion	bicarbonate ion
	HCO_3^-	$\rightleftharpoons$	H^+	+ CO_3^{-2}
				carbonate ion
Ca^{+2} +	CO_3^{-2}	$\rightleftharpoons$	$CaCO_3$(crystalline)	
calcium ion			calcium carbonate	

seawater. Reactions among these constituents result in alterations of the hydrogen ion concentration and hence the pH. If an equilibrium is being maintained between solution and precipitation of calcium carbonate, intermediate ions, and the pressure of carbon dioxide in the atmosphere, then the pH should be about 8—a close approximation to actual measurements. In cases where dissolved carbon dioxide is supplied in great quantities, as in the oxygen minimum zone, the pH may decrease to 7.5. In contrast, high local rates of photosynthesis in shallow water consume carbon dioxide and can raise the pH to 9.

1-4 CIRCULATION IN THE OCEAN

Coriolis Effect

All patterns of global water movement are affected by the earth's rotation. Note that the radius of a circular slice of the earth (perpendicular to the axis of rotation) at a given latitude decreases with increasing latitude. Because the earth rotates once a day on its axis, a particle at rest on the earth's surface at the equator must travel eastward more rapidly than a particle near the pole. A point on the equator has an eastward velocity of ca. 1700 km/hr while points at 30°N or 30°S have a velocity of ca. 1500 km/hr and points at 60° move at ca. 800 km/hr.

Consider a particle of water not attached to the earth, moving north from latitude 30°N. When it is at 30°N, it is moving eastward at a velocity of about 1500 km/hr. But as it moves northward, the earth beneath it is moving eastward at a slower velocity. Thus the particle will deflect toward the east, relative to the earth's surface, because of its initial higher eastward velocity. This process is called the *Coriolis effect* and is proportional to the sine of latitude. It is therefore zero in value at the equator and increases with increasing latitude toward the pole. It causes a deflection to the right for water traveling (in any direction) in the Northern Hemisphere and a deflection to the left for water traveling in the Southern Hemisphere (Fig. 1-9).

The Coriolis effect similarly deflects water moving under the force of the wind. Wind sets water in motion by dragging along the water surface, which is hydrodynamically rough because of ripples and other surface irregularities. Wind then causes water to move in sheets, which drag on layers of water below. The effect of the wind can thus be it is deflected to the right until water flows in balance between gravity and the Coriolis effect. Such flow, known as *geostrophic flow*, is obtained imperfectly in the ocean.

Wind-Driven Surface Circulation

The latitudinal gradient in solar energy capture on the rotating earth drives the planetary wind system. At about 40°N and 40°S the *prevailing westerlies* help move surface water to the east while the *trade winds* move toward the west on either side of the equator.

Figure 1-10 shows the generalized surface circulation in the oceans and planetary wind patterns. The principal feature common to all ocean basins are gyres or large circular current patterns. The Atlantic and Pacific oceans have major gyres centered around the subtropical latitudes of 30°N and 30°S in each hemisphere. Because of the earth's rotation, gyres are displaced toward the west sides of oceans. Boundary currents on the west sides of oceans, such as the Gulf Stream, are stronger and narrower than diffuse eastern boundary currents. The warm north-flowing Gulf Stream transfers heat toward higher latitudes. Originating in the Gulf of Mexico, warm water travels northward along the coast of North America and then eastward toward the British Isles. Coastal waters of Ireland and the United Kingdom are warmed as a result.

Because surface water is transported away from the tropical western coasts of continents by wind, water rises from the bottom to replace it (Fig. 1-11). Such a process is known as *upwelling* and results in the transport of large amounts of nutrients from the bottom and great biological productivity in the surface waters.

Other conspicuous features of oceanic surface circulation are the east and west equatorial currents and the continuous water movement around Antarctica toward the east—the west wind drift.

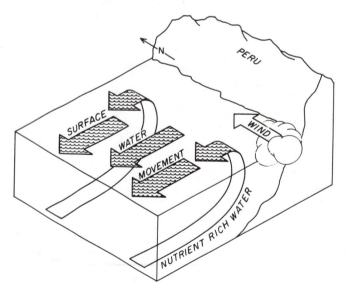

Figure 1-11 Vertical transport (upwelling) of coastal water as induced by winds combined with the Coriolis effect.

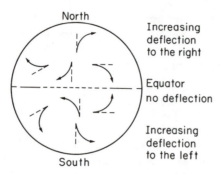

North

Increasing
deflection
to the right

Equator
no deflection

Increasing
deflection
to the left

South

Figure 1-9   The Coriolis effect deflects
moving objects, as shown by this diagram.
Note that no deflection occurs at the
equator.

transmitted to depths of 100 m, forming the surface layer discussed earlier. But the
Coriolis effect deflects water to the right of the wind in the Northern Hemisphere (theoretically at an angle of 45°). Each layer of water is also deflected to the right of the
direction of the immediately higher layer pulling on it. This results in a progressive
deflection of a surface current produced by wind, with increasing depth, known as an
Ekman spiral. Combining the movements of all layers usually results in a net 90° deflection
of the surface layer from the direction of the wind (again, to the right in the Northern
Hemisphere and to the left in the Southern Hemisphere, characterized as *Ekman circulation*). This wind-induced surface transport is crucial to observed patterns of surface
circulation in the ocean.

Because the surface waters are deflected approximately 90° to the right of the wind,
less dense surface water tends to be pushed by the prevailing wind system in toward the
center of clockwise (Northern Hemisphere) or counterclockwise (Southern Hemisphere)
current systems known as *gyres* (Fig. 1-10). The deflection creates a higher sea surface
at the center of, for example, the subtropical North Atlantic. This difference in elevation
(less than 2 m) creates a gravitational pull. But as water flows down the inclined surface,

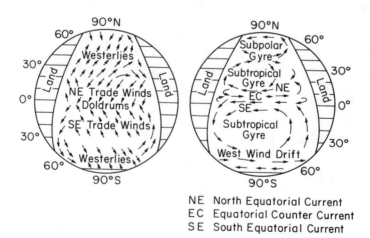

NE   North Equatorial Current
EC   Equatorial Counter Current
SE   South Equatorial Current

Figure 1-10   The relationship of the surface currents of the ocean to the planetary wind
pattern. (After Fleming, 1957, courtesy The Geological Society of America)

Thermohaline Circulation in the Deep Ocean

Although strong currents may exist locally in the deep ocean, deep oceanic circulation is dominated by movement of large *water masses* whose unique temperature and salinity characteristics are acquired during their origin at the surface at high latitudes. Figure 1-12 shows the origin and fate of major water masses of the Atlantic Ocean. The *Antarctic bottom water* originates at the surface in the Weddell Sea in Antarctica. Formation of nearly salt-free sea ice leaves a cold, saline residue water that sinks to great depths, mixing with other water along the way. This water moves northward along the sea bottom and can be traced to the Northern Hemisphere in the Atlantic. It also moves into the Pacific and Indian oceans. The *Antarctic intermediate water* forms at the surface near Antarctica and descends toward the north. It is, however, less dense than the North Atlantic deep water (NADW). NADW originates at about 60°N latitude, sinks and travels southward, overlying the Antarctic bottom water.

A water mass can be traced by its characteristic temperature and salinity acquired at the sea surface. Mixing between any two water masses may be traced as waters whose temperature–salinity characteristics are intermediate between two known water masses. Oxygen, which is consumed by organisms as the water moves, can also be used as a tracer to identify water masses. Finally, decay of radioactive carbon fixed by photosynthesis at the surface can also be used to measure rates of thermohaline convection. For example, C-14 dating has shown that it takes less than 600 years for the Pacific deep water to travel from 60 S to 30 N.

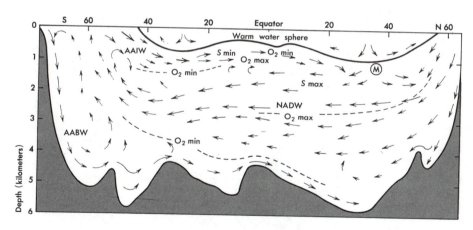

Figure 1-12 Thermohaline deep circulation of the Atlantic and Antarctic oceans, showing water masses identified by temperature and salinity. AABW = Antarctic Bottom Water; AAIW = Antarctic Intermediate Water; NADW = North Atlantic Deep Water. (From Gerhard Neumann, Willard J. Pierson, Jr., *Principles of Physical Oceanography*, © 1966, p. 466. Reprinted by permission of Prentice Hall, Inc.)

1-5 COASTAL PROCESSES

Waves

Ocean waves are generated by wind and deliver kinetic energy to the shoreline. Wave strength depends on the speed and duration of the wind and the distance of sea surface over which it moves. A series of waves passes a fixed point; peaks alternate with troughs. Figure 1-13 shows dimensions of waves. The time taken by two successive crests or troughs to pass a fixed point is the wave period T. L is the wavelength and H is the wave height. Wave speed V is calculated as

$$V = \frac{L}{T}$$

As a wave passes, water particles are set into vertical orbital motion, the diameter of the orbits diminishing to a zero radius deeper than $L/2$ (Fig. 1-13). Water movement in the direction of the wave is negligible relative to the lateral velocity of wave crests and troughs. As the wave approaches shore, vertical orbital movements of water particles strike the bottom when the depth is less than $L/2$. Here the wavelength (L) is shortened and the height (H) increases. Wave steepness (H/L) increases to a maximum of about 1/7 before the destabilization of the wave. At this point the wave "breaks" and expends energy through the formation of small sets of waves and turbulent water that wash on

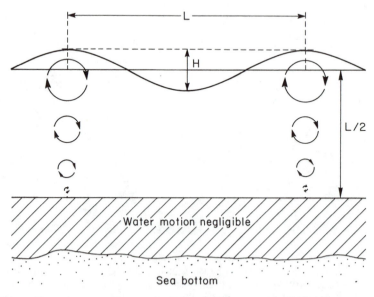

Figure 1-13 Dimensions of ocean waves. (After Shepard, 1963. From *Submarine Geology*, 3rd Edition by Francis P. Shepard. Copyright © 1948, 1963, 1973 by Francis P. Shepard. Reprinted by permission of Harper & Row, Publishers, Inc.)

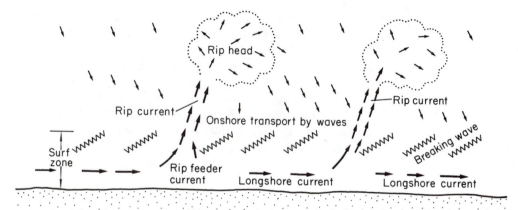

Figure 1-14 Water transport adjacent to an exposed beach. (After Shepard, 1963. From *Submarine Geology,* 2nd Edition by Francis P. Shepard. Copyright © 1949, 1963 by Francis P. Shepard. Reprinted by permission of Harper & Row, Publishers, Inc.)

the beach. This turbulence has enormous force. Organisms living on wave-swept coasts are sturdy and must have strong holdfasts.

As water approaches the shore, some of it is moved parallel to shore in the form of *longshore currents;* some moves rapidly offshore in concentrated *rip currents* (Fig. 1-14). Longshore currents are responsible for the transport of sediment parallel to the shore. The combined processes of breaking waves, wash on the beach, and longshore currents make sandy beach sediments unstable and an extremely rigorous environment for bottom-dwelling organisms.

As a wavefront approaches the shore, horizontal water velocity is less in shallower water relative to deeper water. Thus wavefronts refract as they approach irregular shores

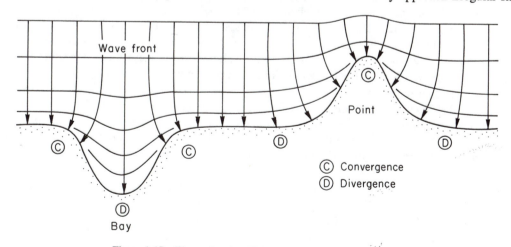

Figure 1-15 Wave refraction. Note that wave action is concentrated at headlands. (After Shepard, 1963)

(Fig. 1-15). Waves tend to concentrate on headlands that extend offshore as shallow protrusions, thereby hastening the smoothing of an irregular coastline. In contrast, when a wavefront enters a bay, the part of the front reaching the deep part of the bay travels faster than at either side of the bay. Thus wave energy diverges at depressions in the shoreline.

Although relatively rarer than typical wave action, large-scale storms, such as cyclones, hurricanes, and even high-force winds, can severely disrupt marine bottoms. Barrier islands are often breached in severe storms. Hurricanes can erode sediment in shallow water and topple large reef-building corals.

Tides

Although tidal motion is measurable in all parts of the ocean, to marine biologists it is only of great interest near the shore. Tidal rise and fall creates gradients of exposure to air and causes *tidal currents* that alter sediment distribution and local hydrodynamic forces.

Tidal movement is the result of the gravitational interaction of the earth, moon, and sun. The moon's proximity to the earth makes it the primary mass responsible for tidal motion. Gravitational attraction (force) between two bodies is proportional to the product of the masses of the two bodies, divided by the square of the distance between them. The attractive force between the earth and moon is balanced by a centrifugal force caused by the rotation of the two bodies about the earth–moon center of mass. Although in balance at the center of the earth, these forces are out of balance on the earth's surface to the extent that there is a net attraction to the moon on the side of the earth facing the moon and a net excess of centrifugal force on the side facing away from the moon (Fig. 1-16). Ideally this should result in a bulge of water (high tide) toward the moon and away from the moon (high tide) at any one time. Corresponding depressions (low tide) will exist on those parts of the earth where there is no net excess of gravitational pull relative to centrifugal force. Because the moon "passes over" any point on the earth's surface every 24 hours and 50 m, or one tidal day, ideally there should be two low and two high tides per day. Because the moon's position relative to the earth's equator shifts from 28.5°N to 28.5°S, the relative heights of high and low water differ geographically from gravitational forces alone.

During times when the sun, earth, and moon are in line (Fig. 1-16) the gravitational force exerted by the sun amplifies that of the moon. *Spring tides* occur at this time of maximal tidal range. When the sun, earth, and moon form a right angle, the two bodies cancel each other and *neap tides* occur, with minimal vertical range. The period between spring tides is 2 weeks.

The earth is not a sphere uniformly covered with water but a complex of oceans and smaller bodies of water interspersed by landmasses. The spatial arrangement of land and sea and the depth and size of basins affect the timing of tides and tidal heights. Moreover, the Coriolis effect also has a great influence on tidal heights and currents. Tidal currents, when affected by the earth's rotation, produce rotating systems around

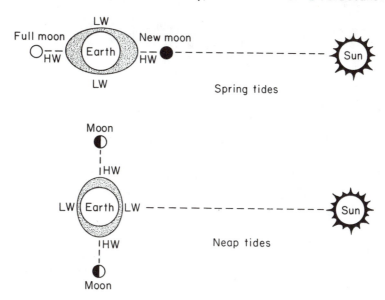

Figure 1-16   Action of tidal forces at different alignments of the sun and moon.

amphidromic points where no change in tidal height occurs over time. The North Sea has three such points.

The size of a given marine basin, relative to the time that a wave can travel from one end of the basin to the other, determines the response of the water in the basin to tide-generating forces. If travel time is short relative to the period of the tide-generating force, then the timing of the tide will be in equilibrium with that expected from an uninterrupted sphere of water. But if travel time is long relative to the tidal period, then the timing of the tide will be out of step with that expected for a uniform water-covered surface. Times of high tide may be quite different or even reversed from expectations based on an equilibrium prediction. One phenomenon commonly observed is the great difference in tidal timing of enclosed basins, such as bays, relative to the adjacent open ocean coast.

The various influences noted earlier, however, tend to cause deviations from the ideal *semidiurnal* tidal pattern. Nevertheless, an even pattern is approximately achieved in such areas as the East Coast of the United States and Southeast Asia. Yet in other parts of the world, such as the northeastern Pacific, the size of the basin seems to respond to a diurnal (solar) component of tidal action and there is only one pronounced low tide per day. In these systems, we speak of lower low water and higher low water. The condition of inequality of tidal highs and lows, where the semidiurnal component interacts with a diurnal component, is known as a *mixed tide*. Figure 1-17 shows some tidal patterns over several days under different tidal regimes.

Because of the various interference effects of landmasses, latitudinal effects, and basin size, some areas have almost no vertical tidal movement. The east coast of the

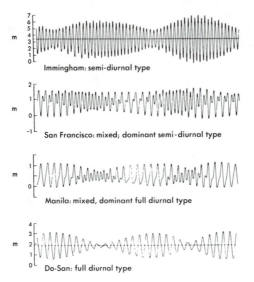

Figure 1-17 Tides from different locations of the world. (From Defant, 1960)

Jutland Peninsula of Denmark, for instance, has hardly any vertical water movement that can be related to tides. Uncoverings of sand flats are related to wind movements that push water toward or away from shore. The Caribbean coast of Panama also has a small tidal range. Consequently, when offshore winds uncover hard surface marine flats for days, mass mortality results from the heat and desiccation of the midday sun (Glynn, 1973b). So although regular tidal motion results in uncovering of marine organisms with concomitant exposure, it at least guarantees immersion some time in the day!

The rise and fall of the tide is accompanied by horizontal movements known as *tidal currents*. Near the coast, a current during the flood, or rise, of the tide is followed by a reverse current generated by the ebb, or fall, of the tide. Such currents can be strong and can transport pelagic larval stages, plankton, and oxygen throughout an embayment.

The vertical range of tides is also affected by the funnel-like nature of some enclosed basins. Usually these blind channels have a much greater vertical tidal range than adjacent open coasts. Dramatic examples include the Gulf of California, the Bristol Channel, United Kingdom, and the Bay of Fundy, Nova Scotia, where tidal range exceeds 10 m.

Ocean Fronts

The substrate facing the sea may be soft sediment or outcrops of rock under active erosion by wave and current action. In quiet water, soft sediment accumulates in *sand* or *mud flats* where sediment movement is minimal and the slope of the flat changes little throughout the year. Such flats can be hundreds to thousands of meters wide, as in the northern part of the Gulf of California, the Bay of Fundy, and the north coast of France. On more

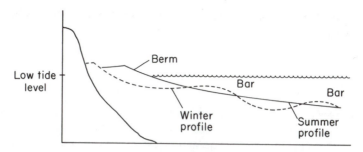

Figure 1-18   Seasonal differences in the profile of an exposed sandy shore.

exposed ocean fronts, *exposed sandy beaches* occur, where extensive wave action causes sediment transport and seasonal change of the beach profile.

Figure 1-18 shows some general features common to exposed beaches. A series of wind-blown dunes, sometimes stabilized by vegetation, lies behind the beach. A relatively horizontal shelf, the *berm,* extends to a break in slope. Descending from this break, the beach increases in slope, especially during winter storms, to the low tide mark. Seaward a complex of troughs and offshore sandbars develop. Some offshore bars become high enough to become islands. Beach complexes may protect lagoons or bodies of water landward of the beach. Longshore migration of sand is responsible for formation of barrier island complexes, such as the barrier beaches of Long Island, New York, which protect Great South Bay.

Rocky coasts develop where outcrops of rock occur in geologically youthful terrains at the sea–land boundary. The nature of such coasts depends on the local lithology (rock type) and wave action. Poorly lithified sandstones, for example, are often weathered and eroded into sand particles, leaving a sandy beach at the base of the rock outcrops (as at Santa Barbara, California). In contrast, highly cemented sedimentary rocks and crystalline rocks maintain their hard surface. Wave attack wears down cliffs at the shore, leaving a slope of debris subtidally beneath the cliff.

Estuaries

An estuary is a body of water nearly surrounded by land whose salinity is influenced by freshwater drainage. A gradient from open marine to freshwater, as in Chesapeake Bay, which extends from Virginia to Maryland, causes dramatic changes in the composition of marine biotas (see Chapter 17). Furthermore, rivers discharge large amounts of dissolved and particulate organic matter, providing nutrients that help sustain substantial estuarine fisheries, such as oysters in Chesapeake Bay.

Water movement in estuaries depends on the amount of river discharge, tidal action, and basin shape. All estuaries share a basic *estuarine flow,* where water of lower salinity moving seaward at the surface is replaced by more saline water moving up the estuary along the bottom. Usually some mixing by wind and tide occurs between the two layers. See Chapter 17 for details.

SUMMARY

1. Oceans cover most of the earth's surface. Ocean basins can be divided into continental margins, ocean basin floors, and oceanic ridge systems. Ridges are the sites of origin of volcanic oceanic rock; the ocean floor moves away from the ridge system in both directions. Such sea-floor spreading makes the earth's crust a dynamic layer; continents move along in crustal plates and change the shapes of ocean basins over millions of years.

2. Water's high specific heat, heat of vaporation, and transparency all permit large amounts of heat to be gathered, stored, and transferred between water masses and the atmosphere. Although the salt in seawater is a conservative property, increases are generated through evaporation and decreases stem from precipitation and river input. Salinity is very variable near coasts, where river input is significant.

3. Oceanic circulation is strongly influenced by the rotation of the earth. Surface circulation is driven by the planetary wind systems. Temperature and salinity variations at the surface influence the movement of water in the deep; cold and saline water moves from the surface at high latitudes into deeper waters and toward lower latitudes.

4. Ocean waves are generated by wind and deliver kinetic energy to the shoreline; orbital motion generated by waves can strongly affect coastal erosion and sediment transport. The influence of the moon and sun's gravitational attraction causes cyclic vertical tidal motion that is especially important at the coastline.

5. Estuaries are bodies of water in which a freshwater river mixes with the ocean. The lower density of freshwater, relative to seawater, generates a two-layered flow when tidal mixing is not too intense.

2 adaptations and responses of marine organisms to the physical and chemical environment

2-1 THE NATURE OF ADAPTATION AND PHYSIOLOGICAL RESPONSE TO THE ENVIRONMENT

Niche

A useful concept in which to frame the environmental responses of marine organisms is that of the fundamental niche (Hutchinson, 1957). The environment can be regarded as a multidimensional space whose component axes (dimensions) are aspects of the environment in which the species lives. It is, of course, only possible to represent pictorially three axes at once. Figure 2-1 shows an example in which two such axes are the median grain size of the sediment and salinity. A hypothetical pair of bivalve mollusk species is able to survive within the dashed ellipses. Each can survive *and* reproduce in a slightly narrower range of environmental variation, the solid ellipses, because reproductive activity is generally curtailed at the extremes of an individual's tolerance range. We here define the space (or hypervolume when more than three axes are considered) within the solid lines as the *fundamental niche*—the hypervolume within which a species can survive and reproduce. Some ecologists would prefer the dashed line as the boundary. The dashed line defines the environmental range within which an individual of a species might occur and thereby interact with a competitor or predator.

Species do not always occur within their whole fundamental niche. Sometimes a competing species requiring similar resources is present and restricts the first species to a narrower range (shaded area in Fig. 2-1) (see Chapter 4). The part of the fundamental niche that is *actually occupied* by the species is the *realized niche*. Implicit in these definitions is the point of view that a prospective range of environments, the fundamental

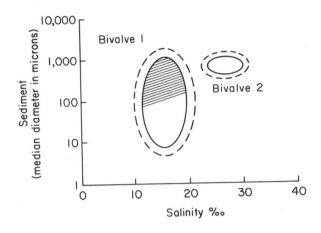

Figure 2-1 Diagrammatic representation of two axes of the niche of two bivalve species. See text for explanation.

niche, can be occupied by the species as the result of evolution and the species' present genetic makeup. In contrast, the realized niche is the narrower range occupied by the species due to presumably ephemeral events.

If a factor limiting the species' fundamental niche to a smaller realized niche were persistent enough, we would expect genetic adaptations to be lost to that range of environment to which the species has not had any success. Thus we would expect a loss of physiological adaptation to varying temperature when a species has lived in a constant environment for many generations.

Time Scales

In examining responses to environmental change, it is useful to distinguish between ecological time and evolutionary time. Ecological time is the time scale within which individuals or populations must respond to environmental change while constrained by each individual's respective genetic makeup. In contrast, evolutionary time is the time scale in which changes in the genetic structure of the species through evolution permit new adaptations to changing environments. This distinction depends on the organism concerned, its generation time, and the time scale of environmental change. If a change of the environment occurs and persists for several generations, then natural selection will favor those genotypes most adapted to the change—that is, those that leave the most offspring in the next generation. This process of population genetic change occurs in evolutionary time.

Adaptations

Any organism is subjected to a range of environmental variation during its lifetime. The ability to survive environmental change is ultimately determined by its genome or genetic composition. When an environmental change, such as a rise in temperature, occurs, the individual must first have *receptors* to sense the change. This information must then be conveyed and translated into an *adaptive response,* such as shivering in the cold or crawling into a cool, wet burrow on a hot, dry day. Sometimes one of a hierarchy of responses can occur, depending on how extreme the environmental change is. Reproductive behavior is often carried out within a much narrower range of temperature for some invertebrates than feeding behavior. Feeding behavior also occurs within a range of temperature narrower than a state of immobility that may permit survival (e.g., a snail's withdrawal into a shell to avoid heat and desiccation shock).

Some environmental changes will bring the organism into a zone of lethality, where *resistance adaptations* must be engaged to prevent death. Thus variation among species in tolerance to extreme temperatures may be thought of as variation in resistance adaptations. Other environmental variations are not so extreme and are well within the organism's range of tolerance. Adaptive responses to such changes are known as *capacity adaptations*. For example, invertebrates can regulate the salt content of their cells (discussed later) in response to a modest change of salinity.

In practice, it is sometimes difficult to distinguish between resistance and capacity adaptations. At low tide, if a mussel does not close its valves to retain moisture, it can die of desiccation. Yet this response is well within the normal range of environmental experience.

Acclimatization, Acclimation, Conformance, and Regulation

Organisms rarely have fixed tolerance limits. For instance, if a group of mussels are collected from Long Island, New York, waters in winter, total mortality soon follows immersion in seawater of 25°C. But no mortality occurs under this experimental condition if summer-collected mussels are transferred to this temperature. Changes in tolerance of organisms coinciding with seasonal environmental change is known as *acclimatization*. Acclimatization usually happens in the context of several correlated environmental changes, such as a summer increase in temperature and food, co-occurring with a decrease in salinity.

An individual collected in the field may be placed in a new laboratory environment and exposed to a shift in one experimental parameter, such as temperature. But, in response, the individual may gradually shift a function, such as rate of oxygen consumption, to a new constant value. Such a compensatory process is known as *acclimation*. With a change of laboratory conditions, an *immediate response* is followed by a *stabilization period* and a new *steady state* [Fig. 2-2(a)]. If the external environment changes, the individual may have a battery of adaptations to keep a factor like body temperature constant. Such organisms are said to be *regulators*. Organisms whose body temperature or cellular salt concentration changes in direct conformance with environmental change are *conformers*. Some species will show regulation with respect to some parameters (e.g., salinity) but conformance to others (e.g., temperature). Figure 2-2(b) shows the responses of complete regulators and complete conformers to changes in salinity.

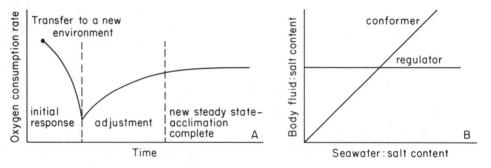

Figure 2-2   (a) Acclimation response of oxygen consumption to a change of temperature. (b) Response of a regulator and a conformer to salinity variation.

Measures of Adaptive Response

How can we systematically examine the significance of an environmental change on the individual? It is a difficult task because responses are generally a complex of genetic, biochemical, physiological, and behavioral components. For an organism that undergoes acclimation in the laboratory or acclimatization in the field, a change in the environment that is beyond that range in which acclimation or acclimatization can occur might be regarded as an *environmental stress*. For conformers, the change in probability of death with a given environmental change might be a measure of stress. The duration of the environmental change might determine whether it raises the probability of the individual's death. If the environmental change is prolonged, the probability of stress from another environmental factor is also increased. In many instances, two factors reaching extremes will act together to cause mortality when either one alone would not.

Extreme environments (a) require an adaptive response that consumes energy at the expense of other functions and (b) increase the probability of death for the individual. The first effect may be quantified by estimating the energy currently available for growth and reproduction relative to the individual's current physiological state. This factor can be ascertained from the following energy equation:

$$E_A = E_G + E_R + E_M$$

where E_A is the energy assimilated per unit time, E_G the energy devoted to growth, E_R the energy devoted to gametes, and E_M the energy required for respiration. We can define scope for growth (see Bayne et al., 1976c) or S as

$$S = E_G + E_R - E_M$$

In order to estimate it, we need to measure the feeding rate, the percentage of food assimilated, and the respiration rate. When surplus energy is available (positive S), energy may be partitioned between somatic growth and gametes. This concept is useful because it represents an integration of behavioral and physiological processes, thereby providing an index of response of the whole organism (Bayne et al., 1976b). Figure 2-3 shows the scope for growth in the mussel *Mytilus edulis* at different temperatures and ration. At

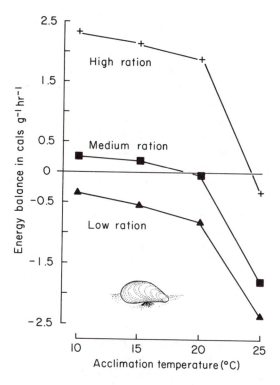

Figure 2-3 Scope for growth of the mussel, *Mytilus edulis,* at different temperatures and rations. (From Bayne et al., 1973, from *Effects of Temperature on Ectothermic Organisms* (ed. W. Wieser), pp. 181–193)

low ration, G is negative at all temperatures whereas G is also negative at 25°C for all food levels. Physiological acclimation, however, results in a stable scope for growth over a wide thermal range.

Scope for activity is the difference in energetic cost between metabolic rate at rest (E_{MR}) and metabolic rate during active motion (E_{MA}). With increasing temperature $E_{MA} - E_{MR}$ changes from positive to negative (Fig. 2-4); there is, as a result, an upper

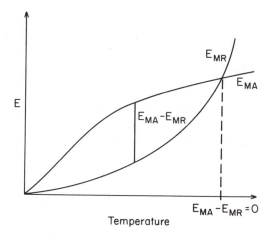

Figure 2-4 Diagram showing scope for activity as a function of temperature for a typical poikilotherm. E_{MR} represents the metabolic rate at rest. E_{MA} represents the metabolic rate when the organism is active.

temperature above which there is no excess energy for activities, such as escape from predators and foraging. Scope for activity is thus a useful measure of physiological state and a predictor of the individual's ability for immediate response in activity.

Lethality of extreme environmental conditions is more commonly assessed than such measures as scope for growth. Experimental populations are usually kept at a standard laboratory condition for a period permitting acclimation. Then lethal temperature, for example, can be determined by (a) a slow decline or rise of water temperature or (b) a rapid transfer of the laboratory-acclimated individuals to a constant, extreme temperature (Kinne, 1970, p. 414). In the latter alternative, the lethal dose required to kill 50% of the experimental population after a specified shock time (24 hours in a common period) is determined. This dose (in the case of a 24-hour shock period) is known as LD_{50} (-24 hours). In order to find the LD_{50}, it is customary to vary the extreme parameter, such as temperature. This step gives a series of points relating temperature to percentage mortality. The LD_{50} can then be interpolated (e.g., Fig. 2-6).

An index of the probability of survival under multifactorial conditions can be obtained by performing experiments under different combinations of more than one factor. For example, one can estimate percent mortality after a given time under different temperature-salinity combinations. This approach permits a more probabilistic means of defining the fundamental niche, especially if we plot the eventual reproductive output as a function of these same variables. Costlow and others (1960) have measured percent survival at different combinations of temperature and salinity for larval stages of the decapod crab, *Sesarma cinereum*. Of particular interest are the substantial changes in mortality as development proceeds. The most notable such change occurs between the larval and adult stages.

An important distinction must be made between the absolute value that a given environmental parameter attains (e.g., 30°C) and fluctuations of that parameter on a time scale that is short relative to the life span of the individual. Fluctuating environments can create rapidly changing degrees of environmental stress to which the individual is subjected. Individuals in such environments must have great physiological behavioral, or morphological plasticity if they are conformers, or an extraordinary ability to regulate.

2-2 THE EFFECTS OF TEMPERATURE

Temperature is probably the most pervasively important and best-studied environmental factor affecting marine organisms. The geographic ranges and biogeographic assemblages of species usually correspond with the latitudinal temperature gradient in the oceans. Because many of our continents have north–south trending coasts, the effect of temperature on geographic range is easily observed. Points on the coast where major latitudinal jumps in temperature occur coincide with major changes in the marine biota (e.g., Cape Cod, Massachusetts, and Point Conception, California).

Marine mammals and birds regulate their body temperature to a constant level and are known as *homeotherms*. In contrast, fishes, invertebrates, and algae do not regulate their temperature to a constant level and are known as *poikilotherms*. In most cases, poikilotherms have the same body temperature as the seawater in which they live. Some active fishes, however, such as skipjack and yellowfin tuna, have slightly higher body temperatures (Barrett and Hester, 1964). This temperature rise is probably related to metabolic heat generated by muscular (swimming) activity and has a role in body temperature regulation. Most fishes, however, are in temperature conformance with their environment.

When exposed to air, intertidal invertebrates also have conspicuously different temperatures than expected from inanimate objects. Tropical intertidal invertebrates have lower temperatures than would be predicted from the temperature acquired by a blackbody exposed to the sun. The body temperature of the gastropods *Nerita tesselata* and *Fissurella barbadensis* and the barnacle *Tetraclita squamosa* were 10, 12.2, and 12.9°C, respectively, below the temperature of an inanimate object (Lewis, 1963). Although these invertebrates were still warmer than air temperatures on a sunny day, evaporative cooling apparently caused some heat loss.

Temperature affects the rate of metabolic processes. Warmer temperatures, within limits, generally increase metabolic and behavioral activity. Oxygen consumption is a convenient expression for overall metabolic activity (Schmidt–Nielsen, 1975) and illustrates the relationship. With an increase of 10°C, the corresponding change in metabolic rate as measured by oxygen consumption is called the Q_{10}. For most poikilotherms, Q_{10} ranges from 2 to 3. Thus there is a doubling or tripling of oxygen consumption with a 10°C rise in temperature. At higher temperatures an increase in oxygen consumption ultimately requires a larger calorie source or food supply to maintain the same scope for growth.

Q_{10} usually decreases as we approach the upper lethal limit in temperature. This result may be due to the depressing effects of high temperature on physiological integration (see later). Cirral activity in barnacles increases with increasing temperature. But Fig. 2-5 shows depressed activity for two species of barnacles near an upper thermal limit. A typically upper intertidal species shows greater thermal tolerance than a typically lower intertidal species (*Chthamalus stellatus* and *Balanus balanoides,* respectively).

The relationship of temperature to metabolic rate causes conspicuous physiological problems for marine species that live in thermally seasonal environments. Cold winter temperatures can depress activity of poikilotherms with no capacity for acclimation. In contrast, summer temperatures can increase oxygen consumption to an extent that metabolic demands exceed energy reserves. Acclimatization to seasonal temperature change therefore serves to maintain activity and to strike a favorable energy balance. With seasonal—that is, relatively slow—changes in temperature, many marine invertebrates acclimatize and adjust the metabolism–temperature relationship to new conditions. This situation can be seen by acclimating individuals to a range of temperatures and then subjecting each one to a sudden change of temperature. Although all individuals will increase oxygen consumption with increasing temperature, those acclimated to cold temperature will have metabolism–temperature curves that are translated to the left of those acclimated to warmer temperatures. So at a given temperature, cold-acclimated individuals

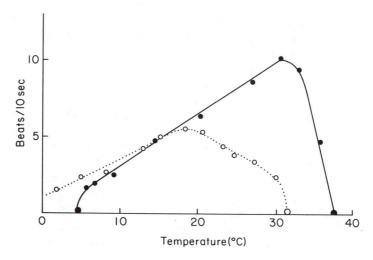

Figure 2-5   Cirral activity as a function of temperature in an upper (solid line) and lower (dashed line) intertidal barnacle species. (After Southward, *Helgoländer wissenschaffliche Meeresuntersuchungen,* vol. 10, pp. 391–403, 1964)

have somewhat elevated metabolic rates relative to individuals acclimated to warm water. This adjustment can result in the metabolic rate being constant for acclimated individuals over a wide range of temperatures (e.g., Newell and Kofoed, 1977). Two favorable results are achieved. The organism depresses the metabolic cost of living in warm water and yet can also adjust metabolism to be active in winter. Acclimation in winter would result in metabolic rates greater than those expected from acute exposure of summer-acclimated animals to winter temperature levels. At 3°C winter-collected burrowing sand crabs, *Emerita talpoida,* consume oxygen at a rate four times greater than summer-collected animals tested at the same temperature (Edwards and Irving, 1943). A seasonal change in tolerance usually accompanies acclimation of metabolic rate; the upper lethal temperature is greater in summer-collected animals relative to those collected in winter.

Similar types of compensatory mechanisms dampen the range of metabolic activity expected within a species living in a latitudinal thermal gradient. At a given low temperature oxygen consumption tends to be greater in animals living at high latitudes relative to conspecifics living at low latitudes. Similarly, such animals as oysters *(Crassostrea virginica)* and sea-squirts *(Ciona intestinalis)* have geographically varying temperature optima for breeding. Populations with compensatory responses, in order to function in different parts of the latitudinal temperature gradient, have been termed *physiological races.* Whether these races are genetically different or merely different in acclimatization is generally unknown (see Chapter 6).

Tolerance to temperature is an important factor regulating the distribution of marine organisms. Because intertidal environments tend to have much greater daily and seasonal temperature ranges, residing organisms are adapted to a broader temperature range than subtidal marine species. Furthermore, acclimatization in seasonal habitats results in greater tolerance to high temperature in summer and lower temperature in winter. The geographic ranges of marine species indicate that natural selection has shifted optimum response of species to that of their native temperature regime. The Antarctic fish genus *Trematomus,* for instance, has representatives that live in water temperatures close to − 1.9°C throughout the year. They will die at an upper limit of only 6°C (Somero and DeVries, 1967). Many Arctic species cannot tolerate the "warm" temperature of 10°C! At any one geographic location the marine biota consists of an assemblage of species whose optimum temperature ranges are different. Thus after the severe winter of 1962–1963 in Great Britain more tropically adapted species showed great mortality whereas Arctic-adapted elements suffered no ill effects (Crisp, 1964).

The upper lethal temperature can be estimated through dosage mortality tests, as discussed earlier. Figure 2-6 shows temperature-related mortality for several crustacean species living in a Maryland estuary (Mihursky and Kennedy, 1967). The two species with the lowest LD_{50}'s (least heat tolerant) are northern species living near the southern end of their range.

The cause of heat death is not easy to explain. In some cases, protein denaturation or thermal inactivation of enzymes may cause death. The lower solubility of oxygen at higher water temperature might limit the individual's capacity for efficient respiration. The most probable common source of heat death, however, is a failure of physiological integration caused by erratically shifting temperature effects among interdependent metabolic reactions (Schmidt–Nielsen, 1975). If various metabolic processes are influenced differently near the upper temperature limit, then coordination among them will be unbalanced.

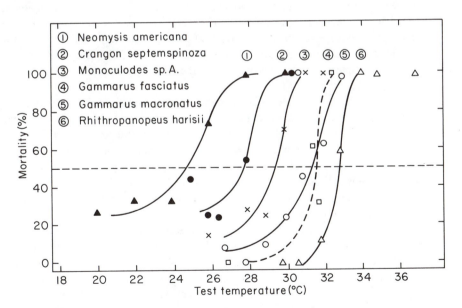

Figure 2-6 Upper lethal thermal limits of adult crustacea acclimated to approximately 15°C. Species 1 and 2 are northern species living near the southern end of their range. (After Mihursky and Kennedy, 1967)

The disruption of physiological integration can be observed at the cellular level through the investigation of *cell thermostability,* as measured by the activity of cilia on ciliated epithelium under different temperatures. Species collected from the upper intertidal in the tropics have greater cell thermostability than lower intertidal species. For any one species, cell thermostability decreases toward the upper lethal limit. In algae, the rate of photosynthesis decreases near the thermal limit (Schramm, 1968).

The effects of cold temperature, particularly in tropical organisms, probably also involve mechanisms similar to those discussed for the upper lethal limit. In tropical fishes cold depression of respiratory systems can lead to anoxia and death. But in many cold marine environments freezing presents a severe environmental problem. Larvae of many fishes, and foraminifera, can be found encased in pack ice in Antarctic waters. Fleshy algae and barnacles often survive the winter under freezing conditions in the intertidal zone. But the formation of ice can shear and distort delicate structures and result in great increases of the cellular salt content of the remaining fluids. It is possible that, in some cases, cellular fluids become supercooled—that is, remain in the liquid state below the freezing point of water. But in most cases investigated tidal animals show varying degrees of freezing under subzero temperatures. At progressively lower temperatures, increasing percentages of the body fluids are frozen. Nevertheless, intertidal fleshy algae can survive extended periods at −40°C (Kanwisher, 1957). Terumoto (1964, in Kinne, 1970) found that some algae can survive a 24-hour shock of −70°C.

The presence of salt in seawater lowers its freezing point. Similarly, salts depress the freezing point of organism cellular fluids. In winter conditions in Labrador temperatures reach the freezing point of both seawater and the cellular fluids of many invertebrates and fishes. Deep-water fishes, out of contact with ice, can remain supercooled. The presence of ice crystals in shallow water, however, enables cellular freezing to begin and precludes the supercooled state (Scholander et al., 1957). The shallow-water fish *Trematomus* counteracts this situation by synthesizing glycoproteins, which are effective in freezing point depression.

Temperature also affects growth and reproduction in marine organisms. Most marine species grow and reproduce over a narrower range of temperature than the range that permits individual survival. Within an intermediate range, growth is usually faster at higher temperatures. In bivalves, members of the same species have been found to grow more slowly, but they survive to older age and reach larger size in high latitudes relative to low latitudes (Weymouth and McMillin, 1931). Growth in seasonal habitats is greater in warmer times of the year, although it should be noted that the increased growth observed in nature in warmer seasons may also reflect the greater availability of food.

Reproduction usually occurs within a geographic range narrower than the range of nonreproductive individuals because pelagic larval stages bring individuals to thermally unfavorable habitats. Every spring larvae of the mussel *Mytilus edulis* reach and metamorphose on beaches around Cape Hatteras, North Carolina. Death from lethal high temperatures, however, ensues in the summer before the age of first reproduction (Gray, 1960). High latitude reproductive margins of species tend to be limited by the maximum summer temperature whereas low latitude extents are limited by minimum winter temperature.

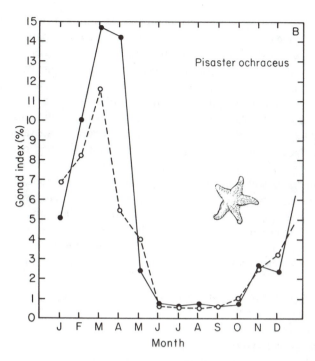

Figure 2-7 Seasonal changes of the gonad development of the starfish, *Pisaster ochraceus,* living in the intertidal of northern California (after Boolootian, 1966). Seasonal variation for 1955 = ●; 1956 = ○.

Temperature often sets the timing and can determine the style of reproduction. Many invertebrates species will spawn only when a given temperature is reached. Switches between modes of reproduction as a function of temperature are known in the coelenterates. At temperatures less than 6 to 7°C, the anthomedusa *Rathkea octopunctata* reproduces asexually but forms gametes above 6 to 7°C. In general, seasonal changes in gamete synthesis and liberation are highly correlated with temperature (as well as food and photoperiod). Figure 2-7 shows seasonal gonad changes in the sea star *Pisaster ochraceus.*

Finally, temperature can apparently affect morphology. Such features as number of ribs on molluscan shells and prodissoconch size in bivalve mollusks often change with latitude. The blue mussel, *Mytilus edulis,* occurs in two color morphs: blue and light brown striped. This difference has a genetic basis (see Chapter 6). Blue mussels absorb more heat on a hot sunny day and have higher body temperatures than light-brown mussels. So we might therefore expect the results found by Mitton (1977); the frequency of blue morphs increases from Virginia to Maine.

2-3 SALINITY

Salinity change in tide pools and estuaries can present problems to marine organisms because of the physical processes of *diffusion* and *osmosis.* Diffusion results when the concentrations of a solute, such as sodium ions, are unequal on either side of a membrane permeable to sodium. Sodium ions diffuse across the membrane until the concentrations are the same. Consequently, marine organisms must actively regulate in order to maintain cellular or circulating fluids at ionic concentrations different from that of seawater. Table 2-1 shows the concentration of common ions in seawater and some marine animals. Note the different concentrations of magnesium maintained by the hagfish and the Norwegian lobster. Regulation by the animal maintains this difference, for neither animal is completely sealed off from its environment. Differences in concentrations of specific ions are presumably maintained for adaptive reasons. Jellyfishes and ctenophores, for example, actively eliminate sulfate, replace it with a lighter ion, and thereby lower their overall specific gravity to enhance flotation.

Osmosis is the movement of water between two solutions with different solute concentrations across a membrane permeable to water but not salts. An osmotic pressure is generated when the solute concentration of body fluids differs from the solute concentration of seawater. If the solute concentration of seawater is less than that of body fluids, water crosses permeable membranes into the body. When salinity changes, marine organisms thus face the danger of water loss or gain, with concomitant changes in body volume.

A solution is said to be hyperosmotic if water will flow into it across a semipermeable membrane; hypoosmotic if water leaves that solution. The osmotic effect generated by dissolved substances of body fluids can be estimated by the depression of the freezing point below 0°C (Δ or Greek capital delta).

Mechanisms of adaptation to changing osmotic conditions and ionic gradients vary greatly among the marine biota. Some marine animals are *osmoconformers* (or are *poikilosmotic*) and regulate neither cell or body volume nor ionic constituents. We emphasize the distinction between osmotic problems and the need to regulate the concentrations of specific ions. Organisms that maintain ionic compositions radically different from that

TABLE 2-1 CONCENTRATION OF COMMON IONS (MMOL PER KG WATER) IN SOME MARINE ANIMALS[a]

	Na	Mg	Ca	K	Cl	SO$_4$	Protein (g liter^{-1})
Seawater	478.3	54.5	10.5	10.1	558.4	28.8	–
Jellyfish (*Aurelia*)	474	53.0	10.0	10.7	580	15.8	0.7
Polychaete (*Aphrodite*)	476	54.6	10.5	10.5	557	26.5	0.2
Sea urchin (*Echinus*)	474	53.5	10.6	10.1	557	28.7	0.3
Mussel (*Mytilus*)	474	52.6	11.9	12.0	553	28.9	1.6
Squid (*Loligo*)	456	55.4	10.6	22.2	578	8.1	150
Isopod (*Ligia*)	566	20.2	34.9	13.3	629	4.0	–
Crab (*Maia*)	488	44.1	13.6	12.4	554	14.5	–
Shore crab (*Carcinus*)	531	19.5	13.3	12.3	557	16.5	60
Norwegian lobster (*Nephrops*)	541	9.3	11.9	7.8	552	19.8	33
Hagfish (*Myxine*)	537	18.0	5.9	9.1	542	6.3	67

[a]After Potts and Parry, 1964, with permission from Pergamon Press, Ltd.

of seawater (e.g., teleost fishes) must adapt to both problems. Most marine organisms show varying degrees of *osmoregulation* and *ionic regulation*. Some organisms, such as crustacea, have relatively impermeable body wall integuments and strongly regulate water and ionic exchange at permeable sites, such as gills and excretory organs.

If we immerse thallus strips of the marine alga *Porphyra tenera* in dilute seawater, cell walls take up water and elongate the thallus strips (Gessner and Schram, 1971). A similar experiment performed with the sipunculid worm *Golfingia gouldii* results in an initial swelling of the body but a subsequent decrease of volume, beginning after 1 day. Volume regulation is achieved by excretion of salts through the nephridiopores. As salts are lost, water moves osmotically across the body wall into the seawater medium.

Bivalve mollusks, among other invertebrate groups, do not osmoregulate extracellular fluids but do regulate the osmotic character of intracellular fluids. They achieve constant cell volume by regulating the concentration of dissolved free amino acids. The free amino acid content increases and decreases to adapt to corresponding increases and decreases of salinity (see Bayne et al., 1976b). Free amino acids function identically to inorganic ions in changing osmotic conditions. Lysozomes, intracellular organelles, have been implicated as sites of protein degradation and amino acid release.

Although volume responses can be fairly rapid, acclimation to salinity change is a comparatively slow process relative to temperature acclimation. When blue mussels *(Mytilus edulis)* are immersed in dilute seawater, valves are closed initially and respiration is low. Then as valves are opened, a new steady-state respiration rate may not be achieved until 4 to 7 weeks (Bøhle, 1972). Furthermore, responses to salinity change vary extensively within a species. If a blue mussel is placed in an experimental medium of a salinity different from its native habitat, oxygen consumption will decrease. Yet when oxygen consumption rates are measured in natural populations collected over a broad range of natural salinities (5 to 30 ‰), they are found to be similar (see Remane and Schlieper, 1971; Theede, 1963). A genetic difference or an acclimatizing process might explain the difference.

Teleost fishes show extensive osmotic regulation, maintaining their body fluids at concentrations of one-third to one-fourth that of normal seawater. Behavioral patterns require further adaptations. The eels *Anguilla rostrata* and *A. anguilla* reproduce in the Sargasso Sea, and juveniles return to salt marshes and other inshore water habitats. They mature and can then live in freshwater. The killifish *Fundulus heteroclitus* can live in freshwater and seawater. Individuals of many salmonid species are born in freshwater, migrate into seawater, and return to freshwater to spawn. Because teleosts are hypoosmotic, they are subject to water loss in normal seawater, and salts must actively be eliminated to maintain their lower (hypoosmotic) salt content. As teleosts drink to maintain water balance, salts are also taken in. The gills maintain the salt balance by actively excreting salts (Fig. 2-8). In elasmobranch fishes, such as sharks and rays, a high concentration of urea is used to maintain osmotic balance, as in the case of free amino acids for bivalve mollusks. Sharks and rays also actively eliminate ions, such as sodium.

Blood concentrations are affected by desiccation as well as by salinity of seawater— a serious problem for organisms of the high intertidal region whose long exposure times can result in extensive evaporation. With water loss, ionic concentrations increase and

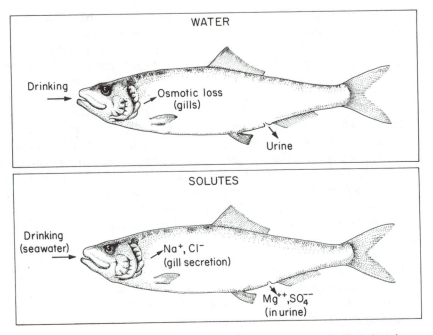

Figure 2-8   Diagram showing the movement of water and solutes by a typical marine teleost fish. Because the teleost is osmotically more dilute than the surrounding seawater medium, it must drink substantial amounts of seawater. (After Schmidt-Nielsen, 1975)

endanger biochemical functions that are influenced by ionic concentration. Organisms living in the upper intertidal show decreased water loss and increased tolerance to desiccation relative to lower intertidal forms. The intertidal salt-marsh mussel *Modiolus demissus* responds to a desiccation-related increase of body fluid osmoconcentration with an intracellular concentration change of organic molecules (Lent, 1969).

Clearly estuaries are the environments most variable in salinity; as a result, they exert the most noticeable changes on the marine biota. In Chapter 17 (Estuaries) the broad ecological effects of salinity variation in estuaries are discussed.

2-4 OXYGEN

The distribution of oxygen in the ocean is controlled through (a) exchange with the atmosphere and (b) the biological processes of photosynthesis (increasing oxygen) and respiration (decreasing oxygen).

Oxygen from the atmosphere dissolves in seawater at the sea surface. The amount that can be dissolved decreases greatly with increasing temperature and, to a lesser extent, with increasing salinity. Such water masses as the Antarctic bottom water are formed at the surface, sink, and carry with them an oxygen concentration acquired through atmospheric interaction. Respiring organisms, however, consume oxygen as the water mass sinks and moves. So although most deep-sea water masses are oxygenated as a result of their origin at the surface, masses like the Pacific deep water can be low in oxygen after their travel time of several hundred years along the bottom.

Active photosynthesis and respiration of abundant planktonic organisms in surface waters can significantly change the oxygen concentration over short periods of time. Photosynthetic plankton in shallow waters can supersaturate the water with respect to oxygen. The process of respiration reverses the direction of oxygen change. Organisms utilize oxygen in the metabolism of energy-rich food, such as carbohydrates, and release carbon dioxide. Schematically,

$$C_6H_{12}O_6 \quad + \quad 6\,O_2\uparrow \quad \underset{\text{respiration}}{\overset{\text{photosynthesis}}{\longleftarrow \atop \longrightarrow}} \quad 6\,CO_2\uparrow \quad + \quad 6\,H_2O \, + \, \text{energy}$$

carbohydrate	oxygen		carbon	water
	(dissolved		dioxide	
	gas)		(dissolved	
			gas)	

During a phytoplankton bloom (see Chapter 10) particulate organic matter, in the form of decaying carcasses of planktonic organisms, sinks and accumulates on the density gradient generated by the thermocline. Bacteria break down the debris and consume oxygen in the process, thereby producing *oxygen minimum layers* at depths of one to a few hundred meters.

In some major basins the supply of organic detritus and its role in oxygen depletion outstrip the mixing of bottom waters of the basin with other oxygenated waters. Thus bottom waters of the Black Sea, the Cariaco Deep of the Caribbean, and some Norwegian fjords are devoid of oxygen. In other cases, the balance is tipped in favor of oxygen exhaustion seasonally, as in summer conditions in deeper parts of the Baltic, and every few years in many restricted basins. In small bodies of water, such as tidal pools, this balance can shift dramatically on a daily basis.

Almost all eukaryotic organisms require oxygen for metabolism. The continued absence or even depletion of dissolved oxygen results in lowering of metabolic activity.

Oxygen consumption is usually expressed as milliliters consumed per unit body weight per unit time ($ml/g^{-1}/hr^{-1}$). Although large individuals consume more oxygen than small representatives of the same species, the oxygen consumption rate per unit body weight is inversely related to body size. Thus on a per unit weight basis, protozoa consume more oxygen than crabs. The relationship of oxygen consumption rate (ml of oxygen hr^{-1}/g^{-1}) the body size can be expressed as

$$\text{ml } O_2 \text{ consumed} = kW^b$$

where b is a fitted exponent, W is body weight, and k is a constant.

Most marine poikilotherms have b values less than 1.0 (0.66 to 1.0), indicating that metabolic rate fails to increase linearly with increasing body weight. There are several possible reasons why metabolic rate does not increase proportionately with body weight. In protozoa, the decrease in surface area to volume ratio with increasing size is a probable limiting mechanism. In organisms equipped with respiratory and circulation systems, other mechanisms may be important. An increase in the proportion of nonrespiring mass (skeletons, fat, and connective tissue), changing activity and growth patterns, and limitations of respiratory organs, such as gills, may be important.

Other trends in oxygen consumption can be briefly noted. Active species tend to consume more oxygen than inactive species. Figure 2-9 depicts such a difference among harpacticoid copepods of differing swimming ability. Sponges, ascidians, and most bivalve mollusks consume much less oxygen than decapod crustacea, cephalopods, and teleosts. Activity may change within the life span of a single individual. Thus species actively feeding during the day concomitantly require more oxygen at this time.

Life-cycle differences also influence oxygen consumption. Larger and older organisms have lower weight-specific metabolic rates, as noted. But relative to actively growing

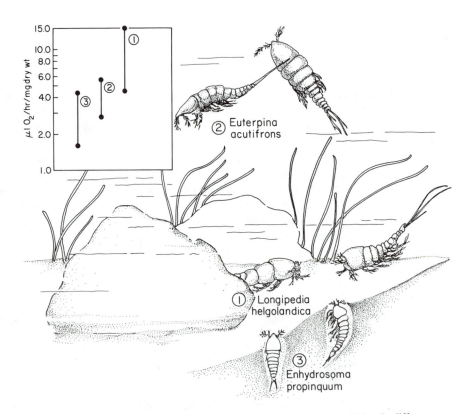

Figure 2-9 Variation in respiratory rates of harpacticoid copepods living in different habitats. *Euterpina* is a swimming species; *Longipedia* is an active benthic form; *Enhydrosoma* is a sluggish benthic form. (After Coull and Vernberg, 1970, from *Marine Biology*, vol. 5)

juveniles, older individuals usually devote a proportionately larger amount of their stored carbohydrate to respiration as opposed to somatic growth.

Mechanisms of oxygen uptake vary with body size, phyletic origin, habitat, age, and activity. Many small organisms, such as protozoa, nematodes, embryos, and larvae, rely on diffusion of oxygen across the body wall. This mechanism of uptake cannot be employed for bodies thicker than a few millimeters. Larger polychaetes, mollusks, and most crustacea use gills and circulatory systems for respiratory exchange. All gills have a large respiratory surface and may be moved rapidly or have water currents passed over them for oxygen exchange. Burrowing species, such as polychaetes, always create a water current to irrigate burrows with oxygenated water. Fishes have respiratory gills on gill arches. In many cases, gills play roles in feeding (e.g., in mollusks) and ion exchange (e.g., in crustacea, fishes), as well as in respiration.

Oxygen availability may periodically diminish and affect metabolism in marine invertebrates. At low tide animals requiring submersion for oxygen uptake are subjected to a protracted period of oxygen depletion. In high latitudes during the winter respiration is depressed and low temperature causes lowered transport rates of oxygen to cells. During these times many marine invertebrates use anaerobic pathways in metabolism (DeZwaan and Wijsman, 1976). At the time of low tide, the end products of anaerobic metabolism (alanine and succinic acid) build up in tissues. In mollusks, a portion of the succinic acid is neutralized by dissolution of calcium carbonate. In winter the inner layer of the shell of the marsh mussel, *Geukenzia demissa,* is pitted due to this dissolution process.

Animals requiring great amounts of oxygen or living in environments where oxygen is difficult to acquire may have *blood pigments* that greatly increase the blood's capacity for oxygen transport (Table 2-2). The Caribbean lucinoid bivalve *Codakia orbicularis* contains hemoglobin and is a resident of shallow soft sediments rich in decaying turtle grass and other organic debris (Read, 1962).

The binding and release of a pigment like hemoglobin (Hb) can be described by

$$HB + O_2 \rightleftharpoons HbO_2$$

At increasing oxygen pressure, more and more hemoglobin is bound to oxygen until the former is saturated. At a given pressure of oxygen, a definite proportion of the hemoglobin present is bound to O_2. Therefore an *oxygen dissociation* curve can be constructed (Fig. 2-10) that portrays the percent saturation as a function of oxygen pressure. If the oxygen pressure is lowered, the hemoglobin gives off oxygen. The pH of the blood and coelomic fluids also affects hemoglobin-binding characteristics. Generally lowering the pH tends to shift the oxygen dissociation curve to the right—the Bohr effect. Thus (a) where CO_2 is abundant, thus lowering the pH, and (b) where O_2 is in lower concentration, oxygenated hemoglobin tends to release oxygen into the medium. This factor is adaptively significant, for as oxygenated blood reaches tissues that are oxygen poor and carbon dioxide rich, oxygen will be made available. We would expect active animals requiring rapid supplies of oxygen to have their dissociation curves shifted to the right of curves for less active forms (Fig. 2-10). This appears to work for invertebrates containing the copper pigment hemocyanin. Cephalopods have high levels of hemocyanin and rely greatly on them for oxygen transport.

TABLE 2-2 COMMON RESPIRATORY PIGMENTS AND EXAMPLES
OF THEIR OCCURRENCE IN THE ANIMAL KINGDOM[a]

Hemocyanin. Copper-containing protein, carried in solution.
   Mol. wt. = 300,000–9,000,000
   Mollusks: chitons, cephalopods, prosobranch, and pulmonate gastropods: not in lamellibranchs.
   Arthropods: Malacostraca (sole pigment in these): Arachnomorpha: *Lumulus, Euscorpius.*

Hemerythrin. Iron-containing protein, always in cells, nonporphyrin structure.
   Mol. wt. = 108,000
   Sipunculids: all species examined
   Polychaetes: *Magelona*
   Priapulids: *Halicryptus, Priapulus*
   Brachiopods: *Lingula*

Chlorocruorin. Iron-porphyrin protein, carried in solution.
   Mol. wt. = 2,750,000
   Restricted to four families of Polychaetes:
      Sabellidae, Serpulidae, Chlorhaemidae, Ampharetidae
   Prosthetic group alone has been found in starfishes, *Luidia* and *Astropecten.*

Hemoglobin. Most extensively distributed pigment: iron-porphyrin protein: in solution or in cells.
   Mol. wt. = 17,000–3,000,000
   Vertebrates: almost all, except leptocephalus larvae and some Antarctic fishes (*Chaenichtys*, etc.).
   Echinoderms: sea cucumbers
   Mollusks: *Planorbis*, Pismo clam *(Tivella)*
   Arthropods: insects *Chironomus, Gastrophilus.* Crustacea *Daphnia, Artemia*
   Annelids: *Lumbricus, Tubifex, Spirorbis* (some species have hemoglobin, some chlorocruorin, others
      no blood pigment). *Serpula*, both hemoglobin and chlorocruorin.
   Nematodes: *Ascaris*
   Flatworms: Parasitic trematodes
   Protozoa: *Paramecium, Tetrahymena*
   Plants: Yeast, *Neurospora*, root nodules of leguminous plants (clover, alfalfa).

[a]From Schmidt–Nielsen, 1975.

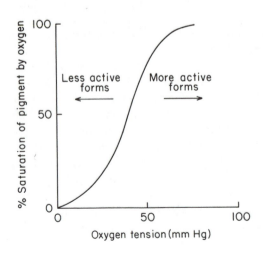

Figure 2-10 Oxygen association-dissociation curve for a typical respiratory pigment, such as hemoglobin.

In contrast, inactive animals living in environments where oxygen is in low concentration should have dissociation curves shifted to the left. Blood pigments may serve as an oxygen reservoir for burrowing animals living in sediments with little oxygen. Bivalves living in tropical soft sediments with little or no oxygen, for instance, tend to have blood pigments. In contrast, the pigments are not present in species living in more oxygenated environments (Jackson, 1973).

High intertidal animals may spend more time exposed to air than immersed in seawater. Several bivalve species open their valves when exposed to air. In the marsh mussel, such gaping allows direct access by the air to a relatively large surface area of water trapped in the mantle cavity (Lent, 1969). This consumption can be meaningful. *Mytilus californianus* consumes oxygen in air at comparable rates to its respiration in water (Bayne et al., 1976a).

2-5 WAVES AND CURRENTS

Distinctions can be made between primary and secondary effects of water movement. Primary effects are properties of water movement, such as current speed. Water velocity and turbulence strongly influence the morphology and taxonomic composition of marine biotas. These effects are discussed in Chapter 13. As an introduction, Table 2-3 summarizes some of the effects of currents on organisms and other environmental factors. Variation in water movement has secondary effects on food, nutrients, and oxygen availability. Stagnant waters will decrease in oxygen content and supply of planktonic organisms for suspension feeders. Sluggish currents will also be inefficient at dispersing pelagic larval stages of marine species (see Chapter 8).

2-6 LIGHT AND TURBIDITY

Light

In Chapter 10 the effects of sunlight on phytoplankton are described. Light energy is attenuated exponentially with water depth. The magnitude of attenuation restricts photosynthesis to the upper 100 to 200 m. Because of the harmful effects of ultraviolet radiation, maximal levels of solar energy depress photosynthesis near the sea surface.

Although more light may permit more photosynthesis, intertidal algae do not increase photosynthesis above a saturation light intensity. Photosynthetic rates of the seaweeds *Ascophyllum nodosum* and *Fucus vesiculosus* exposed to air in salt marshes are constant over a wide range of light intensities. Seaweeds probably acclimatize to the broad range of light intensities found throughout the year in temperate habitats (Brinkhuis et al., 1976).

Algae may acclimatize to conditions of low light. Both the spectral distribution and the total energy of light decrease with water depth. Increase in pigment concentration is a response to the decrease of irradiance while changes in proportions of pigment types adapt the algae to the changing spectral distribution of light with depth. Phytoplankton living at depth acclimatize to low light conditions. At a given irradiance, deep-water

TABLE 2-3 THE INFLUENCE OF WATER CURRENTS
ON OTHER ENVIRONMENTAL FACTORS[a]

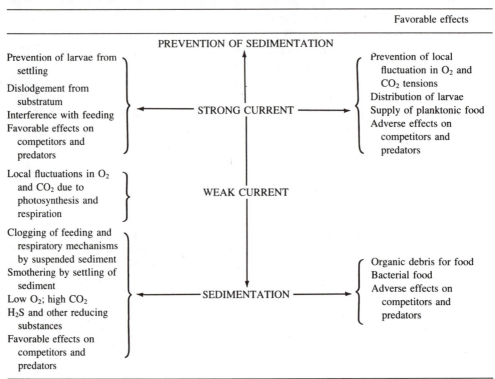

	Favorable effects

PREVENTION OF SEDIMENTATION

Prevention of larvae from settling
Dislodgement from substratum
Interference with feeding
Favorable effects on competitors and predators

STRONG CURRENT

Prevention of local fluctuation in O_2 and CO_2 tensions
Distribution of larvae
Supply of planktonic food
Adverse effects on competitors and predators

Local fluctuations in O_2 and CO_2 due to photosynthesis and respiration

WEAK CURRENT

Clogging of feeding and respiratory mechanisms by suspended sediment
Smothering by settling of sediment
Low O_2; high CO_2
H_2S and other reducing substances
Favorable effects on competitors and predators

SEDIMENTATION

Organic debris for food
Bacterial food
Adverse effects on competitors and predators

[a]From Lilly et al., 1953. Blackwell Scientific Publications Ltd.

phytoplankton show higher photosynthetic rates than those of shallow-water phytoplankton (Scott and Jitts, 1977). Acclimatization in seaweeds to dim light involves changes in total photosynthetic pigment concentrations and in the relative proportions of the component pigments. Algae kept at greater depths show increased pigment concentration and increased proportions of the accessory pigments chlorophyll *b* and phycobiliproteins. The differences are reversible if seaweeds in shallow and deep water are reciprocally transplanted. Seaweeds found only in the intertidal, however, change their pigment concentration but not pigment proportions (Ramus et al., 1976). They acclimatize as "sun" or "shade" forms by changing pigment concentration only.

Seasonal acclimatization with respect to photosynthesis also occurs in seaweeds. King and Schramm (1976) studied photosynthetic rates for 22 species of Baltic seaweeds and found a general increase of photosynthetic rate in spring and summer, corresponding to growth-rate increases in the algae. In winter all species showed an acclimatization of the "light compensation point," the irradiance at which oxygen consumed in respiration equals oxygen evolved in photosynthesis. The light compensation point was less in winter than in summer, allowing net photosynthesis to proceed at lower light intensities in winter.

Seasonal temperature change was also probably involved. In addition, there was a trend for saturation to occur at lower light intensities in winter.

Light is an important cue in behavioral adaptation of marine organisms. Many mobile intertidal animals use positive and negative responses to light to adjust their optimum position relative to tidal height (see Chapter 15). Diurnal vertical migrations in response to diurnal light changes are well known in zooplankton and are discussed in Chapter 10. Intertidal fishes, such as blennies, often use well-developed vision to navigate excursions and returns to preferred moist shelters. Some fishes and invertebrates are believed to use the sun as a compass to accomplish migrations on tidal flats or to and from feeding grounds. Parrot fishes living in Bermudan coral reefs migrate from offshore caves to feeding grounds nearshore by maintaining a constant orientation between their direction of movement and the sun (Winn et al., 1964). This behavior is depressed at night and on cloudy days.

Many salmon species and sea turtles accomplish spectacular migrations over thousands of kilometers. The role of solar navigation in these migrations is questionable, however. Salmon accomplish their long journey from freshwater spawning grounds to the open sea and back under any and all conditions of cloud cover. Green turtles are extremely myopic in air and can only see stars from under water in perfectly calm seas (Ehrenfeld and Koch, 1967). Chemoreception is a more likely mechanism to explain green turtle migrations (Koch et al., 1969).

Turbidity

Turbidity reduces the transparency of seawater and therefore reduces the amount of light available for phytoplankton and benthic algal photosynthesis. This topic is discussed further in Chapter 10.

Because much of the particulate matter in most marine waters is organic (POM), heterotrophic organisms like bacteria increase in abundance with turbidity. Thus vertical profiles of suspended particles (or *seston*) correlate positively with bacterial abundance.

The concentration of suspended particles also affects suspension feeding. This problem is further discussed in Chapters 14 and 18. In summary, sufficiently high turbidity inhibits the feeding efficiency of suspension feeders and deters growth. Reef-building corals require light and are also greatly inhibited by turbidity (see Chapter 20). Corals survive poorly where suspended sediment is deposited from the water column (Loya and Slobodkin, 1971). Particles in turbid waters often clog the gills of fishes.

2-7 CYCLES: PHYSIOLOGICAL AND BEHAVIORAL RESPONSE

Throughout this chapter we incidentally mentioned the influence of various cycles on the physiology and behavior of marine organisms. Clearly annual-seasonal, tidal, and diurnal changes exert both direct and indirect effects on organisms; organisms correspondingly adapt to these changes in a variety of ways. It is easy to understand the extent to which cycles influence living creatures.

Seasonal changes are more pronounced in middle to higher latitudes and in shallow water. As might be expected, tropical shallow water shows less temperature variability throughout the year than in higher latitudes. Polar regions also show depressed seasonal temperature variation. Thus midlatitudes present the maximum amount of seasonal contrast in temperature. Light, however, obviously shows the greatest variation in polar regions between winter and summer. Consequently, tropical regions are, on balance, the seasonally most constant environments for marine organisms.

The interaction of seasonal changes of temperature and light can have important consequences for the physiology of marine animals. In northern latitudes winter is a time of cold temperature and diminished feeding. Growth is therefore reduced. Reduced feeding, however, may also lead to altered metabolism; many invertebrates rely more on anaerobic pathways of metabolism because of the lack of food necessary to fuel aerobic metabolism. At the other end of the scale, summers in midlatitudes may be stressful because there is not enough food to fuel the metabolic requirements enforced by the high temperature. So a switch to a predominance of anaerobic metabolism may occur.

Seasonal variation in temperature and weather results in a rapid increase of temperature and available food in spring in midlatitudes. Therefore somatic growth and reproduction are usually coupled to seasonal cycles (Fig. 2-7). The transfer of food to deeper water may result in a seasonal component even at depths where temperature variation is damped. Thus growth cycles can be coupled with seasonality in deeper water as well; growth cyclicity, however, is not closely related to season in the deep sea (see Lutz and Rhoads, 1980).

Tidal cycles similarly exert a pervasive effect on marine organisms. Obviously the intertidal zone is the habitat most affected by tidal variation (see Chapter 15). Exposure to air at each low tide presents a challenge and an opportunity to organisms, depending on morphology, behavior, and physiology. At low tide, for instance, animals that must draw oxygen from the water or must have a moist gill to take up oxygen are subjected to a long period of oxygen depletion. The bivalve *Mercenaria mercenaria* actively respires oxygen when it is submerged at the time of high tide. At low tide, however, anaerobic pathways predominate and end products of anaerobic metabolism, such as succinic acid and alanine, are found (see preceding discussion on oxygen). The cyclicity in metabolism is recorded in the shell; during anaerobic periods some calcium carbonate in the shell is dissolved, leaving a zone of organic matrix (Lutz and Rhoads, 1977). In contrast, the marsh mussel, *Geukenzia demissa*, is capable of air breathing at low tide (Lent, 1969). Presumably exposure to air has not nearly as much a physiological impact for *G. demissa* as for *M. mercenaria*.

Exposure to air similarly presents contrasting challenges to burrowing and feeding behavior. For example, at low tide many burrowing polychaetes and crustacea retreat into moist burrows to avoid desiccation. At this same time, however, predators and grazers are less active as well. Thus certain bivalves of the family Tellinacea actively protrude their inhalent siphons and feed at the sediment–water interface; bottom-feeding fishes would easily spot and consume the siphons if the bottom were covered by water. Many species of benthic diatoms migrate a distance of about 2 mm from within the sediment to the sediment surface at the time of low tide. In some cases, the presence of both air and daylight is required for this rhythmic behavior. In others, the rhythmic behavior is coupled strictly to the tides and is reinforced by an endogenous rhythm that decays if the diatoms are brought into the laboratory and maintained under constant submersion (see review in Lewin and Guillard, 1963). This general behavior may be adaptive in that it brings diatoms to the surface for photosynthesis when light is maximal and grazers are inactive. Apparently diatoms are capable of rapid withdrawal below the sediment–air interface on mechanical disturbance, which might be a response to avoid surface grazers.

The general increase of feeding activity at low tide suggests a cycle of growth; this factor is well established for mollusks. Bivalve mollusks actively secrete calcium carbonate plus organic matrix when submerged. Yet when exposed to air, clams produce acidic products of anaerobiosis and the most recently deposited calcium carbonate is dissolved. Because the organic matrix remains, we can see an alternating series of growth increments consisting of organic matrix and calcium carbonate–organic matrix complex (Lutz and Rhoads, 1977; Gordon and Carriker, 1978). The fortnightly component of the tides results in a cyclicity of thickness of growth increments as well.

Finally, diurnal changes are of great importance. Clearly daily changes in light have a strong influence on photosynthetic organisms, animals dependent on light for detection of prey, and prey whose visibility is heightened during the day. In Chapter 10 one of the most fascinating and mysterious behavioral responses is discussed—the daily vertical migrations of planktonic animals.

SUMMARY

1. Marine organisms possess a variety of means to adapt to environmental change. The internal state of an organism—for example, body temperature—may fluctuate directly with the environment or the organism may be able to regulate the body's internal state. Organisms may suffer effects as a result of the environmental change, effects that may be measured in increased mortality, reduced growth, and reduced activity.

2. Temperature affects rate processes in marine organisms. High temperature can denature proteins or promote a failure of integration of biochemical reactions and cellular function. Freezing is especially damaging to delicate cells. Most marine organisms fail to regulate body temperature; adjustments of metabolism to seasonal change or local climate permit compensation to some extent, however.

3. Salinity change causes movement of both water (distilled) and solutes across permeable membranes in order to equilibrate cell fluids with the external medium. Most marine organisms regulate the volume of cells with water transport or regulation of cell constituents. Those that cannot tend not to occur in regions of fluctuating salinity. Many marine organisms regulate specific ions as well as volume.

4. Oxygen in seawater is determined as a balance between solution of atmospheric oxygen and photosynthesis opposing respiration. Although some marine organisms

rely only on diffusion of oxygen across a body wall, most have some form of gill adapted for oxygen uptake.

5. Waves and currents directly affect organisms by presenting them with mechanical stress. They also exert indirect effects, however, such as determining sediment type, affecting phytoplankton abundance, and influencing oxygen supply.

6. Light decreases exponentially with increasing depth; it obviously limits photosynthesis but also serves as a cue for many behavioral adaptations of animals as well.

7. Marine organisms are strongly influenced by cycles in the sea. Daily, tidal, and seasonal cycles in temperature, light, and water currents exert crucial effects on marine biota.

II

Some Models and Principles of Populations in Marine Ecology

In Part II we focus on the principles necessary to understand population growth and interspecies interactions in marine habitats. Part II serves partially as a resource for the chapters devoted to specific environments (Chapters 9 to 11, 15 to 21). We discuss controls on populations of a single species and extend the inquiry to interactions within assemblages of species. Species interact as competitors, symbionts, predators, and prey and as parasites and hosts to determine the fabric of a community. We relate the discussion of these interactions to possible evolutionary consequences. We also introduce evidence for the pattern of evolution of the world marine fauna and its ecological significance.

3 populations and their limiting resources

3-1 POPULATION DYNAMICS

Population size and change are touchstones of ecological effort. First, population density determines the resources available per individual in a generally resource-limited world. Secondly, the age structure and probability of death as a function of age reflect selective forces that determine the evolution of reproduction, physiological tolerance, morphology, and habitat preference. Finally, population density and size determine encounter probabilities of mating, predator–prey, and parasite–host encounters and the probability that an environmental perturbation will cause a local extinction.

By *population size* we refer to the number N of individuals in an area under consideration at time t. *Density* refers to the number of individuals per unit space. In the case of plankton, this may be numbers per cubic centimeter or cubic meter whereas in two-dimensional habitats, such as rock surfaces, numbers per square centimeter or square meter would be of interest. Because the environment can support only a finite number of individuals of a species, we say that the environment has a *carrying capacity, K*.

Population Growth

Consider a large population. If the overall death rate (number dead per capita per unit time) is d and the birthrate is b, then

$$\frac{dN}{dt} = (b - d)N$$

The expression $(b - d)$ is the instantaneous rate of change, or r, of the population. Thus on integration

$$N_t = N_0 e^{(b - d)t}$$

$$= N_0 e^{rt}$$

where N_t is population size at time t and N_0 is an initial population size. If r is positive, then N_t will increase exponentially over time. This situation is idealized in planktonic systems by diatoms that reproduce by splitting into two daughter cells over a mean generation time [Fig. 3-1(a)]. If r is negative, the population will decrease to zero [Fig. 3-1(b)].

If we measure the change in population size over one generation period, T_G, additions due to reproduction and subtractions from mortality will yield a change in population by a factor R_0 over the time T_G. Thus

$$R_0 = e^{rT_G}$$

R_0 is a net gain due to reproduction over mortality. Rearranging and taking logarithms give

$$r = \frac{\ln R_0}{T_G}$$

So rapid population increases could be accomplished by (a) decreasing generation time or (b) increasing R_0. Such selection is probably favored in species adapted to colonize and proliferate in newly opened habitats.

If r were measured in an unlimiting optimal* environment, we could estimate r_{max}, the intrinsic rate of increase of a population. Given optimal growth, we would shortly be knee deep in bacteria, ciliates, and diatoms. Limited resources would not be expected to allow population size to exceed carrying capacity or K for very long. The *logistic model* assumes that $(b - d)$ or r decreases linearly as N_t approaches K and becomes negative when N_t exceeds K. We assume no immigration or emigration.

$$\frac{dN}{dt} = r_{max} N \frac{K - N}{K}$$

So as a diatom population consumes nutrients from the water column, population growth declines as N approaches K [Fig. 3-1(c)].

The logistic model is oversimplified and the resources available at a given time might result in a reproductive increase a discrete time interval later. Such a time lag might cause an overshoot of K with a subsequent decline. Note that adjustments of N to K can be through an increase of d or a decrease in b; we only require that as K is reached, r tends to zero. Thus with a limiting resource, population growth is a *density-dependent* process.

---

*Note that r_{max} is only useful conceptually. In real-world conditions, an r_{max} might be defined under specific physical conditions.

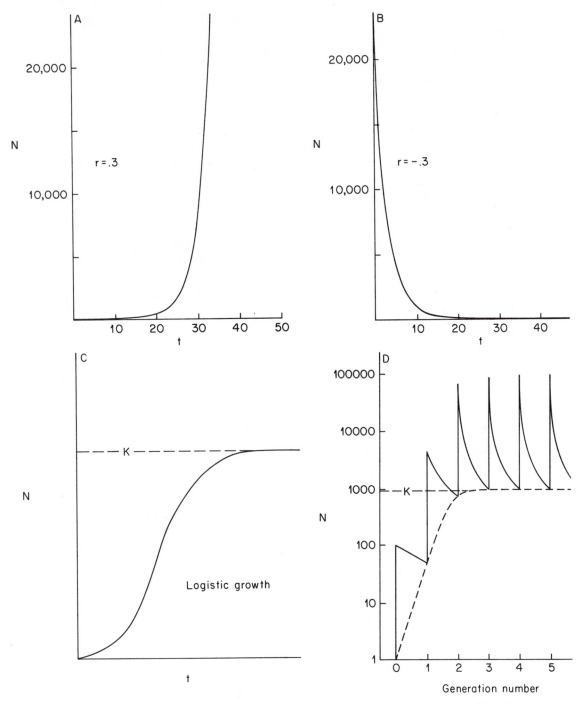

Figure 3-1 Different types of population growth: (a) Exponential growth, where $r = 0.3$; (b) Exponential growth, where $r = -.3$; (c) Population growth according to the logistic equation; (d) A net process of logistic growth, with overshoots of juvenile recruitment and subsequent mortality, adjusting the population to the logistic curve.

Reproduction in the marine environment is often confined to discrete periods, followed by recruitment of large numbers of larvae that settle and die off throughout the year. Seasonal recruitment of larvae can be traced through successive year classes that are detectable as discrete size classes in a living population (Fig. 3-2). The survivors breed at the year's end and produce another generation. If a density-related death factor is introduced with discrete breeding periods, a population growth is obtained, such as Fig. 3-1(d) (Christiansen and Fenchel, 1976). Juveniles are produced in great numbers but die back to a level determined by carrying capacity. The line connecting the adult population sizes, therefore, is similar to the logistic growth curve.

The depression of population growth might be due to the following limiting aspects of the carrying capacity.

1. Available food per consumer. The reduction in food per individual means that the costs of maintenance and behavior exceed the point where devotion of resources to reproduction is possible.

2. Space available to consumer. A given amount of space may support a finite number of organisms. In crowded populations reproduction is suppressed due to interindividual interactions.

3. Density-dependent mortality.

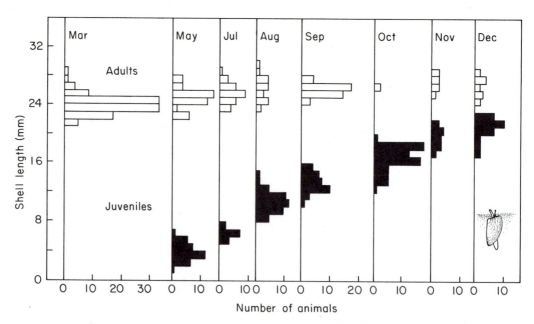

Figure 3-2 Size-frequency histograms of the sandy beach bivalve, *Donax incarnatus*, showing change in body size of 2 year classes as a function of season. (After Ansell et al., 1972, reprinted from *Marine Biology*, vol. 17)

(a) Disease and parasitism may spread more easily in a crowded population where individuals are crowded and short of resources for maintenance.

(b) Cannibalism might increase under crowding.

Population size and change are often controlled by factors that differ greatly from those included in our model. Marine benthic algae and invertebrate populations often export all their reproductive products in the form of pelagic gametes or larvae. Consequently, local changes in population size can be explained by immigration of settling larvae. We can expect frequent overshoots of K, with an influx of more migrants than the environment can support. Furthermore, with increasing density, emigration might be increased because selection would favor colonization of areas where available resources yield more reproductive output. Thus we might alter our basic population model to

$$\frac{dN}{dt} = rN \frac{(K - N)}{K} + i - e$$

where i represents immigrants per unit time and e emigrants per unit time.

Age Structure and Population Growth

In the simplest case, a population "begins" when a large number of colonists of zero age arrive at a habitat free of the same species. Most cases are far more complex, with overlapping generations and simultaneously occurring emigration and immigration. Settling of pelagic larvae on a newly opened substrate satisfies this arbitrary example. To determine the subsequent fate of the population (ignoring emigration and immigration), we need to know the probability of death with age and the number of young produced as a function of age (females are usually only considered in such analyses). Our settling larvae are defined as a cohort and we can follow their history on a *life table* that records, for a specified time interval x (units of time are defined by the investigator),

1. l_x, the number of animals alive at the beginning of interval x.
2. d_x, the number dying during interval x.
3. L_x, the average number of individuals alive during interval x ($= 0.5 \{l_x + l_{x+1}\}$).
4. e_x, the life expectancy from time period x of an individual of age x ($=$ sum of L_x from time x to the end, divided by l_x).
5. m_x, the number of young produced by a female of age x to $(x + 1)$.

Thus the change in population size from the beginning to the end of the cohort, R_0, is the sum of the probability of death at time x times the fecundity per individual, m_x, or

$$R_0 = \sum_{x=0}^{n} l_x m_x$$

More specifically, the sum represents the total number of offspring per female produced in a single generation. Because

$$r = \frac{\ln R_0}{T_G}$$

an R_0 of 1 means that $r = 0$, and the population size is stationary.

The life table approach, in toto, rarely has been used effectively to predict population change in marine organisms. It is particularly difficult to estimate mortality in the larval stage or even the percent fertilization between pelagic eggs and sperm to get an idea of the initial size of the cohort. Moreover, age-specific fecundity is difficult to measure in natural or laboratory populations. Table 3-1 shows a partial life table for the Gem clam, *Gemma gemma*, a resident of beaches (Sellmer, 1967).

TABLE 3-1   LIFE TABLE FOR THE INTERTIDAL DIMINUTIVE CLAM, *GEMMA GEMMA*, LIVING IN A NEW JERSEY SAND FLAT (MEAN LENGTH OF LIFE = 1.13 MONTHS)[a]

x	x^1	d_x	1_x	1000_{qx}	e_x
Age (months)	Age as % deviation from mean length of life	Number dying in age interval out of 100,000 liberated	Number surviving at beginning of age interval out of 100,000 liberated	Mortality rate per 1000 alive at beginning of age interval	Expectation of life, or mean lifetime remaining to those attaining age interval (months)
0–1	− 100.0	79,000	100,000	790.0	1.09
1–2	− 11.5	8,800	21,000	419.1	2.32
2–3	+ 77.0	3,700	12,200	303.3	2.63
3–4	+ 165.5	2,850	8,500	335.3	2.56
4–5	+ 254.0	1,850	5,650	327.4	2.59
5–6	+ 342.5	1,240	3,800	326.3	2.61
6–7	+ 431.0	850	2,560	332.0	2.55
7–8	+ 519.5	550	1,710	321.6	2.70
8–9	+ 608.0	355	1,160	306.0	2.75
9–10	+ 696.5	191	805	237.3	2.74
10–11	+ 785.0	116	614	188.9	2.43
11–12	+ 873.5	137	498	275.1	1.88
12–13	+ 962.0	188	361	520.1	1.40
13–14	+1050.5	92	173	537.6	1.38
14–15	+1140.0	43	81	530.1	1.39
15–16	+1227.5	20	38	526.3	1.40
16–17	+1316.0	9	18	500.0	1.39
17–18	+1404.5	5	9	555.6	1.28
18–19	+1493.0	2	4	500.0	1.25
19–20	+1581.5	1	2	500.0	1.00
20–21	+1679.0	1	1	1000.0	0.50

[a]From Sellmer, 1967.

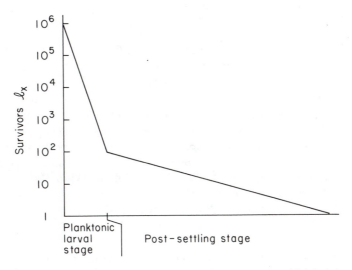

Figure 3-3   Expected survival curve for a marine invertebrate species with pelagic larvae.

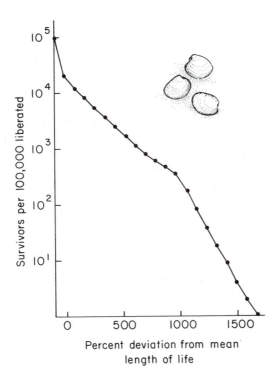

Figure 3-4   Survivorship curve for the protected beach bivalve mollusk, *Gemma gemma*. Age is expressed as percentage deviation from mean length of life; the curve gives the number surviving at the beginning of each age interval, starting with an initial cohort of 100,000 clams. (After Sellmer, 1967)

Survivorship

Although a complete life table is desirable, the determination of l_x (or d_x, the number of animals dying in age interval x) allows the construction of a *survivorship curve*, which gives an idea of the probability of death with increasing age. Survivorship analysis can therefore be used to assess changing ecological risk as a function of age.

If the l_x's are plotted as log (l_x) with time (plotted arithmetically), any change in slope indicates a change in *mortality rate*, the percent mortality per unit time. Thus a straight line indicates that there is no difference in probability of death between young and old age. Figure 3-3 shows the presumed survivorship curve for a fish or invertebrate with pelagic larvae. An initially high mortality rate (probably over 95%) is followed by a much lower adult mortality rate. The female gem clam broods its young and releases them on the sand flat. But Fig. 3-4 shows that there is still a high probability of death early in life. Although the clam can be as much as 2 years old, the mean length of life is only 1.1 months!

Measurement of survivorship can be done in several ways. The least reliable is to examine the distribution of individuals among age classes in a living population. A decrease in relative abundance from one age class to the next can be interpreted as an estimate of the amount of mortality between those two corresponding ages. This process is possible with organisms that reproduce during a short period at a certain time of year and survive for several years. But the method must assume that the number of juveniles entering the habitat each year is the same; otherwise the starting point determining the mortality sequence differs. So if recruitment of juveniles was ten times the number for previous years, then we would wrongly infer a high mortality between the juvenile $(0+)$ and subsequent year classes.

A second, more reliable way is to follow a cohort through its history and measure the diminution of numbers. Figure 3-5 shows a postmetamorphosis survivorship curve for the clam *Tellina martinensis*, constructed from sequential samples from a mud flat. The probability of death is constant with age (straight-line mortality). Following marked individuals increases the power of this technique. Seed (1969) examines survivorship of marked cohorts of the mussel *Mytilus edulis* at three levels of an exposed rocky shore. Low in the intertidal zone, mortality is high due to predation. Mussels high on the shore escape predation and often reach ages of 15 to 20 years.

A third possible way of estimating mortality is to estimate d_x by collecting the accumulated carcasses of dead animals. If the relation of age to size is known, a dead body or skeleton is a record of the age of death (see Deevey, 1947). This procedure is only possible with animals that have a skeleton, such as mollusks. With a collection of N shells taken from a mud flat, the number of shells of age 0 to age t, divided by N, is the mortality rate from zero to age t.

In order to estimate mortality rates from an accumulation of skeletons, a means must be available to estimate the age of death. Annual growth rings can be used in some mollusks. Without the availability of growth rings, a logarithmic correction yielding the proper age–size relation is still possible (Hallam 1967; Levinton and Bambach, 1970).

Several problems are inherent in this approach.

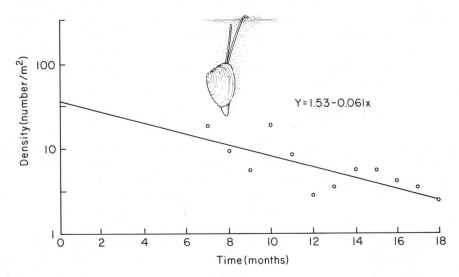

Figure 3-5   Survivorship of the bivalve, *Tellina martinicensis,* estimated from successive samplings of a beach. (After Penzias, 1969)

1. The death assemblage of shells is probably an amalgam of several generations, thus blurring year-to-year variations.
2. Alterations of the size-frequency distribution by currents and destruction of shells by bacteria, predators, and waves can bias the estimate of mortality.
3. The shell must be aged or the relative age must at least be known.

Figure 3-6 illustrates a survivorship curve for the bivalve *Mulinia lateralis,* based on shell collections in Long Island Sound subtidal muddy sediments (Levinton and Bam-

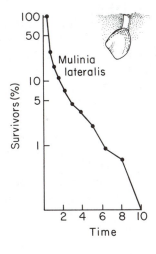

Figure 3-6   Survivorship curve for the bivalve, *Mulinia lateralis,* constructed from dead shell collections taken from Long Island Sound. (After Levinton and Bambach, *American Journal of Science,* vol. 268, pp. 97–112, 1970)

bach, 1970). *M. lateralis* feeds by filtering suspended particles from the water. Suspension feeders have difficulty stabilizing their living position within the muddy sediment. Furthermore, juvenile suspension feeders are easily clogged by suspended sediment. These two factors combine to explain the observed mortality pattern of declining mortality rate with increasing age and size.

Population Strategies

We can imagine a case in which a population has two genetic variants. One variant efficiently locates and feeds on rare resources. The other variant is inefficient at locating and feeding on scarce resources. It can, however, proliferate rapidly in an unlimited environment. We further presume a tradeoff between the ability to exploit food efficiently and the ability to grow a large population. If population density is high, the first genetic type will take over because it is superior at exploiting scarce resources. Yet if the environment fluctuates and population size is repeatedly diminished, the second type will be favored (see MacArthur and Wilson, 1967; MacArthur, 1972; Roughgarden, 1971). Selection for the efficient variant is known as k selection; selection for the fast-reproducing variant is known as r selection. In this discussion r corresponds to r_{max} as defined earlier.

It is important to remember that the distinction between r and k selection relies on the assumption of a tradeoff between efficiency and reproductive capacity. Is this assumption valid? What data led to this inference? The latter question is probably easier to answer. When food is abundant, efficiency at exploiting the food may decrease. When phytoplankton are dense, copepods do not digest a considerable fraction of the food that they ingest (e.g., Harvey, 1926). Above a critical upper density of diatom cells, copepods no longer can successfully increase ingestion rate (Frost, 1972). When food is typically scarce, however, organisms may be able to adapt to the efficient processing of the rare food. In deep-sea bottoms bivalves have elongated guts for efficient assimilation of scarce food (Allen and Sanders, 1966). Thus the rarity of food may influence the evolution of increased efficiency.

The tradeoff between efficiency and reproductive capacity can be understood in the following way. In an unpredictable environment, where environmental change causes frequent population crashes, there is an advantage in having a large r_{max}—to recover from the crash and exploit the unlimited environment. But such a large r_{max} probably results in frequent overshoots of carrying capacity. In contrast, species with low r_{max} have a longer response time after a crash. Overshoots of k will be less dramatic (May, 1976), however, r-selected species would presumably be at a disadvantage in a constant environment. k-selected species would not increase as rapidly as r-selected species in a newly opened environment.

One more aspect of a tradeoff between efficiency and population increase is important. Consider a newly opened habitat that is invaded by a species. Since all resources are available to that species, we expect the population increase to be maximized if all resources are exploited. Thus we expect species adapted to colonize new habitats to be generalists. Nevertheless, if other species are present and resources are rare, such generality might be a disadvantage because many of the resources will be monopolized by other species. So there might be an advantage to specializing in locating and exploiting a particular resource.

Can we observe a gradient in r_{max} among a spectrum of adaptive types? Freshwater and oceanic zooplankton habitats provide an environmental framework within which to look for such a trend. Allan (1976) shows that the spectrum of life histories of major zooplankton groups (rotifers, cladocera, and copepods) conforms to a gradient from unpredictable, often food-rich small lakes and inshore waters of large lakes to more predictable and food-limiting open waters of lakes and open-ocean habitats. Rotifers and cladocera have higher r_{max} (Fig. 3-7) than copepods. This factor allows rotifers and cladocera to proliferate rapidly into their favored habitats of highly seasonal lakes and inshore marine waters (e.g., cladocera occur in estuaries, not in open marine waters). But in nutritionally dilute environments, such as open oceans, copepods win out because an r_{max} strategy is not favored. Copepods seem to be more specialized feeders and more effectively resist predation than the other two groups.

Some caution must be taken in interpreting r_{max} patterns because of an overall inverse correlation between r_{max} and body size. Fenchel (1974) showed that r_{max} is related to body weight over a broad range of species from all phyla. Consequently, Allan's correlation might be explained only on the basis of the small body size of rotifers and cladocera relative to copepods.

The total energy fraction allocated to reproduction (reproductive effort) relative to that energy used for nonreproductive functions might better reflect adaptation to repeatedly disturbed environments (Gadgil and Solbrig, 1972). Three species of the mud snail genus *Hydrobia*, for example, occur in Danish waters. *Hydrobia neglecta* occurs commonly in temporary ponds that often dry up or are diluted to freshwater. The other two species live in more permanent habitats. *H. neglecta* correspondingly devotes more of its available energy to reproduction than either of the other two snail species (Lassen and Clark, 1979).

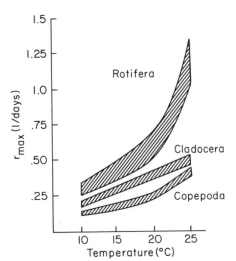

Figure 3-7 Maximal intrinsic rates of increase (r_{max}) for major zooplankton taxa. (After Allan, 1976)

It is a fugitive species, adapted to invasion of new and often temporary habitats (Fenchel, 1975a).

H. neglecta is a brooder and produces fewer eggs per female than *H. ulvae,* which has pelagic dispersal. Therefore dispersal mode is not necessarily linked to *r* selection, but reproductive effort is. This finding is further confirmed by the brooding Pacific sea star *Leptasterias hexactis,* which devotes a greater precentage of its energy to reproduction than a coexisting sea star *Pisaster ochraceus* with planktonic dispersal (Menge, 1974). *Leptasterias* seems adapted to invasion of newly opened habitats. Grahame (1977) measured reproductive effort in the snail genus *Lacuna* as the ratio: total spawn weight/body weight. The planktonic disperser *L. vincta* devotes twice as much effort as the direct-developing *L. pallidula. L. vincta* is a more generalized species.

3-2 THE NATURE OF LIMITING RESOURCES

Interindividual Interactions

As a population increases to carrying capacity, available resources per individual steadily decrease. Below *K,* resources are still sufficient to permit a population increase. Resources are plentiful and interindividual contacts are relatively rare. As the population size approaches *K,* intraspecific competition for the limiting resource intensifies. Above *K,* there are not enough resources per individual to sustain the population at that level.

Two types of interaction for the resource are possible. First, direct *interference competition* can occur (Miller, 1967), where individuals actively prevent other individuals from utilizing the same resource. Such interactions are most frequent when space is the limiting resource. After a dense larval settlement of the barnacle *Balanus balanoides,* individuals overgrow or undercut other individuals (Connell, 1961a). Amphipods, stomatopods, and other crustacea vigorously defend hiding places against intrusion by members of the same species, often with fierce battle tactics. Fights between individuals of *Gammarus* (Amphipoda) for hiding places often result in loss of limbs and sometimes death.

Exploitation competition is less direct and implies that individuals of a population are each exploiting the limited resource at a given rate. Those capable of exploiting the resource more rapidly or efficiently than others will grow faster and leave more progeny than less successful exploiters. A good example of this form of interaction would be diatoms taking up such nutrients as nitrogen and phosphorus from the water column. The success of a given diatom clone, relative to another, will be determined by the relative rates of nutrient uptake and utilization efficiency (see Chapter 10 for discussion of the kinetics of nutrient uptake).

Resource Renewal

Some resources, such as the space occupied by a cementing sponge, must be utilized for the entire life of the individual. Other resources are utilized less permanently. As temporary shelters are deserted, they become available to others. A predator might reduce

TABLE 3-2 EXAMPLES OF NATURAL MARINE SYSTEMS
WITH RENEWABLE RESOURCES

Exploiter	Renewable resource	Habitat
Copepods	growing diatom population	plankton
·Sediment ingestors	bacteria and other microflora attached to particles	marine soft bottoms
Diatoms	nutrients	plankton
Herbivorous snails, urchins	benthic microalgae and seaweeds	marine hard substrate

the population size of a prey population (the resource). Rapid population growth of the prey, however, might maintain a steadily abundant resource for the predator. Such are renewable resources.

The amount of renewable resource available to an exploiter at any time depends upon the rate of resource renewal and the rate of resource exploitation. The latter is clearly a function of exploiter population density and rate of exploitation per individual. If resource renewal is rapid, then resources are always plentiful. If exploitation is intense, then even rapid renewal may be insufficient to maintain the resource. For example, plaice *(Pleuronectes platessa)* are active visual-hunting bottom-feeding fishes that nip off siphons of the bivalve mollusk *Tellina tenuis* (Trevallion et al., 1970). Still, a *Tellina* individual can regenerate a siphon, thereby renewing the resource. If fish feeding is intense, then *Tellina* does not have the opportunity to use its siphon to take in food in the form of phytoplankton. As a result, it may starve or have insufficient energy reserves to produce gametes. Thus the resource will be collapsed. A similar argument can be made for fisheries exploited by humans (see Chapter 11). Table 3-2 lists other examples of renewable resources.

In some cases, grazing of renewable resources can enhance the recovery rate of the resource. Such an effect can be seen when bacterial grazers consume bacteria in the process of mineralizing particulate organic detritus. Grazing increases the rate of detrital breakdown, perhaps by changing the physiological state of the bacteria (Barsdate et al., 1974; Lopez et al., 1977). Thus more microorganisms are available to the grazers as more nutrients are being released from the detritus and converted into microbial biomass. This effect can also be seen in animals grazing on phytoplankton.

3-3 BEHAVIORAL ADAPTATIONS TO THE RESOURCE SPECTRUM

Resource Abundance

The availability of a resource will determine the effect of the exploiter on the resource and perhaps determine the search behavior and the degree of specialization on the resource. If several resources are available but the consumer does not distinguish among them and

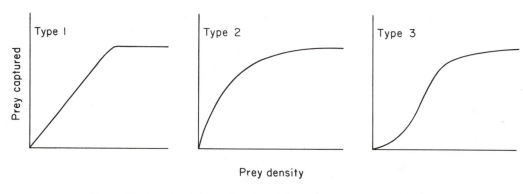

Figure 3-8 Functional response types, relative to prey density. (After Holling, 1959)

consumes them proportionately to their abundance, exploitation is *fine grained* (MacArthur and Levins, 1964). Examples are suspension feeders feeding indiscriminately on several phytoplankton species or deposit feeders that swallow sediment without sorting for grain size. *Coarse-grain* exploitation occurs when this situation does not hold and specialization on a particular resource occurs.

The number of units of a resource consumed (e.g., number of prey taken) as a function of the density of the resource (e.g., density of prey) is known as the *functional response* (Holling, 1965). Responses can be of three types (Fig. 3-8).

1. Consumption increases linearly with resource density until an upper density, beyond which consumption does not increase. Copepods feeding on diatoms as a function of diatom density are a good example (Frost, 1972).
2. There is an increase of the resource taken with increasing resource density but with a logarithmic deceleration to a plateau. Some invertebrates and fishes who exploit by random encounter can take progressively smaller increments of resource with increasing resource density.
3. A sigmoid exploitation curve shows an increase in consumption, over a threshold density. At the threshold, the exploiter can locate the resource more efficiently than just by random encounter. With increasing resource density, consumption reaches a plateau. In some instances (see section on optimal foraging) the exploiter may be capable of avoiding a given resource type unless it surpasses a threshold density that permits economical exploitation.

Niche Width and Resource Spectra

Some species will exploit a greater spectrum of resources than others. The starfish *Pycnopodia helianthoides,* for instance, feeds voraciously on almost any invertebrate prey that it can capture. In contrast, some species of the gastropod genus *Conus* feed on a narrow range of prey items, such as fishes or polychaetes only.

Niche width can be defined as the "spread" of exploitation over a given number of resources. It has two components: (a) a range due to the utilization of the resources by each individual and (b) a range due to variation among individuals in a population (Roughgarden, 1972). The second component may be due to differences of exploitation with age (as when the larva is a suspension feeder and the adult a herbivore), polymorphism (as when dark morphs live on dark substrates and light morphs live on light substrates, due to cryptic behavior), or sexual dimorphism.

Optimal Diet

Optimal foraging theory predicts behavioral rules that maximize an animal's rate of food intake. The environment consists of an assortment of food types of varying nutritional value, abundance, and spatial arrangement. The importance of optimizing foraging behavior becomes clear when we consider that increased nutrition will result in healthier animals with more resources available for growth and reproduction; optimal foraging should maximize fitness.

If two different prey types are available, we would thus predict that the forager would choose the nutritionally favorable prey. To test this hypothesis, we would require a criterion for nutritional quality; this problem is usually more than superficial. We might measure the calorific content of various prey types (e.g., seaweed species) and then measure preference of a forager (e.g., a sea urchin). This approach is naive (e.g., Paine and Vadas, 1969a), for calories are often vested in relatively indigestible structural carbohydrates. So it would be more reasonable to estimate nutritive value in terms of a more realistic measure of calories that can be readily assimilated. A study by Vadas (1977) demonstrates that urchins prefer seaweeds that are relatively easy to assimilate. Thus the forager responds positively to higher quality food.

Prey-size variation presents another situation of choice to the forager. Larger prey items yield more nutrition than small prey items. The larger payoff afforded by a larger prey item, however, might fail to compensate for the additional time required to subdue and consume the prey. Consequently, the food consumed per unit capture and handling time might increase with increasing prey size until a maximum, beyond which capture time and handling time reduce the payoff. There would, as a result, be an optimal prey size that maximizes food consumed per unit capture, plus handling time.

Crabs snip and crush bivalve shells and then consume the soft flesh within. A larger prey item will yield more food than a small one. The time required to cut open a mussel, however, increases with valve size; so we might expect an intermediate size of mussel to provide the maximal amount of food per unit handling time. Elner and Hughes (1978) studied consumption of mussels by the crab *Carcinus maenus* in Great Britain. They measured handling time and food value as a function of mussel size. Food obtained per unit handling time reached a maximum at an intermediate shell length. Preference experiments showed that the crab preferred an intermediate-sized mussel slightly larger than that predicted by the optimization model. Furthermore, as predicted by expectations based on energy gain, larger crabs preferred larger mussels.

Food density also plays a role in choosing among prey types. When food is scarce, it is less profitable to spend the additional time necessary to locate only valuable food items. When profitable prey are common, however, more food will be obtained by passing over less valuable items and restricting the diet to a reduced set of valuable prey types (MacArthur and Pianka, 1966; Charnov, 1976). Charnov (1976) shows that we can calculate the number of prey types that should be consumed, given (a) the food value per unit handling time for each prey type and (b) the cumulative reduction in travel time between prey items, as successively less valuable prey types are included in the diet. As food density decreases, an animal's optimal diet should consist of a broader range of prey types; less profitable prey types should be included in the diet because the reduction in prey quality is compensated by the reduced travel time between successive prey items. The reader should consult Krebs (1978, and references therein) for a more complete treatment of this model.

Field and laboratory tests tend to support the model. Goss–Custard (1977) studied feeding on mud-flat polychaetes by redshank (*Tringa totanus*) and found that the shorebird increased in selectivity for food-rich large worms in direct proportion to the increased density of the worms. Smaller polychaete species were ignored when large worms were common. When presented with two food types of differing profitability, great tits (*Parus major*) were unselective at low food density but became strongly selective when the density of the more profitable prey was increased to the point that the birds could do better by ignoring the less profitable prey type (Krebs et al., 1977).

Although optimal foraging theory has predictive power, several factors suggest that it is not always possible to make simple predictions based on the profitability and the density of prey. First, other features might outweigh a predator's "decision" to choose the most profitable prey. Although seemingly it might make sense to choose the best prey, what if the organism does not have the neurological integration needed to remember the series of feeding experiences that are necessary to assess its environment properly? Furthermore, the predator might be equipped to detect only certain types of prey; profitable prey might simply be harder to spot. Redshanks, for instance, prefer the amphipod *Corophium volutator* to more food-rich polychaete worms—perhaps because of their ability to spot the rapid motions typical of the amphipod. Alternatively, the selectivity might indicate a nutritive value to the amphipod that is not understood by the investigator but *is* understood by the bird! In either case, it is clearly not possible to make an unambiguous prediction that the preference optimizes food intake for a definite reason. A set of prey types with varying predator defenses, digestibility, and spatial distribution might make a prediction of the optimal diet difficult indeed.

Optimal behavior may be a compromise determined jointly by the available diet and the level of predation. If the forager risks attack in feeding on "good" food, a balance must be struck between the food reward and the risk of mortality. Therefore, information on the available food spectrum may often be insufficient to predict the diet in a field population.

3-4 SPATIAL STRUCTURE OF POPULATIONS

The spatial distribution of a population determines the pattern of resource exploitation. It is at once determined by (a) behavioral interactions, (b) colonization, and (c) mortality.

If a population can be considered a series of points in space (e.g., a two-dimensional sediment surface), three patterns are possible. If the probability of a given point's location is the same throughout the space, then the spatial distribution is *random*. If more points occur in a given subarea than expected by chance, then other areas will be depleted of points and *aggregation* or *patchiness* occurs. If every equal subarea of the space contains a constant number of individuals, or a number more even than expected by chance, then the pattern is *uniform* (Fig. 3-9). Only territoriality or the uniform spatial distribution of a preferred substratum could generate the last pattern.

The scale at which an investigation is done affects the interpretation of pattern. Planktonic organisms may occur in large patches of 0.5- to 10-km diameter. But spatial variation might be random within the patches. Moreover, population heterogeneity can affect results. Animals may space evenly on a surface, but if the distance to the nearest neighbor is a function of body size, a population with body-size heterogeneity might appear aggregated if they were considered a series of points in space (Pielou, 1969).

Spatial pattern can be analyzed in two ways. If the density of points (individuals) per unit area is known, the average expected distance from an individual to its nearest neighbor in a random distribution, $\bar{r}_e$, is

$$\bar{r}_e = \frac{1}{2\sqrt{\lambda}}$$

where λ is the number of individuals per unit area. If the measured mean distance $\bar{r}$ is greater than $\bar{r}_e$, then a tendency toward uniformity is demonstrated; if $\bar{r}$ is less than $\bar{r}_e$, aggregation is present. An index of difference from a random pattern (Clark and Evans, 1954) is

$$R = \frac{\bar{r}_A}{\bar{r}_E}.$$

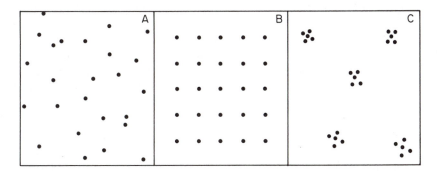

Figure 3-9 Patterns of spatial distribution: (a) Random; (b) Uniform; (c) Aggregated.

The spatial pattern can also be evaluated by dividing the space into a series of contiguous *quadrats* of equal area. The number of points can then be counted in each quadrat and the variance in number of individuals per quadrat compared with the mean number per quadrat λ. In a random distribution where the total number of individuals is large and the population sparse relative to what the total space can contain, the fraction of quadrats with x individuals, P_x, is

$$P_x = \frac{\lambda^x e^{-\lambda}}{x!}$$

The actual frequency distribution of quadrats with number x can then be compared with the expected random distribution, using a chi-square test (see Pielou, 1969). If there is an excess of quadrats with low and high numbers, then the pattern is aggregated; an excess in intermediate classes indicates uniformity. Other analyses of spatial distribution are covered in Pielou (1969).

Uniform distributions are of interest because they usually indicate that interindividual interactions maximize spacing among neighbors. Animals living on intertidal rock surfaces are often territorial. Clones of the anemone *Anthopleura elegantissima* display aggressive behavior with neighboring clones and the owl limpet *Lottia gigantea* forcibly pushes neighboring limpets from its territory. It is conceivable, however, that the spatial pattern might be uniform if the animal depends on resources that are uniformly distributed. Regularly spaced cracks (as in volcanic rocks) might be an example.

Uniform distributions may occur in small-scale populations that are themselves patchily distributed. The tellinacean bivalves *Tellina tenuis* (England; see Holme, 1950) and *Tellina agilis* (New England; see Gilbert, 1970) both space evenly on the scale of ca. 100 cm^2 or less. But *T. agilis* is concentrated in certain areas by currents and both species show considerable spatial variation in density on tidal flats.

Random distributions in mobile species imply independence of individuals in a population. In sessile species, random distributions can result from random larval settlement or uniform settlement with subsequent random mortality.

Aggregations in mobile species are caused by (a) attraction to a common resource, such as a plant, (b) attraction to form breeding pairs or swarms, and (c) aggregations formed by accumulation of individuals moving randomly in a microtopographically complex environment. Random movement of snails might still yield an aggregated distribution if a crack prevents lateral movement.

In sessile species, or on a scale too large for an individual's mobility to matter, aggregations may be due to (a) patchy larval settlement (see Chapter 8) and (b) environmental variation (e.g., a sediment change) that affects abundance by selective mortality or preferential larval settlement.

Spatial Autocorrelation

When the density of a population can be predicted from the density of neighboring populations, then that population exhibits spatial autocorrelation. A simple example would involve the regular decline in abundance of a species with distance from a given point.

Many marine populations exhibit this type of spatial structure. The density of deposit-feeding invertebrates, for example, usually decreases along a spatial gradient from mud to sand.

Explanations for spatial autocorrelation may fit into the following classes.

1. A concomitant spatial variation in some environment parameter may be important to the organism and may regulate population density.
2. A spatial pattern may be observed that reflects the movement of a population in a defined direction. The regular change of density with distance may reflect the tail end of a migrating swarm (of seabirds, for example). Alternatively, a unidirectional current might move a population of plankters in one preferential direction, with a diffuse leading edge mixing with water devoid of plankters.
3. Random processes might, nevertheless, occasionally generate a nonrandom pattern.

Spatial autocorrelation need not entail a continually increasing or decreasing population density with distance from a point. Cyclic increases and decreases of density may be correlated with cyclic environmental variables. Ripple marks on a beach might generate a spatially cyclic series of microenvironments favorable to benthic invertebrates. The interested reader should consult Eckman's (1979) study of larval settlement for an interesting case and relevant literature on techniques.

SUMMARY

1. Population size and growth determine the availability of resources to individuals and the time scales necessary for one species to become dominant. Populations grow until limiting resources slow reproduction or increase mortality.
2. Species might evolve to change reproductive investment, generation time, investment in protection against predators, or specialization to particular resource types. To some degree, there exists a tradeoff between investment in rapid population increase and investment in mechanisms to increase specialization and adult longevity.
3. Resources may be either fixed (attachment site for a seaweed) or renewable (feeding on a bacterial population). Consumers may be well adapted to locate and exploit their resources optimally; optimality involves behavioral decision rules that have evolved in the species.
4. Spatial structure of populations is determined by behavioral interactions, colonization patterns, and mortality. Organisms may be randomly distributed in space. More often, a patterned environment causes aggregation. Negative interactions among individuals may lead to uniform distributions.

4 interspecies interactions and the structure of marine communities

4-1 INTERSPECIFIC COMPETITION

Theory

Interspecific competition occurs when two or more species inhibit or interfere with one another as a result of the common use of resources. In Chapter 2 we introduced the concept of *fundamental niche,* the hypervolume plotted in multidimensional space within which the organism can live, modifiable by natural selection. The *realized niche* is that part of the fundamental niche within which the organism exists at any given time. Interspecific competition might determine the size of the realized niche. In some cases, competitive pressure results in evolutionary change such that the species exploits a new resource where competition is less intense (given that sufficient genetic variance is initially present). Such evolutionary displacements are difficult to study and good examples are rare (Grant, 1972).

If we consider a group of species and their utilization of resources along one niche axis, say size of diatom eaten by a group of copepod species, we might get a spectrum as in Fig. 4-1. Note the bell-shaped resource utilization curves for each species. Also note specialist species 3, which exploits a narrow range of the resource as opposed to more generalist species 6, which exploits a broader range of the resource. Finally, we see that species 6 and 5 overlap to a significant degree and so will compete if resources are limited. The bell-shaped resource exploitation curve and an additional potential bell-shape resource availability curve yield complications beyond the scope of this book but suggest theoretical limitations to the similarity of coexisting competing species (MacArthur and Levins, 1964).

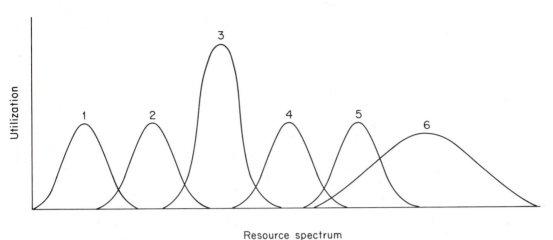

Utilization

Resource spectrum

Figure 4-1 Hypothetical resource utilization curve along a spectrum of available resource types. Numbers refer to different species.

Figure 4-2 shows two species with a degree of niche overlap in two niche dimensions. Note that although there is only a small amount of displacement in niche axis 2, the displacement along both axes shows that the two species, in fact, overlap very little. Two deposit-feeder species, for example, might require nearly identical sedimentary grain-size spectra, but this overlap would be diminished if the two species lived at somewhat nonoverlapping depths below the sediment-water interface. This situation illustrates the importance of a multidimensional approach to niche studies. Species pairs with similar diets may separate spatially whereas spatially concordant pairs may have divergent diets due to competition. Such a pattern has been observed in snails of the genus *Conus* (Kohn, 1971) and lizards of the genus *Anolis* (Schoener, 1968).

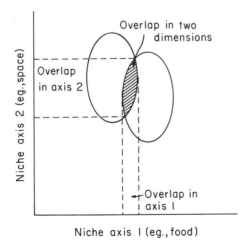

Figure 4-2 Overlap in resource utilization between two species along two niche axes (e.g., food and space). Dashed lines indicate apparent overlap in niche axis 2, while shaded area indicates much smaller actual overlap.

If we reconsider the logistic population growth model, we can see the effect of a competitor on population growth of a species. If we have two competitors, their growth equations will be

$$\frac{dN_1}{dt} = r_1 N_1 \frac{(K_1 - N_1 - \alpha_{12}N_2)}{K_1} \tag{1}$$

$$\frac{dN_2}{dt} = r_2 N_2 \frac{(K_2 - N_2 - \alpha_{21}N_1)}{K_2} \tag{2}$$

where N_1 and N_2 are the population sizes of species 1 and 2, r_1 and r_2 their respective r_{max}, and K_1 and K_2 their respective carrying capacities. The factor α_{ij} is the effect of species j on species i. Thus if α_{12} is 0.5, an individual of species 2 has half the effect on species 1 that species 1 has on itself. Therefore the change of N_1 with time will be zero if $N_1 = 0.5K_1$ and $N_2 = 2N_1$. At this point $(K_1 - N_1 - \alpha_{21}N_2)$ is zero. Thus the α_{ij} is a *competition coefficient* scaling the effect of species j on species i. If $\alpha_{12} = \alpha_{21} = 0$, then the species have no effect on each other.

At equilibrium, $dN_1/dt = 0$, we can set the expressions in parentheses equal to zero by multiplying both sides of Eq. (1) by K_1 and dividing both sides by $r_1 N_1$ and doing similarly for Eq. (2). Thus we get two linear equations that describe lines on coordinates of N_1 and N_2.

$$0 = K_1 - N_1 - \alpha_{12}N_2 \tag{3}$$

$$0 = K_2 - N_2 - \alpha_{21}N_1 \tag{4}$$

These are the lines, or isoclines, shown in Fig. 4-3 for $dN_1/dt = 0$ [Eq. (3)] and $dN_2/dt = 0$ [Eq. (4)].

Four possible outcomes of competition are possible, given variations in α_{12}, α_{21}, K_2, K_1 (Fig. 4-3). In two cases ($K_1/\alpha_{12} > K_2$ and $K_1 > K_2/\alpha_{21}$, or reverse of the inequalities), one species will win out irrespective of the starting values of N_1 and N_2. In a third case ($\alpha_{ij}/K_i < 1/K_j$), the two species will coexist at the intersection of the two lines. Here the effect of a species on itself is greater than on its competitor. A fourth case results in one or the other winning ($K_1 > K_2/\alpha_{21}$ and $K_2 > K_1/\alpha_{12}$, and reverse of inequalities), depending on the starting frequencies of the two species.

Several important points emerge. First, if α_{21} and α_{12} are much less than one, the effects of one species on the other is minimal and coexistence is possible. If $K_1 = K_2$, the α_{12} and α_{21} must both be less than one for coexistence.

We can rewrite conditions for coexistence as

$$\frac{1}{\alpha_{21}} > \frac{K_1}{K_2} > \alpha_{12}$$

(MacArthur, 1972, p. 35). As α_{12} and α_{21} both approach one, the ratios of K_1 and K_2 must be very precise to allow coexistence. If $\alpha_{12} = \alpha_{21} = 0.9$, then we must have

$$1.1 = \frac{1}{0.9} > \frac{K_1}{K_2} > 0.9$$

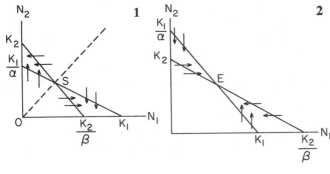

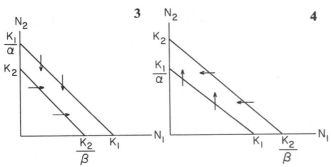

Figure 4-3 Possible outcomes of interspecific competition between two species, following the competition equations described in the text. Lines are isoclines, or lines where population growth will be zero. *Case 1:* Unstable equilibrium is possible at point *S*, however one species or another will persist, depending upon whether the initial mixture of species lies to the left or right of the line *O-S*. *Case 2:* A stable equilibrium occurs at point *E*, irrespective of the initial abundance of the two species. *Case 3:* Species N_1 always wins in competition because species 1 continues to grow even when species 2 is at its maximum carrying capacity. *Case 4:* Species N_2 wins for reasons similar to case 3.

whereas if $\alpha_{12} = \alpha_{21} = 0.2$,

$$5 = \frac{1}{0.2} > \frac{K_1}{K_2} > 0.2$$

So because very similar species have α_{12} close to α_{21}, each depresses each other's growth as much as its own. We conclude that very similar species coexist only under a precise K_1/K_2 ratio whereas dissimilar species coexist more easily (MacArthur, 1972).

The Lotka–Volterra competition equations imply linear isoclines determining the outcome of competition, but other models generate curvilinear isoclines whose form often predicts differing outcomes with given values of *K* and competition coefficients.

Although the Lotka–Volterra model has the limitations of assuming the logistic growth model, nonvarying α's, and ignoring stochastic variation and threshold effects, it provides a conceptual framework for understanding the competitive effect of one species on another. Figure 4-4 shows the effect of snail density of *Hydrobia ulva* as a function of density of *H. ulvae* alone and as a function of density of a competitor, *H. neglecta*. Note that $\alpha_{ulvae, neglecta}$ must be close to one. This allows an appreciation of the intensity of competition. The interested reader should consult MacArthur (1972), Christiansen and Fenchel (1976), and references therein for more on the subject.

If we have several competitors living with species N_1, then

$$\frac{dN_1}{dt} = r_1 N_1 \frac{K_1 - N_1 - \sum_{j=2}^{n} \alpha_{ij} N_j}{K_1}$$

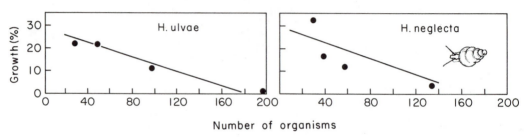

Figure 4-4   Somatic growth of the mud snail, *Hydrobia ulvae,* as a function of density of its own species and of another species *(Hydrobia neglecta).* (From Fenchel and Kofoed, 1976)

where there are *n* total species, including N_1. Then the equilibrium population density N_1^* of species 1 is

$$N_1^* = K_1 - \sum_{j=2}^{n} \alpha_{ij} N_j^*$$

Consequently, when many competitors are present, competition has a dramatic effect on the population size of species 1.

Migration can change the trend of competitive interaction if an inferior competitor is losing out to a superior one. If such an interaction is occurring in a habitat receiving migrants of the inferior competitor, it can be shown that this migration is sufficient to prevent competitive exclusion (Fenchel, 1975a). With high migration rates into the habitat, it is theoretically possible for the otherwise inferior competitor to displace the superior competitor. This effect can be important in marine habitats like lagoons, which experience considerable immigration from outer habitats. Fenchel (1975a) shows that transitions from *Hydrobia ulvae* to *Hydrobia ventrosa* dominance can be best explained from fjord to fjord on the basis of migration distance interacting with the effects of salinity. In fjords of shorter length, the open marine *H. ulvae* penetrates into areas of lower salinity.

Resources Reconsidered

As noted, comparisons of competing species along one niche axis (e.g., food) are easy but potentially misleading in understanding the magnitude of niche overlap. Two species requiring identical substrates may be found at different salinities; two species both consuming diatoms may be found at different levels in the intertidal. Similarly, the spectrum of resources available, relative to potential specialization possible, must be considered in comparisons of competing species.

Recall the definition of *fine-grained* exploitation, where consumption is proportional to the relative abundances of the resources of the resource spectrum. Such a species is effectively exploiting only one resource, for no distinction is made among the components. When *coarse-grained* or specialized exploitation occurs, opportunities for competitive shifts to given resources are possible. MacArthur and Levins (1964) show that no more competing species can coexist than the number of resources. Furthermore, if two species compete for two resources, the more similar the exploitation of the species on the two resources (a) the more the initial proportion of resources determines the outcome of competition between the species and (b) the more likely that the initial resource combination will result in displacement of one species by another. The less similar the species, the greater the opportunity for stable coexistence. This result fits intuitively with our definition of α_{ij} and the justifiable conclusion that very different species can coexist by exploiting different resources. There is currently a great deal of controversy surrounding competition coefficients that account for variation in resource abundance (e.g., see Hurlbert, 1978; Schoener, 1974b).

Competition Theory Is Not Simple

The interested reader will soon discover that theoretical models of interspecific competition are far more complex than presented here. Of greatest significance are the studies demonstrating cases in which n competing species can coexist even on fewer than n resources (e.g., Armstrong and McGehee, 1980). These systems persist because of internally generated cyclic behavior of competing species populations. Because the theory is so variable in outcome, depending on assumptions behind the models and the nature of the environment, we can view the current theoretical structure of ecology as a series " of what may be" predictions rather than a theory that makes singular predictions about the structure of communities.

Evidence for Competition

Theoretically extensive niche overlap of two species should result in competitive displacement. In practice, however, detecting competition in the marine environment is a more difficult matter. In fact, Hutchinson (1961) posed the paradox of the plankton based on the enigmatic coexistence of many phytoplanktonic species with the same apparent nutrient requirements. Great environmental variability, causing frequent local extinctions and shifts of competitive advantage, might explain this situation. It is difficult, however, to formulate criteria for asserting coexistence of two species in defiance of the expectations of competitive theory. Many species of the bivalve genus *Macoma*, for example, coexist in subtidal soft sediments (Dunnill and Ellis, 1969) of British Columbia waters. Some feeding differences among the species exist (Reid and Reid, 1969), but our ignorance of the biology of the genus precludes any hasty judgements as to coexistence with significant niche overlap. This problem leads to assertions that one has not looked hard enough if niche differences among coexisting species are not observed.

Consequently, niche dimensions must be defined, good criteria must be established for demonstrating competitive effects, and suitable criteria must be used to select a group of species for study. It is popular to select a group of closely related species, for we expect that niche overlap is probably maximal (e.g., Kohn, 1959; Fenchel, 1975b; Harger, 1968). However, it is better to consider a *guild* (Root, 1967) or group of species known to exploit overlapping resource spectra. Such an approach recognizes clearly the point that competition for a resource like space might occur among algae, territorial fishes, and urchins.

Evidence for competition may be of the following types.

Experimental manipulations.   If we remove a competitor and then observe the ecological expansion of another competing species, competition is demonstrated (Connell, 1975). This situation is probably the strongest evidence for competition and provides the opportunity for controlled experiments. Connell's (1961a) pioneering study of competition between the barnacles *Balanus balanoides* and *Chthamalus stellatus* was accomplished through transplants and removal experiments. *Chthamalus* survived successfully in the intertidal occupied normally by *Balanus* when the latter was removed (see Chapter 16). When the sea star *Pisaster ochraceus* was removed from a rocky intertidal locale, the body weight of the sea star *Leptasterias hexactis* increased relative to a control locality where both species coexisted (Menge, 1972). Additions of *P. ochraceus* individuals to another locale resulted in significant loss in mean body weight of *L. hexactis* relative to the control. Such manipulations are now commonplace in marine benthic studies (see Paine, 1966; Woodin, 1974; Peterson, 1977).

Laboratory experimental demonstrations of interference or niche overlap.   Here guilds are inferred from field evidence. Laboratory experiments are then designed to elucidate the mechanisms of competitive interaction. This approach has the advantage that behavior and interaction can be closely examined. It has a disadvantage in that interactions observed in the laboratory must be extended by correlation to patterns of nonoverlap in the field. Fenchel (1975b) noted that sympatric (co-occurring) populations of the mud snails *Hydrobia ulvae* and *H. ventrosa* always differed in size whereas allopatric (non-co-occurring) populations showed no size difference. He inferred that competitive displacement occurred in localities of sympatry through the evolution of body size. In laboratory studies he showed that small *Hydrobias* ate small particles whereas large *Hydrobias* ate large particles. All three species in Denmark conform to the same body-size, particle-size relation. Thus the change in body size served to reduce overlap with respect to food.

Displacements in nature.   This approach assumes that a "natural experiment" has been performed when a competitor is missing from a habitat, relative to other habitats. If species *a* expands its ecological range in the absence of competitor *b*, then competition may be inferred. This evidence is less satisfactory than the evidence discussed earlier, for it lacks the opportunity to understand the mechanism, and usually lacks the appropriate control situation, because some other aspect of the environment generally changes along with the removal of competitor *b* (Connell, 1975).

The most dramatic marine expansions of ecological range in the absence of competitors can be observed in large bodies of brackish water. The bivalve *Macoma balthica* is restricted to intertidal muddy sediments in open marine habitats of normal salinity. In brackish areas, such as Chesapeake Bay and the Baltic Sea, however, it is commonly found subtidally and in a much broader range of sedimentary types. Species diversity decreases in brackish waters, thereby reducing the occurrence of normal marine subtidal competitors of *Macoma balthica* (see Segerstråle, 1962; Remane and Schlieper, 1971).

Contiguity of niche space of coexisting field populations.   This approach assumes that competition has been a structuring force in the assemblage of species when a set of species coexists and has contiguous, nonoverlapping resource requirements. This type of evidence, although circumstantial, is strengthened by laboratory experiments and can provide useful baselines for the design of field-manipulation studies. It is also strengthened by trends in niche width with number of coexisting members of a group of ecologically similar species. *Conus* is a gastropod genus consisting of species that poison prey with a specially adapted radula. In southern California *C. californicus,* the single species of *Conus,* feeds on a wide variety of prey types. There are almost 30 species of *Conus* in Hawaii, however, and food specialization is greatly increased (Kohn, 1966). A similar trend can be seen when habitat diversity increases. Whitlatch (1981) found that the number of coexisting deposit feeders increased with particle-size diversity of the sediment.

Contiguity of niche space has also been observed along the niche axis of food resource. The amphipods *Neohaustorius schmitze* and *Haustorius* sp. coexist in upper intertidal beaches of the southeastern United States. But *Haustorius* has long maxillae and can filter larger food particles than *Neohaustorius*. Niche division is probably accomplished by selective particle-size feeding (Croker, 1967). Similarly, Fenchel (1968) found four typically coexisting ciliates in marine sea bottoms. All four species of the genus *Remanella* showed differences in the size range of diatoms eaten.

Evolution of Competitive Assemblages

If competition is so strong a force, we would expect interspecific competition to favor evolution to more specialized forms. Unfortunately, such a process probably would occur over a longer time span than a Ph.D. study or even the lifetime of a patient investigator. The specialization of congeneric species of *Conus* is perhaps evidence, for a *Conus* specialized in fish hunting is precluded from feeding on many other prey types. Still, we cannot be sure that the specializations were acquired because of competitive pressure. The adaptive radiation of Darwin's finches in the Galapagos Islands is a case in which a single lineage has diverged into a remarkable array of ecologically specialized types (see Lack, 1947).

The size difference of *Hydrobia ulvae* and *H. ventrosa* in sympatry (Fenchel, 1975b) may be a good example of such evolutionary divergence. In several places, size diverges between the species in sympatry, with complete overlap in allopatry. There is no evidence at present that the size difference is fixed genetically, but size is a character that can rapidly change over a few generations in selection experiments on shellfish, *Drosophila*, agricultural plants, and domestic mammals.

Yet it is surprising that intense competitive interactions, such as between *Balanus balanoides* and *Chthamalus stellatus,* have not resulted in evolutionary changes of the two species' fundamental niches. Otherwise we could not successfully perform manipulation experiments. This area of evolutionary aspects of ecological interaction deserves special attention in the future. Comparisons of structure among marine guilds in different parts of the world in similar habitats should be helpful in scaling the potential amount of competitive interaction and evolutionary specialization that can occur (see Cody, 1974).

Styles of Competition

We implicitly assume that most competitive exclusion is hierarchical—that is, the order of competitive superiority in a three-species system would be

$$A \xrightarrow{\text{superior}} B \xrightarrow{\text{superior}} C$$

Under this scheme, A always beats B and C.

But what if the competitive system is nonhierarchical, with the following set of relationships?

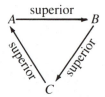

Here we cannot assume the overall superiority of A because C is superior to A. Such a system can only develop when there are differing mechanisms of competitive displacement for all three relationships. For example, A might be able to overgrow B whereas C might be able to poison A.

Nonhierarchical competitive relationships are known as *networks* and may depress the rate at which one species can competitively displace the rest (Buss and Jackson, 1979). Epibenthic invertebrates growing as sheets (sponges, ectoprocts) under *Agaricia* coral plates in Jamaica cover nearly 100% of the available space. Yet nearly 300 species coexist. Buss and Jackson (1979) suggest that the presence of networks depresses the rate of competitive exclusion enough to permit extensive coexistence. Several networks among common species have been inferred from records of overgrowth of one colony by another (see Chapter 21).

Indirect interactions among species may also exert unexpected effects within a community of coexisting competing species. Consider a pair of species A and B that have similar resource requirements; these species would clearly be in direct competition. Now consider a third species C that overlaps substantially with species B but overlaps only to a small degree with species A. This situation may be represented as

where the length of the arrows represents the strength of interspecific competitive effects. The presence of species *C* would clearly affect species *B* negatively; species *A* would be affected to a minor degree. So even though the introduction of species *C* implies competition with both *A* and *B*, the net effect will be to facilitate the increase of species *A*. Species *A* benefits from the depression of *B* by species *C*.

Such indirect effects suggest strong complexities in communities of more than two species. Great variation in patterns of species abundance may therefore be due to indirect effects, as well as nonhierarchical relationships among competing species. The prediction of indirect effects further suggests that the population growth of a given species is not simply depressed when more and more competing species are added to a community. Indirect effects can lead to the facilitation of population growth of some species.

4-2 PREDATION, MUTUALISM, AND COMMUNITY STRUCTURE

By *predation*, we mean the consumption of one living organism by another. The herbivorous snail *Littorina littorea* is a predator on benthic diatoms attached to rocks on the shore. Such a definition allows a comprehensive view of species interactions.

Because a predator is consuming a renewable resource, the rate of renewal of the prey relative to the rate of predation will determine the fate of the prey population. We can qualitatively illustrate this statement by going to a locale where urchins are feeding on algae. Where urchins are dense, rocks will be clear of algae. At high rates of grazing, the recovery of algae is too slow to balance consumption. However (assuming other herbivores are unimportant), when urchins are rare or absent, algae will be abundant.

There is an inherent time lag in predator–prey interactions. When prey is most common, predators consume the most prey. A time lag in reproduction, however, may cause a new generation to appear, when prey are at a minimum due to overexploitation. This lag will increase as the generation time T_G increases. The lag can result in one of the following patterns: (a) predator–prey oscillations and (b) overexploitation with extinction of prey and subsequent extinction of the predator.

In more northerly latitudes, where fewer generations per year of the genus *Calanus* can be generated, we see an increasing lag between the seasonal peak of diatom productivity and the peak of copepod abundance. Copepods feed on diatoms at the peak of the diatom bloom, but the lag in reproduction time yields a large new generation when much less food is available. The new copepod generation appears when the diatom population is overgrazed and nutrients sinking from surface waters further depress diatom growth.

Stabilizing Forces

In a sub-Arctic planktonic environment, a single common copepod, *Calanus finnmarchicus,* feeds on a very few diatom species in a relatively homogeneous water column. Such a system can result in overexploitation and great instability. Similarly, sea stars can devastate lower intertidal zones, preventing the establishment of invertebrate populations.

The sea star *Asterias rubens* causes such intense predation that the lower rocky intertidal of Northeast England is almost free of potential prey (Seed, 1969). Still, with the following conditions, overexploitation can be damped and prey extinctions can become less common.

Rapid recovery rate.   If the prey can rapidly reproduce, a proportionately greater amount of predation is needed to eliminate the prey. This situation can be accomplished through high fecundity or short generation time. Many sediment-ingesting invertebrates, for example, digest bacteria (see Hargrave, 1976). Lopez and Levinton (1978) showed that the digestion efficiency of *Hydrobia* feeding on bacteria is about 50%. So only one doubling period, probably less than a day, would completely restore the resource.

Predators limited by other factors.   A predator may itself be limited by still other predators. Overexploitation of prey is thus prevented. A predator population may also be limited by space to an extent that overexploitation of prey is not possible. For instance, *Octopus* individuals require crevices and holes as shelters. If such holes are in short supply, then its population will be too small to affect typical invertebrate prey populations significantly.

Refuges.   If prey can escape overexploitation through occupation of predator-proof refuges, then prey devastation can be avoided. Refuges can be in (a) habitats unsuitable to the predator, (b) rapid growth to sizes beyond the predators' killing ability, (c) shifts of habitat to temporarily cryptic conditions, and (d) shifts of polymorphism to morphs capable of crypsis or predator resistance.

Many intertidal predators are incapable of surviving the desiccation of the upper intertidal zone (Connell, 1961a; Seed, 1969; Paine, 1974). Thus a prey species, such as the mussel *Mytilus californianus*, finds a spatial refuge in the upper intertidal, above the foraging limit of the predator sea star *Pisaster ochraceus* (Paine, 1974). But the mussel can also escape predation by passing a size threshold, for sea stars of a given size have an upper limit to the size of prey that can be handled (Paine, 1976).

Prey switches.   Fisher–Piette (1934) found that the drilling snail *Thais lapillus* shifts its prey preference between barnacles and mussels, depending on which is more common. Such prey switching prevents overexploitation by allowing a prey species to recover its population when it decreases in numbers below a threshold.

Evolutionary shifts.   If sufficient genetic variance exists, a prey may be able to elude the predator permanently through an evolutionary change. A shift in habitat can be acquired or the production of poisonous substances or possession of unwieldy morphology can depress predation. The brown seaweed *Desmarestia* produces sulfuric acid, which discourages grazing (Meeuse, 1956). The black tunicate *Ascidia nigra* has converged on acid production as an effective deterrent to predators (Stoecker, 1978).

Predation and Competition

In some habitats, predation is a pervasive force in controlling population size. Connell (1970) suggested that intense predation in the rocky intertidal of the San Juan Islands is the major force in population control of many attached invertebrates whereas rocky intertidal populations in Scotland experience less intense predation and competition regulates population size of attached barnacles and mussels.

Slobodkin (1964) examined competitive interactions between two species of *Hydra* in laboratory cultures, where one species always excluded the other. In order to simulate a nonselective predator, he repeatedly removed a constant percentage of each species. Both prey species persisted in coexistence. Using our equation for competition derived earlier, we can add the effect of predation:

$$\frac{dN_1}{dt} = r_1 N_1 \frac{(K_1 - N_1 - \alpha_{12} N_2)}{K_1} - m N_1$$

$$\frac{dN_2}{dt} = r_2 N_2 \frac{(K_2 - N_2 - \alpha_{21} N_1)}{K_2} - m N_2$$

where m is the predation rate. Under such conditions the species with the greatest r is favored under competition. If r_1 is greater than r_2, with intermediate values of m, a competitive advantage of species 2 will be dampened and both species will coexist. If $r_1 > m > r_2$, then species 1 will displace species 2 (Slobodkin, 1961). So the effect of predation is to dampen competition and perhaps even to shift competitive success toward species with higher r's.

A field experimental approach can demonstrate the ability of a predator to suppress the outcome of competition normally expected without the predator. This effect is especially pronounced when the successful competitor is the preferred prey item. Paine (1966, 1974, 1976) manually removed the sea star *Pisaster ochraceus* from a rocky locale containing a variety of species. Subsequently the preferred prey of *Pisaster—M. californianus*—outcompeted coexisting species for space and a virtual *Mytilus* monoculture developed.

If the predator prefers or selectively feeds on a competitively inferior prey, predators can actually hasten the outcome of competition (see Van Valen, 1974). Such an effect is likely when prey are competing for a resource that protects against the predation itself. We can imagine two species of amphipods competing for a limiting number of shelters that afford protection against a fish predator. The inferior competitor will lose out in competition for shelter and will be exposed more frequently to predators. Exposure to predators will hasten the extinction of the inferior competitor.

Trophic Structure and Competition–Predation Equilibria

A community may have one or more *trophic levels*. Each trophic level consists of one or more species, all consuming another species or group of species in the trophic level beneath it. The primary trophic level fixes carbon through photosynthesis, making the

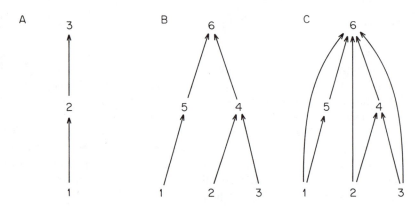

Figure 4-5 Some possible configurations of a food web.

next trophic level the herbivorous level. Because there is some inefficiency in food capture and all food consumed is not converted calorifically into growth, energy is lost from one trophic level to the next. Consequently, food availability decreases with increasing trophic level and less production is supported (see Chapter 11).

Figure 4-5 shows three types of *food webs* or associations of trophic levels. In case (a), energy is transferred through a food chain consisting of single-species trophic levels. Case (b) shows several species at lower trophic levels that enter into competition. In both cases, a species cannot eat anything but the trophic level immediately beneath. If we relax this assumption, we might allow a predator to consume all species at lower trophic levels [case (c)]. The top predator is a Keystone species; its removal results in competitive displacement at all lower trophic levels. Carnivores consuming a wide variety of prey at all trophic levels, such as the intertidal *Pisaster ochraceus* mentioned, are good examples.

With fewer trophic levels, the probability of predation-mediated communities decreases. With more trophic levels, the limitation of food with increasing trophic level causes predators to crop lower trophic levels and alter the course of competition of species sharing a lower trophic level. In communities with only two trophic levels, chance events reducing the top level would ensure that competition goes to completion at the lower trophic level. This pattern seems to fit the occurrence of competitive equilibria in the rocky intertidal of the northeast coast of the United States, with few trophic levels, versus the rocky intertidal of the San Juan Islands, Washington, with more trophic levels. Predation prevents competition from going to completion in lower trophic levels more frequently in the West Coast locality (Menge and Sutherland, 1976).

Mutualism

Interspecies interactions that benefit all participants are said to be *mutualistic*. Small invertebrates living attached to larger invertebrates may represent such a mutualistic assemblage. The larger invertebrates provide an attachment site while the smaller invertebrates may camouflage their hosts. Such interactions tend to stabilize species associations in the short run; the presence of one species increases the survival potential of the other. In the long run, a stereotype dependency might result in extinction for one species if its obligate mutualist is not present.

Ecologists have tended to emphasize antagonistic species interactions (competition, predation) in studies of community structure. Nevertheless, the extent of mutualisms may be underemphasized and may play a decisive role in community structure.

4-3 DISTURBANCE

Disturbance

Disturbance refers to the local eradication of one or more species as a result of severe modification of the structural environment or other factors that suddenly change the species composition of a local habitat. The dumping of dredge spoils on a subtidal bottom is an example of a severe modification of structural habitat. All the resident species are smothered and colonization of the newly deposited sediment will subsequently occur. Populations of adult horseshoe crabs *(Limulus polyphemus)* migrate inshore and devastate local mud-flat invertebrates. The disturbance effect is due to the active burrowing of the crabs, which disrupts the sediment.

Disturbance can have effects on communities similar to the effects of predation. A sudden disturbance can eliminate the competitive dominant and permit colonization by competitive inferiors. Fugitive species are those species adapted to colonize newly disturbed environments.

The Intermediate Disturbance–Predation Hypothesis

Disturbance and predation both mediate interspecific competitive exclusion by favoring the persistence of competitively inferior species. Strong disturbance or intense predation, however, will obviously eliminate the majority of the disturbed or prey species populations; so we would expect species richness to decline as either predation or disturbance increases in intensity. This factor suggests that intermediate levels of disturbance or predation tend to conserve the maximum number of species in a system; under low levels of predation–disturbance, competitive exclusion will go to completion whereas under high levels all species will be eliminated. We would thus predict that, along a gradient of increasing predation–disturbance, species richness will first increase and then decrease (Paine and Vadas, 1969b).

4-4 SUCCESSION

Succession refers to an orderly sequence of species observed in a habitat following a large disturbance that eliminates the local biota. Three major processes are conceived by Connell and Slatyer (1977).

1. Some species might alter the habitat in such a way as to facilitate the entry of other species. By implication, this facilitation makes the habitat unsuitable for species earlier in the successional sequence.

2. Some species might modify the habitat so that it becomes less suitable for earlier species in the successional sequence. No adverse effect is exerted on "late succession" species, however.

3. Early occupants monopolize the habitat at the expense of all other species. As long as early occupants survive, they suppress colonization by all other species. Thus species with the highest colonization rate would appear earlier in succession. More slowly colonizing forms would appear later, as earlier colonists die off.

All three processes can be imagined for marine environments. In subsequent chapters we illustrate the interactions that can result in successional sequences. It is important, for now, to remember that such sequences are initiated by disturbances.

Odum (1969) has attributed many properties to early and late stages of succession. Species occurring early in succession are envisaged as adapted for rapid colonization, early reproduction, and devotion of resources to growth and reproduction. Late successional species are slower to colonize, devote more resources toward maintenance rather than growth, and are superior competitors for space. Thus early successional species are adapted for quantity while later species are adapted for quality. Table 4-1 summarizes a broad range of biological attributes of early and late successional species.

TABLE 4-1   CHARACTERISTICS OF EARLY AND LATE STAGES OF SUCCESSION[a]

	Successional stage	
Structure	Early	Late
Biomass	variable	variable
Species diversity	low	high
Energy flow		
Number of trophic levels	few	many
Primary production per unit of biomass	high	low
Individual populations		
Fluctuations	more pronounced	less pronounced
Life cycles	simple	complex
Feeding relations	generalized	specialized
Size of individuals	smaller	larger
Life span of individuals	short	long
Population control mechanisms	abiotic	biotic
Exploitation by humans		
Potential yield	high	low
Ability to withstand exploitation	good	poor

[a]Modified after Odum, 1969, "The Strategy of Ecosystem Development," *Science* Vol. 164, pp 262–270. Copyright 1969 by the American Association for the Advancement of Science.

The basis for this theory of succession is the premise that species with strong colonizing and reproductive ability are most capable of invading newly disturbed habitats. But as time passes, these early successional species are outcompeted by species more capable of devoting resources to structural body features that contribute to competitive superiority for available space. There must be a tradeoff between (a) the ability to colonize, grow, and rapidly reproduce and (b) the alternative adaptations related to being a successful competitor.

Littler and Littler (1980) performed successional manipulations by disturbing intertidal seaweed communities on San Clemente Island, California. The alga *Ulva californica* was found to be a typical early colonizer whereas *Pelvetia fastigiata* typically appeared in late successional stages. Algal productivity and nutritive value (kcal per ash-free unit dry weight) decreased from early to late successional species. Late successional seaweeds tended to allocate more energy to the synthesis of structural elements that supported the seaweed, provided a firm attachment to the substratum, and served to resist grazing by urchins. Later successional macroalgae showed greater toughness, as well as greater resistance to wave-shearing forces, than the fugitive species, thus implicating selection for persistance in late successional stages.

SUMMARY

1. Interspecific competition occurs when exploitation of resources by one species exerts negative effects on another. If the two species are similar in resource requirements, the superior species will displace the inferior one. Coexistence is possible if resource requirements are sufficiently different or if the species evolve to minimize resource overlap.

2. Predation can completely eliminate prey populations; however, alternating cycles of abundance of predators and prey may also occur. Predation may ameliorate competition via the removal of a competitive dominant.

3. Disturbance opens new space and initiates a sequence of colonization in newly opened habitats. Disturbance can also be a mechanism of removal of the competitive dominant. A disturbance is often the initiation of succession, a predictable, ordered temporal array of dominance by different species.

5 marine biotic diversity

5-1 DIVERSITY AND COMMUNITY STRUCTURE

Implications of Species Richness

The number of species in a region is the end product of a long evolutionary process of speciation events balanced by extinction events. If all the ecological interactions discussed earlier are potent selective agents, then the variety of species in a habitat must be the result of species origins and subsequent evolutionary specialization mediated by ecological processes. The study of diversity therefore includes both ecological and evolutionary questions.

There are less than ten species of benthic foraminifera in the northeastern U.S. shallow subtidal but more than 80 living on the abyssal plain of the North Atlantic (Buzas and Gibson, 1969). What evolutionary and ecological processes have generated a *diversity gradient* of such magnitude? Are there differences in ecological structure and among coexisting species in assemblages poor and rich in species? Surely characterization of differences of such communities will hint at factors regulating their development. But only a good fossil record will permit analysis of the evolutionary dynamics inherent in generating differences in species variety.

Two types of among-species change intuitively seem to accompany the evolution of diverse communities. First, variety can be increased through the multiplication of trophic levels. This is a limited process because energy is lost as it passes from one trophic level to the next. A general estimate of 10% efficiency of transfer (Slobodkin, 1961; but see Slobodkin, 1970) decreases the energy available to a fifth trophic level by a factor of 10^5 over the energy available to the first trophic level. Inshore low-diversity planktonic communities usually have about three trophic levels. But "blue-water" high-diversity planktonic communities rarely exceed five trophic levels.

The second type of change to be expected is an increase of ecological specialization with increasing diversity. Given that resources are limiting, the evolution and migration of species into communities should be accompanied by ever-greater levels of specialization accompanying interspecific competition. An increase of the variety of predators and potential symbionts further permits specialization to ever-increasing biological variety. Is there a limit to how diverse a community can get? Theoretically the number of species cannot exceed the number of resources. But how to gauge resource number when the appearance of a new species itself creates a new complexity?

Problems arise in comparing communities of different diversities. Characterizing differences on the basis of specialization or development of mutualistic associations is not as easy as it first seems. Most naturalists, for example, are struck by the number of "oh my!" species and mutualistic associations in the tropics relative to higher latitudes—the giant clam *Tridacna,* colorful fishes, orchids, frogs, and so on. But such species and associations are expected to increase in large communities for stochastic reasons alone (Jumars, 1974). With N species, there are $N(N - 1)/2$ possible associations between pairs of species. With two communities of N_j and N_k ($N_j > N_k$) species, we expect a factor of $N_j(N_j - 1)/N_k(N_k - 1)$ more species interactions on a random basis alone. If $N_j = 100$, $N_k = 10$, the ratio is 110 for two-species interactions; 1361.4 for three-species interactions.

The same problem develops when comparing the degree of average species specialization of two such communities. If ecological specialization is normally distributed, then the probability of finding a specialized species more than 2.326 standard deviations more specialized than the "mean" species is $1 - 0.99^{10} = 0.09$ if $N_k = 10$ and $1 - 0.99^{100} = 0.63$ if $N_j = 100$. Thus it is easier to find a specialized species when more species are present. Multispecies approaches, however, do support the hypothesis of a greater range of specialization in diverse communities. Schoener (1965) showed that the range of both insect prey and bill lengths of bird predators is greater in tropical, diverse forests than in temperate and less diverse forests.

Measuring Diversity

Diversity is divided into two components: (a) richness, S, is the number of species present; (b) evenness, J, is a measure of the distribution of population sizes of the respective species. J is maximized as the population sizes of the species approach equality. The literature is groaning under the weight of various proposed indices of diversity.

Indices that combine S and J have been developed. The simplest is the Shannon–Wiener measure H':

$$H' = \sum_{i=1}^{s} p_i \ln p_i$$

where p_i is the proportion of species i and s the number of species. H' increases with increasing S. For a given S, H' is maximized (H'_{max}) when $p_1 = p_2 = \cdots = p_s$, and H' equals $\ln S$. Therefore a measure of evenness is the following:

$$J' = \frac{H'}{H'_{max}} = \frac{H'}{\ln S}$$

The measure H' is not strongly affected by the abundances of the common or rare species, only by the middle species. It is often used in studies of biological diversity, but the emphasis away from the common and rare species reflects no biological model. Rare species are especially important in disturbed communities in the process of recovery. Formerly rare species—the ones not emphasized in H'—will become important and should not be so deemphasized. Nevertheless, this index is fashionable because of its combination of S and J and its weak links with information theory. The reader will encounter it often. It is useful when we wish to measure the evenness of distribution of species in a community.

Some general patterns in H' can be related to disturbance. Newly disturbed environments have low species richness, high dominance, and hence low H' and low J'. With further succession, species richness increases, but dominance may still be high due to the competitive superiority of a few species. Generally, H' increases in later successional stages.

Buzas and Gibson (1969) surveyed S, H', and J' as a function of water depth in benthic foraminifera of the western North Atlantic.* S and H' parallel each other with strong peaks at 150 m and 4000 m (Fig. 5-1). There is a further peak of H' at 40 m, coincident with a peak in J'. Thus the species-richness component is maximal at 150 and 4000 m whereas the evenness component is maximized at 40 m. Explanations are not obvious.

Another index commonly used is *Simpson's index,* the probability that two individuals drawn at random from the same community are the same species:

$$S_I = \sum_{i=1}^{s} \frac{n_i(n_i - 1)}{N(N - 1)}$$

where n_i is the number of species i and N is the total number of individuals of all species. This index emphasizes the degree of dominance by one or a few species.

A third measure is

$$D_s = \frac{1}{\sum_{i=1}^{s} p_i^2}$$

This measure has biological meaning because it can be related to the diversity of resources available to the s species present when competitive interactions are considered (see MacArthur, 1972).

Probably species richness is the only biologically tractable measure readily applicable to natural communities. It is true, however, that evenness shows trends, such as a reduction in disturbed and polluted environments.

*Buzas and Gibson (1969) actually calculate H, a parameter closely related to H'.

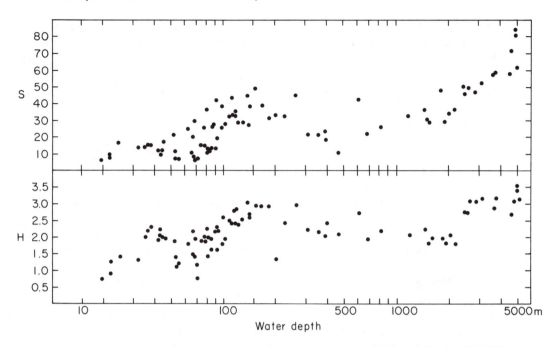

Figure 5-1   Plots of species richness and H with increasing water depth in benthic foraminifera. (After Buzas and Gibson, 1969, *Science*, vol. 163, pp. 72–75. Copyright 1969 by the American Association for the Advancement of Science.)

5-2 PATTERNS OF SPECIES RICHNESS

Within-Habitat and Between-Habitat Comparisons

In any comparative study of diversity, an homogeneous habitat with few resources or microhabitat types will support fewer species than an heterogeneous habitat with many resources, topographic complexity, and many substrate types. A comparison of diversities between two habitats of differing structural complexity would be a *between-habitat* comparison (MacArthur, 1965). Therefore a *within-habitat* approach is preferable in comparing diversity between different regions—that is, measuring diversity in similar habitats (e.g., muddy bottoms of the deep sea and muddy bottoms of shallow-water lagoons). There is an inherent logical flaw in the definition, for the differences in *S* themselves make the habitats different. Furthermore, rarely are all variables kept constant—as in mud-bottom communities of *deep* and *shallow* water. So the statement of a within-habitat comparison is insufficient without mentioning which variables do and do not remain constant.

Within-habitat comparisons are useful because they heighten the probability that we will see patterns of diversity change among groups of species that are ecologically related and competitively interactive. In this context, the pattern of diversity of a guild is preferable to that of one taxonomic group, which might be replaced by an ecologically equivalent taxonomic group in a different biogeographic realm (MacArthur, 1972). Dominance of Buzzards Bay, Massachusetts, region mud-bottom communities may switch among bivalve mollusks of differing superfamilies and polychaetes, for instance, and yet the number of dominant species changes little (Sanders, 1958, 1960; Levinton, 1977).

Gradients of Species Diversity

A parsimonious theory of diversity would assume that all well-defined gradients of species richness (which we equate with diversity in what follows) have developed as the result of similar forces. Nevertheless, the history of an individual region may locally alter diversity. The impoverished biota of Australia is due to isolation from mainland Asia.

Latitudinal gradient.   The most well-known gradient is an increase of S from high to low latitudes in continental shelf and planktonic organisms (Fig. 5-2). This pattern has been recorded in detail for bivalve mollusks (Stehli et al., 1967), gastropods (Fischer, 1960), planktonic foraminifera (Stehli et al., 1972), and many terrestrial groups. Such a consistent trend must have a pervasive explanation, independent of the details of given dispersal types, feeding adaptations, and trophic level. Some groups, such as hermatypic corals (Chapter 20), are tropical by nature, but many groups showing this trend are well represented in all latitudes.

Few detailed studies have compared species richness to habitat diversity at differing latitudes. Spight (1977) examined prosobranch gastropod diversity in Washington and Costa Rica. For Costa Rica, the regional species list contains five times as many species as in Washington. And yet typical tropical beach quadrats contained no more species than temperate beach quadrats. Consequently, tropical species must be more patchily distributed than temperate species.

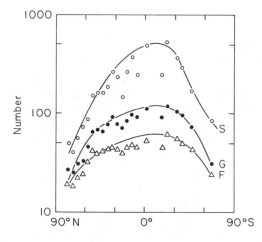

Figure 5-2 Plot of bivalve mollusk taxon richness versus latitude. Points are average numbers of species, S, genera, G, or families, F. (After Stehli et al., 1967, courtesy The Geological Society of America)

Habitat structure-diversity relationships differed in the two latitudes. High-shore habitats supported more species in the temperate community. Low-shore cobble habitats, however, supported many more species in the tropical site, further accounting for the overall richness of the tropical site.

Spight finds that the tropical site contains more habitat specialists than the temperate site. But most of this specialization is due to the relative richness of the tropical low-shore cobble microhabitat. Substrate diversity in tropical low-shore cobble microhabitats is greater than in the temperate site. Therefore the tropical community is more diverse because there are many available substrates with no temperate counterparts, not because competing species have divided up the available resources more finely than in the temperate zone.

Between ocean basins.   The Pacific Ocean has more species than the Atlantic. This fact has been documented for hermatypic corals (Stehli and Wells, 1971), bivalves (Stehli et al., 1967), fishes (Goldman and Talbot, 1976), and is probably true for most other groups. A specific case is the greater diversity on the Pacific side of the United States relative to the Atlantic. Table 5-1 shows differences, with some important exceptions, such as polychaetes. The diversity difference has been related to the more constant year-round water temperature (maritime climate) versus strong seasonal fluctuation on the East Coast (continental climate; Sanders, 1968).

Continental shelf, deep-sea gradient.   Sanders (1968) collected samples from mud bottoms ranging from shelf to deep-sea (continental slope) depths. He found that diversity of the polychaete and bivalve fraction increased dramatically with water depth. A similar pattern has been shown for benthic foraminifera (Buzas and Gibson, 1969) and gastropods (Rex, 1973). An interesting detail is that diversity decreases again from continental rise depths to the abyssal plain (Rex, 1981). Food shortage is especially

TABLE 5-1 SPECIES RICHNESS OF VARIOUS INVERTEBRATE GROUPS IN THE VICINITY OF WOODS HOLE, MASSACHUSETTS, AND FRIDAY HARBOR, WASHINGTON. NUMBERS COMPILED FROM KEYS TO INVERTEBRATES OF THE MARINE BIOLOGICAL LABORATORY AND THE FRIDAY HARBOR LABORATORIES.

Taxonomic Group	Friday Harbor	Woods Hole
Shell-less opisthobranchs	61	23
Shelled gastropods	88	51
Bivalve mollusks	114	55
Asteriod starfish	19	6
Polychaetes	174	218
Isopods	42	27
Amphipods	76	47

acute on the abyssal plain, which is far removed from a shore or down-continental slope source of organic debris. This stressful situation may permit the survival of few species.

The deep sea was characterized by Sanders (1968) as being thermally constant relative to shelf depths, which are seasonally variable in temperature. Sanders (1968) drew an analogy between the shelf–deep-sea gradient and the species richness of deeper waters of Lake Baikal (Siberia) relative to shallow waters of the lake. Both diversity gradients pass from variable (shallow) to constant (deep) environments (see Chapter 19 for more detailed discussion).

Inshore–offshore plankton communities.   Temperate-zone planktonic communities near shores, as in bays, support fewer species than offshore assemblages. Furthermore, fewer trophic levels are usually present in inshore waters. Seasonal variation in plankton abundance is also of greater amplitude near the shore. A similar pattern can be seen from species-poor upwelling areas, such as in the Humboldt Current off Peru, relative to high-diversity "blue water" plankton communities at the same latitude (Ryther, 1969).

Estuary–open marine gradients.   Estuaries are generally impoverished in species relative to adjacent open marine waters. Decreases of diversity into estuaries are often accompanied by ecological expansion of the species that penetrate the brackish water, due to the relaxation of competition (as discussed earlier).

Area.   Habitat area also contributes to diversity. This effect was first noticed on islands (see MacArthur and Wilson, 1967); at a given distance from the mainland, larger islands support more species than smaller islands. This relation, however, also holds for species on continents (Flessa, 1975). Area may contribute substantially to explaining the latitudinal diversity gradient (Schopf et al., 1978) and the continental shelf, deep-sea diversity gradient (Abele and Walters, 1979a,b). Area complicates matters when comparing the larger Pacific coral reef province with that of the smaller Caribbean province. We discuss the role of area in extinction in Chapter 6, Section 6-1.

5-3 MODELS EXPLAINING DIVERSITY GRADIENTS

Unlike experimental manipulation of field populations, processes involved in generating diversity gradients can only be approached through correlation and the search for parsimony. The implications of these gradients for community structure and evolutionary ecology make this weaker attempt at explaining causality entirely worthwhile. The following models explaining diversity gradients are therefore constructed inductively. They may all contribute in some measure to the generation of differences in diversity. Pianka (1966) discusses likely and unlikely diversity models. Here we exclude temperature as a likely factor, for diversity increases into the deep sea (cold) and toward the tropics in shallow water (warm). We must also remember that area effects might often be the most parsimonious explanation for some diversity gradients.

Stability-Time Hypothesis

Communities in physically stable and geologically ancient environments accumulate more species than variable environments (Sanders, 1968). The influence of environmental variability lies in the adaptations that organisms must have to deal with increasingly fluctuating environments. If the environment varies often from state *a* to state *b,* then a species will go extinct unless it can survive and reproduce in both states. Given that resources are limited, however, the environment can only support a few species of such broad adaptability. Perturbations are rare in a constant environment and specialized species can survive. The habitat can therefore support more competing species. The age of an environment is thought to determine the extent to which more specialized species have been added to the system.

This hypothesis finds support in the high species richness to be found in large, stable, and ancient lakes (e.g., rift valley lakes of East Africa; Lake Baikal, Siberia). Furthermore, the deep sea harbors communities of high species richness, relative to the shallow shelf (see Chapter 19, Section 19-3); of note is the thermal stability of the deep sea.

The high diversity of the Pacific coast versus the Atlantic coast of the United States is therefore explained in the context of the more constant maritime climate of the Pacific relative to the fluctuating, continentally dominated, seasonal climate of the Atlantic (Sanders, 1968). Temperature fluctuation alone cannot be held accountable, for both Arctic and tropical latitudes are much less thermally variable throughout the year than mid-latitudes. Despite this, species richness increases with decreasing latitude.

Unpredictable environments are thought to be more important in depressing diversity than predictably variable environments, where a change in the environment is cyclic. In predictably variable environments the species can presumably adapt to the change by evolving the ability to acclimatize to the cyclic change (e.g., daily change in tidal height). Unpredictable changes, however, will often be at or beyond the limit of the species' tolerance. In such environments only broadly adapted species will avoid extinction.

Sanders (1968) calls the fluctuating-environment, low-diversity communities *physically controlled* and the constant-environment, high-diversity communities *biologically accommodated.* These labels are somewhat misleading, for we assume that biological factors, such as competition, are less common in habitats of low diversity. It is probably not so, as documented in Chapters 16 and 18. As we argued earlier, the frequency of interspecies encounters increases exponentially with increasing species richness. Thus the opportunity for symbioses, elaborate antipredator adaptations, and unusual competitive "solutions" increases in diverse communities.

The time aspect of the hypothesis is difficult to consider because it is based mainly on the difference between large ancient lakes, such as the rift valley lakes of eastern Africa and Lake Baikal (Brooks, 1950), and large younger lakes, such as the Great Lakes of the United States and Great Slave Lake of Canada. But these latter lakes are only 11,000 years old or less (last glacial retreat), making them much too young to expect evolutionary events to occur at all. Consequently, it is hardly a fair test of the hypothesis. Furthermore, there is no compelling evidence that the deep sea is older than the shelf or intertidal zone. Shallow-water platforms have been present in varying abundance through-out Phanerozoic geologic time (see Valentine, 1973, for a thoughtful discussion). Abyssal faunas are, if anything, younger than shelf faunas (Menzies and Imbrie, 1958). So the stability aspect seems to be the dominant factor.

Resource Stability

Similar to the preceding hypothesis, this explanation emphasizes the fluctuation of primary production and its role in selecting for generalized and specialized species (Valentine, 1971, 1973). Seasonal variations in food availability increase from the poles to the tropics because of photoperiod changes. As a result of seasonally rich-nutrient supplies, upwelling regions are more variable than gyre centers. Inshore waters are variable relative to the open ocean because of increased environmental fluctuation inshore.

With relative trophic constancy, species may become food specialized and persist at low densities, thereby allowing complex food webs to evolve, with efficient and complete utilization of energy. In contrast, in environments with fluctuating trophic resources, species must be more generalized in order to survive the fluctuations (e.g., a phytoplankton blooms in the Arctic for only a few weeks, with only organic detritus and rare phytoplankton available for the rest of the year). Species in such environments have high r_{max} and can increase during favorable times. Thus a trophically fluctuating habitat will support few species of generalized feeding habit and communities will tend to be similar over broad ranges of habitats (Valentine, 1973, p. 306).

Species evolving in such habitats will tend to resemble parent species and be generalized forms. In trophically stable habitats, daughter species will differ more from their parent species, for they can "afford" to specialize with lower environmental vari-ability and extinction rate. Newly formed species may be forced to diverge morpho-logically due to competitive pressure. Evidence for such evolutionary change is presented later.

Competition

Competitive pressure is greater in areas where high diversity develops, thus forcing specialization. This hypothesis probably derives from the two preceding hypotheses.

Predation

Cropping of prey species prevents competitive displacement and allows the coexistence of more species. Thus cropping in the deep sea by predators would permit more species to coexist (Dayton and Hessler, 1972). As diversity increases, the number of trophic levels increases, resulting in a greater incidence of predatory depression of competition in lower trophic levels (e.g., see Menge and Sutherland, 1976). The predation hypothesis, however, ignores the question of how more trophic levels evolve in diverse habitats in the first place.

This hypothesis is attractive but difficult to test, for we must measure predation levels in habitats of differing diversity. No current evidence proves that predation is more important quantitatively in tropical shallow waters than in the temperate zone. It is true, however, that tropical gastropods seem morphologically superior in resisting predation, relative to temperate forms (Vermeij, 1977). The presence of more specialized or capable predators (e.g., Vermeij, 1977) is difficult to evaluate because the probability of finding a specialized species increases disproportionately with S, as discussed earlier (Jumars, 1974). Jackson (1977) presents compelling evidence that the co-occurring array of nearly 300 species of invertebrates living under colonies of the foliaceous coral *Agaricia* do not experience much predation at all and occupy nearly 100% of the available space.

Environmental Stress

An extreme environment can be successfully colonized by fewer species than a less extreme environment. So although some species may inhabit hot springs (e.g., bacteria), most of the phyla have not evolved representatives capable of such an invasion. Polluted environments, estuaries, and intertidal areas scoured by sea ice might be examples. The logical difficulty is obvious. If one species got there, why haven't others? Furthermore, extreme environments are unstable as well. But the regular decrease of species richness up a seasonal estuary might be cited as a stress diversity gradient (see Chapter 17, Section 17-2).

5-4 COMMUNITY EVOLUTION

Given our static, present-day picture of diversity gradients, we might expect patterns of community evolution to differ in unstable and stable environments. Such hypotheses might suggest that in stable environments more specialized species evolve and more interdependencies among species of the community develop. Such dependencies would endanger the long-term prospects for community survival, for extinction of a given prey species might endanger the predator species that depends on it. Species that have evolved mutualistic associations are likely to disappear together (Bretsky, 1969; Futuyma, 1973).

The eel grass *(Zostera marina)* "epidemic" of the 1930s provides evidence for the long-term danger of such evolved dependencies (see Rasmussen, 1973, 1977, for a review). *Zostera marina* forms shallow subtidal grass beds in soft sediments of the North Atlantic and Pacific. The mass destruction and recovery were restricted to the Atlantic (excluding the Mediterranean) and can be chronicled in the following phases (Rasmussen, 1973):

1. 1931–1932: the years of destruction, catastrophic and sudden decline in the United States and Europe.
2. 1933–1940: slight and local but temporary recovery.
3. 1941–1944: fluctuations in abundance.

4. 1945–1950: noticeable recovery, especially in places where *Zostera* had been absent since 1932.

5. 1951–1953: still more recovery, nearly complete recovery with some exceptions.

The reasons for this mass mortality remain obscure but have been related to disease and temperature change (see Rasmussen, 1973, 1977, and references therein).

Many species of benthic invertebrates had apparently evolved habitat preferences for *Zostera* beds. The eastern American bay scallop *Argyropecten irradians* decreased dramatically and shifted habitat to bare bottoms. The bivalve *Cumingia tellinoides* virtually disappeared from Cape Cod eel grass beds and has only partially recovered. The urchin *Psammechinus miliaris* disappeared in Danish waters and has not returned to the Holbaek fjord, where it was abundant before the epidemic.

The loss of eel grass depleted surrounding benthic communities due to a lowered influx of eel grass-derived detritus. Grass beds also protected areas from bottom erosion by currents (Rasmussen, 1973). Thus the loss of an important element of the community resulted in direct effects on coadapted species and indirect effects on associated species. An expected drastic decline of fisheries dependent on *Zostera* detritus, however, never took place (Rasmussen, 1977).

Bretsky (1969) inferred that fossil communities of clastic sediments living in offshore stable regimes should consist of specialized species, perhaps with high interspecific dependency. They should be more prone to extinction than onshore communities, which are more adaptable to change, less specialized, and less coadapted mutualistically with other species. An examination of the Paleozoic fossil record shows that onshore communities persisted in much the same taxonomic configuration ("linguloid-molluscan") while offshore communities experienced several major taxonomic reorganizations. The suggestion that the offshore and onshore habitats differ in stability and evolutionary characteristics is further strengthened by the data of Bambach (1977), who demonstrates that species richness stays the same in onshore assemblages since the Ordovician (ca. 400×10^6 years ago) whereas offshore associations have progressively, if episodically, increased in diversity (Table 5-2). (See Chapter Appendix for Geologic Time Scale.)

TABLE 5-2 MEAN NUMBERS OF SPECIES IN FOSSIL COLLECTIONS FROM "HIGH STRESS," "VARIABLE NEARSHORE," AND "OPEN MARINE" ENVIRONMENTS OF THE PHANEROZOIC[a]

Median number of species	High stress environments	Variable nearshore environments	Open marine environments
Cenozoic (most recent)	8.5	39	61.5
Mesozoic	7.5	17	25
Upper Paleozoic	8	16	30
Middle Paleozoic	9.5	19.5	30.5
Lower Paleozoic (most ancient)	7	12.5	19

[a]After Bambach, 1977.

A more global historical context has been proposed for the evolution of marine benthic communities (Valentine and Moores, 1972; Valentine, 1971). As mentioned in Chapter 1, sea-floor spreading and continental drift have resulted in movements and rearrangements of continental blocks throughout geologic time. Valentine (1971, 1973) argues that climatic stability is enhanced when continents are fragmented and surrounded by seas. In this case, climate is dominated by the ocean, a stabilizing force because of the high heat capacity of water. When continental blocks are associated into one super-continent, however, environmental stability is lowered, for climate is dominated by the continent. Seasonal changes are then exaggerated. Because an increase in seasonality would result in seasonal trophic resource fluctuations, there would be selection for generalized species and hence a reduction in diversity. Patterns of diversity change of continental shelf invertebrates through Phanerozoic time do correlate well with independently inferred periods of continental assembly and fragmentation (Valentine and Moores, 1972).

The Permian extinction that so pervasively affects marine invertebrate faunas is therefore related to a period of continental assembly in the Permian period. There are reasons to believe that the fossil record is biased and gives misleading estimates of diversity that may be concordant with the abundance of exposed lithified sediments (Raup, 1972). Consequently, this hypothesis requires further examination but provides an integral framework within which to view community evolution.

Earlier we mentioned the species–area relationship. Increased area probably reduces extinction and thereby raises species richness. Valentine and Moores (1972) mention the possibility that during active sea-floor spreading extensive development of midoceanic ridges provides enough topographic displacement to raise sea level and increase the area of shallow seas on continental platforms of the several continents. Thus when continents are joined, the marginal continental area available for epicontinental sea development is reduced. Furthermore, an expected diminution of midoceanic ridge development when continents are assembled might also lower sea level and reduce the area of shallow epicontinental seas. Such a reduction would decrease habitat complexity and increase environmental variability of the now more localized water bodies. Schopf (1974) shows that the area of shallow seas, as inferred from geologic maps of Permian sediment distribution, is reduced in the late Permian to 15% of the area present in the early Permian.

Some of the dynamics of evolutionary change in areas of differing species richness can be understood from an examination of taxonomic overturn in different parts of a diversity gradient. In general, it seems to be the case that forms living in low diversity areas are geologically more ancient than forms living in diversity centers (Stehli and Wells, 1971; Stehli et al., 1972). The tempo of evolution seems to be greater in diversity centers.

The Cenozoic history of planktonic foraminifera is of interest in this regard (Cifelli, 1969). A major Paleocene evolutionary radiation of planktonic foraminifera occurs from ancestors associated with cooler water and looking much like modern-day *Globigerina*. A large diversity of ornamented forms evolves until the Eocene, when a major extinction takes place. All but the cold-water globigerinoid forms go extinct. A subsequent radiation from these same ancestors occurs, yielding forms indistinguishable from the first radiation. The extinction periods favoring global occurrence of globigerina types are times of cold water. During radiations globigerina types are found in the species-poor higher latitudes. So we see a pattern similar to that described by Bretsky (1969), where a generalized form persists through an extinction of specialized forms. This generalized form, however, is also the progenitor of an adaptive radiation. A similar process can be seen in the middle Cambrian of the western United States, where invasions of a few trilobite species from cooler, open marine realms give rise to an adaptive radiation. Extinction is followed by another cycle (Palmer, 1965).

The translation of static diversity trends to dynamic processes in the fossil record leaves us with more general questions of the balance of speciation and extinction and equilibria of species numbers that might be generated by such processes. If there is an upper bound to S in a habitat, speciation rate might decrease (or, more properly, survival rate of newly isolated protospecies) or extinction rate of all species might increase as new species approach the "carrying capacity." Raup and co-workers (1973) argue that evolution could well be viewed in the context of stochastic evolutionary processes, with an upper limit set by the world resource level on the total number of coexisting species. Flessa and Levinton (1975) present evidence of significant departures from randomness. This complexity is in too early a stage for obvious conclusions, but the time dimension afforded by the fossil record telescopes the question of community evolution into the proper context of evolutionary time.

A species growth curve much like a standard population growth can be written

$$S_t = S_0 e^{(B - D)t}$$

where S_t is the number of species at time t, B the speciation rate, D the extinction rate, and S_0 the starting number of species (Stanley, 1975). At the start of an adaptive radiation, species may occupy broad geographic areas and have wide ecological tolerance and low values of D. But as the radiation succeeds, species become more specialized and restricted in geographic range, thus increasing D. Variations in B during this period are difficult to predict, but if speciation is a random process of selecting for peripherally isolated populations (Mayr, 1963), then it might be a constant, characteristically determined by ecology, morphology, and population size. As D increases, there may be a point where $B = D$ and an evolutionary equilibrium is generated. Such a prediction is supported by short geologic ranges of species in diverse species assemblages (Spiller, 1977). Sepkoski (1978) shows that the number of orders of fossil invertebrates increases to a plateau in the early Paleozoic.

SUMMARY

1. The numbers and relative abundance of species are important in characterizing the complexity of interactions in communities. Because of differences in complexity, differing habitats may have differing species richness (between-habitat differences). However, strong differences in species richness are noticeable with habitats of similar type—for example, mud bottoms—as well (within-habitat differences).

2. Several well-known geographic patterns of diversity have been described, including
   (a) latitudinal,
   (b) among-ocean basins,
   (c) water depth,
   (d) inshore-offshore plankton,
   (e) estuarine-open marine.

3. Diversity is probably regulated by the following parameters:
   (a) environmental stability,
   (b) age of the habitat,
   (c) resource stability,
   (d) competition,
   (e) predation,
   (f) environmental stress,
   (g) area of habitat coverage.

4. Communities may have evolved dependencies that increase the chance for large catastrophes following the extinction of key species. Diversity of communities over time may be viewed as a balance between speciation and extinction. The balance may be struck at different equilibrium levels, as determined by the factors cited under (3).

Chapter Appendix: Abbreviated Geological Time Scale

Era	Period	Years Before Present (millions)
Cenozoic	Pleistocene	2
	Tertiary	65
Mesozoic	Cretaceous	
	Jurassic	
	Triassic	225
Paleozoic	Permian	
	Carboniferous	
	Devonian	
	Silurian	
	Ordovician	
	Cambrian	600

6
biogeography, speciation,
and evolution
within populations
of marine organisms

6-1 BIOGEOGRAPHY

Provinces

The arrangement of the continents and oceans, combined with the influence of the latitudinal temperature gradient on oceanic circulation and shallow-water temperature, organizes the world oceans into a series of areas differing in thermal and salinity characteristics, environmental stability, and nutrient supply. Land barriers and water mass differences intensify isolation of the areas and local adaptations of resident organisms also enforce this isolation.

To illustrate a restriction of geographic range, consider the following situation. A marine urchin restricted to shallow water can only occur between some shallow depth and an upper limit in the intertidal zone. If the urchin lives on a north–south trending coast—the common case in our present-day world—some thermal limitation will also restrict the urchin's north–south extent. A pelagic larva may be the means of dispersal. Unless the larva is very long lived, however, the deep ocean is an effective barrier to transoceanic migration (see Scheltema, 1971, and Chapter 8, Section 8-4).

Because of geographic barriers and thermal gradients, the ocean can be divided into *provinces* or biogeographic regions with characteristic species assemblages. Boundaries may be water masses, major thermal discontinuities, points of land coincident with thermal discontinuities, or boundaries between water bodies of differing salinity (e.g., North Sea–Baltic). Thus on the East Coast and West Coast of North America we can delineate provinces that characterize latitudinal ranges (Fig. 6-1). Some species may have ranges

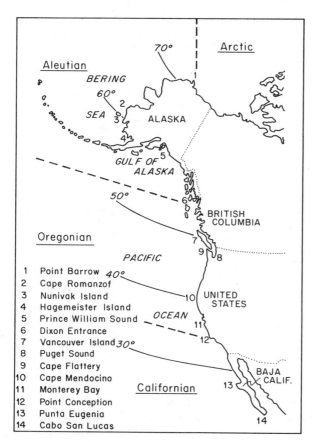

Figure 6-1 Biogeographic provinces for the mollusca of the northeastern Pacific continental shelf. (After Valentine, 1966).

1 Point Barrow
2 Cape Romanzof
3 Nunivak Island
4 Hagemeister Island
5 Prince William Sound
6 Dixon Entrance
7 Vancouver Island
8 Puget Sound
9 Cape Flattery
10 Cape Mendocino
11 Monterey Bay
12 Point Conception
13 Punta Eugenia
14 Cabo San Lucas

of more than one province (e.g., the bivalve *Mytilus californianus* extends from Alaska to Baja California).

Classifications of groups of species belonging to a province are usually based on qualitative judgments. We can organize quantitatively, however, and delineate provinces by grouping localities based on species content. A between-locality similarity index can be calculated and similar locales then grouped into provinces. A simple matching coefficient for two localities is Jaccard's coefficient

$$M_{ij} = \frac{C_{ij}}{N_i + N_j - C_{ij}}$$

where C_{ij} is the number of species the two samples have in common, N_i the number of species in sample i, and N_j the number of species in sample j.

From a matrix of n localities and m species we get a matrix of Jaccard coefficients (M_{ij}'s). These samples can then be clustered hierarchically, based on the M_{ij}'s (e.g., one can use the unweighted-pair group method in Sneath and Sokal, 1974). A trellis diagram

TABLE 6-1   COEFFICIENTS OF SIMILARITY (SEE TEXT)
BETWEEN ADJACENT PAIRS OF MOLLUSCAN
BIOGEOGRAPHIC PROVINCES AND SUBPROVINCES,
PACIFIC NORTH AMERICAN SHELF[a]

Provincial or subprovincial pair	Similarity, total mollusks[b]	Similarity, less endemic forms
Aleutian–Arctic	33	42
Oregonian–Aleutian	31	38
Columbian–Aleutian	45	57
Mendocinan–Columbian	54	61
Montereyan–Mendocinan	58	63
Californian–Montereyan	46	55
California–Oregonian	38	48
Surian–Californian	38	45

[a]After Valentine, 1966.

[b]Shelled gastropods and bivalves.

has been calculated for samples of mollusks from each degree of latitude for the North American West Coast shelf (Valentine, 1966). Breaks in clusters correspond to qualitatively established provinces but now have an objective basis.

Table 6-1 shows similarities between adjacent provinces. Provincial boundaries correspond well to major thermal discontinuities: (a) the 27–28° north latitude break is a transition from the Pacific equatorial water mass to the colder California current; and (b) the 34–35° break at Point Conception, California, divides cool water to the north from warm water (brought by geostrophic flow from offshore) to the south.

Imbrie and Kipp (1968) used a multivariate technique known as *factor analysis* to construct assemblages of Atlantic planktonic foraminifera that behave statistically independently of other so-constructed assemblages. Figure 6-2 shows the occurrence of these independent groups and their excellent fit with water mass distribution in the Atlantic. Other groups, such as diatoms, reveal similar trends (e.g., Maynard, 1974). Thus objective clustering techniques produce provinces that generally match the surface thermal structure of the oceans.

Guillard and Kilham (1977) summarize major aspects of the biogeographic distribution of planktonic diatom assemblages in relation to climate and hydrography. They conclude that major climatic zones and hydrographical systems, such as gyres and currents, contain characteristic groups of species. Latitudinally, temperature is the overriding factor. For example, different clones of the same species grow optimally in their respective native temperature regions (Guillard and Kilham, 1977). Physiological differences also exist between nearshore and offshore assemblages, however, diatoms inhabiting nearshore waters are more resistant to environmental stress, such as pollutants, than offshore species (e.g., Fisher et al., 1973). Genetic differences may explain differential physiological responses of conspecific populations living in different environments (Guillard and Kilham, 1977).

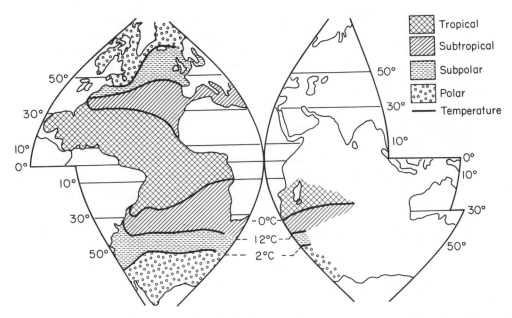

Figure 6-2   Distribution of deep-ocean core top samples dominated by climate-diagnostic assemblage defined through factor analysis. (After Imbrie and Kipp, 1968, *Late Cenozoic Glacial Ages*, edited by Karl K. Turekian, Yale University Press. Copyright © 1971 by Yale University.)

Adaptability and Biogeographic Range

Species living in variable environments should also have the flexibility to occur over a broad biogeographic range. Fugitive species adapted to invade newly open patches might be expected to fit this pattern. The fugitive polychaete *Capitella capitata* (see Grassle and Grassle, 1976, for some taxonomic revision) has worldwide distribution from the boreal to the tropics. Similarly, the fugitive bivalve (see Chapter 7) *Mulinia lateralis* has one of the longest latitudinal ranges of any primarily subtidal bivalve: New Brunswick to the Yucatan Peninsula.

Jackson (1974) tabulated biogeographic ranges for marine shallow-water bivalves and demonstrated that bivalves adapted to variable shallow waters have broader geographic ranges than deeper-shelf bivalves. Several problems arise in extrapolating the results. First, deep-sea bivalves, living under constant conditions, can have very broad geographic ranges (see Sanders and Hessler, 1969). This factor is probably due to the environmental homogeneity and probable lack of barriers of the deep sea. Secondly, high intertidal gastropods often have highly endemic, or localized, distributions. Selection against dispersal mechanisms occurs because of the difficulty of pelagic larval location of the high intertidal.

The significance of biogeographic range lies in its possible inverse relation to extinction rate. Species spread over a wide area might be less prone to extinction when the environment deteriorates in one region. Thus the deep-sea biota might, through broad geographic occurrence, survive a localized environmental change.

Bretsky (1973) analyzed extinction before the Permian major marine extinction. Bivalve molluscan genera with cosmopolitan (worldwide) ranges go extinct much later than geographically restricted genera. Levinton (1974) shows that cosmopolitan bivalves of the Paleozoic and Mesozoic have much longer geologic time ranges than genera with more restricted geographic range. Boucot (1975) marshalls a great deal of evidence to show that Paleozoic brachiopods with broad geographic range are less prone to extinction than those of narrow range.

Thus biogeography creates a mosaic on which the environment may exert differential extinction. We presume that a broad geographic occurrence, whether due to adaptability or to a broadly homogeneous environment, will reduce extinction. Note the problem of distinguishing between adaptability and environmental homogeneity over a broad geographic extent. For example, does the mussel *Mytilus edulis* have a broad occurrence because it is adaptable to a changing environment or because the rocky intertidal is similar everywhere?

Dispersal and Range Extension

New colonizations of marine organisms, often facilitated by humans, provide insight on the rapidity of spread and effects of the colonists on the local marine biota. Three types of colonizations are possible: (a) planktonic larvae (see Chapter 8) may traverse great oceanic expanses and colonize new coasts; (b) attached forms (e.g., barnacles) may raft across oceans on floating logs, ships, and so on; and (c) species may be introduced by humans for culture, or species associated with the cultivated species may be accidentally introduced.

The rocky shore periwinkle, *Littorina littorea,* was noticed in Nova Scotia in the nineteenth century. Since that time, it has spread southward and has become the dominant shore gastropod in New England. The spread southward was probably facilitated by coastal shipping and the construction of rock jetties along the southern New England and middle Atlantic coast. Its current southernmost extent in Maryland is probably controlled by an inability to evolve tolerance to high summer temperatures. Currently, it is the major herbivore on rocky shores and probably has displaced native species of *Littorina.*

Dispersal and Extinction

Biogeographic range can be extended through chance migration across a barrier, such as an ocean, with subsequent establishment of a growing population. We define propagule as the minimum number of individuals that can establish a new reproducing population. If the species is dioecious, we need one of each sex, or the hoped for pregnant female. But if a species arrives at a site, stochastic events may result in extinction, particularly if the number of migrants are few or r_{max} is very low. If the colonization target is small, the absence of a suitable habitat or the chance of extinction due to predation or competition by residents will increase.

If immigrants of a species invade a target area devoid of the species at constant rate i, the expected number of individuals, N_t, at time t is

$$E(N_t) = \frac{i}{b - d} [e^{(b-d)t} - 1]$$

(Bailey, 1964), where b and d are birth and death rates on the target area. If $b \leq d$, population size is maintained or increased (if $b = d$) only through immigration. As b/d surpasses 1, the contribution of individuals through population growth on the area of colonization increases. If we have an exponentially growing population with initial size a number of propagules, the probability of extinction by some time t is

$$P_0(t) = \left[\frac{d(e^{(b-d)t} - 1)}{be^{(b-d)t} - d} \right]^a$$

If b is greater than d, the probability increases asymptotically to $(d/b)^a$. With $b = 0.7$, $d = 0.2$, the limiting probability of extinction with one propagule is 0.29; with two propagules it is 0.08; with four propagules it is only 0.006. The probability of ultimate extinction is therefore greatly decreased with increasing $b - d$ and increasing a (Poole, 1974, p. 441).

The significance of this calculation applies to the study of gastropods living on both sides of the Atlantic, many of whose pelagic larvae are found commonly in open-ocean plankton (Scheltema, 1971). Therefore a is probably high. We know that cross-ocean immigration is not the major source of population growth in many cases, for between-coast morphological differentiation is apparent in some of the species.

Island Biogeography

A simple but elegant theory proposed by MacArthur and Wilson (1967) explains the species richness of islands as a balance between immigration, extinction, and species interactions. It clearly has applicability to all communities receiving colonists, but patterns on isolated targets are liable to be more definable.

Consider a species pool P, all of which are capable of eventually reaching an island. We can plot the immigration rate I of new species as a function of species already present, S (Fig. 6-3). As S approaches P, the rate of addition of new immigrant species must decrease, for the species pool is exhausted. Because some species are better migrants than others, the rate will be high at first but lower as S approaches P. Consequently, the curve is concave up. The extinction rate E, as a function of S, is similarly constructed (Fig. 6-3). With $S = 0$, E must equal 0; but as the number of species on the island increases, the rate at which species go extinct increases because there simply are more species to go extinct. Because more species are on the island, competitive interactions are likely to be more intense. Thus the extinction curve is concave upward, increasing as S approaches P. As seen in Fig. 6-3, there is a point where the two curves cross, and extinction is balanced by immigration. This is a stable equilibrium, at $\hat{S}$.

If we consider near islands and far islands, curves for immigration must be greater

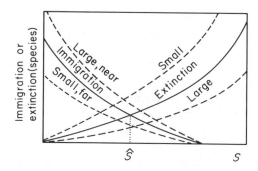

Figure 6-3   Immigration and extinction (expressed as numbers of species per unit time) as a function of species numbers already present on an island. The preceding exists for islands of different sizes and distances from the colonization source. (From *The Theory of Island Biogeography* by Robert H. McArthur and Edward O. Wilson © 1967 by Princeton University Press. Reprinted by permission of Princeton University Press.)

in slope for near islands, as when S is low, I must be greater due to the island's proximity to shore. Thus $\hat{S}$ should be higher for near islands than far islands. This relationship holds well in oceanic island bird faunas (MacArthur and Wilson, 1967). Island size also influences the extinction curve, for E should be greater on small islands than larger islands, as S approaches P. Thus $\hat{S}$ should be greater for larger islands than small islands.

The area effect is probably related to two processes.

1. Large islands present larger targets for migrants and large continents or ocean basins present larger areas with more ecological heterogeneity. Isolation barriers are probably more abundant in larger areas and allow the evolution of more species.
2. Extinction rate is probably higher in small areas due to stochastic events acting on small population sizes in a probably more ecologically homogeneous habitat.

This approach has been tried successfully in lakes (Lassen, 1975) and coastal river systems (Sepkoski and Rex, 1974). Lassen (1975) showed that the species–area relationship for freshwater snails is different for oligotrophic (nutrient-poor) and eutrophic (nutrient-rich) lakes in Denmark (Fig. 6-4). Oligotrophic lakes have a steeper species–area relationship. An analogy may be drawn with small and large islands. The extinction curve is likely to be steeper in oligotrophic lakes, for population size will be less than in food-rich eutrophic lakes. Also, the E curve might show increased concavity due to intense interspecific competition for food as S approaches P.

The theory suggests that a newly created island would increase in species to an equilibrium. A later equilibrium lower than $\hat{S}$ might be reached because of competition. Simberloff and Wilson (1969) examined the recolonization by arthropods of defaunated islands in Florida and found a return to predefaunation levels within 200 days. There is some evidence for a competitive effect on species richness.

A. Schoener (1974) followed the colonization of marine invertebrates onto plastic mesh sponges. Large and small sponges were placed near and far from an algal bed (a

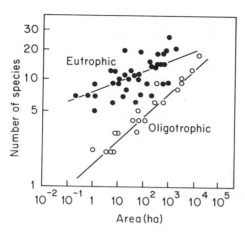

Figure 6-4 The species area curve for freshwater snails in oligotrophic and eutrophic ponds in Denmark. (After Lassen, 1975, from *Oecologia*, vol. 19)

presumed source of colonists). The experiment was performed at Bimini Lagoon, Bahamas; so P was probably large but indeterminate. The following results were obtained.

1. A stabilization of species number did occur with time.
2. There were more species and individuals on large than on small sponges.
3. There was no difference in species number on "near" or "far" sponges, probably because the source area was not the algal bed.
4. Early colonizing species were suspension or detritus feeders and other types followed.

Sutherland and Karlson (1977) followed the colonization of epibenthic invertebrates onto ceramic tiles suspended off a dock at Beaufort, North Carolina. Although S stabilized after a time, the composition of the community moved to no predictable species assemblage. Random events of colonization and complex competitive outcomes precluded simple conclusions.

Types of Colonizing Species

We mentioned fugitive species as those that colonize newly opened habitats and are soon outcompeted by superiors, in the absence of predation or disturbance. Thus a community is a mosaic in various stages of movement to a competitively determined equilibrium. Grassle and Grassle (1974) suggest an ordering of species on the basis of colonization rate. They observed that some polychaetes always precede others on newly established habitats. Similar hierarchies of colonization have been established for subtidal benthic invertebrates colonizing boxes of sediment in Long Island Sound (McCall, 1977). One might suggest that earlier colonizers will probably lose out to later, more superior competitors. The assumption is that colonization rate is greater in poorer competitors or that in a group of newly opened patches poorer competitors will appear more frequently.

Diamond (1974) proposed that the survival prospects of a particular species may be quantified on a series of islands by determining the fraction of islands of a given small range of S on which it occurs. The frequency spectrum might include three alternative types. Type a is a "supertramp" found only on small islands of low S and is always competitively excluded by other species on large islands. Type b is a "tramp" that appears in islands of intermediate S and can always be found on large islands. Type c is a "high-S" species, probably found only on large islands that have diverse microhabitats and low extinction rates. Such species may be viewed in the context of population parameters as having decreased r_{max}. Such an approach is probably useful in marine assemblages, but it remains untried.

6-2 SPECIATION AND SPECIES IN MARINE HABITATS

Speciation in the Sea

The origin of a new species probably requires the isolation of a daughter population from a parent population. Sufficient subsequent genetic divergence prevents interbreeding and complete mixing (introgression) when the two isolated populations are subsequently brought into contact (see Mayr, 1963). Isolation of populations followed by speciation is characterized as *allopatric speciation*. The genetic divergence during the period of isolation can develop in the following ways.

1. The habitat encountered by the isolated daughter population is different and selects for morphological, physiological, or reproductive change.
2. The genetic variation of the colonizing daughter population is only a small and unrepresentative sample of the parent population. Thus similar selection agents operate on a population that is genetically different, causing divergence from the parent population.

If such changes have occurred, the fitness (proportionate contribution of offspring to the next generation) of progeny of crosses between newly merged parent and daughter isolated populations might be significantly less than crosses within each population. Thus there will be selection for any genetic variants that permit displacement of reproduction of the two populations. A displacement could occur through separation of reproductive periods. Steele and Steele (1972) examined the reproductive season in species of the amphipod *Gammarus*. In this group males cannot select females of the same species in precopulatory behavior when more than one species is present (Fenchel et al., 1978). Interspecific crosses result in no progeny, however, suggesting strong selection for reproductive displacement. Co-occurring species of *Gammarus* do not have overlapping breeding periods (Fenchel et al., 1978).

In the Dorvilleid polychaete *Ophryotrocha*, species have evolved various patterns of antihybridization mechanisms. Co-occurring species are morphologically indistinguishable, but avoidance of pseudocopulatory activity can be demonstrated. Species occurring as isolated populations (e.g., Mediterranean versus western Atlantic) mate freely, but offspring are inviable (see Åkesson, 1973).

Isolation in the sea requires barriers that are of common occurrence. The open ocean is an effective barrier for most shelf benthic invertebrates. North–south provincial boundaries are often locations of between-water mass isolation. Continents, midocean ridges, isthmuses, and broad expanses of unsuitable substrate type also provide isolation. Many virtually identical pairs of bivalve mollusk species are separated by the Isthmus of Panama and islands or oceanic ridges in the Caribbean.

Although allopatric speciation is an obvious explanation for speciation on either side of a geographic barrier, such as an isthmus, some species' ranges end at thermal discontinuities. Point Conception, California, and Cape Hatteras, North Carolina, are two examples of major north–south breaks in temperature. It is possible that genetic divergence can occur on either side of such a discontinuity even in the presence of some gene flow. Thus species might originate because of strong divergent selection in two geographically adjacent but environmentally different habitats. This model of divergence and speciation is known as the *parapatric model*.

Species Problems

Although species recognition usually depends on qualitative and quantitative attempts to characterize a reproductive and population genetic phenomenon with morphology, several pairs of "species" in the sea are so close as to have stimulated much controversy as to their status. A taxonomic revision of any closely related group of species is likely to result in changes of the number of recognized species. But several famous question marks remain.

Eels.   Two species of eel, *Anguilla rostrata* (American eel), and *Anguilla anguilla* (European eel), are now recognized, but the species difference is the subject of hot dispute (see Rodino and Comparini, 1978). Both species migrate from freshwater rivers, ponds, and brackish waters of salt marshes to spawn in open marine waters of the Sargasso Sea. Schmidt (1923, 1925) collected evidence suggesting overlapping but nonidentical spawning grounds for the two species (Fig. 6-5).

Although geographic variation of vertebral number within the range of the American eel is not observed, there is a large difference in vertebral number between the two species (Fig. 6-6). Because vertebral number in fishes is often a function of ambient temperature during the larval stage, it is possible that each species originates as a single randomly mixed (panmictic) population in its respective spawning grounds in the Sargasso Sea. The two spawning grounds might be at differing latitudes and hence differing temperatures. Then the Gulf Stream and the West Wind Drift might carry larvae that leave the Gulf Stream to reach shore (American eel) or remain in the current until Europe is reached (European eel). It is possible that European eels do not survive the trip to the Sargasso Sea (Tucker, 1959). All European eels would therefore originate as progeny of American eels reproducing in the Sargasso Sea.

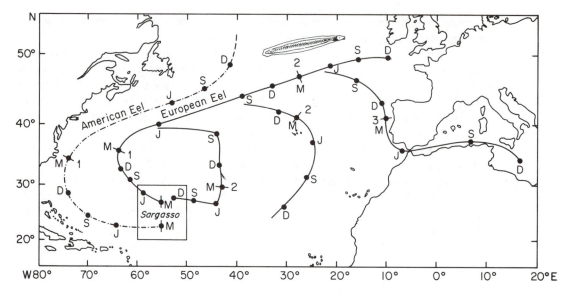

Figure 6-5 The drift of larvae (leptocephali) from the spawning areas of the American eel and the European eel. The geographic positions of larvae hatched in March (M) in the two spawning areas are plotted at quarterly intervals: June (J), September (S), December (D), and March (M). (From Harden Jones, 1968, *Fish Migration,* with permission of Edward Arnold (Publishers) Ltd.)

A difference in chromosome number has been found between the two species (Ohno et al., 1973), but only a large allele frequency difference at one biochemical enzyme locus was detected in an extensive investigation of many loci (G.C. Williams, unpublished; De Ligny and Pantelouris 1973). Furthermore, the frequency of vertebral number of eels from Iceland is intermediate between the European and American samples, further blurring the distinction between the species (G.C. Williams, unpublished).

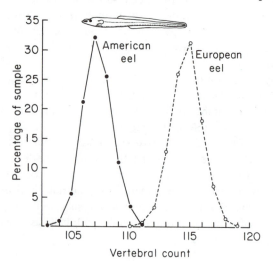

Figure 6-6 The vertebral count distributions of eels caught in America and Europe.

Mussels.   *Mytilus edulis* is a species with subpopulations in boreal and temperate waters of the Northern Hemisphere and with closely related representatives in South America, New Zealand, and Australia (Seed, 1976). In European waters *M. edulis* extends southward from the Arctic to the English Channel, where overlap with the southerly distributed *M. galloprovincialis* occurs. In the Mediterranean *M. galloprovincialis* occurs without *M. edulis*. The species status has been questioned for *M. galloprovincialis,* but morphological and physiological differences do exist between the two named species (Seed, 1971, 1978). In an investigation of biochemical genetic differences, Ahmad and Beardmore (1976) found an allele frequency difference at only one of several enzyme loci investigated, with all alleles identical in the two species. In the English Channel the two species are difficult to tell apart and progeny can be obtained from interspecific crosses (crosses are less successful, however, than within-species crosses). Skibinski and co-workers (1978) present evidence of natural hybridization between the two species.

Barsotti and Melluzzi (1968) believe that *M. galloprovincialis* is a rather recent derivative from *M. edulis,* favored by (a) the warm climate that developed in the Mediterranean since the last glacial maximum and (b) the geographic separation between the Atlantic and the Mediterranean. The speciation process is therefore now in progress (Seed, 1976). A similar distributional problem exists for the species pair *Cardium edule* (northern) and *Cardium glaucum* (Mediterranean); these cockles have overlapping distributions in southern Britain and Jutland.

6-3 EVOLUTION AND GENETIC VARIATION WITHIN MARINE POPULATIONS

Genetic Variation

In diploid organisms, the process of meiosis and zygote formation ensures the generation of a large number of genotypes if there is genetic variation in the population. New variants can be introduced into the population through mutation or immigration of genetically different individuals. Genetic analysis of traits involves controlled breeding programs, using phenotypic variation of genetic origin.

If we consider a genetic locus with two different variants, or alleles, *a* and *b*, we define their respective frequencies as p and q. There can be three possible diploid genotypes:

$$aa, ab, bb$$

where *aa* and *bb* are the homozygotes and *ab* is the heterozygote. If mating is random, population size is infinitely large, and no selection, immigration, or emigration occurs, we can predict the frequencies of the genotypes in a population. Because of meiosis and gamete formation, p is the frequency of gametes carrying the *a* allele; q is the frequency of the gametes carrying the *b* allele ($p + q = 1$). The probability of getting *aa* homozygotes is $p \times p$ or p^2. The probability of getting *ab* heterozygotes is $2pq$ (for you can get *ab* and *ba* gamete unions). Thus the frequency of the genotypes for any future generation are as follows:

$$f(aa): p^2$$

$$f(ab): 2pq$$

$$f(bb): q^2$$

This is known as the Hardy–Weinberg law and holds as long as p and q remain constant, given the preceding conditions.

Several processes can cause deviations from the expected genotypic frequencies. Selective mortality might preferentially remove an unfavorable genotype. Heterozygotes might exhibit superior performance (*overdominance*), or homozygotes might be preferentially favored (*underdominance*). Genotypes bearing a specific allele might exhibit preferential survival; this would impose *directional selection,* resulting in the fixation of the favored allele in the population. Assortative mating would also cause genotypic frequencies to deviate from expected proportions as progeny would no longer be the result of random combinations of gametes. Finally, small population size increases the probability that chance processes may generate deviations from the Hardy–Weinberg ratios.

Mixing of two populations with very different allele frequencies can cause apparent deviations from Hardy–Weinberg ratios. This may happen when two swarms of planktonic larvae from genetically different source populations mix at a single site. The two suites of genotypes are mixed and have an aggregate set of allele and genotypic frequencies that do not represent adequately either source population.

Genetic Drift, Population Size, and Selection

Changes in allele frequencies may occur because of the finite size of the population. In a finite population, the transition from one generation to the next is accomplished through a finite number of gametes. There will usually be a difference of allele frequency between generations because of random sampling of gametes of the parent generation to form the offspring generation. This random difference will, on average, increase with decreasing population size and is the explanation for the *random genetic drift* of allele frequencies. Drift is the accumulation of such sampling bias over generations. Without selection, mutation, or migration, this process will cause the increase of one allele to a frequency of one, with loss of variability. If population size decreases drastically, the potential for biased sampling of the parent generation's genetic composition by the daughter generation is greatly increased. Such population bottlenecks are therefore times when random genetic drift is maximized.

The *founder principle* states that a small isolate of a species migrating to a peripheral location is bound to have a differing genetic constitution than the parent population. Such an effect may be important when only a few larvae survive a trip of ca. 100 km to a site, originating from a parental population that produced several million larvae. With strong selection, this process will be countered and allele frequencies will be shifted by environmental change. The dominance of selection over drift is proportional to the amount of selection per generation and to population size of individuals producing progeny. (See Li, 1955, or Cavalli–Sforza and Bodmer, 1971, for a quantitative introduction to these processes.)

Quantitative Genetics

Frequently we have information that cannot be interpreted as genetic variation at a single or at a few loci and yet we know that there is a genetic component. For example, the length obtained by a snail may have a genetically determined component and an influence due to environmental factors, such as food and temperature. The character variance V in a population can be ascribed* to the following, given that several genes, the environment, and random error determine the variation:

1. V_A, the *additive genetic variance,* or variance due to differences between homozygotes.
2. V_D, the *dominance variance,* or variance resulting from effects of specific alleles in heterozygous form—for example, when a heterozygote is identical in morphology to one of the homozygotes.
3. V_E, variance due to the environment and error.

Heritability is the fraction of V ascribed to V_A, or V_A/V. It is a measure of the variation available on which selection can act to change the character. Thus when heritability is substantial, selection for the largest-sized individuals will result in a substantial genetically based increase in size in the next generations (see Falconer, 1960, for more technical definitions). This is the principal type of approach possible in most variation because most characters are polygenically determined and difficult to partition among a few interpretable genetic loci. Mother–offspring correlations (Fig. 6-7) are useful in determining the heritability of characters because of their typical availability for analysis in natural populations (Christiansen and Frydenberg, 1973).

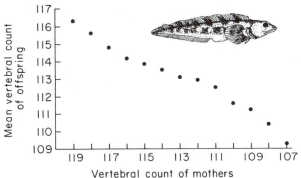

Figure 6-7   The mean vertebral count of *Zoarces viviparus—a* comparison of vertebrae in mothers with those of their offspring. Results suggest that the vertebral number is genetically determined. (Data from Schmidt, 1920).

*Note that this is a simplified discussion.

The methods of quantitative genetics have rarely been applied to marine species. Bradley (1978) has shown a strong genetic component in local adaptation of copepods to different thermal regimes.

Types of Variation Measurable in Marine Organisms

The data useful for studies of genetic variation are as follows.

Chromosome number and polymorphism.   Some marine species have variable average numbers of chromosomes per individual among populations and variation in morphological characteristics of given chromosomes within a population (as in inversion polymorphisms of *Drosophila*). In the drilling gastropod *Nucella (Thais) lapillus,* variation in chromosome number has been observed along the Breton coast of France (Staiger, 1957). A haploid number of $n = 13$ was found on wave-exposed shores but $n = 18$ in sheltered localities. The cause of this difference is unknown, but it could have been accomplished through fusion of chromosomes of the sheltered population. Any genetic changes that can be ascribed to various gene position effects due to chromosome rearrangement would apply here. Some chromosomal polymorphism has been observed in the genus *Mytilus* (Ahmed and Sparks, 1970). Chromosomes differ in form, but not in number.

Color polymorphisms.   Color spot patterns have been successfully interpreted as products of one or a few loci. The harpacticoid copepod *Tisbe reticulata* is a benthic species found in shallow water on *Zostera* (eel grass), seaweeds, and bare substrates. Color spots on the cephalothorax and thoracic segments vary in size, pattern, and color and can be ascribed to several independent genes (Bocquet et al., 1951). In Venice populations, one polymorphic locus with several alleles explains the variation (Battaglia, 1958). In crowded laboratory cultures there is evidence of heterozygote superiority vis a vis a significant excess of heterozygous survivors over expected Hardy–Weinberg proportions. The basis of this feature is unknown, but color may be "marking" a portion of the genome responsive to crowding effects (oxygen depletion, and so on). Such color variation has been also observed and genetically interpreted through crossing tests of the isopod *Sphaeroma* (see Bocquet et al., 1951; West, 1964).

In populations of the mussel *Mytilus edulis,* a brown shell phenotype co-occurs with a (usually) more common black phenotype. Crosses between different phenotypes show that a simple one-locus, two-allele genetic model governs the variation (Innes and Haley, 1977). The brown allele is dominant over black. A second gene controls the presence of black stripes on the shell.

Mitton (1977) studied the geographic variation of light-colored, striped morphs and found an increase of black morph frequency toward the north (Fig. 6-8). Dark mussels gain heat more rapidly on a sunny day and hence are superior in habitats with freezing.

Heritable morphological characters.   Although we have the least-detailed genetic information on such characters as size and shell thickness, ecological interpre-

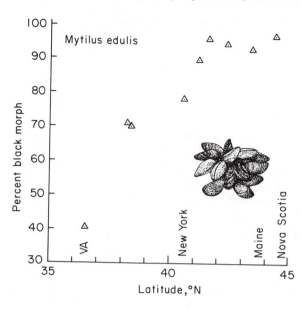

Figure 6-8 contents: Mytilus edulis; y-axis "Percent black morph" (100, 90, 80, 70, 60, 50, 40, 30); x-axis "Latitude, °N" (35, 40, 45); location labels VA, New York, Maine, Nova Scotia.

Figure 6-8 Latitudinal variation in black and brown shell color morphs of the blue mussel, *Mytilus edulis*. (Data from Innes and Haley, 1977; and Mitton, 1977)

tations are often more readily obtained. The sculpture pattern of the Hawaiian intertidal snail *Littorina picta* varies from smooth to coarsely sculptured forms (Struhsaker, 1968). Cultivation experiments of larvae of *Littorina* show that sculpture patterns are inherited from parents. In the field, smooth forms occurred on low-angle beaches where wave intensity was highest whereas more strongly sculptured forms were more common on high-angle rocks and beaches. Presumably smooth shells present the hydrodynamically most stable surface in strong current situations. In one case, the diversity of sculptural types of one locality decreased from juveniles to adults, indicating selection for optimal sculptural types from a polymorphic population. This factor is important in a species where pelagic larvae often settle on substrates that will cause shifts from the morphological diversity of the source of the larval population. The intertidal drilling snail *Thais lamellosa* shows a similar variation from smooth shells in rough outer-coast environments of Washington to delicately sculptured forms in the more protected waters of the San Juan Islands, Washington (Spight, 1973).

Physiological differentiation. We might expect two populations living in environments of differing salinity or temperature to differ genetically due to adaptation to local conditions. Mussels *(Mytilus edulis)* collected over a wide range of ambient salinities show inhibition of oxygen consumption when salinity is changed from that of ambient conditions of their native habitat (Bayne et al., 1976b). The genetic difference involved here is unknown but potentially great.

Although physiological differences among populations may have a genetic basis, individuals may irreversibly acclimatize to their native environment. Therefore a genetic difference cannot be inferred simply because individuals collected from two different environments show different physiological response. Raising the populations under identical laboratory conditions into the next generation should eliminate acclimatization differences. If individuals of the next generation respond differentially as a function of habitat, then genetic differentiation is more probable.

Protein polymorphisms.   The techniques of electrophoresis have permitted the discovery of many genetic markers in marine organisms without the use of crossing programs. Although any genetically interpreted variation should be confirmed by controlled crossing experiments, the Mendelian variation commonly observed in protein polymorphism as staining patterns on starch, acrylamide, or cellulose acetate gels lends strength to their genetic significance. The technique of electrophoresis involves separating proteins on gels of differing supportive media and pH, in the presence of an electrical potential. Proteins move at different rates as a function of surface charge, molecular weight, and conformational differences. Gels are then stained by using cofactors, substrates and stains specific for a given enzyme or protein.

Band patterns thus observed can often be simply interpreted as gene products of a single locus. In the simplest case, a single-chain-molecule (monomer) produced by a homozygote stains as a single band and a heterozygote occurs as a double band. We assume that there is no variation that is not electrophoretically detectable (not always true).

This technique has permitted many genetic loci to be screened in marine populations where no genetic information was previously available (see Gooch, 1975, for an exhaustive review of the marine literature). The adaptive nature of variation at enzyme and other protein loci (e.g., hemoglobin), however, is poorly understood. Geographic patterns of variation may be measured and correlations with environmental factors may be made. An integrated study of natural populations, biochemistry, physiology, and genetics is essential in order to understand fully the adaptive nature of enzyme polymorphism.

Patterns of Variation

Genetic polymorphism.   The amount of polymorphism maintained in a species is of great interest because it suggests the scope of variation on which selection may act. If an environment is heterogeneous, we might expect polymorphism to increase. The increased variation of color polymorphism in the limpet *Acmaea digitalis* on substrates of varying colors supports this hypothesis.

Species of *Macoma* living on different varieties of sedimentary substrata were examined for degree of genetic polymorphism at two enzyme loci (Levinton, 1975). At a leucine aminopeptidase locus, no difference was found among species, but a direct correlation between polymorphism and variety of substrate occupation was found at a phosphoglucose isomerase locus. Thus heterogeneity within the genome in response to environmental heterogeneity is expected. Similar results for the latter enzyme locus were obtained in studies of variation of bivalves living in Irish waters (Wilkins and Mathers, 1975; Mathers, 1975).

The correlation of genetic polymorphism with heterogeneity of the environment might suggest that a variable environment may support more genetic variants than a more stable environment. This hypothesis would predict that shelf environments would support more polymorphism than the deep sea (Bretsky and Lorenz, 1970). In fact, this prediction turns out to be untrue (Schopf and Gooch, 1971). Valentine (1976) reviews the literature on enzyme polymorphisms and variable environments and concludes that total polymorphism tends to increase in species living in trophically stable environments. Valentine suggests that species living in variable environments must have only a few variants shared by most of the population in order to be able to cope with strong environmental change, such as seasonal change in temperature. Thus proteins maintained in such polymorphisms will be biochemically more flexible (active over a greater range of temperature for example) than proteins of species living in constant environments. Although a flexible protein may not be as efficient as a specialized protein in any one environmental state, it might be favored when the environment varies over a great range. A similar hypothesis was suggested by Levinton (1973). In a more constant but heterogeneous environment consisting of a series of patches, each of which does not vary much over time, more specialized alleles might be maintained. The entire question of polymorphism and environmental variation remains a mystery. We cannot exclude the possibility that there are no adaptive differences among protein variants and that random processes maintain polymorphism.

Geographic variation.   Trends in the distribution of allele frequency have been commonly observed for marine species, particularly in protein polymorphisms. Variation along a geographic distance is known as a *cline*. Clines have been found for shell morphology (Struhsaker, 1968), chromosome number (Staiger, 1957), shell color (Mitton, 1977), and protein allele frequencies (Gooch, 1975). O'Gower and Nicol (1968) investigated the geographic variation of alleles at a hemoglobin locus in a marine bivalve along the east coast of Australia. There was a continuous change of frequency along the coast, with a local reversal consistent with temperature as the correlative environmental factor. Johnson (1971) showed that a two-allele lactate dehydrogenase locus in the intertidal crested blenny, *Anoplarchus purpurescens,* showed latitudinal clinal variation in Puget Sound, Washington, consistent with temperature variation. The *A'* allele is associated with warm temperature and increases from less than 0.05 from the Strait of Georgia southward to over 0.25 near Tacoma, Washington (Fig. 6-9). At some sites, the *A'* allele decreased into deeper and colder water localities and an increase of the *A'* allele in 1967 coincided with an unusually hot summer.

Enzyme polymorphisms have been mapped geographically for various ectoproct species living on rocks and wood pilings (Gooch and Schopf, 1970). At two loci, geographic variation of allele frequencies in the Cape Cod region correlated with local water temperature for the warmest season (Schopf, 1974). Allele frequencies were found to be stable between years. An interesting aspect of this study is the parallel variation found at two differing biochemical loci.

Differing among-locus patterns in clinal variation have been found for other species. Christiansen and Frydenberg (1974) found that two unlinked loci of the eel pout, *Zoarces viviparus,* show parallel clines in Danish waters whereas two other unlinked enzyme loci maintain constant allele frequencies over this same geographic range. Similar patterns

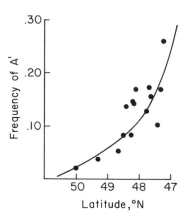

Figure 6-9 Latitudinal variation of an allele at the lactate dehydrogenase locus in the intertidal crested blenny, *Anoplarchus purpurescens,* in Puget Sound, Washington. (After Johnson, 1971)

have emerged in geographic variation in enzyme polymorphisms of the mussel *Mytilus edulis* along the eastern coast of North America. One locus shows abrupt step-clines, one shows smooth linear changes with latitude, and others show no variation at all (Koehn et al., 1976). Among-locus differential geographic variation cannot be compatible with any model postulating isolation among populations or genetic drift, for a different pattern would need to be postulated for each locus, all carried by the same organism. Thus selection must maintain observed geographic variation at some of the loci.

Latitudinal genetic variation in the American eel, *Anguilla rostrata,* provides further evidence of selective maintenance of genetic variation along clines. Samples from five widely spaced localities from Florida to Newfoundland show marked clinal variation in gene frequency for three loci (Williams et al., 1973). Because populations apparently come from one randomly mixed population in the Sargasso Sea, large-scale selection within one generation must apparently explain the differentiation.

Differences of geographic variation as a function of dispersal type have also been found for intertidal gastropods. Snyder and Gooch (1973) showed that the pelagic disperser *Nassarius obsoletus* shows less regional differentiation of allele frequencies than the brooding snail *Littorina saxatilis.*

Levinton and Suchanek (1978) investigated among-locality genetic differences in the mussels *Mytilus edulis* and *Mytilus californianus. M. edulis* lives in a wide variety of habitats ranging from outer-coast exposed rocks to mud flats in estuaries. *M. californianus,* however, lives only in open-marine exposed rocks. At two enzyme loci, *M. edulis* showed more among-locality differentiation than *M. californianus.* Greater local differing selective pressures over *edulis* habitats relative to *californianus* habitats was invoked to explain the difference. A comparison was also made of variation at two enzyme loci in *Mytilus edulis* along the East Coast of North America, with variation in *M. californianus* along the West Coast at the same loci. Latitudinal variation on the East Coast was pronounced, but it was insignificant along the West Coast. The greater latitudinal thermal gradient along the East Coast, relative to the homogeneous thermal regime of the Pacific Coast, explains this difference.

Microgeographic variation.   Variation among patches in a local habitat is common in marine species. In the case of color variation of limpets (Giesel, 1970), it can be explained on the basis of behavioral selection for appropriate substrates in combination with selection against genotypes that are unsuited for specific habitats. Color morphs that are conspicuous against a rocky substrate of contrasting color are easily located by birds. Where individuals are sessile, microgeographic variation may be explained by selective mortality among genotypes.

In mussels, several cases of small-scale clines as a function of intertidal height are known (Balagot, 1971; Levinton and Fundiller, 1975). Levinton and Fundiller found an association between size and genotype at an enzyme locus. A subsequent study, however, showed no growth rate differences among genotypes of mussels at the same locus (Levinton and Lassen, 1978).

A somewhat larger-scale variation has been found at a leucine aminopeptidase locus in the mussel *Mytilus edulis* south of Cape Cod. In several estuaries, the frequency of the LAP^{94} allele decreases from about 0.55 to 0.1, from open marine waters into the estuary. Given the dispersal powers of this species (pelagic planktotrophic larva of 3 to 7 weeks), such differentiation requires strong selection at the locus unless the estuarine populations are isolated in some fashion.

The variation in natural populations observed suggests the interaction of migration and selection as a mechanism for maintenance of polymorphisms in nature. The strong selection observed in shell morphology of a benthic snail at a locality can be constantly diluted by pelagic migrants from adjacent areas. In planktonic species, mixing of water masses may continuously blend populations, thereby precluding much geographic differentiation into clines. We do, however, observe geographic variation in the cod and the eel as a function of latitude. This variation may be explained by strong selection or the maintenance of separated "stocks" of fishes that genetically differentiate (Møller, 1969).

The question arises as to whether such selection operates directly at the loci being examined. It is possible that the loci under investigation are markers for other parts of the genome. It is unlikely that viability differences among color morphs of *Tisbe*, for example, are related directly to the color differences themselves. Instead the color loci are probably closely linked to parts of the genome implicated in the viability differences.

Physiological races.   Different physiologically adapted populations of the same species are known as *physiological races*. The presence of physiological races can also influence geographic variation. If different races exist within estuaries and in open marine waters, pelagic larvae entering the estuary might be poorly adapted to low salinity and have greatly lowered viability, thus ensuring the isolation of the estuarine population.

Battaglia (1959) compared populations of *Tisbe reticulata* living in an open-marine salinity habitat (Roscoff, France) and a low-salinity lagoon (Bay of Venice). Laboratory cultures were kept for at least ten generations from the two localities at 23 and 35 ‰. Raising populations in the laboratory eliminates field differences in irreversible acclimatization. After this period, fecundity of the Roscoff population was found to be depressed by a factor of nine when raised at 23 ‰ relative to Roscoff stocks raised at 35 ‰. No difference was observed when the Bay of Venice stocks were raised at the two salinities. Thus the difference in physiological tolerance can be ascribed to genetic differences between the two populations.

Physiological races of diatoms adapted to different temperatures have been found (Guillard and Kilham, 1977). Strains of the same species of diatom collected from tropical coastal and northern coastal Atlantic waters have differing temperature optima and ranges for growth. Inshore clones of the diatom *Thalassiosira pseudonana* have less genetic variation, but higher reproduction rates, than clones derived from oceanic waters (Brand et al., 1981).

Seasonal variation.   Species with short generation times might be expected to display rapid changes in allele frequency. Marine protistans and bacteria have short generation times (time between fissions). The possibility arises that one genetically distinct clone might rapidly displace all others within a season. Natural selection among clones would therefore continually change the relative abundance of genotypes. Allele frequencies at any one time would presumably reflect selection for superior genotypes.

Gallagher (1980) investigated allozyme polymorphisms in clones of the diatom (see Chapter 9, Section 9-2) *Skeletonema costatum*. Populations collected in winter differed substantially from those collected in summer. These differences may have reflected a form of seasonal (cyclic) natural selection.

SUMMARY

1. The configuration of the ocean basins, the latitudinal temperature gradient, and ocean currents combine to determine the geographic distribution of the marine biota. The species richness of any one area may be considered as being the result of a balance between speciation colonization and extinction.

2. Marine species probably arose as the result of isolation of subpopulations of parent species. The open ocean is an effective barrier for most continental shelf invertebrates. Current systems can also serve as barriers. Strong selection gradients may also result in local adaptation and subsequent isolation.

3. Marine species, like others, display a large amount of genetic variability. Of special note is the presence of physiologically distinct subpopulations, adapted to local environments.

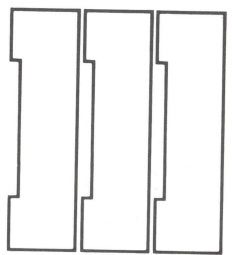

Reproduction, Dispersal, and Larval Ecology

In Part III we describe the means by which marine organisms reproduce and disperse to new habitats. We relate these adaptations to the organism's environment and discuss the environmental cues that permit marine species to disperse to or migrate between favored habitats. We discuss some theories explaining the origin of reproductive and dispersal strategies and relate them to marine organisms. Although an evolutionary subject, reproductive and dispersal strategies are the keystones on which ecological success is translated into evolutionary longevity.

7

reproductive strategies

7-1 THE COSTS AND BENEFITS OF REPRODUCTION

Allocation of Resources

The three component processes of maintenance, growth, and reproduction compete for the limited resources that any given individual has at its disposal. The growth rate of the common intertidal barnacle *Balanus balanoides,* for example, decreases during the reproductive or spawning period (Barnes, 1962). Energy is consumed in egg production, defense of reproductive territories, such as nesting areas, and defense of eggs against predators. If a marine organism broods its young, then more energy will be devoted to reproduction per offspring than in those cases where larvae are sent off into plankton to fend for themselves.

The profit obtained for a given reproductive investment is the number of offspring produced by a parent up to a given age. The allocation of resources between reproduction, maintenance, and somatic growth should be optimized to produce the maximum number of successfully surviving offspring. A low-somatic and high-reproductive investment indicates that continued existence of the reproductive individual beyond the first reproductive period will not increase the profit to the individual. In contrast, a higher investment of resources to growth, relative to reproduction, indicates that the body is necessary to support future reproductive efforts.

It is profitable for Pacific migrating salmon to invest heavily in one reproduction because, once a successful migration has occurred, the production of large numbers of young is relatively easy. Increasing the reproductive effort will increase the profit gained in offspring. But utilization of equivalent energy reserves for a return migration is not favored because the probability of a second successful return is too small.

The relation of effort to profit determines whether a species reproduces only once *(semelparous)* or repeatedly *(iteroparous)*. Gadgil and Bossert (1970) treat semelparous and iteroparous modes of reproduction as selected responses to costs and profits associated with reproductive activities at a certain life-history stage. The diversion of energy from somatic growth and maintenance to reproduction increases parental mortality and lowers somatic growth rates. If the quantity of resources available to an organism is fixed, an increase of resource allocation to current reproduction results in a decrease of available resources for somatic maintenance and growth necessary for future reproductive activity. Therefore current reproductive activity and future reproductive success are inversely correlated (Williams, 1966).

If a current investment is so high that it jeopardizes future reproductive success, then the residual reproductive value for the rest of the individual's life span is minimized. Consequently, an optimal reproductive strategy is determined by the profit gained by allocating resources to current reproduction as opposed to saving some energy for future reproduction through investment in somatic growth (see Gadgil and Bossert, 1970, and Pianka and Parker, 1975, for some graphical theory).

Most marine animals have a life cycle that can be divided between larval and adult stages. The larval stage includes the period of development from the formation of the zygote to the point where the larva is recruited to the habitat of the adult and metamorphoses to adult morphology. Fish larvae that drift in currents from a spawning ground to a nursery ground for feeding are typical examples.

In such life cycles the risk of being a larva may be compared with adult mortality to suggest the strength of selection for reproductive changes. In Chapter 4 we show that under high adult mortality the species with the highest r_{max} will win in competition. So where there is significant danger to the adult population we expect selection to favor early reproduction and one-time spawning (semelparity). The probability of mortality is so great that any genetic variants allowing a second reproduction in some genotypes would be reduced in favor of those that reproduce early and invest a great deal of energy in that spawn, thereby lowering the probability of survival to a second reproduction.

An increase in r_{max} must be assessed in terms of reproductive effort because many young produced per female may each have a lower probability of survival than when only a few young are produced per female. If resources are limited, then the production of many eggs per spawn must be accompanied by little investment of yolk or mother care per egg. Thus an *"r"* strategist may increase r_{max} with either many or few eggs per female, but reproductive effort must increase.

The selection for early maturity and single reproduction may be reversed if there is uncertainty in juvenile recruitment. Using simulation models of populations, Murphy (1968) showed that iteroparity is favored if a random component to juvenile mortality is introduced. Under these circumstances reproduction more than once increases the probability that a year class of adults will successfully recruit another year class of juveniles. If the random component is removed, the situation is reversed and forms with high reproduction early in life are favored.

If the adult has a low mortality rate and the larval stage is uncertain, then iteroparity is even more favored because the favored adult form will probably devote more energy to survival and competitive interaction in order to survive. Under conditions of low adult mortality the genotypes devoting much energy to early reproduction will lose in competition and not reproduce as well again. Because juvenile recruitment is uncertain, early reproducers will not last in the population.

Murphy tested this hypothesis by comparing variation in spawning success with age of first reproduction in several populations of planktivorous fishes (Table 7-1). As can be seen, those populations in which spawning success is consistent have earlier ages of first reproduction. Some benthic invertebrates seem to exhibit low recruitment and relatively low adult mortality. Many asteroids are slow in reaching sexual maturity, sporadic in reproductive success, and have a long reproductive life (Birkeland, 1974).

TABLE 7-1 SUMMARY OF REPRODUCTIVE PARAMETERS OF HERRING, SARDINE, AND ANCHOVETA[a]

Population	Age at first maturity	Reproductive span	Variation in spawning success (highest/lowest)
Herring (Atlanta–Scandian)	5–6	18	25x
Herring (North Sea)	3–5	10	9x
Pacific sardine	2–3	10	10x
Herring (Baltic)	2–3	4	3x
Anchoveta (Peru)	1	2	2x

[a]After Murphy, 1968.

The mussels *Mytilus edulis* and *M. californianus* seem to represent two alternative reproductive strategies consistent with the mortality patterns of the adults. *Mytilus edulis* usually colonizes open patches in *M. californianus* beds, estuaries, and protected bays. Generally, however, it suffers high mortality from predation and competition for space with *M. californianus*. Consequently, it shows a strong spawning peak in the winter and an extensive amount of reproductive investment. *M. californianus* is less affected by predation and competition and shows no strong seasonal spawning peak (Suchanek, 1978).

Reproductive Strategy and Colonization Potential

With successful reproduction, a population will increase in a favorable environment. Reproductive output of a population must balance mortality and emigration if the population is to persist. The replacement rate R_0 equals unity when the population is stable, is greater than one when the population increases, and less than one as it declines (see Chapter 3).

In Chapter 3 we show that

$$r = \frac{\ln R_0}{T_g}$$

The instantaneous growth rate r of the population, which can be compared to a compound interest rate, is therefore directly proportional to the log of the replacement rate R_0 and inversely proportional to generation time T_g. Thus population increase can be accomplished through an increase of surviving juveniles per female and a generation time decrease. Species living in environments where the probability of local extinction is high are subject to selection on those parameters that increase r_{max}: reproductive effort and generation time.

The polychaete *Capitella capitata* is an indicator of polluted environments throughout the world in shallow-water soft sediments. During an episode of pollution *Capitella* replaces the normal benthic fauna after that fauna has gone extinct (Grassle and Grassle, 1974). New environments are colonized rapidly by planktonic larvae that are mainly produced in summer but occur in the plankton all year round. Mean egg clutch size is about 130 and generation time is 30 to 40 days. Instead of producing large numbers of young through high fecundity, short generation time allows rapid population increase.

A second example of a marine organism capable of rapidly increasing population size is the diminutive, mactracean bivalve *Mulinia lateralis*. Pelagic larvae can rapidly colonize a muddy bottom and achieve densities of 10^4 to 10^5 m^{-2} (Levinton, 1970; Calabrese, 1969, 1970). Rhoads (personal communication) observed a large colonization of the species on muddy bottoms after an oil spill in 1969 off West Falmouth, Massachusetts.

Mulinia lateralis has a high fecundity of 3 to 4 $\times$ 10^6 eggs at a single spawning (Calabrese, 1969). The greatest observed number of eggs released by a single female was about 7 $\times$ 10^6 eggs. Not all the female's eggs had even been released in this very large spawning. It has a longevity of about 2 years, with repeated spawning possible. Finally, *M. lateralis* has a remarkably short mean generation time of about 60 days (Calabrese, 1969). Under ideal conditions a new generation can be established in less than 39 days.

In Long Island Sound, *M. lateralis* successfully colonized newly disturbed subtidal muddy bottoms. Areas eroded by storms or bottoms covered with dredge spoils are usually soon colonized. Having established a brief refuge, *Mulinia* individuals soon produce large numbers of young. As other species colonize the bottom, *M. lateralis* generally loses in competition to superior competitors and may succumb to predators (e.g., Virnstein, 1977).

Body-Size Limitation

Many invertebrates are thought to have indeterminate growth; there is no necessary upper limit to body size. Constraints related to the balance between the efficiency of food intake and the cost of maintenance metabolism, however, may influence the maximum size that an animal might reach and the optimal size at which the production of eggs might occur. Consider an anemone. The animal could continue to grow indefinitely. Does not larger body size permit more gametes to be produced? Why not?

Consider the following argument (Sebens, 1977, 1979). An animal's food intake can be related to the surface area of its feeding organ. In the case of the anemone, let us consider the dorsal surface as the site for food capture. The animal must face a problem as it grows larger; the surface-to-volume ratio of a geometrical solid must decrease with increasing size. Because the surface area increases with the square of length and the weight increases with the third power of length, the relation between surface area S and weight W should be a power function:

$$\text{Surface } (S) = aW^c$$

where a is a constant and c is an exponent equal to 0.67 (corresponding to the surface area/volume relationship). Therefore energy intake might be envisioned as a constant K multiplied by surface:

$$\text{Intake } E_i = KaW^c$$

Metabolic cost is also related to body weight: A power function with an exponent of 0.7 to 1.0 is generally found (Zeuthen, 1953):

$$\text{Cost } E_m = bW^t$$

where b is a constant and t is the exponent. Therefore we can calculate the energy surplus as the difference of intake and cost:

$$\text{Surplus } E_s = KaW^c - bW^t$$

If the exponent for the cost function is greater than 0.67, then at some upper level of W the cost and intake curves must cross—that is, above that size the cost of metabolism will be greater than the possible intake (Fig. 7-1). So there is a maximum size that an animal can reach, given these energetic considerations.

The surplus, E_s, is plotted in Fig. 7-1. As can be seen, E_s reaches a maximum and then falls to zero, with increasing body weight. It thus follows that there is a body size at which a maximum amount of surplus energy is available. Consider the animal with indeterminate growth at the time of reproduction. If the animal's potential for leaving

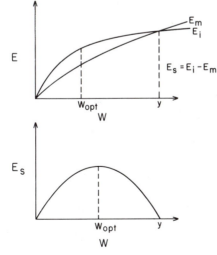

Figure 7-1 Diagrammatic representation of Sebens's theory of optimal allocation of resources to reproduction as a function of body size. As discussed in the text, the difference between energy taken up (E_i) and energy consumed in metabolism (E_m) is maximal at an optimal body size (W_{opt}). See text for details.

offspring is maximized by the maximization of output of gametes, then it would be advantageous to cease growth at the body size that maximizes energy surplus available for the production of gametes. Consequently, the optimal body size for reproduction should be less than the maximum size that the animal could reach before energetic limitations reduced the value of E_s to zero. The animal should therefore grow to the body size corresponding to maximum E_s and then reproduce.

This argument should not be taken literally, for other factors might contribute to the determination of optimal body size. A given body size, for example, might be optimal with regard to surplus energy available for reproduction. This body size, however, might be prone to successful predation. So it might "pay" to grow larger to escape the capabilities of a size-limited predator.

Sebens (1977) has tested some assumptions of the model in a species of anemone. The value of c is close to the predicted value of 0.67; the value of t is somewhat greater than c. Therefore the model probably holds some promise for understanding the energetic limits to growth and reproduction.

7-2 MODES OF SEXUALITY

Many marine organisms are capable of reproduction without the aid of meiosis, gamete formation, and zygote production. Asexual reproduction is common in both plants and animals but is rarely the exclusive means of reproduction. Most asexually reproducing species have sexual adaptations as well. Apparently the generation of new genotypes through recombination and the maintenance of genetic variability through sexuality are strongly favored in living organisms.

Asexual Reproduction

The advantages of asexuality over sexuality are twofold. First, asexual reproduction permits the proliferation of a genotype that successfully colonizes a given microhabitat. Secondly, the proliferation of this genotype is not hampered by the suite of adaptations necessary for gamete union. But all individuals have sprung from one individual and are therefore of identical genotype. Thus an asexually (clonally) generated population may be poorly suited to survive if the favorable environment suddenly changes.

Three types of asexual reproduction are discussed: binary fission and fragmentation, parthenogenesis, and vegetative reproduction. Binary fission allows a protistan to increase in numbers from a single parent cell. The diatoms, for example, are a dominant group of phytoplankton in the open ocean. These algae are usually unicellular and have a siliceous skeleton composed of two valves. At cell division, each daughter cell inherits one valve. The number of cell divisions per day can vary dramatically, depending on culture conditions (e.g., Davis et al., 1973). Diatom species, however, have doubling rates ranging from 0.6 to 6 doublings day^{-1} under optimal conditions (Eppley, 1977). So the capacity for population increase is enormous.

Because a diatom population can double so rapidly, many generations are traversed during a season. A large number of generations will allow for the selection of the genetically most favorable type and asexual reproduction preserves the dominance of the superior genotype. Rapid turnover has apparently led to strong genetic differentiation of diatom populations living in seawater of different temperatures (Guillard and Kilham, 1977). Diatoms also can reproduce sexually, but sexuality seems to be a response to a critically minimum cell size (Bold and Wynne, 1978) or a sudden increase in nutrient supply (Davis et al., 1973).

In multicellular organisms fragmentation can serve the same function as binary fission. Fragments of seaweeds, such as *Enteromorpha,* can develop into new individuals. The archianellid *Dinophilus* can also reproduce by breaking into pieces that develop into complete worms. In some cases, special buds are designed to break off from the "parent" individual. Such buds are known in both seaweeds and sponges.

A second asexual type of reproduction in multicellular organisms is parthenogenesis, where unfertilized eggs develop into normal individuals. In some cases, such as the rotifers, sperm are required to initiate cleavage but do not contribute their genes to the offspring. This mode of reproduction is relatively rare in the marine environment.

A most important type of asexual reproduction in marine invertebrates is vegetative reproduction or division of a given animal into many individuals who may or may not be connected. It is utilized by such colonial animals as encrusting sponges, coelenterates, ectoprocts, and some other invertebrate groups. Marine algae and angiosperms also have vegetatively reproducing species. In the invertebrate groups, a single larval individual settles on a substratum and metamorphoses. It then reproduces its own genotype, covering the substratum by vegetative reproduction. Thus a single planula larva of a coral species can settle on a hard substratum and cover an open hard surface with a colony.

The adaptive nature of such reproduction is illustrated by the settlement and subsequent vegetative reproduction of the intertidal anemone *Anthopleura elegantissima,* a common resident of intertidal rocky coasts of western North America. Contiguous aggregations are composed of individuals from a single clone and are the products of asexual reproduction (Francis, 1973a). A single planula larva settles, metamorphoses, and then divides asexually, producing hundreds or thousands of genetically identical individuals. Francis inferred the clonal nature of aggregations by noting that all individuals had the same color pattern and sex. Sustained contact between individuals within an aggregation resulted in no interindividual interactions whereas contact between individuals from different aggregations resulted in an intraspecific aggressive response, resulting in the subsequent separation of these individuals (Francis, 1973b).

This mode of reproduction allows the anemone the advantage of a single individual colonizing an open environment with the ability subsequently to reproduce and exploit the whole local environment without the need for a sexual reproductive phase. The population size of particularly successful genotypes can thus be amplified (e.g., Shick et al., 1979). *Anthopleura elegantissima* also has separate sexes and normal sexual reproduction. The need for genetic variability for adaptation to changing environments is thus maintained. Localized success of certain genotypes will lead, however, to spatial variation in genotypic frequency (Shick et al., 1979).

Other types of colonial organisms have more structured colonies and therefore the arrangement of individuals in the colony may serve a function beyond their origin as an asexually generated population. Many sponges and corals have a stereotyped colony shape, as in the elkhorn coral *Acropora palmata.* This shape is designed to minimize current damage. In the ectoprocts and some hydrozoan coelenterates, asexual reproduction produces different types of individuals in a colony, resulting in polymorphism. There are feeding individuals, reproducing individuals, and individuals that either clean or protect the colony. Thus a group of asexually produced individuals may be organized into a colony to perform a series of interrelated functions or to produce a colony shape that is adaptive to water turbulence.

Hermaphroditism

In sexual reproduction, gametes of two different sexes must unite to form a zygote that subsequently develops into the adult individual. In animals, (a) the sexes may be separate. Alternatively, (b) a single individual may contain gonads producing gametes for both sexes or may produce gametes of either sex at different stages of its lifetime. The latter conditions are known as *hermaphroditism.*

Typically an hermaphroditic individual will have gonads for both sexes and so is capable of reproducing jointly with any other individual of its species. Self-fertilization is unknown or rare in most species. An example of this mode of reproduction is the common acorn barnacle *Balanus balanoides,* although hermaphroditism is the rule in all barnacles. Barnacles are the only large group of hermaphroditic crustaceans and cross fertilization is the most common condition. The ovaries of sessile barnacles lie in the bases and walls of the mantle whereas testes are usually located in the cephalic region. The penis is protruded out of the body, is long, and may be inserted into the mantle cavity of another individual for the deposition of sperm. Given that a copulatory organ is necessary for successful reproduction, hermaphroditism ensures that an individual of complementary sex will be near by.

In *sequential hermaphroditism,* individuals change sex either from male to female (protandry) or from female to male (protogony). Sequential hermaphroditism is the result of selection when an individual reproduces more efficiently as a member of one sex when small or young but efficiency of reproduction increases with age or size as a member of the other sex (Ghiselin, 1969). Sequential hermaphroditism can convey a selective advantage to an individual by increasing its reproductive potential relative to non-sex-changing members of the population. Warner (1975) concludes that male-to-female sex change should be selected where female fecundity increases with size and age and female-to-male sex change should occur when male reproductive output is depressed at early ages for various reasons, such as inexperience, territoriality, or mate selection.

Sequential hermaphroditism is common in many invertebrate groups. Protandry is the general rule in these forms. In the Eastern Oyster, *Crassostrea virginica,* smaller and younger males transform into females after a few years. Functional sperm and ova can be found in the gonads of transitional forms; self-fertilization is possible as well (Galtsoff, 1964). A similar transition occurs in the suspension-feeding gastropod *Crepidula fornicata.* Individuals usually occur in stacks, with larger and older females below and smaller

Figure 7-2 A stack of the sequentially hermaphroditic snail, *Crepidula fornicata*. Females are at the bottom (right).

and younger males on top (Fig. 7-2). Members of such stacks are oriented with the right anterior margin in contact with the same margin of the preceding member, facilitating insertion of the penis into the female gonopore (Hyman, 1967).

Some of the most interesting examples of sequential hermaphroditism come from fishes, particularly many of the small grazing fishes inhabiting coral reefs. The cleaner wrass, *Labroides dimidiatus* (Robertson, 1972), usually occurs in a small school of about 15 individuals on coral reefs. The largest member of the school is male, all other members being female. If the male dies, then the largest member of the group of females will change into a male. The male-to-female change is thus totally dependent on environmental factors, such as predation. The female-to-male transition can be experimentally shown by removing males. This form of sex change is common among reef fishes (Robertson, 1972).

Separate Sexes

Although asexual and hermaphroditic species abound in the sea, separate sexes are the general rule among marine invertebrates, fishes, and algae. Sexuality promotes genetic variation through recombination, but it necessitates a series of adaptations designed to ensure the transfer of gametes from male to female. Sperm transfer may be accomplished through the following modes: (a) both eggs and sperm may be shed in the water column; (b) sperm may be shed in the water column and fertilize eggs held by the female; (c) sperm may be mechanically transferred from male to female via a copulatory organ or passed through the male gonopore and shed on eggs laid by the female.

Pelagic sperm transfer normally involves simultaneous spawning of males and females. In cases where males and females both shed gametes into the water, spawning by one of a few individuals induces *epidemic spawning* in the whole local population. This process ensures union of gametes. In some cases, spawning may be correlated with lunar phases and occasionally whole populations spawn within a matter of a few hours. In some reef sponges all individuals spawn at precisely the same time, resulting in a fog of sperm over the reef. Adult nereid, syllid, and eunicid polychaetes change morphology into individuals filled with gametes (epitokes) that swim to the surface (usually timed with lunar phases) and spawn. A nuptial dance may occur in which males and females swim rapidly in small circles while releasing gametes into the water. By shining a strong light in the water in July in temperate Atlantic or Pacific waters during the first or last quarter of the lunar cycle, nereid polychaetes can be observed swarming near the surface.

The deployment of pelagic eggs and sperm engenders a problem when more than one species is considered. In seasonal habitats, more than one species may spawn at the same general time, increasing the possibility that many sperm will encounter eggs derived from another species. The result would be to reduce the frequency of successful fertilizations, especially if sperm were attracted to eggs of closely related species. Apparently eggs can have highly species-specific attraction for sperm of the same species. Miller (1979) presents data for 32 species of marine hydromedusae, siphonophores, and sessile hydroids. Of the 752 egg-sperm interspecific combinations tested, only 13 cases demonstrate interspecific attraction of sperm. Sperm are highly specific for eggs of the same species. Species-specific sperm chemotaxis may be a significant mechanism both for ensuring fertilization in an environment that subjects the gametes to massive dilution and for reducing the chance of hybridization.

Although numerous invertebrates shed sperm and eggs openly into seawater, many species use modified means of copulation or spermatophore transfer. This mode of sexual contact increases the probability of fertilization and allows for mate selection. Breeding swarms are often formed at the time of courtship to maximize mate contact in species requiring interindividual contact or copulation for sperm transfer. Several species of intertidal prosobranch gastropods form breeding swarms (e.g., *Thais lamellosa*—Spight, 1974).

Where the sexes are separate and mate location is difficult, males may be small and either attach or reside very close to larger females. Good examples are found in the barnacles, where tiny dwarf males attach within the mantle cavity of normal individuals (e.g., the pedunculate genera, *Scalpellum* and *Ibia*, and the boring Acrothoracica). Many bathypelagic fish species have dwarf males; presumably this factor guarantees sperm transfer in an environment where mates might be difficult to find when eggs are ripe.

Males show degrees of degeneracy of various structures or are sometimes simply miniatures of the host individual. This type of adaptation also occurs in the Ophiuroids or brittle stars, where dwarf males cling to the much larger females (Hyman, 1955). The male remains permanently in this position, with his mouth pressed against the female's and his arms alternating with hers (Fig. 7-3). In a gorgonocephalid from the Antarctic, females are mostly found with dwarf males attached to their aboral surface.

Figure 7-3   A female brittle-star, *Amphilycus androphorus,* carrying dwarf male mouth-to-mouth. (From Hyman, 1955)

Sexual Dimorphism and Sexual Selection

Differences between the sexes can be dissected into three components. First, the mechanisms of sex determination tell us how one gets to be male or female in the first place. In most species, either sex chromosomes or single genetic loci determine sex. More rarely, sex may be determined by environment. In some invertebrates and fishes, the environment may determine the age at which a sequential hermaphrodite switches sex, or even may determine the permanent sex of a nonhermaphrodite.

Secondly, morphological differences between the sexes (sexual dimorphism) can relate to their respective reproductive function. Such primary sexual characteristics as male and female gametogenic organs may generate a strong sexual dimorphism. Distensible body walls, brood pouches, and other structures contribute to the distinct morphological differences between sexes. As mentioned, the relatively greater cost of egg production, relative to sperm production, may result in female reproduction at larger sizes, relative to the male.

Sexual dimorphism with respect to secondary sexual characteristics is by far the most difficult component of sexual difference to understand. Strong differences between sexes are strongly correlated with behavioral interactions accompanying reproductive activities. Male and female secondary sexual structures are often used to facilitate displays calculated for mate attraction. In many fish species, males acquire strong coloration during the reproductive season. In other cases, males are permanently and differently colored than females. Some of these colorations are genetically polymorphic within populations. Presumably, there is a benefit for the male to be brightly colored: More females are attracted. However, the bright coloration might also make the male more conspicuous to predators, thus selecting for pale coloration. The extent of the coloration may therefore be a balance between sexual selection for conspicuous coloration, as opposed to selection for cryptic coloration to avoid predation.

Sexual dimorphism reaches its peak in fiddler crabs of the genus *Uca*. In the males, one claw is very large and makes up approximately 40% of the body weight of the crab! The claw serves only in displays used for mate attraction and is used in male–male territorial defense. The claw seems to serve no other purpose; indeed, males must feed for longer periods than females to make up for the deficiency of having one less feeding claw.

Sexual Reproduction in the Benthic Algae

Sexuality and life histories in benthic algae are complex and may consist of alternations of diploid and haploid generations. Vegetative stages (attached macroscopic forms) may be either haploid or diploid. The plant producing gametes is called the gametophyte and may bear gametes in gametangia. Gametes may be motile (as in the brown algae) or nonmotile (as in reds). The zygote produced from the fusion of two gametes develops into a diploid sporophyte that may be a vegetative phase or a motile flagellated form. The sporophyte stage then produces a meiospore that usually develops once more into the gametophyte. Structures on the sporophyte that produce meiospores are known as sporangia. In the green and brown algae, meiospores are usually motile and flagellated.

Although gametangia and sporangia in many algae are elaborated from single cells, many seaweeds have elaborate reproductive structures. The green algae *Codium* produces gametangia in the form of special lateral branches. Male and female gametes may be produced on the same plant (monoecious) or on different plants (dioecious). In the seaweed group Fucales (e.g., *Fucus*) both monoecious and dioecious species are known.

Life histories vary, especially with regard to the morphology of the sporophyte and gametophyte stages. In some seaweeds they are morphologically identical (isomorphic), as in the sea lettuce, *Ulva*. In other cases, the two stages are quite dissimilar (heteromorphic); in the Laminariales (e.g., *Laminaria*) the sporophyte is the noticeable large vegetative stage while the gametophyte is a small filamentous form. In some seaweeds sporophyte and gametophyte generations have been erroneously assigned to different species (see Bold and Wynne, 1978, and Chapman, 1979, for extensive discussion).

Life histories and reproduction in benthic algae are complex and usually bewilder the beginning student. Figure 7-4 diagrams some of the common life histories of benthic seaweeds. As can be seen, almost anything is possible. Most notable is the lack of consistent relationship between haploid or diploid condition and morphological form of the vegetative stages.

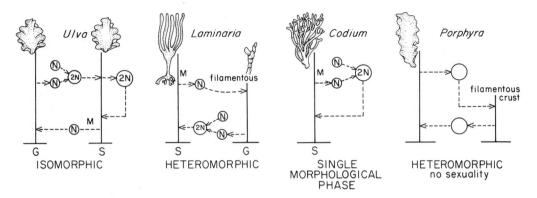

Figure 7-4 Some life histories of benthic algae. S = sporophyte (2N chromosomes); G = gametophyte (N); M = meiospore (N). Size of t-bar denotes size of sporophyte relative to gametophyte stage (after Chapman, 1979). Note that genera used as examples may contain species of different life history type.

7-3 ENVIRONMENTAL CUES TO REPRODUCTION

In general, the timing of reproduction in marine invertebrates correlates with seasonal changes in temperature (Fig. 7-5) and with lunar cycles. The former sets the pace of the seasonal cycle of gametogenesis and the approximate time of spawning. The latter factor permits the organism to spawn at a time favorable to appropriate tidal conditions or serves as a mechanism for a whole population to spawn simultaneously and ensure maximal fertilization.

The role of temperature and season is well established in the reproductive cycles of marine invertebrates. Gametogenesis and spawning are both usually correlated with

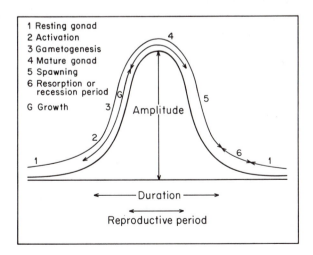

Figure 7-5 A generalized seasonal gonad cycle for a marine invertebrate.

seasonal change and an approximate critical temperature that contributes to the induction of spawning. As spring develops, the onset of the spring diatom increase in temperate–boreal habitats favors egg production due to food availability. The availability of phytoplankton also favors spawning and release of plankton-feeding larvae. Although spawning may be correlated with a specific temperature, it is not always obvious that this temperature is necessarily critical; it may simply correlate with other factors, such as light and food availability. It had been believed that all populations of a species—for instance, the oyster *Crassostrea virginica*—spawned at the same critical temperature. We now know that spawning occurs at progressively lower temperatures with increasing latitude on the East Coast of North America. Therefore a local "critical" temperature cannot be extrapolated to other populations at different latitudes. Differences in spawning temperature may be part of a large number of physiological factors that divide populations of invertebrates into genetically distinct physiological races (Loosanoff and Nomejko, 1951).

Within a region, strong differences in response to the same temperature regime may be found. We mentioned the winter spawning peak of West Coast North American *Mytilus edulis,* as contrasted with the less peaky spawning of sympatric *M. californianus.*

Lunar cycles permit some organisms to time spawning to correlate with favorable environmental circumstances. The marsh snail *Melampus bidentatus* is an air breather but lays eggs that develop into pelagic larvae. Because its typical habitat is in marshes above the mean tide level, it is crucial that larvae are released at high water springs when the bottom is immersed. Similarly, larval settlement must be timed with high water springs so that the appropriate habitat can be found (Russell Hunter et al., 1972).

The grunnion *Leuresthes tenuis* is similarly adapted to tidal cycles. Fishes come onto sand flats to spawn at high water springs at night. Females dig a pit in the sand at the highest level of the tide; males curl around the females and deposit sperm. The fertilized embryos then have a little less than 2 weeks in which highest tide levels are below the level at which the eggs were laid. The next spring tide washes the eggs out and embryos hatch into swimming larval fishes.

As noted, many species time spawning with lunar cycles to maximize the probability of fertilization. Many polychaetes have swimming epitokes that swarm at full or new moon to spawn. Yet many invertebrates do not show any correlation at all with lunar cycles. The oyster *Crassostrea virginica,* for instance, does not spawn in synchrony with lunar cyles in Long Island Sound (Loosanoff and Nomejko, 1951).

SUMMARY

1. The three component processes of growth, maintenance, and reproduction compete for the limited resources that a given individual has at its disposal. Patterns of reproductive investment strike a balance between the payoff in offspring and the reduction in adult viability that might reduce the probability of another reproduction.

2. High adult mortality should select for increased investment in reproduction at an early age. Low adult mortality should favor delayed reproduction and repeated reproduction.

3. Asexual reproduction permits the spread of an advantageous genotype in a given environment; the costs of mating are also circumvented. Sexuality promotes genetic variation and the consequent potential to adapt to new environments. Sexual species show great variation in mate location and life cycles; in the case of the benthic algae, life cycles can be especially complex.

4. Environmental cycles are important in reproductive timing. Cycles permit the maximization of mating, breeding in special habitats, and so on.

8 migration, dispersal, and larval ecology

8-1 FISH REPRODUCTION, DISPERSAL, AND MIGRATION

Many abundant, commercially exploited fish species conform to the spawning and migration cycle shown in Fig. 8-1. Juveniles drift from a spawning area to a nursery area and are recruited to an adult stock that feeds on a feeding ground. Adults then migrate back to the spawning area. In some cases, this pattern is complicated by a migration to a wintering area.

The general pattern of spawning, drift, and active migration back to the spawning area reflects a maximization of reproductive success by feeding and spawning in the most optimal sites. The gradual movement of fishes from nursery grounds to different adult feeding grounds may reflect competition between juveniles and adults for limited food resources. The nursery ground may not be rich enough in food to maintain both adults and juveniles (Nikolsky, 1963).

The spawning area is the region where eggs are laid and inseminated. Eggs may be laid in midwater (pelagic) or on the bottom (demersal). In salmonids, spawning grounds are located on the bottom of freshwater streams and young smolts then migrate downstream toward the sea. In Atlantic salmon, migration toward the sea occurs after 2 to 3 years whereas in Pacific salmon it happens only a few months after spawning. In the Atlantic herring, *Clupea harengus*, eggs are attached to stones and gravel by the female while the male swims around and deposits sperm. In contrast, the cod *Gadus morhua* spawns in the water column—male and female fishes swimming belly to belly as eggs and sperm are released.

The marine migratory pattern can be as follows. *Anadromous* fishes (salmon, shad, sea lamprey) spend most of their time in the sea and return to freshwater to breed.

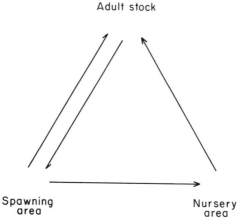

Figure 8-1   The pattern of fish migrations. (After Harden Jones, 1968, *Fish Migration,* with permission of Edward Arnold (Publishers) Ltd.)

Catadromous fishes (eels of the genus *Anguilla*) spend their adult lives mainly in freshwater and migrate to the sea to reproduce. *Oceanodromous* fishes (herring, cod, plaice) live and migrate solely in the ocean. Migrants and differentiation into fish stocks have been studied with tagging, aging techniques (e.g., scale-growth rings, otolith-growth bands), and biochemical identification (e.g., electrophoretic variants of proteins).

Migratory fishes vary widely in their respective degrees of homing. Salmon are capable of homing to the same stream from which they departed as juvenile smolts. Most of the entire migratory sequence involves active swimming. By contrast, the early larval life of the herring (Fig. 8-2), *Clupea harengus,* involves drifting in currents toward the shore from offshore spawning grounds. As they grow, herring move offshore to deeper

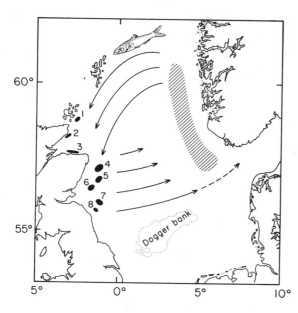

Figure 8-2   The migration of the herring, *Clupea harengus,* in part of the North Sea. (After Harden Jones, 1968, *Fish Migration,* with permission of Edward Arnold (Publishers) Ltd.)

water and feed upon larger zooplankton. They return to the spawning ground after the first year, but homing is probably not exact. Like other species, such as cod, herring seem to occur in distinct subpopulations that each maintain a unique migratory route and are genetically distinct from other subpopulations (see Sick, 1961, 1965).

8-2 INVERTEBRATE LARVAL DEVELOPMENT

Planktotrophic Development

In this mode of development (Fig. 8-3) the larva feeds on either phytoplankton or organic detritus and may live for extended periods of time in the plankton. Larvae may develop from planktonic gametes (mussels), emerge from egg cases (the mud snail, *Illyanassa obsoletus*), or emerge from a parent (barnacles). Planktotrophic larvae are the most capable of long-distance dispersal. Scheltema (1971) demonstrates that some tropical plankto-trophic larvae survive many months in the plankton and may traverse geographic distances as large as the width of the Atlantic Ocean.

In many invertebrate species, the time between release of larvae in the plankton and final metamorphosis is probably programmed genetically and mediated through a series of developmental stages. Scheltema (1971) demonstrates that metamorphosis may not occur if no suitable substratum exists for adult life. The lack of a suitable substratum (true, of course, in the open Atlantic Ocean) may result in delay of metamorphosis with continued feeding and long larval life. Otherwise the lack of suitable substratum results in the larvae metamorphosing or degenerating, with subsequent death.

A good example of a typical planktotrophic larva is that of the common blue mussel *Mytilus edulis*. Sexes are separate and eggs and sperm are shed into the open water. Within 9 hours after fertilization larvae are completely ciliated (pretrochospheres) and are strong swimmers. On the fifth to seventh days fully developed veligers are present, capable of feeding on phytoplankton (Field, 1922). Normal larval life may be 4 to 5 weeks (Chipperfield, 1953). Thus larvae have over a month to disperse from their parental starting point.

Stage-one nauplii of the barnacle *Balanus balanoides* appear in the Clyde Sea plankton about the time of the spring plytoplankton outburst (Barnes, 1956). The cypris stage appears in about 3 to 4 weeks and settles and attaches to a suitable substratum.

Planktotrophic dispersal involves a compromise between the risk of mortality in the plankton (see later) and the benefits of dispersal away from the source area. This risk is measurable in comparisons of populations of short- and long-lived larvae. Limnfjord populations of *Corbula gibba*, a species with a long-lived (several weeks) planktotrophic larvae, strongly fluctuated in adult population size from 1910 to 1924. *Nucula nitida* has a very short pelagic stage and maintained stable populations over this same period. Planktotrophic dispersal has probably evolved because there is some measurable danger in maintaining young at a single place that outweighs the presumable advantage of depositing young in the habitat that was suitable enough to permit the previous generation

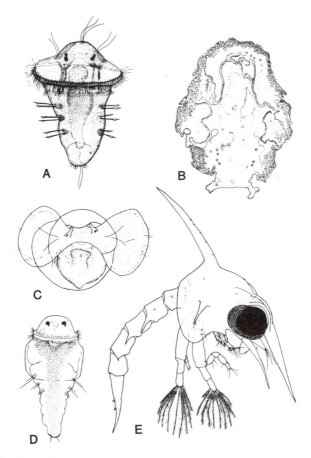

Figure 8-3 Planktonic larvae: (a) the polychaete *Spirobranchus giganteus:* plankto-trophic–planktonic for 1 to 2 weeks; (b) the sea cucumber *Auricularia nudibranchiata:* planktotrophic–teleplanic larva capable of many months' existence in the plankton; (c) the gastropod *Ilyanassa obsoleta:* planktotrophic 12 days to 6 weeks; (d) the polychaete *Spirorbis spirorbis:* lecithotrophic–planktonic for less than 1 day; (e) the crab *Portunus sayi:* planktotrophic–zoea larva; adults live on pelagic sargassum weed. (Drawings (a)–(d) by R. S. Scheltema; (e) by I. P. Williams)

to grow and reproduce. Local population extinction due to disturbances, predation, or overcrowding might provide enough selection pressure for dispersal.

Lecithotrophic Development

Here (Fig. 8-3) the larva has a short planktonic stage and uses a large egg yolk as a source of energy during development in the plankton. Such development occurs in the majority of ctenostome and cheilostome ectoprocts. Ectoprocts may brood eggs in specially modified chambers called ovicells, as in the common bryozoan *Bugula flabellata*. The embryo develops in this chamber until it is a fully ciliated larva. A digestive tract does not develop in this early swimming embryo and yolk reserves only permit swimming periods ranging from about 1 hour to about 1 to 3 days (Hyman, 1959). Long-distance dispersal is thus precluded.

Although various styles of development are largely controlled by taxonomic differences between the various groups, the general production of a nonfeeding larva dependent on yolk from the original egg is a common mode of development. Even though the dispersal phase is short, lecithotrophy must have evolved for the purposes of dispersal because even the existence for as short a period of time as 1 hour in the plankton will permit a set of dispersing larva to travel several hundred yards and escape local crowded conditions.

Direct Release from Egg Cases

The planktonic phase is completely absent and development in an egg case is often supplemented by some form of parental protection. Young escape into the immediate and local environment instead of dispersing hundreds or thousands of meters away. Such a strategy could be employed by species that live in a predictable and non-resource-limiting environment where it is not necessary to get juveniles away from the parental population. Newly released young would be ensured an environment that was suitable to their parents. It can also be used by a fugitive species to proliferate rapidly within an area that becomes open to colonization.

Direct release occurs in species of the drilling gastropod genus *Thais*. In *Thais lamellosa*, a Pacific American rocky intertidal species, the sexes are separate and fertilization is achieved through copulation. During the breeding season adults gather in groups of 50 to 1000 snails, an individual snail preferentially returning to a breeding site that it has previously used (Spight, 1974). Each female encloses approximately 1000 eggs in 40 to 60 vase-shaped capsules, attaching them to a rock in the same area of other females of the breeding aggregation. About three-quarters of 1% of the hatchlings survive the first year while 8 to 13% of this group live to the age of sexual maturity at 4 years. This rate of mortality is probably much lower than the rate of mortality of larvae in a planktonic stage.

The restriction of gene flow between populations of a species with direct release is manifested in genetic differentiation. Regional allozyme differentiation of the mud snail *Ilyanassa obsoleta*, a species with a long-dispersing planktotrophic larval stage, was contrasted with the rocky intertidal snail *Littorina saxatilis,* which has viviparous development (Snyder and Gooch, 1973). *Littorina saxatilis* has more local geographic differentiation than does the long disperser *Ilyanassa obsoleta.* Thus these different reproductive strategies also determine the extent of gene flow, which, in turn, influences the development of inter- and intrapopulation genetic differences.

Within-Parent Direct Development

Complete development occurs within the parental animal up to the point that the embryo is capable of escaping to the bottom as a fully developed juvenile. A number of prosobranch gastropods brood their eggs, retaining them in the female body until the hatching stage. A good example is the rocky intertidal snail *Littorina saxatilis.* Eggs arrive in the brood chamber already fertilized and are provided with an albuminous layer covered externally by a resistant hull representing the capsule. Brooding occurs in an altered part of the oviduct, the "uterus" containing young in various stages of development (Hyman, 1967). Young snails then emerge as fully shelled juveniles. Viviparous development is found in the suspension-feeding Comatulid crinoids, or feather stars. Most feather stars produce pelagic larvae, but many Antarctic species (true of many Antarctic invertebrate benthic species) develop their eggs in brood chambers located near the gonads in the arms at the pinnule bases (as in *Notocrinus*—Mortensen, 1920, cited in Hyman, 1955). The brood chamber is formed from the body wall invagination next to the gonad, expanding into a sac that retains an external aperture. Development in the marsupium continues to the "stalked larva" stage, which is found projecting from the marsupium (Hyman, 1955).

Egg production of viviparous species is generally reduced relative to other comatulids that produce eggs that develop into pelagic larvae. This general relationship of egg number to mode of reproduction is common to all invertebrate groups. Because of the production of large amounts of yolk, there is an energetic cost in producing eggs that must be a long-term energy supply for developing embryo. So given a limited energy source, fewer of these eggs can be produced relative to eggs that do not have large yolk reserves, as in the case of planktotrophic larval development.

8-3 DISTRIBUTION OF LARVAL TYPES

One means of determining the relative adaptive significance of different modes of reproduction and dispersal would be to establish the relative abundance of each mode of dispersal in different environments. The most well-known trend (Fig. 8-4) is a latitudinal correlation discussed extensively by Thorson (1950), who elaborated on the findings of late nineteenth-century workers, such as J. W. Murray (1895, 1898, cited in Thorson, 1950). Thorson estimates that about 70% of contemporary species of marine-bottom invertebrates have planktonic development. In the high Arctic (East Greenland), however, well over 90% of all marine species of bottom invertebrates develop without a planktonic phase and with large yolky eggs.

Arctic planktotrophic larvae confront two problems: a short phytoplankton season

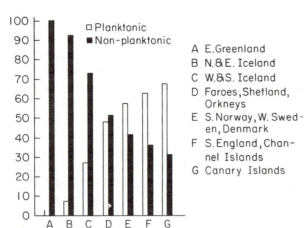

A E. Greenland
B N. & E. Iceland
C W. & S. Iceland
D Faroes, Shetland, Orkneys
E S. Norway, W. Sweden, Denmark
F S. England, Channel Islands
G Canary Islands

Figure 8-4 Variation in percent planktonic dispersers as a function of latitude in the Prosobranch gastropods. (After Thorson, 1950)

and cold temperatures. Timing the release of larvae with a plankton bloom of 1 to 1½ months is chancy indeed. Furthermore, cold temperature lengthens the time needed for larvae to develop. Consequently, the chance of starvation and the exposure of larvae to predation increase. Because even lecithotrophic larvae are rare in the Arctic, temperature must be an important contributing factor.

Similar conditions hold for the Antarctic Ocean as well and many of the well-known examples of brooding organisms come from this region. Over half the Antarctic species of brittle stars are viviparous and 14% of asteroid starfish protect their developing young. But 44% of the total number of species of asteroids do not have brood protection. Of the 30 species of urchins known from Antarctic and sub-Antarctic seas, 37% protect their offspring. This trend is also seen in bivalve mollusks, gastropods, actinians, and polychaetes. Although rich plankton production occurs in Antarctic seas, the production takes place at the very surface of the open ocean while bottom organisms inhabit partly shallow-water shelves in coasts where food reserves are known to be poor.

The percentage of planktonic development increases toward the tropics. It might be predicted that planktotrophic larval development should be best developed in midlatitudes where primary productivity is greatest. Primary productivity is greatest in mid-latitude open oceans, however, and not in inshore shelf regions where bottom invertebrates are most common. In these latter regions, primary productivity is just as high in tropical areas as in midlatitudes.

In addition, midlatitude primary productivity is confined to periods of warm temperature and long photoperiods. In the tropics primary productivity is at a continually moderately high level all year round. Thus an organism producing planktotrophic larvae has the opportunity to reproduce all year. Observations of many different types of bottom invertebrates demonstrate that seasonality in reproduction is depressed in the tropics. Reproductive seasonality is not entirely eliminated due to other aspects of seasonality, such as storms, regional salinity, and current patterns. Although several species do have restricted breeding seasons, the breeding season is, on average, much longer than in temperate seas (Thorson, 1950).

A second trend in reproductive types first suggested by Thorson (1950) is a variation in reproduction with depth of water. Thorson claimed that the dominance of nonplanktonic development in the Arctic and Antarctic oceans also held for the deep sea as well. He felt that there was a consistency here because of low food conditions of the deep sea (discussed in Chapter 19) and low temperatures. The data he had were spotty and in later years many examples have been presented of planktonic development in abyssal environments.

Schoener (1968) studied development in the abyssal ophiuroid *Ophiura jungmani*. She presents convincing arguments for the presence of free planktonic development in this species in contrast to a previous study showing direct development in the deep-sea-dwelling sea star family Porcellanidae. Schoener (1967) found that none of five deep-sea species of brittle stars showed brooding.

Among the bivalve mollusks, most evidence points to lecithotrophic development as the most common in abyssal bivalve mollusks rather than direct development (Ockelmann, 1965; Scheltema, 1972; Knudsen, 1967). Of the 26 deep-sea bivalve mollusk species collected on the John Murray expedition, only 2 species were believed to have planktotrophic development, 6 species had direct development, and 15 species had lecithotrophic development. Later studies by Knudsen (1967) show that even larger percentages of abyssal bivalve species have lecithotrophic development. Scheltema (1972) evaluated reproductive types, using Ockelmann's criterion of prodissoconch size and determined that of the four deep-water species of the protobranch bivalve genus *Nucula* off the northeastern coast of the United States only one had direct development. On the continental shelf, one species of three developed directly. Scheltema also found a change of egg size and fecundity with depth. Deeper-water species of the genus *Nucula* produced fewer eggs per female. The number of eggs per millimeter of shell length also decreased with depth.

8-4 LONG-DISTANCE DISPERSAL OF LARVAE

Scheltema's (1971) discovery of the large-scale occurrence of long-lived planktonic larval stages in the tropics sprang from a large-scale survey of larvae of benthic stages occurring in the Gulf Stream and North Atlantic Drift and equatorial currents. He found larval gastropods of the families Cymatiidae, Tonnidae, Cassidae, Muricidae, Architectonicidae, Neritidae, and the family Ovulidae. Scheltema termed such ocean-going larvae of shelf invertebrates *teleplanic* larvae.

Larvae of some species were found in all three (Fig. 8-5) trans-Atlantic surface currents: the North Atlantic Drift and the north and south equatorial currents (Scheltema, 1971). Other species also occurred only in the equatorial currents and were absent from the North Atlantic Drift. The first distributional pattern includes forms having wide temperature tolerance, permitting them to survive the low temperatures of the North Atlantic Drift. The latter distributional pattern must be characteristic of larvae of narrow temperature (stenothermal) tolerance.

Scheltema estimates a range of 42 to 320 days of possible life in the plankton for

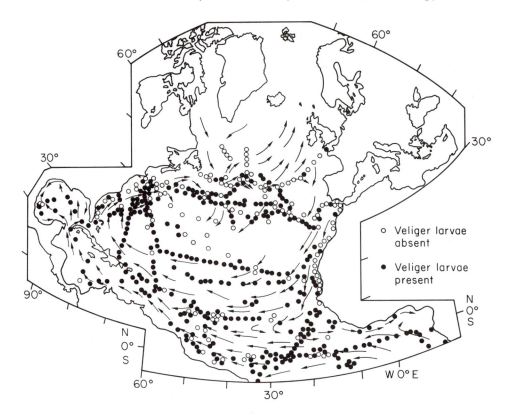

Figure 8-5 Geographical distribution of Teleplanic gastropod veligers taken in the tropical and North Atlantic Ocean.(After Scheltema, 1971, *The Biological Bulletin,* vol. 140, pp. 284–322)

planktotrophic gastropod larvae. Comparing the duration of larval development and the velocity of North and tropical Atlantic surface currents shows that transoceanic dispersal is possible without any delay in settlement in many species. The frequency of long-distance dispersal across ocean basins depends chiefly on the probability that larvae will be carried offshore into the major ocean surface currents rather than retained in the inshore waters near the parent population and that the size of parent population from which larvae originate is large enough to provide sufficient larvae to compensate for the very high mortality rate that occurs in the plankton.

One result of long-distance dispersal is the possibility of large-scale geographic ranges of tropical marine invertebrates. Larvae commonly found in open-ocean surface currents are those species that have amphi-Atlantic distributions. Amphi-Atlantic species that show the largest amount of geographic differentiation on either side of the Atlantic Ocean have shorter larval stages. Because these species depend on shallow-water marine bottoms for adult life, a deep oceanic barrier is just as formidable as a terrestrial barrier for dispersal. This barrier can only be traversed through favorable currents and an extensive larval life.

The presence of phytoplankton, together with equitable temperature conditions throughout the year, seems to select for species with planktotrophic larvae and therefore long-distance dispersal methods. Yet why do organisms need long-distance dispersal mechanisms? And why is it that, given the "opportunity," most tropical benthic species have opted for planktonic dispersal? Given that tropical climates are supposedly more benign and have less environmental stress, we might expect tropical species to show selection for reduced dispersal stages because the environment is so favorable. But the probability of local population extinction may be so great that there is always a premium for selection for pelagic dispersal. It is only in high latitudes that severe conditions select against planktonic dispersal (although the few planktonically dispersed Arctic forms can be very abundant). This conclusion is admittedly speculative, but the latitudinal variation in reproductive types is probably the only clearcut information we have on trends in reproduction in marine benthic invertebrates.

8-5 PROBLEMS FACED BY SETTLING PELAGIC LARVAE

Larvae face many significant dangers before successful settlement and metamorphosis. We must conclude that, despite the risks of mortality, natural selection has chosen planktonic dispersal strategies because there is an overall benefit in getting away from the adult population that produces the larvae.

Thorson (1966) summarizes three major stages through which larvae must pass before successful settlement: (a) successful development, (b) retention nearshore, and (c) substratum selection. Bayne (1964) documents the disappearance of a larval population of *Mytilus edulis* larvae in the Menai Straits. Although three peaks of larval production were observed, the first pulse disappeared and never survived to successful settlement in intertidal regions. The reason for this disappearance is unknown. The technological difficulties in tracing a larval population from the parent to settling precludes a reasonable assessment of the cause and amount of mortality. Several particularly important sources of mortality for pelagic larvae are discussed in the following pages.

Food Shortage in the Plankton

Because planktotrophic larvae must feed for a period on plankton, a pulse of larvae appearing asynchronously with phytoplankton abundance may cause starvation and a failure of larval recruitment. Bad years for plankton may also be bad years for planktotrophic larvae. In the spring of 1950 Clyde Sea phytoplankton blooms were lush and cyprids of *Balanus balanoides* were abundant and settled densely on intertidal rocks. But in 1951 the spring diatom increase did not materialize (Fig. 8-6) and few cyprids were observed to settle (Barnes, 1956).

Shelbourne (1957) examined the myotomal musculature of plaice larvae *(Pleuro-*

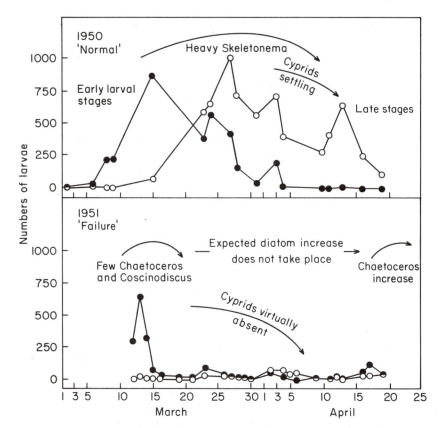

Figure 8-6 Successful settlement and failure in the barnacle, *Balanus balanoides,* as a function of phytoplankton abundance from year to year. (After Barnes, 1956)

nectes platessa) and found that larvae were in better condition in March, when plankton were more abundant, relative to the plankton-poor time of January. Wyatt (1972) showed that myotomal musculature is better developed under conditions of abundant food. If plaice larvae are starved of zooplankton for more than 8 days, a point of no return is reached and feeding does not occur even when food is then presented to them. Older larvae can stand at least 25-days duration before the point of no return is reached. Therefore food shortage may be particularly important in early larval stages. Thus food shortage should make survivorship curves appear concave, with a decrease of mortality rate with increasing larval age (Wyatt, 1972).

Wastage of Larvae

Although planktonic dispersal permits colonization of sites not available to adults of the previous generation, it carries a risk of encountering totally unfavorable settling sites. What happens if a large population of planktonic larvae adapted to metamorphosis on sand is transported by a current into a broad basin of mud? "The larvae cannot know that if they continue to drift over the bottom for perhaps 10 to 20 kilometers more, they might meet a much more attractive substratum" (Thorson, 1966). The brittle star *Amphiura filiformis* has a seasonal settling rate of 3500 larvae per square meter of sand although adults occur strictly in mud (Thorson, 1966).

Loss of larvae may be common from the shelf toward the open sea (Thorson, 1950). Fourth zoeal stages of the intertidal sand crab *Emerita analoga* were commonly found in plankton tows over a hundred miles from the mainland of California (Johnson, 1939). Earlier we discussed the presence of larvae of shallow-shelf invertebrates in trans-Atlantic currents. Although larvae are capable of crossing the Atlantic, the majority of these larvae probably die before reaching a suitable adult habitat. Some of this presumed loss is less than might be supposed. For example, along the European Atlantic coast the predominant wind is from the west, especially during the summer when most invertebrates spawn. At this time, currents tend to move water toward the shore or along the shore, thus keeping larvae near shelf habitats. Some species may have larvae that swim below the depth of seaward surface currents; washout to sea is avoided in this way.

If a colonizing population reaches a distant oceanic island, reproductive effort invested in long-distance dispersal would generate propagules that would probably never reach a suitable site because of the remoteness of the island. Therefore selection will operate against dispersal into these inevitably unfavorable habitats. If wastage of larvae is an important factor, we should expect such a reduction in marine organisms as well (Vermeij, 1972a). Abbott (1966) observed a prevalence of direct development of shallow-water shelf marine benthic organisms in the Galapagos Islands and on the coasts of California and Peru, where currents would carry larvae offshore during a large portion of the year. Vermeij (1972a) demonstrated that high intertidal gastropods living in Ghana and Fernando de Noronha, Brazil, show higher rates of endemism than species of gastropods living lower down on the shore. Suitable high intertidal habitats are normally more discontinuous than low intertidal habitats. Gastropods with nonplanktonic dispersal are usually more common in the high intertidal.

Estuarine circulation may also result in wastage of planktonic larvae of estuarine benthic forms. Surface waters of low salinity leave the estuary, thereby carrying larvae away from their optimal habitat. To avoid this loss, selection may have favored behavior that kept larvae within the estuary. Larvae of the oyster *Crassostrea virginica* do not have the same distribution as coal particles of comparable size and density in the James River estuary. Oyster larvae are most abundant in waters of high salinity at flood tide whereas coal particles were found in areas with the highest current velocity at all tidal stages. There was a net movement of larvae upriver, perhaps being facilitated by selective swimming behavior (Wood and Hargis, 1971). Selective swimming behavior by the larvae of spionid polychaetes keeps the larvae at the bottom during tidal ebb but brings them into the water column at flood stage (Bird, 1972, quoted in Lockwood, 1976).

Predation on Larvae

Marine benthic pelagic larvae are very small, usually less than 250 μ, and must combat many predators. Predation in the plankton is probably a major source of the demise of whole larval populations. Many planktivorous fishes and invertebrates feed effectively on larvae of benthic invertebrates. Herring in the North Sea feed on larval stages of benthic invertebrates.

Settling larvae face similar dangers from benthic predators. The benthic foraminiferan, *Astrorhiza limicola,* is a voracious predator of small benthic organisms in subtidal British waters. A typical population of *Astrorhiza* can sweep away half the bottom of small animals (Thorson, 1966). Deposit feeders (organisms feeding on sediments) may ingest newly settled larvae that attach to sand grains.

Benthic predation on larvae has been implicated in the disjunct distribution of the bivalve *Macoma balthica* and the amphipod *Pontoporeia affinis* in the Baltic Sea. *Pontoporeia* prefers fine-grained sediment and tends to occur in the deeper parts of the Baltic Sea. *Macoma* fails to colonize deeper habitats where *Pontoporeia* is dominant. If sediment suitability alone were the major determining factor, there would be no reason for *Macoma* to fail to co-occur with *Pontoporeia.*

Segerstråle (1962) hypothesized that benthic populations of *Pontoporeia* exerted an adverse effect on the settling larvae of *Macoma,* either by suffocating *Macoma larvae* by stirring the sediment or by ingesting newly settled larvae. *Pontoporeia* is an efficient consumer of bottom material. Because it was impossible to secure *Macoma* larvae, Segerstråle used larvae of the blue mussel *Mytilus edulis* for experiments on successful settlement in the presence or absence of *Pontoporeia.* In aquaria in which *Pontoporeia* was present, all *Mytilus* larvae died within 6 weeks. Although substantial mortality occurred in the absence of *Pontoporeia,* however, there were still large numbers of larvae that had successfully survived. Unfortunately, we do not know from Segerstråle's investigations whether predation on the larvae directly or disruption of the sediment was the major source of mortality for *Mytilus* larvae. Using slightly larger juveniles of *Macoma* (about 1 mm in length), Segerstråle found no adverse effect by *Pontoporeia.* Thus the negative effects were exerted at or near the time of settling.

Grazing snails and urchins may be a significant biological deterrent to settling larvae. Although accidental ingestion may occur, grazers kill settling larvae while scraping hard surfaces for benthic algae. Limpets inhibit barnacle larvae by bulldozing aside newly settled cyprids (Dayton, 1971). The tropical urchin *Diadema antillarum* can scrape off new scleractinian coral colonies and may be a determining factor of coral patch reef species composition (Sammarco, 1977, 1980).

Avoidance of Crowding

Settling larvae face a potential shortage of space for successful settling or adult growth. Cyprid larvae of the common barnacle *Balanus balanoides* settle in great densities each spring on hard substrata in Long Island Sound. A carpet of cyprid larvae appears on rock surfaces with no open space available. Crowding is inevitable when we consider that these larvae will settle, metamorphose, and grow. Consequently, there should be strong selection for various types of behavior patterns to avoid crowding.

Settling larvae may enter a crawling stage when a suitable microenvironment is selected and the presence of nearby larvae is detected. Such behavior is particularly important when the species in question remains permanently attached to the bottom (e.g., barnacles and serpulid polychaetes). Spacing has been elegantly investigated in the tube-constructing worm *Spirorbis borealis* by Wisely (1960). Larvae settle on fronds of the algae *Fucus serratus,* enter an exploratory crawling stage, and establish a uniform spatial distribution. Larvae are about 0.5 mm long and do not settle closer than 0.5 mm from another individual. The mean distance between settled individuals is about 1 mm in a crowded population.

Newly settled cyprids of *Balanus balanoides* exhibit an early exploratory stage in which contact with other newly settled larvae results in movement away from these larvae (Crisp, 1961). Early spacing allows the establishment of basal plates so that the barnacle can firmly attach to the substratum (Crisp, 1961). Yet the degree of territoriality and spacing exhibited by cyprids of *Balanus balanoides* is not sufficient to avoid overgrowth and undercutting between individuals as they grow to full adult size. Such overgrowth and undercutting are of major significance in barnacle mortality (Connell, 1961a). Deevey's (1947) analysis of Hatton's (1938) data on settling success versus crowding in the barnacle *Balanus balanoides* shows expectation of life versus potential interindividual contacts that develop as barnacles grow in size. Figure 8-7 shows the calculated expectation of life in months versus potential crowding that might occur. Crowding has a pronounced detrimental effect on life expectation.

Barnacle morphology and growth as a function of crowding strongly influence subsequent survival. Acorn barnacles growing under crowded conditions tend to be tall and thin, with a relatively low surface area of attachment to body weight ratio. Consequently, it is easy to break off a barnacle with only a small amount of lateral pressure. Under uncrowded conditions barnacles have large basal areas of attachment and are relatively low in verticle profile, thereby making it difficult to dislodge. This effect alone

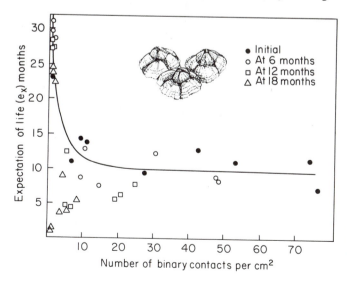

Figure 8-7   Relation between expectation of life of *Balanus balanoides* at selected times and the crowding coefficient (contacts per square centimeter) at the same time. (After Deevey, 1974)

is sufficient to explain why crowding is disadvantageous. Competition for food can also be more intense under crowded conditions. Moreover, crowding probably limits growth rates and thus provides prey for predators limited to eating small animals.

8-6 ADAPTATIONS FOR LARVAL SUBSTRATUM SELECTION

Planktotrophic and lecithotrophic development larvae must find a substratum compatible with adult survival. Almost all species with dispersing pelagic larval stages exhibit varying degrees of selectivity for an adult substratum. A planktonic larval stage must find some physical or biological clue attracting it to a favorable adult habitat. The problems involved in reaching an adult habitat cannot be underestimated. Imagine being a free-swimming larva. It must develop in the plankton for a specified time and then find an appropriate benthic substratum on which to metamorphose. It must have adaptations to leave the surface of the water column, recognize a suitable benthic substratum, and have an environmental cue to induce metamorphosis.

Figure 8-8 is a schematic diagram of the stages through which a free-swimming larva goes before metamorphosis. After being released into the plankton, the larva usually

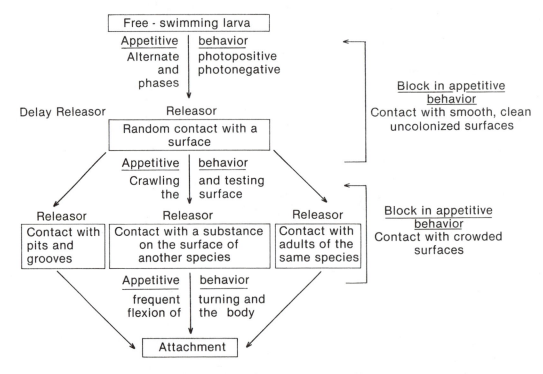

Figure 8-8   Stages in selection of a suitable substratum by planktonic larvae. (After Newell, 1979)

spends anywhere from a few hours to a few months near the water surface, guided by positive phototaxis. If the adult form lives typically within the intertidal zone, the larva remains positively phototaxic. A good example are the pelagic larval stages of the intertidal barnacle *Balanus balanoides*. Those larvae whose adult environment is subtidal are first photopositive and then photonegative at the time of settling.

A larva may not have access to a suitable substratum at the time when metamorphosis should occur. In this case, it may undergo a so-called delay period (Scheltema, 1971). Suitable substratum will cause movement toward that surface, however, with subsequent random contact. Then the larva tests the surface by either "feeling" the quality of the surface or "searching" for various types of biological cues. If these cues exist, the organism will commit itself further to the substratum, perhaps either by metamorphosing or by producing various types of cements (as in the case of the tube worm *Spirorbis*).

The cues that elicit metamorphosis of the larva on the substratum can be classified as follows:

1. Physical characteristics of the substratum, such as the presence of pits, grooves, or the presence of sand grains of suitable diameter.
2. Presence of adults of the same species (gregarious settling).
3. Contact with a substance produced by a species that predictably co-occurs with adults of the larva in question.
4. Contact with some generalized biological substratum feature, such as bacterial films.

In most cases studied, one of these factors is often not sufficient to induce settling and metamorphosis. Instead, a combination of factors, often arranged in a hierarchy of importance, is involved. For discussion purposes, however, we will separate these various factors.

Physical Substratum Characteristics

Both the presence and the nature of the substratum strongly influence the settling and metamorphosis. The presence of a substratum is often required for settlement to occur at all. Larvae of the echinoid, *Mellita sexiesperforata*, for instance, will not metamorphose unless they are in contact with natural sand. Planktonic larvae of many polychaete species undergo a period ranging from a few days to several weeks, during which the fully developed larva metamorphoses whenever it comes into contact with a favorable substratum (Wilson, 1937). If no substratum is available, some delay is usually possible.

Planktonic larvae also favor certain physical attributes of the substratum. Larvae of the sand-dwelling polychaete, *Ophelia bicornis*, settle preferentially on well-rounded sand grains similar in size to those in natural sand flats where the species abounds. The mud snail, *Ilyanassa obsoleta*, selects and requires fine-grained sediments for settlement and metamorphosis. This corresponds well to the sediment-ingesting habits of the adults (Scheltema, 1961). Barnacle larvae are particularly selective (Fig. 8-9) for grooves and

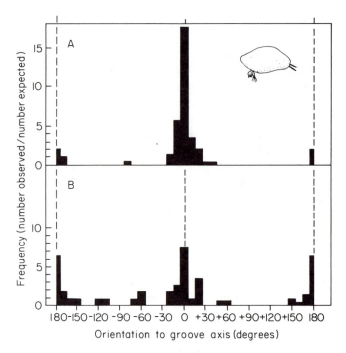

Figure 8-9   Orientation frequency of cyprids (football shaped) of the barnacle, *Elminius modestus,* to the grain of wood in various positions: (a) Light parallel to grain; (b) Light diffuse. (After Crisp and Barnes, 1954)

pits in hard surfaces. Through a series of feeling movements with the antennae and rotations of the abdomen, cyprids sense whether a groove or pit is present (rugotropism). Larvae metamorphosing in these crevices obtain a firm purchase on the substratum and thus fare well in intraspecific competition with neighbors (Crisp and Barnes, 1954).

Gregarious Settling

Many planktonic larvae seek adults of their own species. The value of such gregarious behavior is obvious. When a vigorous population of a species is established, juveniles of the same species can find the microhabitat that has been successfully colonized by previous generations. One disadvantage, however, is the possibility of being eaten by adults of the same species, especially when the adults are suspension feeders capable of ingesting planktonic larvae or deposit feeders capable of ingesting and digesting newly settled larvae on the bottom.

Gregarious settlement usually requires the direct contact of larvae with adults of the same species, as in oysters and barnacles (Knight-Jones, 1953). Typically, a relatively insoluble and species-specific substance induces settlement (e.g., arthropodin in barnacles). Experimental manipulations demonstrate that oyster and barnacle settlement is always heavier on substrata with adults of the same species as the larvae.

Response and Settling to Biological Cues
Other Than Adults of the Same Species

Planktonic larvae often seek biological cues other than conspecifics before settlement and metamorphosis. Another species may serve as the attachment site for the larvae. Larvae of many ectoproct species are particularly selective for certain macroalgae. One of the prime attractive Atlantic algal species, *Fucus serratus,* contains an extractable product that can be transferred to an abiotic substratum, rendering it equally attractive to settling larvae (Ryland, 1959). Adults of the hydroid family, Proboscidactylidae, live on the tubes of certain sabellid polychaetes. At settling time, the planula larvae of the hydroid can detect the prostomial tentacles of the sabellid and settle on them. After settling, they move down toward the opening of the tube and finally assume an adult life position (Campbell, 1968).

Planktotrophic larvae of the blue mussel, *Mytilus edulis,* first settle preferentially upon filamentous algae. Following primary settlement, the early plantigrades (as they are called at this stage) pass through a migratory phase during which they are transported by water currents and may attach and detach themselves from filamentous algae or hydroids. Upon reaching the size of between 0.9 and 1.5 mm, mussels then attach to the main mussel beds (Bayne, 1964). This secondary settlement occurs presumably when the juvenile mussels are large enough to avoid being ingested by adults or to be smothered in the copious feces and pseudofeces produced in the mussel bed.

Influence of Bacterial Coatings on Inorganic Surfaces

The presence of organic coatings and bacterial films on the surfaces of substrata is often essential in inducing settlement and metamorphosis of marine invertebrate larvae. Wilson (1954) showed that acid-cleaned sand kept in seawater becomes increasingly attractive to settling larvae of the polychaete *Ophelia bicornis* with time. The factor inducing metamorphosis and settlement of *Ophelia bicornis* larvae must be the presence of living microorganisms on sand grains (Wilson, 1954, 1955). Similar results were obtained in Scheltema's (1961) study of experimental settling of the pelagic larvae of the mud snail *Illyanassa obsoleta.* Treatment of natural substrata with heat and ultraviolet light greatly reduced the rate of larval settling and metamorphosis.

8-7 SOME THEORETICAL CONSIDERATIONS

The multifactorial control of reproductive success suggests that it is appropriate to formulate deductive models predicting the stable domain of different reproductive/dispersal modes. Vance (1973a,b) attempted to predict optimal reproductive and dispersal strategy from a few simple assumptions. He regarded the amount of energy invested in egg production as a rate-limiting process influencing dispersal. Embryos that feed in the plankton require less yolk investment per egg relative to those embryos that must rely on yolk reserves to nourish their whole development.

Two questions are raised by Vance's theoretical study.

1. Why do we observe only exclusively lecithotrophic or exclusively planktotrophic development in most species? Few species have an extended period of dependence on yolk reserves, with a subsequent dependence on a feeding planktonic larva.

2. Under what conditions should a larva feed in the plankton or the benthos, develop in the plankton living off a yolky egg, or simply develop within an egg case or parent and hatch as a free-living juvenile?

Vance's model considers the following parameters (Fig. 8-10):

s = egg energy content in units of relative energy investment $(0 < S \leq 1)$

L = length of the prefeeding period

P = length of the feeding period

T = the total larval developmental period $(T = L + P)$

a, b = proportionality constants

d = planktonic mortality rate

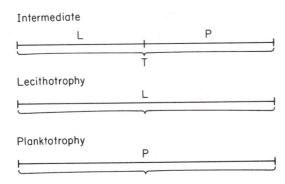

Figure 8-10 Representation of three larval feeding types, according to Vance (1973). Total larval development occurs over time, T. (a) Intermediate, where a prefeeding stage of time, L, depends upon yolk reserves, followed by a stage of time, P, feeding in the plankton $(L + P = T)$. (b) Lecithotrophy, where all of larval development is occupied by the prefeeding stage $(L = T, P = 0)$. (c) Planktotrophy, where all of larval development is taken up by the feeding stage $(L = 0, P = T)$.

Vance assumes that L is linearly proportional to egg size and that P decreases linearly with egg size. Thus the more time spent feeding, the less the parental energy investment per egg. More specifically, $L = as$, $P = b(1 - s)$. So

$$T = L + P \tag{1}$$

$$T = b - (b - a)s \tag{2}$$

The constants a and b are probably functions of environmental conditions and larval development pattern. Water temperature, salinity, availability of suitable substrata for metamorphosis, and concentration of dissolved nutrients all contribute to the determination of a and b.

Consider the case where the prefeeding stage is planktonic. If N_t is the number of larvae produced by a single female that remains alive at some time t after release into the plankton, then for $t \leq T$,

$$N_t = N_0 e^{-dt} \tag{3}$$

N_0 represents the number of larvae released by the female. If the female devotes C energy units to eggs and each egg contains s units (as defined earlier), then $N_0 = C/s$. From this, we can substitute into Eq. (3) and evaluate at $t = T$:

$$N_T/C = \frac{1}{s} \cdot e^{-d[b-(b-a)s]} \tag{4}$$

This equation specifies the relationship between the number of larvae per unit energy devoted to eggs and egg size when both prefeeding and feeding periods are planktonic.

Using calculus, we can determine that the curve is concave upward throughout the range of s. Therefore the only possible values of s that maximize reproductive efficiency are at $s = 0$ and $s = 1$. So the only stable states are complete planktotrophy ($T = P$) or complete lecithotrophy ($T = L$).

Another result can be summarized as follows. If a species has a fixed lower limit of egg size, then the relative efficiencies of planktotrophy and lecithotrophy depend on the product of the planktonic predation rate and the difference between feeding and prefeeding development time. Development times can be assumed to be related to food concentration, which, in turn, influences the values of a and b. With abundant food and low planktonic mortality rate, selection will favor planktotrophic development; under the reverse conditions lecithotrophic development will be more efficient.

Vance attempts to explain the latitudinal gradient in benthic invertebrate dispersal in terms of these results. In the Arctic, a short phytoplankton season and cold temperatures would reduce food availability for planktotrophic larvae and increase the time needed to complete development. These two factors would increase the planktonic mortality rate. Vance's model predicts that lecithotrophy should be favored when planktonic mortality is very great and when the planktonic development is long. The model is thus in conformance with Thorson's (1950) intuitive arguments discussed earlier.

In summary, this model predicts that

1. Only the extremes of the possible ranges of egg size and method of nutrition are stable in an evolutionary sense.
2. Over a certain range of environmental parameters the two developmental types are both evolutionary stable states.
3. Planktotrophic development is more efficient than lecithotrophic development when planktonic food is abundant and planktonic mortality is low and lecithotrophic development is more efficient when either or both conditions are reversed.
4. Benthic prefeeding development results in greater efficiency when lecithotrophic development time is long and/or planktonic predation more intense than benthic predation. Planktonic prefeeding is more efficient when these conditions are reversed.

An extension of these arguments (Vance, 1973b) shows that the introduction of larval starvation changes none of the qualitative conclusions of these arguments. Larvae that can withstand only short periods of food shortage during the feeding stage are more likely to evolve lecithotrophic development than larvae that can survive longer periods.

The effect of starvation is probably greater in forms where the prefeeding stage is plank-tonic.

Unfortunately, despite some intuitively satisfying results, there is not a large body of data that can be easily used to prove or disprove Vance's models. Underwood (1974) presents data showing a lack of correlation between mean egg diameter and mean length of either the prefeeding or the feeding developmental period. So one of the main as-sumptions of the model—namely, the inverse relationship between feeding period and egg size—seems to be violated. Furthermore, many invertebrate larvae have delays of metamorphosis when no suitable substratum is available for settlement. This factor adds a potentially long delay period, which extends the period during which larvae are vul-nerable to planktonic predation, negating the assumption that the feeding period of de-velopment is linearly related to egg size (Underwood, 1974). These two points could be disputed, for present data are not sufficient to substantiate or falsify the first assumptions of Vance.

More importantly, Underwood points out that several factors might far outweigh energy limitation in the production of one or another type of larval stage. The "need" for dispersal, for instance, might be so paramount in marine benthic invertebrates that even if some energy inefficiency is encountered, pelagic modes of dispersal might be selected anyway. The population might become extinct by staying in essentially the same place with no dispersal. Sacrificing a little energy and even taking the risks of predation in the plankton might permit survival of the species.

Another theoretical consideration concerns the reasons for long-distance dispersal. Lecithotrophic dispersal generally involves periods of only hours and probable dispersal distances of only a few hundred to a few thousand meters. But what of larvae that can disperse for distances as much as thousands of kilometers and survive for 6 to 12 months? Planktotrophic larvae can disperse to distances greater than needed to reach open habitats, for a traverse of only a few hundred meters would probably result in the colonization of an uncrowded area. Typically planktotrophic larvae have a feeding stage nurturing growth and development, during which it is impossible to settle and metamorphose. Therefore it is incorrect to say that planktotrophic larvae are traveling until they find a suitable settling site. Instead they are ready for settlement and metamorphosis at some specified time after being released in the plankton. This time will be lengthened if the temperature is low or food is scarce, thereby further separating the larval population from that of the parent.

This seeming paradox led Strathman (1974) to suggest that the function of pelagic larvae is not simply for dispersal from parental habitats. Planktonic dispersal also spreads larvae over a large number of potential environments. Genotypes that spread larvae more evenly between temporarily favorable and unfavorable patches would have a higher average fitness over several generations. This situation occurs because the relative rate of population increase is directly related to the product of the reproduction of successive generations. The relative rate of population increase is highest when variation from one generation to the next generation is minimal.

If a female produces several hundred thousand eggs that are fertilized and the subsequent developing larvae happen to disperse a long distance and all land in a single patch, the probability of this patch's long-term success controls the future of the population. If this same group of larvae, however, is spread over many different environments whose potential for population increase and decrease tends to cancel out over the long run, then the female producing these eggs encumbers a smaller risk that her genotypes will go extinct. So if we define the larvae that develop from eggs of a single female as siblings, then this type of selection is to spread sibling larvae over as many habitats as possible.

It seems possible that this mechanism would work not only for individuals deriving from a single female but also for individuals originating from a local population. Larvae originating from a single area that are spread over a wide area would probably have a higher long-term probability of survival and contribution to succeeding generations relative to a large group of larvae that all land in a single patch. Thus Strathman's limitation to sibling larvae of a single individual is probably overly restrictive.

A piece of circumstantial evidence comes from a study of the intertidal snail *Thais lamellosa,* a species with no pelagic dispersal. Here there is a long-term advantage of having many subdivided populations. Some might go extinct, but others might survive and carry on the species (Spight, 1974). With a pelagic larval stage, it is not possible to maintain a series of subdivided populations as in *T. lamellosa.* Local selection favoring optimal phenotypes would be diluted by dispersal and gene flow from neighboring populations. But then we may ask "Why is this mode of dispersal favored in the first place?" The simple answer might be found in the survival prospects of widely dispersed species. It must invite eventual extinction to maintain a continuous population in the local site of origin of the adults. Crowding, predation by adult conspecifics, parasites, and the rapid extinction rate of local habitats all contribute to selection for planktonic dispersal. Species with planktonic dispersal may therefore survive longer than species with more restricted movements. Evidence from the molluscan fossil record suggests an overall pattern of greater geographic spread and species longevity for pelagic dispersers (e.g., Spiller, 1977).

SUMMARY

1. Migrating fishes divide time among spawning grounds, nursery grounds, and adult feeding grounds. A combination of spectacular adaptations and drift in currents permits the migrations.

2. Invertebrate dispersal ranges from long-distance planktonic feeding larvae, to short-lived planktonic larvae with large yolks and no feeding, to directly released larvae, to embryos nurtured within the parent. This sequence involves an increasing investment of energy per young and a decreasing produced number of young per female.

3. Planktonic dispersal with small-yolked larvae increases in frequency toward the tropics. This fact may suggest that cold climates prolong the developmental time

to an unfavorable degree. The short period of the phytoplankton bloom in high latitudes may also select against feeding larvae in the plankton.

4. Food shortage in the plankton, wastage of larvae through loss to the open sea, predation on larvae, and crowding on settlement all increase the mortality of planktonic larvae substantially.

5. Planktonic larvae possess a variety of adaptations to locate and select optimal adult habitats. Both chemical and tactile cues are important.

6. The selective value of various dispersal types is a difficult question. Short-distance dispersal permits development of localized, perhaps well-adapted, populations. Longer-distance dispersal, however, may minimize the chance that a local catastrophe will eliminate the species population.

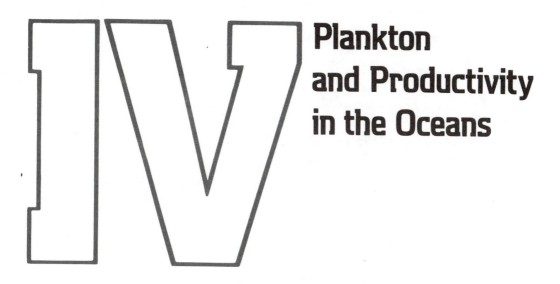

Plankton and Productivity in the Oceans

We turn next to the creatures and processes that regulate the dynamics of the plankton. After introducing the organisms, we examine the factors that cause the birth, peak, and demise of the spring phytoplankton increase—the preeminent feature of most planktonic communities outside the tropics. We then examine plant–animal interactions and the nature of planktonic food webs. Finally, production in the sea and the theory devised to explain variation in production are described.

9 introduction to the plankton

9-1 INTRODUCTION AND TERMINOLOGY

In the surface waters of the ocean dissolved nutrients and sunlight are taken up by photosynthetic organisms (phytoplankton) that, in turn, provide most of the organic matter for the rest of the marine food web. Most shallow-sea-bottom faunas depend on organic matter derived from photosynthesis in surface waters. Remote deep-sea-bottom communities probably rely almost exclusively on this source.

In Part IV we describe the interactions between nutrients, photosynthesizers, and consumers. We show that a delicate balance between photosynthesis, nutrient exchange, and grazing usually determines the dynamic properties of planktonic biotas. We also discuss the physical and biological factors that determine significant geographic and local variation in production and food web properties. We begin with some definitions.

Neritic environments (Fig. 9-1) occur in waters above the continental shelf; the oceanward limit is the continental shelf–continental slope break. *Oceanic* or *pelagic* environments are found oceanward of the shelf–slope break. Because of circulation patterns, effects of insolation, and differences in sedimentary regimes, we often find a completely different biota in an oceanic environment living adjacent to a given neritic one. Pelagic environments can be divided into epipelagic (0 to 150 m), mesopelagic (150 to 2000 m), bathypelagic (2000 to 4000 m), and abyssopelagic (4000 to 6000 m) realms.

Plankton consist of the group of organisms lacking the means to counteract transport by water currents. Plankton can be divided into the bacterioplankton (bacteria), zooplankton (animals), and phytoplankton (plants). (A single specimen is a plankter.) Plankton may be floating, approximately neutrally buoyant in seawater, or live at the air–seawater interface. *Neuston* are those plankters that live attached to the air–sea in-

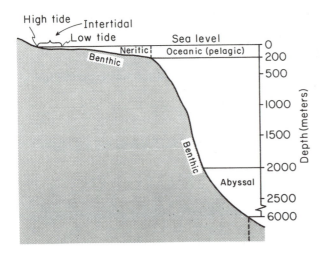

Figure 9-1 A water depth profile, traversing oceanward from shore, showing various significant biotic environments.

terface; the group includes some bacteria, protozoa, phytoplankters, and insects. Members of the *nekton* are capable of counteracting currents through various swimming mechanisms. The group includes adult fishes, some crustacea, cephalopods, and marine mammals.

Plankton are often classified on the basis of size. This classification is related to the collection of plankton in towed nets of various mesh sizes. These size classes are ultraplankton (less than 2 μm), nannoplankton (2 to 20 μm), microplankton (20 to 200 μm), macroplankton (200 to 2000 μm), and megaplankton (> 2000 μm).

Holoplankton are organisms that spend essentially all their active life stages in the open waters; some holoplankters have resting stages that reside in the bottom temporarily. *Meroplanktonic* species spend alternate parts of their life cycles in the water column and on the seabed. They are mainly invertebrates with planktonic dispersal stages.

Organic detritus refers to particulate organic substances not transferred up the food web by predation (i.e., not living cells or organisms). Detritus may be derived from the breakdown of land plants, benthic algae and grasses, phytoplankton, and zooplankton. There is undoubtedly a continuum from particulate organic detritus of various sizes to *dissolved organic matter* (those constituents that are organic in origin but are dissolved in seawater). The range of types of organic matter in the sea is complex and hence difficult to classify. Dissolved organic matter may include urea, soluble proteins, amino acids, fatty acids, carbohydrates, and simple sugars.

The term *nutrients* refers to those constituents required by plants. *Limiting nutrients* are potentially in short supply and may regulate plant growth or reproduction. Nutrients include dissolved inorganic substances, such as the nitrate and phosphate ions, and complex organic compounds, such as vitamins. *Autotrophic* members of the phytoplankton can fix carbon through photosynthesis and produce all necessary constituents of the cell simply with light and inorganic nutrients. *Auxotrophic* phytoplankton species, however, require one or more organic nutrients, such as vitamins, in order to survive. Finally, almost all zooplankton (all but some protozoa with algal symbionts) and phytoplankton under certain conditions consume organic matter for energy and thus are *heterotrophic*. It is important to remember that autotrophy and heterotrophy are not necessarily unique characteristics of a given species. Some organisms can live autotrophically in the light and heterotrophically on organic substrates in the dark.

The *thermocline,* discussed in Chapter 1, is the region of relatively rapid change of temperature with depth, usually due to solar heating of surface water (Fig. 1-8). Because cold water is denser than warm water, the establishment of a thermocline helps prevent the turnover and mixing of the water column by wind-driven circulation. In estuaries and bodies of water adjacent to large masses of melting ice, density gradients called *haloclines* are due to salinity rather than temperature. Any change of density with depth, whether caused by temperature, salinity, or a combination of them, is known as a *pycnocline*.

Productivity refers to the amount of living tissue produced per unit time. It is often estimated as carbon contained in living material and generally expressed as grams of carbon produced in a column of water intersecting one square meter of sea surface per day (g C/m^2/day) or as grams of carbon produced in a given cubic meter per day (g C/m^3/day). *Primary production* is that part of the productivity ascribed to photosynthetic plankton. *Secondary production* refers to the production of organisms that consume the

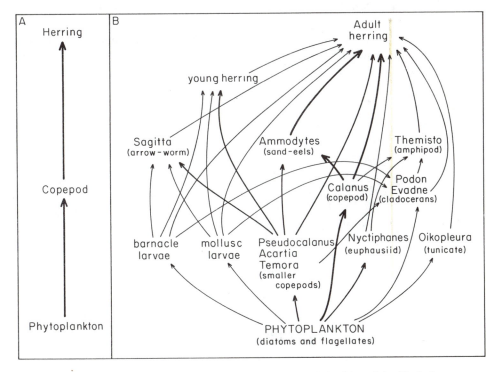

Figure 9-2 Food web constructed from the feeding relationships of the North Sea herring, *Clupea harengus,* during different life history stages. A simplified food chain leading to adult herring is diagrammed at the left. (Modified after Hardy, 1924)

growth products of primary production. Similarly, *tertiary production* refers to the consumption of the secondary producers. We can construct a *food chain:* a transfer scheme of primary, secondary, and higher-level producers. More realistically, a *food web* describes an array of species with more complex food transfer patterns (Fig. 9-2).

Standing crop or *biomass* is measured as dry weight of living tissue, either in a given volume of seawater (mg/liter, g/m^3) or contained in a whole column of seawater over a square meter of bottom (g/m^2). In order to estimate biomass, instead of dry weight we might measure a related parameter, such as chlorophyll (when considering the primary producer level). Biomass measurements do not estimate productivity. For example, a dense phytoplankton population (high biomass) might be photosynthesizing at a low rate; hence productivity would be low. In contrast, phytoplankton with a low biomass might be photosynthesizing rapidly, with cell division occurring several times a day; thus productivity would be high. Zooplankton grazing might immediately consume this production, keeping biomass at low levels, but the productivity would remain high.

9-2 COMPONENTS OF THE HOLOPLANKTON

Phytoplankton

Diatoms.   Diatoms (class Bacillariophyceae) are dominant members of the phytoplankton, particularly in the Arctic and Antarctic and in temperate and boreal inshore waters (Fig. 9-3). They may occur as single cells or as cell chains. Each cell is encased

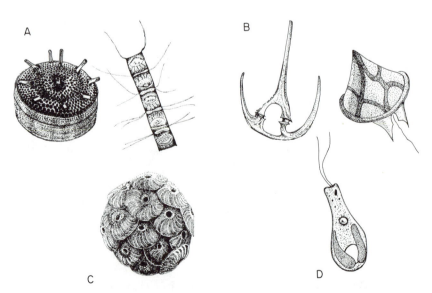

Figure 9-3 Some members of the phytoplankton: (a) Diatoms; (b) Dinoflagellates; (c) Coccolithophores; (d) The microflagellate, *Isochrysis.*

in a silica cell wall, the frustule, which is composed of two valves—one large and the other slightly smaller. The valves fit together in a form that can be likened to a box. The valves are usually flanged and overlap, forming a boundary known as the girdle. Some species have valves ornamented with rather long spines (e.g., the genus *Chaetoceras*). Pores and ridges are present to varying degrees in patterns that are taxonomically significant. *Centric* diatoms have a frustule that is radially symmetrical about an axis whereas *pennate* diatoms are commonly bilaterally symmetrical. In some places, diatom shells accumulate in great deposits of silica on the sea floor.

Binary cell division is the usual means of reproduction in diatoms. One valve of the parent cell goes to each daughter cell and serves as the larger valve for each respective daughter. Therefore the average cell size decreases in most (but not all) species over several cell generations. Restoration of a large cell size is accomplished in most diatoms through sexual reproduction. Sexual reproduction can be induced through changes in light, salinity, or nutrient conditions (see Drebes, 1977). Gametogenesis varies among diatom groups; the resulting zygote usually enlarges to form a large cell called an *auxospore*. A new large-sized frustule is formed from the auxospore and asexual reproduction can commence. Resting spores are produced by many neritic species during unfavorable conditions. Generally they are not sexual spores.

Diatoms rapidly take up nutrients and have optimal cell doubling rates in the range of 0.5 to 6 doublings per day (Eppley, 1977). Diatoms may be autotrophic, but many species are auxotrophic instead. Some diatoms are facultatively heterotrophic and can subsist on sugars, amino acids, and other compounds in the dark (see Hellebust and Lewin, 1977). A few species have no photosynthetic pigments and hence are obligate heterotrophs. Because of its importance for the synthesis of the frustule, silicon is an essential element and is required by diatoms in relatively large amounts. Silicon depletion inhibits cell division and disrupts metabolism.

Dinoflagellates.   The dinoflagellates (class Dinophyceae) are biflagellated, unicellular forms that often dominate the subtropical and tropical phytoplankton and are important in the phytoplankton of temperate and boreal autumn assemblages. Motile cells, which are haploid, have two unequal flagella (Fig. 9-3). One flagellum is located in a groove (the girdle) that divides the cell into two subequal parts. The other flagellum is oriented perpendicularly to the transverse flagellum and extends posteriorly. Cells may be covered only with a series of membranes or may also be armored with a taxonomically diagnostic and contiguous array of cellulose thecal plates. Thecate forms include many of the common genera found in phytoplankton blooms *(Ceratium, Peridinium)*.

Dinoflagellates normally reproduce asexually through binary division. In some armored forms, the theca is shed before mitosis. In others, however, a part of the theca is retained by each daughter cell. The missing part is then synthesized. Sexual reproduction has been observed in many species as well and leads to the formation of special cysts (hypnocysts), a response that seems to depend on environmental conditions.

Dinoflagellates are particularly noteworthy as the cause of harmful *red tides*. "Red tide" is the name given to a dense phytoplankton population appearing suddenly and coloring the water red or red brown. Various algae may contribute to red tides. Species of the genera *Gonyaulax* and *Gymnodinium*, however, are responsible for a variety of toxic effects ranging from fish and invertebrate mass mortality to a deadly neurological disorder known as paralytic shellfish poisoning. In the latter, the toxic agent saxitoxin, derived from dinoflagellates and filtered from the water by bivalve mollusks, accumulates in the siphons and hepatopancreas of the mollusks, some of which are commercially exploited. Saxitoxin interferes with sodium transport and therefore depresses synaptic function. Within 12 hours of ingestion of a toxic bivalve, human respiration is inhibited and cardiac arrest ensues (Steidinger, 1973).

Red tides seem associated with sudden influxes of nutrients through upwelling, tidal turbulence, or washout of nutrients into the sea from land sources (see Steidinger and Ingle, 1972; Hutner and McLaughlin, 1958). Dark uptake of nitrate, influxes of vitamins from shore, and shore-derived sources of iron chelated by humic substances may all contribute to dinoflagellate success.

The availability and transport of a seabed pool of cysts may be important in the origin of dinoflagellate red tides. Turbulence generated by storms or seasonal increases in runoff may transport large numbers of cysts to nutrient-rich waters. In New England cysts germinate in response to the spring increase in temperature (Anderson and Wall, 1978). The great New England red tide of 1972 may have been caused by flushing of cysts from estuarine bottoms to adjacent coastal waters after a hurricane (Anderson and Wall, 1978).

Bioluminescence is common in dinoflagellates and the flickering of shallow waters at night can usually be attributed to *Noctiluca*. The luminescence is of the luciferin-luciferase type and is produced according to an endogenous circadian rhythm. Light production reaches a maximum at night.

Coccolithophores and other golden brown algae (class Haptophyceae).

The coccolithophores are unicellular, lie in the size range to be nannoplankton, and are important in the pelagic phytoplankton, dominating at times in tropical waters. They are approximately spherical and covered with a series of calcium carbonate buttons or plates known as coccoliths (Fig. 9-3). Vast deposits of coccoliths occur in places on the sea floor.* Some coccolithophore species have complicated life cycles involving more than one life form (and type of coccolith). A few have benthic stages. The class Haptophyceae also contains many flagellated species that occur in both inshore and offshore waters. These forms are mainly among the nannoplankton and are difficult to preserve and study taxonomically—fortunately, they can usually be cultured easily. Many are of considerable importance in the primary productivity of phytoplankton assemblages. The colonial form *Phaeocystis pouchetii* is of worldwide distribution but only important numerically in cold waters. It is curious in that it produces the poisonous acrylic acid (Guillard and Hellebust, 1971).

---

*Dr. R. R. L. Guillard informs me that the plankton group, Thoracosphaera, long thought to be a coccolithophorid, is actually a dinoflagellate! This form predominates in what is often identified as coccolith-dominated sediment (see Honjo, 1978).

Blue green algae.   (class Cyanophyceae; also called Cyanobacteria because of their prokaryotic condition) They occur in blooms in certain restricted nearshore and brackish waters. Also, the filamentous genus *Trichodesmium* (called *Oscillatoria* by some) is characteristic of the nutrient-poor waters of the warm oceanic gyres. It has been discovered (Waterbury et al., 1979) that tiny unicellular members of the class (ca. 1 μm) are widespread and numerous. They may be important as foods of the smallest zooplankters.

Green-colored algae.   Members of the true green algae (class Chlorophyceae) and the closely related class Prasinophyceae are grass green in color and are encountered in small numbers in most marine waters but may occur in blooms in estuarine or enclosed bodies of water, especially in late summer and fall. There are both flagellated and nonmotile forms, generally of nano- or ultraplanktonic size. The Eustigmatophyceae is a class newly created on the basis of pigmentation and ultrastructural properties of certain species (Norgård et al., 1974). So far only one form is known from the sea, but it can occur in enormous abundance in estuaries or enclosed regions (such as Great South Bay, Long Island, or the Lake of Tunis, Tunisia). This marine small species was for years misidentified as a species of the true green algal genus *Stichococcus;* the taxonomy is by no means clear yet, but it is currently assigned to the genus *Monallantus* (Antia et al., 1975).

Cryptomonad flagellates (class Cryptophyceae).   These plants are unique in pigmentation, having chlorophylls *a* and *c* and phycobilins also. They are widespread and locally abundant in estuaries.

Zooplankton

Copepods.   Copepods are by far the largest group of crustacea in the world zooplankton fauna. They range from less than 1 mm to several millimeters in length. Most species are free-living, mobile zooplankters, usually ingesting phytoplankton as their principal food source. The calanoid copepods dominate oceanic systems. *Calanus finmarchicus* is widespread in all temperate to Arctic seas. Harpacticoid copepods are locally abundant as well, however. A large number of marine species of the order Lernacopodoida are ectoparasitic on fish and marine mammals and have simplified morphologies suitable for ectoparasitic life. Species of the parasitic genus *Penella* may be over 1 foot in length.

The body of calanoid free-living copepods is usually cylindrical and segmented, composed of a head, thorax, and abdomen. A median nauplear eye is prominent and compound eyes are absent. Figure 9-4 diagrams the locomotory appendages and feeding apparatus. Movement is accomplished by strokes of the large first antennae or by coordinated movement of the five posterior pairs of thoracic appendages.

The calanoid copepods employ a filtering mechanism to obtain particles from the

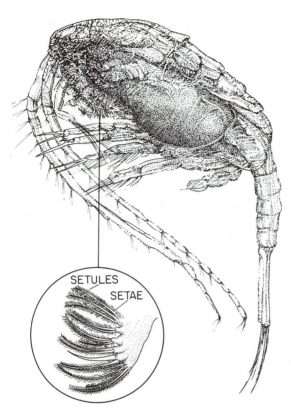

SETULES

SETAE

Figure 9-4 Side view of *Temora longicornis,* a common copepod of inshore waters of New York. Second maxillary appendage is shown in detail. (From Ninivaggi, 1979)

water. An array of mouthparts form a boxlike apparatus. Particles are entrapped on hairlike maxillary setules after the movement of maxillipeds. Particles are removed and are moved forward by long setae on the maxillipeds or by the endites of the maxillules (see Marshall and Orr, 1955a).

In some species of *Acartia,* the two maxilli spread apart to form an open basketlike structure and are then quickly drawn together. This filtering apparatus efficiently retains particles less than about 30 to 50 μm in diameter. The lower limit of particle retention is usually 10 μm. Calanoids may also feed by scraping or seizing larger food (Conover, 1960). Thus larger diatoms and even smaller zooplankton are subject to raptorial attack by Calanoid copepods, such as *Calanus*. We discuss copepod feeding further in Chapter 10, Section 10-5.

In the genus *Calanus* the female lays eggs in clutches of about 50, with an interval of 10 to 14 days between each clutch. Egg production takes place in a series of bursts, each lasting about a week, and clutch size is determined by food availability for females. Both this periodicity and food limitation have an influence on lag time of copepod population response to phytoplankton availability (see Cushing, 1975).

Copepods have large depth distributions and undergo vertical migrations involving active movement toward deeper water during the day with movement toward the surface at night. Copepods may also change their depth distribution seasonally. Depth distribution can vary with latitude. *Calanus finnmarchicus* occurs near the surface in sub-Arctic latitudes but lives in deeper waters near the Gulf of Maine (Bigelow, 1926). An equatorward deepening of depth distribution is known as tropical submergence. Tropical submergence keeps cold-adapted animals in preferred cold-water conditions.

Euphausids.   Euphausids are small shrimplike planktonic creatures that dominate Antarctic seas and are common in pelagic waters of high productivity throughout the world. They are the most important food of baleen whales in the Antarctic and also are the most abundant zooplankton in areas of high productivity, such as the Benguela Current off Africa. Euphausids often occur in great concentrations and occasionally are prominent members of *deep scattering layers* that reflect sonar. Body lengths may reach 5 cm in length and many species are red in color.

In the euphausids [Fig. 9-5(a)], six long limbs of the cephalothorax form a basket that can function as a filtering device but may also be used for seizing larger individual prey, especially copepods (Jørgensen, 1966). In the common Antarctic euphausid, *Euphausia superba*, the basket is closed on all sides through interlimb contact. The segments of the thoracopods have setae that form filtering walls of the basket, filling in the spaces between adjacent limbs. Interbristle distance is 7 μm on the segments closest to the body wall, 20 μm on the median segments, and 35 to 230 μm on the distal segments. During swimming water is pressed through the coarse filter of the distal segments into the basket and out again laterally and posteriorly through the filters of the proximal segments. So smaller zooplankters and the largest phytoplankton cells are prevented from passively entering the basket. The smallest particles that can be retained efficiently must be larger than 7 μ. Stomach contents of *Euphausia superba* agree with these predictions (Barkley, 1940). Diatoms seem by far the most important food of euphausids, although selection of larger particles (e.g., zooplankters) is possible.

Sperm are transferred in spermatophores and eggs are carried in a basket formed by thoracic limbs, setae, or in ovisacs attached to the ventral thorax.

All but a single genus of euphausids are luminescent. The luminescent material is intracellular and located within special light-producing organs called photophores, commonly located on the upper end of the occular peduncle.

Cladocera.   Although a major constituent of freshwater lakes, cladocera are only important in inshore and particularly estuarine zooplankton communities. The genus *Podon* [Fig. 9-5(b)], for example, occurs in estuaries and preys on other zooplankton.

Other crustacea.   A few mysids, ostracods, and cumaceans are truly planktonic but rarely dominate the zooplankton. Some mysids rise into the plankton at night but live on the bottom most of the day. A few amphipods are truly holoplanktonic (such as the genera *Euthemisto* and *Hyperia*) and are at times important members of the zooplankton community in almost all parts of the world (Raymont, 1963).

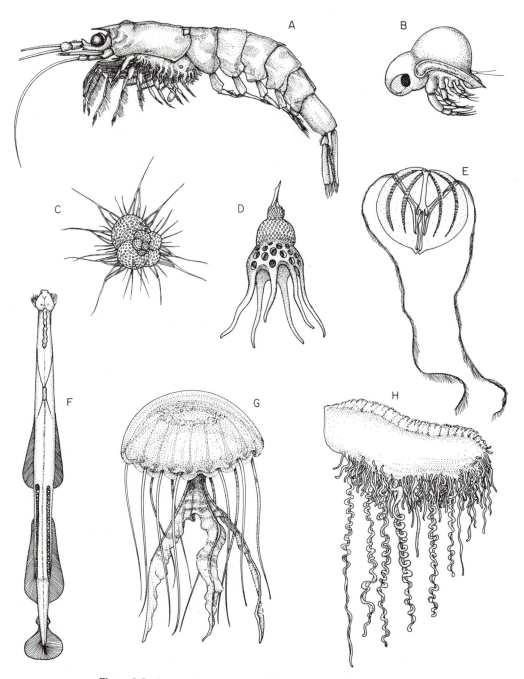

Figure 9-5 Some other components of the zooplankton: (a) euphausiid shrimp; (b) the cladoceran *Podon;* (c) a foraminiferan; (d) a radiolarian; (e) a ctenophore; (f) a chaetognath, or arrow worm; (g) a scyphozoan jellyfish; (h) the Portuguese man-of-war *Physalia,* a siphonophore.

Protistan groups.   Foraminifera [Fig. 9-5(c)] are a major protistan group dominating the open-ocean zooplankton. Typical foraminiferans secrete calcium carbonate shells having one or more chambers, enclosing cytoplasm that streams out through a main aperture and many perforations in the shell. Forams are slow moving and depend on nearby particles for food. Pseudopodia are capable of extending, contracting, and bringing in small particles of food.

Reproduction can be either asexual, through cell division, or sexual, involving gamete production. Gametes fuse and the zygote thus formed produces a shell (the microspheric form) that is usually smaller than the test of the stage that reproduces by cell division (the megalospheric form).

In general, foram tests are less than a millimeter in size. Morphologies range from simple arrangements of spherical chambers to elaborate spinose and branching shells. High-latitude forms are usually simple (e.g., *Globigerina)* whereas low-latitude restricted forms have more spinose sculpture. Tests of the more common genera, such as *Globigerina,* fall to the bottom at such great rates and in such quantity that they form sediments composed strictly of foraminifera, the famous *Globigerina* ooze. Foraminiferal species assemblages coincide with water mass distributions.

The Radiolaria [Fig. 9-5(d)] range from less than 50 μm to a few millimeters in size. Colonial forms can attain several centimeters. They are common members of the zooplankton, particularly in tropical oceanic waters. A membrane of pseudochitin separates the body into a central capsule and extracellular cytoplasm (calymma). Straight, threadlike pseudopodia (axopods) radiate from the central capsule. The silica skeleton is generally a combination of radiating spines and spheres, producing a complex latticework of great beauty.

Nutrition is probably similar to that of foraminifera and involves capturing any particles within reach of the axopods. In many forms, symbiotic algae known as zooxanthellae live within the calymma. These radiolarians probably derive some nutrition from the photosynthate of their algal symbionts.

Asexual reproduction by binary fission occurs, as well as a form of sexual reproduction. Radiolarian tests also settle down from the plankton in large enough numbers to form radiolarian oozes on the sea floor.

Ctenophora.   The comb jellies [Fig. 9-5(e)] are common in most plankton and may occur at depths of 3000 m. Generally they are rather transparent, gelatinous, and egg shaped. The most distinguishing feature is the presence of eight external rows of meridional plates. Tentacles may also be present.

Ctenophora are exclusively carnivorous and are major predators on copepods. In the common genus *Pleurobrachia,* food is caught by the tentacles, which are spread out and moved in a variety of looping turns. When food is captured, the tentacles are retracted and food is appressed against the mouth rim. In another common ctenophore, *Mnemiopsis,* ciliary action brings the prey into grooves, where it becomes entangled by a row of short tentacles. These tentacles pass particles to the labial trough, which leads directly to the mouth (Hyman, 1940). *Mnemiopsis leidyi* chiefly eats molluscan larvae, copepods, and other small zooplankters. Failures of oyster larval settlement may be related to comb jelly predation (Nelson, 1925, cited in Hyman).

Gametes are usually shed into the water. The newly formed embryo is a free-swimming larva that closely resembles the adult. To avoid sinking, many comb jellies lower their specific gravity by reducing their content of the heavy sulfate ion to about half the concentration in seawater, replacing it by a lighter ion.

Chaetognaths.   This group [Fig. 9-5(f)] occurs throughout the world oceans and includes species that are restricted to given water masses and depth zones. Normally they are shallow-water animals but may also occur in the deep sea.

Arrowworms are torpedo shaped and bear one or two pairs of lateral fins. The body terminates anteriorly in a rounded head armed with grasping spines. The head is also equipped with a pair of eyes. Body length is usually less than 4 cm but may range up to 10 cm.

Chaetognaths generally swim by rapid contractions of longitudinal trunk muscles, resulting in a rapid forward motion. This feature aids their carnivorous habit and they feed voraciously on other plankton, such as copepods. Open-ocean forms, such as *Sagitta*, can consume large prey, sometimes as large as the worm.

Arrowworms are hermaphroditic and eggs are shed freely into the water or attached to floating objects. Development proceeds directly to a free-living juvenile.

Coelenterates.   The class Scyphozoa—the true jellyfish—is divided into several orders with varying stomach and tentacle morphologies. Scyphozoans swim by rhythmic pulsations of the bell [Fig. 9-5(g)], allowing the individual to stay near the surface. The group is carnivorous and feeds on animals captured through the tentacle-nematocyst feeding apparatus. The Scyphozoan jellyfishes are found abundantly throughout the world oceans, and may be important local predators on other zooplankton groups. They themselves are often eaten by such carnivores as fishes.

The Siphonophores are polymorphic swimming or floating hydrozoan colonies commonly found in the plankton. Modified medusoid and polypoid individuals are adapted for differing functions (Hyman, 1940). Most important among these polymorphic individuals is the pneumatophore or float, floating the colony on the surface of the ocean. A series of tentacles dangles beneath.

The nematocysts may be very large and extremely toxic. For example, stinging by the Portuguese man-of-war, *Physalia,* can sicken a normal adult human being and has been known to be fatal in some cases. *Physalia* may reach a large size. The float is as much as 10 to 30 cm long [Fig. 9-5(h)] and tentacles may dangle for many meters. A small fish (*Nomeus*) lives among the tentacles of *Physalia* as a commensal, perhaps attracting prey into the deadly grasp of the tentacles. Most Siphonophores are not as dramatic as the very large man-of-war, but others, such as the sail-floated *Velella,* are often abundant and usually of great beauty.

Pteropods.   Pteropods are highly modified holoplanktonic gastropods, swimming with lateral projections from the side of the foot (parapodia). The ventral part of the foot is reduced to three small lobes. Pteropods can occur in great numbers under appropriate conditions of tides and currents. The shells of one group of pteropods, thecosomes, sink to the bottom in great abundance and form sediments consisting almost exclusively of pteropods (pteropod ooze).

Polychaetes.   A few families of polychaetes are exclusively adapted for holoplanktonic existence. Most famous is the genus *Tomopteris*. Both parapodia and sense organs are well developed.

Salps.   The Thaliacea are tunicates specialized for a free-swimming planktonic existence. They differ from the benthic ascidians in having the buccal and atrial siphons at opposite ends of the body. In the genus *Salpa,* the body is barrel shaped and the organism is solitary. The genus *Pyrosoma,* however, is colonial and has the form of a cylinder that is closed at one end. The colony reaches over 2 m in length, and individuals are oriented to the wall of the colony so that the buccal siphons open to the outside and the atrial siphons empty into the central cavity.

Salps strain phytoplankton and fine particulate matter on a ciliary mucus net. Salps feed efficiently on particles ranging from 1 μm to 1 mm in diameter (Madin, 1974). They are also known to be important predators on fish larvae and can be an important inhibiting factor in the development of a fish population.

Larvacea.   Like the salps, the larvacea are also very specialized free-living tunicates found in the plankton. Usually tiny, reaching a few millimeters in length, they are found in marine plankton throughout the world. The organism is neotenic and has retained some of the typical ascidian tadpole characteristics, such as a tail. The animal constructs a house in which the body is enclosed or to which it is attached. The beating of the tail of the animal creates a water current that passes through the house. The orifice through which water enters is covered by a grid of fine fibers that keep out all but the finest plankton. The water is then filtered a second time and plankton are delivered to the anterior of the body. Particles enter the mouth with water and are strained a final time through the pharynx.

SUMMARY

1. We define terms important in discussing the plankton. Plankton lack the means to counteract transport by water currents. Nekton swim and are not simply carried passively by currents.

2. Phytoplankton are those organisms capable of photosynthesis. Most groups are unicellular; the diatoms are of particular importance in temperate and boreal seas. Dinoflagellates tend to dominate more tropical habitats.

3. Zooplankton consist of a wide variety of animal groups. Copepods are of particular importance in temperate and boreal seas. They are the principal consumers of phytoplankton. Euphausids are common in the Antarctic and in localized regions of high productivity. Gelatinous zooplankton, such as comb jellies, salps, and larvacea, are important predators in open seas.

10 plankton dynamics and spatial structure

10-1 PLANKTON DYNAMICS IN A SEASONAL CYCLE

Plankton standing stock often occurs in a predictable seasonal pattern that is explainable in terms of a few parameters. The pattern differs geographically as follows. In the Arctic a single summer peak of phytoplankton abundance is followed by a peak of zooplankton abundance [Fig. 10-1(a)]. In neritic temperate–boreal locales a spring phytoplankton increase is followed by a decrease, concomitant with a zooplankton increase [Fig. 10-1(b)]. In late spring and summer the zooplankton decline and a peak of phytoplankton follows in the fall. In the tropics no obvious alternate pattern of phytoplankton and zooplankton abundance occurs [Fig. 10-1(c)].

Figure 10-2 shows an idealized diagram tracing changes in plankton, light, and nutrients during the year in a temperate–boreal inshore locale. The onset of spring is marked by an enormous increase in phytoplankton, the *spring diatom increase*. In the following paragraphs we describe the rise and decline of the phytoplankton and the onset of the zooplankton increase. We attempt to explain these changes with reference to the factors that regulate phytoplankton and zooplankton populations.

Phytoplankton

Long Island Sound is one of the most extensively studied inshore bodies of water in the world. From mid-December to early January 1952, a steady increase of the phytoplankton population took place (Conover, 1956). The peak of this increase—the annual maximum—occurred in February or March. The initial increase was dominated by centric diatoms,

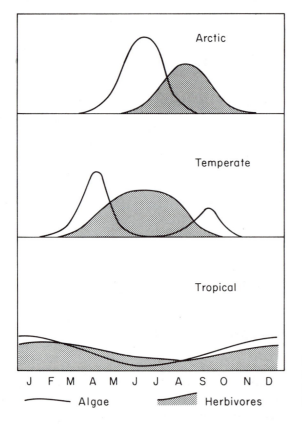

Figure 10-1 Phytoplankton and zooplankton is a seasonal cycle: (a) Arctic; (b) temperate-boreal neritic; (c) tropical. (After Cushing, 1975)

although pennate diatoms and silicoflagellates (members of the golden-brown algae) were also present.

Chlorophyll and cell numbers paralleled each other reasonably well until summer, when chlorophyll values were higher. This situation is probably related to the summer dominance of microplanktonic flagellates, which preserve poorly. Much of the summer microplanktonic flagellate standing stock consisted of cells too small to be filtered by the zooplankton. Consequently, much of the production sank to the bottom as organic detritus. The fall was another period of phytoplankton abundance, although this latter peak was not as dramatic as the spring diatom increase.

A similar pattern of spring diatom increase, decline, and fall buildup has been documented for Loch Striven, a sea loch in Scotland (Marshall and Orr, 1958). In March the diatom *Skeletonema costatum* increases dramatically, followed by blooms of other diatom species in the summer. A fall peak is dominated by still other diatom species.

A predictable change in taxonomic composition occurs in the phytoplankton during the production season. This change conforms to our definition of *succession* given in Chapter 4. Margalef (1958, 1962) and Guillard and Kilham (1977) have summarized three main stages of succession in the spring diatom increase. Stage I is initiated by a complex of factors that permit sudden nutrient availability under favorable conditions for

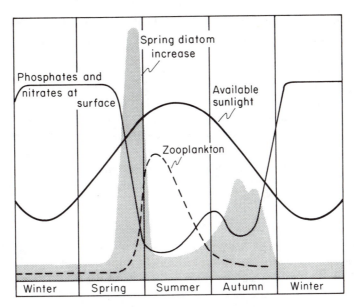

Figure 10-2   Idealized diagram tracing changes in plankton, light, and nutrients during the year in a temperate-boreal inshore body of water. (Modified after Russell Hunter, 1970)

phytoplankton population growth (discussed in next section). Upwelling, the accumulation of nutrients from winter, or a relaxation of zooplankton grazing may contribute to this initiation (see Conover, 1956; Pratt, 1965, for examples). A diatom increase is dominated by species with small cell size, high surface-to-volume ratio (1 in terms of $\mu m^2/\mu m^3$), high division rates (greater than one doubling per day), and high standing crop (10^6 to 10^7 cells 1^{-1}). Small cells have, relative to large cells, the advantage of greater surface area per unit volume. This, combined with rapid division, permits them to (a) absorb nutrients rapidly, and (b) bloom under initial nutrient enrichment.

Stage II is characterized by larger cells with reduced surface/volume ratio (0.2 to 0.5), an increased diatom species richness, and reduced standing stock (order of 10^4 to 10^5 cells 1^{-1}). Dissolved nutrients are now on the decline.

Stage III diatoms are those capable of doing well under conditions of low dissolved nutrients. Species in this stage are survivors of Stage II. Standing crop decreases and diatoms that have dominated previous stages produce resting stages and sink out of the nutrient-poor water column.

The decline of the diatoms is followed by a phytoplankton flora that varies, depending on locale. In Long Island Sound in summer and fall, microflagellates, and particularly dinoflagellates, become abundant. In Loch Striven, Scotland, the fall increase is characterized by a few diatom species (Marshall and Orr, 1958).

The pattern of seasonal succession seems generally to reflect the geographic distribution of phytoplankton. Phytoplankton species that flower early in succession characterize nutrient-rich (eutrophic) coastal waters whereas those that occur later are prominent in nutrient-poor (oligotrophic) offshore environments (Guillard and Kilham, 1977). Diatom succession is most pronounced in mid- to high latitudes and indistinct in tropical locations.

Conceivably, groups like dinoflagellates are dependent on exudates and nutrients produced in the excretion and decomposition of species earlier in the successional sequence. For example, diatoms early in the successional sequence may be autotrophic, requiring only inorganic nutrients for their survival, whereas species later in the successional sequence might be auxotrophic, requiring nutrients like vitamins that they cannot produce themselves. Later species therefore cannot reach great abundance until the flowering of earlier species. Dinoflagellates are known typically to require more nutrients that they cannot manufacture themselves than diatoms, perhaps explaining the successional sequence in the plankton.

Allelopathy—the production of toxic compounds by one organism to inhibit another—may play a role in succession. In eutrophic lakes blue green algae can become dominant and inhibit the development of diatom populations. Cell-free filtrates of blue green algal cultures inhibited the growth of diatoms isolated from the same lake (Keating, 1977, 1978). Blue green algal blooms may thus alternate with diatom outbursts in lakes and some polluted estuaries.

Although prominent changes in relative abundance of phytoplankton species occur during succession, it is important to remember that all species are present at all times of the year. Otherwise there would be no progenitor population from which a population explosion of a given species could develop. In some cases, a population of cysts in the bottom might help initiate blooms of some species.

Zooplankton

Zooplankton in Long Island Sound become abundant and reach their yearly maximum after the spring diatom increase begins to decline. The initial dominance is primarily by calanoid copepods, the major grazers of diatoms. Depending on varying conditions, the genera *Acartia* or *Temora* may dominate the Long Island Sound zooplankton. Meroplanktonic larvae of benthic invertebrates are common in late spring and early summer. Maximal copepod standing stock is reached in the summer months (Fig. 10-2). These copepods consist mainly of small forms that are not readily available as food to planktivorous fishes (Deevey, 1956). Menhaden are the only successful plankton-feeding fishes found abundantly in the Sound.

In Long Island Sound ctenophores become abundant in the zooplankton and prey on calanoid copepods. These gelatinous creatures clog plankton nets in summer, when arrowworms and tunicates may also become abundant.

Depending on latitude, one or several generations of copepods may occur during an inshore zooplankton bloom. In Arctic waters only one generation of *Calanus finnmarchicus* occurs during the season. In the warmer waters of temperate–boreal latitudes, however, several generations can be produced in the spring and summer. Three broods of *Calanus* are produced during the spring and summer in Loch Striven (Marshall and

Orr, 1955). At any one time several different developmental stages of copepods may be found in the plankton.

Variations in the mean size of different generations of copepods within a season is well known (Raymont, 1963). Marshall, Nichols, and Orr (1933 to 1935, cited in Raymont) found that there were large size differences among broods of *Calanus*. In Loch Striven maximum brood size and body weight is typical of the first brood that matures at the end of March. The availability of phytoplankton food for feeding adults may determine the size of individuals and clutch size within a brood. Phytoplankton fluctuations may be responsible for large between-generation population size differences.

Unfortunately, because of the problems of correlating many changing parameters with zooplankton, it is difficult to assess the quantitative effect of food limitation in natural waters (see Raymont, 1963, for extensive discussion of seasonal changes in the zooplankton). In the laboratory Slobodkin (1954) examined the control of population size of the freshwater cladoceran *Daphnia obtusa* and found that population size was linearly related to food supply and that no significant interaction occurred between *Daphnia* in the population except with regard to their competition for food. Time lags in population adjustment to new conditions were also found significant in this species. Moreover, reproduction was affected by food supply.

10-2 WATER COLUMN PARAMETERS: GENESIS OF THE SPRING DIATOM INCREASE

Light intensity decreases exponentially with increasing depth (Fig. 10-3) and becomes a limiting factor to photosynthesis. The *compensation depth* is that depth at which the amount of oxygen produced in photosynthesis (see later) equals the oxygen consumed in respiration. We can estimate the compensation depth by placing phytoplankton cells in a clear bottle. At depths shallower than the compensation depth, there is a net increase of oxygen over time whereas at depths deeper than the compensation depth, there is a net decrease of oxygen over time. The compensation depth is thus an indicator of the potential of a photosynthesizing cell to be a net producer. These relationships would hold if there were no mixing with waters of differing oxygen content. The light intensity corresponding to the compensation depth is the *compensation light intensity.*

The compensation depth is controlled by season, latitude, and transparency of the water column. As the temperate–boreal spring progresses, the increasing photoperiod tends to increase the compensation depth, to an eventual maximum. The Arctic winter photoperiod is zero, however, and therefore the compensation depth is at the surface. Suspended matter in coastal waters reduces the compensation depth relative to the open sea. Similarly, as a phytoplankton bloom develops and as suspended matter (seston) becomes trapped in the water column, the compensation depth decreases due to light absorption and shading by these particles. A yellow pigment originating from rivers and other terrestrial sources is also important in the extinction of light with depth (Jerlov, 1951).

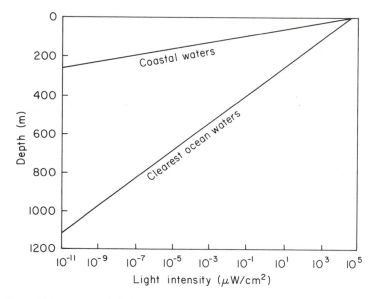

Figure 10-3   Decrease in light intensity with increasing depth in coastal water and clear ocean water.

Consider the state of the plankton and of the water column in the winter before the spring diatom bloom occurs. At this time the water column is isothermal, with little or no density variation with depth. Near the shore temperate–boreal winters are times of high wind stress, resulting in extensive overturn of the water column. Because there are no density differences, the water column is unstable and winds cause extensive vertical mixing.

The *mixing depth* is the depth above which all water is thoroughly mixed under the wind's influence. Because the winter mixing depth is great due to storms, phytoplankton cells can be easily swept down to great depths, where there is not enough light for photosynthesis. A phytoplankton bloom cannot start because any potential profit in photosynthesis is lost through mixing to greater depths. So even though the photoperiod may increase, the instability of the water column may preclude the development of a phytoplankton bloom. Water column stability is thus an essential part of the development of phytoplankton bloom, along with the increase of photoperiod.

If water is mixed in winter, plankton are uniformly distributed and hence respiration must be approximately constant as a function of depth. Because a population increase of phytoplankton requires that total production must exceed total respiration, a phytoplankton bloom can only occur when the volume of water in which photosynthesis occurs has a net excess of production over consumption (respiration, in this case). The depth above which total production in the water column equals total consumption (respiration) is known as the *critical depth*. Above this critical depth, phytoplankton should increase, but if the mixing depth is greater than the critical depth, phytoplankton will be swept down below the area of optimum light, preventing a phytoplankton bloom (Fig. 10-4).

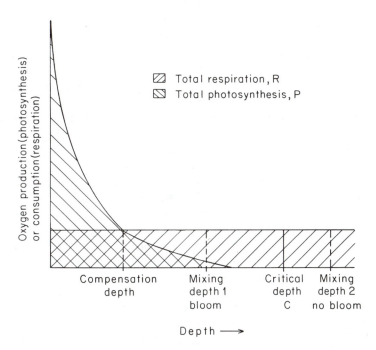

Figure 10-4   The relation of critical depth (C) and mixing depth (M). In the absence of vertical mixing, P = R at the critical depth. If the mixing depth, M, is less than the critical depth, C (e.g., at mixing depth 1), then *P* > *R* and a bloom develops. If M > C (e.g., at mixing depth 2) a bloom fails because some phytoplankton cells are swept below to waters of sufficiently low light intensity to yield the condition P < R.

A spring thermocline generated by solar heating results in less dense water at the surface, thereby impeding wind mixing and overturn. The water column then stabilizes, decreasing the mixing depth. The onset of relatively quiet summer conditions also further enhances the decrease of mixing depth. In inshore waters the water column may be further stabilized by lowered salinity due to influxes of freshwater from terrestrial sources. The late spring and early summer runoff from the Fraser River in British Columbia, for instance, results in salinity minima and typically high production throughout the northern Puget Sound and the Strait of Georgia region. The earlier timing of inshore Norwegian fjord phytoplankton blooms over open-water phytoplankton blooms may be related to the greater stability of the water column inshore.

The stabilization of the water column causes both the birth and the eventual demise of the spring diatom bloom. The spring stabilization of the water column maintains phytoplankton in the upper layer, thus precluding its removal from the zone of active photosynthesis. As the water column stabilizes, however, and the thermocline is established, phytoplankton die or are ingested and egested by zooplankton, sinking below the compensation depth. Because of the stabilization of the water column, these materials and other nutrients are not returned to the surface from greater depths and from the bottom. In a shallow area like Long Island Sound there is extensive exchange between the bottom and the overlying water in terms of resuspension of detritus and dissolved nutrients. Once the thermocline is established in the spring and summer, however, this exchange is greatly diminished, trapping large amounts of detritus in the water column because of its stability (Rhoads, 1973). Toward the end of summer, with the advent of fall storms, the thermocline may be disrupted, bringing nutrients toward the surface from the bottom in shallow water. This may result in a fall maximum of phytoplankton of a different species composition.

In the case of phytoplankters denser than seawater—for example, diatoms—the stabilization of the water column in spring and summer probably creates a situation in which populations rapidly sink out of the photic zone. Riley, Stommel, and Bumpus (1949) calculate the turbulence necessary to counteract the sinking of diatoms. Such considerations do not hold for phytoplankters whose swimming abilities (e.g., dinoflagellates) circumvent this problem.

If these ideas are correct, then we may conclude that the hydrographic conditions tied to seasonal variation play the primary role in the birth, development, and demise of the spring phytoplankton increase. The stabilization of the water column in spring initially permits the development of the spring increase. The stability of the water column, however, prevents nutrients lost from the surface waters from returning to the surface where light is available for photosynthesis. Furthermore, dense phytoplankters sink out of the water column when spring–summer stability sets in. The poor nutrient situation prevails until the fall and winter overturn of the water column.

10-3 LIGHT

Energy from solar sources can be expressed in terms of energy units, such as langleys per minute (g cal/cm^2/min).* The angle of the sun at different times of day, the latitude, and other factors contribute to the spectral distribution of light that strikes the sea surface and the amount of backscattering. The spectral distribution of light striking the sea surface includes a large part of the ultraviolet and infrared spectrum; however, only visible parts of the spectrum penetrate to great depths. At temperate latitudes in clear weather during the summer, the maximum energy striking the sea surface is about 1.4 langleys per minute. About one-half the total radiant energy is in the infrared region of the spectrum and so is not available to marine photosynthetic organisms.

Light is attenuated in the water column through absorption and scattering. Scattering can be accomplished by water molecules, dissolved organic matter, particulate organic and inorganic material, and living plankton themselves.

Solar radiation is exponentially attenuated with increasing depth in seawater. If λ is the wavelength of a monochromatic source, and $I_{\lambda o}$ the illumination at the surface, and $I_{\lambda d}$ the illumination after the radiation has passed through the depth (d), then

*One langley min^{-1} = 0.0698 watt cm^{-2}.

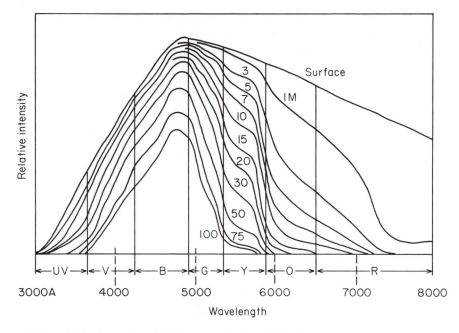

Figure 10-5 Attenuation of different wavelengths of light with increasing depth below the sea surface. Wavelength is in angstroms ($\times$ 0.1 = nm). (From Clarke, 1939)

$$I_{\lambda d} = I_{\lambda o}e^{-Kd}$$

where K is the absorption coefficient for that wavelength. Figure 10-5 shows attenuance values for different wavelengths of light. In the clear open ocean the attenuation spectrum of light transmission results in a maximum transmission at about 480 nanometers (nm). In turbid inshore waters, however, a more pronounced maximum occurs at longer wavelengths, approximately 500 to 550 nm. Because ultraviolet light has detrimental effects on DNA, its penetration is of great interest. In moderately turbid coastal waters incident light with a wavelength of 380 nm or less is almost attenuated at a depth of 1 to 2 m, but in very clear parts of the ocean 20 m may be required to remove 90% of the incident or radiation (Strickland, 1965).

The intensity of incident radiation is so great near the surface that photosynthesis may be inhibited through bleaching of photosynthetic pigments, such as chlorophyll *a,* or the arresting of pigment production. Photosynthetic phytoplankton use chlorophyll *a, c,* and a variety of "accessory" pigments, such as protein-bonded fucoxanthin and peridinin, to utilize fully all the incident light in the visible spectrum. Within the usable wavelengths of 400 to 700 nm, the light absorbed by phytoplankton pigments can be divided into (a) light of greater than 600 nm, which is mainly absorbed by chlorophyll, and (b) light of less than 600 nm, which is mainly absorbed by accessory pigments. The combined absorption of chlorophylls and accessory pigments allows the yield of photosynthesis to be constant from 530 to 680 nm in the diatom *Navicula minima* (Tanada,

1951). The extent of utilization of different wavelengths can be studied by determining the action spectrum, attained by using different monochromatic sources of light and determining the amount of oxygen evolution per phytoplanktonic cell.

Figure 10-6 illustrates a theoretical photosynthesis/light curve showing photosynthetic rate as a function of light intensity. Photosynthesis increases logarithmically with increasing light intensity up to some maximal value, P_{max}, at which the system becomes light saturated.

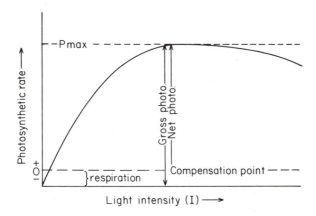

Figure 10-6   Theoretical photosynthesis versus light intensity relationship.

At the compensation light intensity, the photosynthetic rate (in this case, perhaps measured by oxygen evolution) equals the amount of oxygen consumed in respiration. Because the amount of light reaching a phytoplankton cell varies over a day, the compensation light intensity is usually expressed on a 24-hour basis. An average 24-hour compensation light intensity is in the range from 3 to 13 langleys/day in temperate seas (see Parsons and Takahashi, 1973, p. 65). It is assumed that respiration is the same in the light and the dark, a shaky assumption because of light-accelerated respiration (photorespiration). Figure 10-6 shows photoinhibition at very high light intensities.

The physiological adjustment to surrounding light conditions involves some of the following morphological and biochemical changes:

1. Change in total photosynthetic pigment content.
2. Change in pigment proportions.
3. Change in the morphology of the chloroplast.
4. Change in chloroplast arrangement.
5. Change in availability of dark reaction enzymes.

For example, in diatoms, chloroplasts shrink and aggregate under strong light conditions (Brown and Richardson, 1968). Adaptations to changes in light intensity usually occur within 1 day. Deep-water phytoplankton can acclimatize photosynthetic rate to low light intensities.

Physiological adjustments to changing light intensity are best illustrated by changes in light intensity–photosynthesis curves (Jørgensen, 1977, and references therein). We can distinguish between the *Chlorella*-type and *Cyclotella*-type adaptations. The *Chlorella* type is characterized by changes in cell chlorophyll content—for instance, an increase in chlorophyll when a culture is transferred to low light. Figure 10-7(a) shows the effect on a photosynthesis-light intensity diagram. Because the culture acclimated to low light has an increased cell chlorophyll content, photosynthesis is greater at any light intensity than in a culture acclimated to high light.

In the *Cyclotella* type of adaptation, cellular chlorophyll is about the same in cultures grown at low and high light intensity. The photosynthetic rate, however, is higher in cells grown at high light intensity [Fig. 10-7(b)]. Jørgensen (1977) suggests that increased concentrations of photosynthetically active enzymes in the dark reaction steps of photosynthesis cause the rise of photosynthetic rate. The photosynthetic physiology behind these adaptations is beyond the scope of this text; the reader might consult Prézelin and Alberte (1978) and Prézelin and Sweeney (1978).

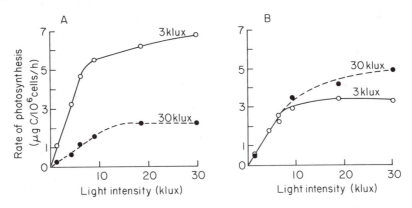

Figure 10-7   Two types of adaptions to increasing light intensity. (a) The "Chlorella-type" of adaption; (b) The "Cyclotella-type" of adaption. Curves labeled with acclimation intensity.(After Jørgensen, 1977, © Blackwell Scientific Publications, Limited)

Little is known about the distribution of these two adaptive types. It must be assumed that some phytoplankton species that often experience changes in depth distribution (e.g., seasonally) may have the *Chlorella*-type adaptation. Most species studied by Jørgensen (1977) seem to belong to the *Cyclotella* type.

10-4 NUTRIENTS

As defined earlier, nutrients are all those constituents required by plants. We can speak of required *nutrient elements* (e.g., nitrogen) that occur in dissolved inorganic form (e.g., ammonia, nitrate) or in organic form (e.g., amino acids). *Major nutrient elements* are required in great amounts and include carbon, nitrogen, phosphorus, oxygen, silicon, magnesium, potassium, and calcium. *Trace nutrient elements* are required in far smaller amounts and include iron, copper, and vanadium. *Organic nutrients* include vitamins. Recall that autotrophic and auxotrophic uptake refer to the uptake of inorganic and organic nutrients, respectively, in association with photosynthesis. Heterotrophic uptake refers to uptake of organic substances for nutrition in the absence of photosynthesis.

Many elements essential for phytoplankton nutrition can be found in particulate as well as dissolved forms. Carbon, nitrogen, and phosphorus may be found in carcasses of phytoplankton and zooplankton sinking in the water column. Microturbulent water motion may also create particles from collision of dissolved organic molecules (Baylor and Sutcliffe, 1963). This resulting fragile organic aggregate material is known as *marine snow* and is often found to be enriched by a variety of planktonic organisms and detrital products of plankton (Silver et al., 1978). Finally, influxes from rivers and seaweed beds also contribute to particulate organic matter (see Marshall, 1970).

Dissolved organics, particulate organics, and living biomass occur, on average, in the approximate proportions 100 : 10 : 2 (Strickland, 1965). The ocean is a solution of nutrients in the dissolved state, with a depleted surface layer due to uptake by living forms in the photic zone. Exchange processes, such as upwelling and wind mixing, may balance the loss of surface nutrients to greater depths via sinking carcasses and zooplankton fecal pellets. The turnover of the ocean's dissolved organic matter is estimated to be 30 to 300 years (see Fenchel and Blackburn, 1979), but some molecules (e.g., glucose) have much more rapid rates while others probably turn over slowly.

Nitrogen

Nitrogen is required for the synthesis of proteins. It occurs in three principal inorganic dissolved forms: ammonia, nitrate, and nitrite. Nitrogen also occurs in dissolved organic forms, such as urea, amino acids, and peptides.

Ammonia is usually the preferred form of nitrogen (Dugdale and Goering, 1967) because it can be used directly for amino acid synthesis without a change in oxidation state. Some phytoplankters—certain euglenids, cryptomonads, and green algae—require reduced nitrogen (ammonia, amino acids) for growth. Nitrate and nitrite must be reduced by the enzymes nitrate reductase and nitrite reductase, making ammonia use a kinetically favored process. Dissolved ammonia can inhibit the uptake of nitrate (Dugdale, 1976). Furthermore, ammonia can be taken up more efficiently at low light levels than nitrate (Dugdale and Goering, 1967).

The highest concentrations of dissolved nitrogen that occur in the ocean are of nitrate (ca. 100 μM), and nitrate is often the most abundant form of nitrogen in eutrophic coastal waters. Upwelling and storm-induced turbulence bring nitrate to the euphotic zone. Nitrite is generally the rarest of the three and behaves similarly to nitrate in phytoplankton nutrient uptake. Under certain circumstances ammonia can surpass nitrate in abundance (usually when the nitrate is used up). In the Gulf of Maine ammonia was found to be more abundant than nitrate at depths of 25 to 100 m (Redfield and Keys,

1938). The dissolved concentrations of all three forms of nitrogen increase in the temperate–boreal winter and decrease in the spring and summer when phytoplankton populations build up.

In shallow coastal bays and estuaries, coupling with the benthic system may influence phytoplankton nutrient dynamics. Nixon and co-workers (1976) showed extensive transfer of nitrogen in the form of ammonia from the bottom to the overlying water. Ammonia was released during the process of microbial decomposition of detritus in the bottom. Half the return of nitrogen from the bottom to the water column was probably in forms of organic nitrogen, such as urea. Some phytoplankton species are capable of taking up urea, uric acid, and amino acids (Carpenter et al., 1972; Wheeler et al., 1974). Benthic–pelagic coupling of nutrients is discussed further in Chapters 17 (Section 17-4) and 19 (Section 19-1).

Recycling of different forms of nitrogen depends on the habitat and the nature of the nutrient regeneration cycle. In coastal areas of high upwelling, as off the coast of Peru, nitrate is regenerated from the bottom and is the main currency in which nitrogen enters the planktonic system. Tracer studies employing N-15 show that over half the nitrogen uptake in upwelling areas is in the form of nitrate. The remainder is in the form of ammonia that recycles from zooplankton excretion and decomposition back to the phytoplankton (MacIsaac and Dugdale, 1972). Excretion of the anchoveta may be the principal source of regenerated nitrogen in the Peru upwelling region (Walsh, 1976). In contrast, in the nutrient-poor gyres, less than 10% of the measured nitrogen uptake is in the form of nitrate (Dugdale and Goering, 1967; Eppley et al., 1973). Most nitrogen uptake must therefore involve efficient recycling of ammonia and organic nitrogen between the zooplankton and phytoplankton.

Nitrogen may be incorporated into marine food chains through the process of nitrogen fixation (accomplished by some bacteria, blue green algae, and yeasts). Many diatoms living in nutrient-poor waters form symbiotic associations with a blue green algae suspected of nitrogen fixation (see Guillard and Kilham, 1977). A nitrogen-fixing blue green species, *Oscillatoria (Trichodesmium) thiebautii*, is found in tropical gyre centers and may possibly contribute nitrogen to the phytoplankton (Carpenter, 1972). Denitrifying bacteria may return much nitrogen to the atmosphere.

Phosphorus

Phosphorus occurs in the ocean as inorganic phosphate (orthophosphate ion), dissolved organic phosphorus, and particulate phosphorus. It also occurs in polluted areas in the form of polyphosphate that may be used occasionally by phytoplankton. Orthophosphate is the preferred form for phytoplankton and exchanges rapidly between phytoplankton and seawater. Grazing and excretion by the zooplankton allow rapid regeneration in the plankton (Pomeroy et al., 1963).

The absolute amount of phosphate in seawater probably limits standing crop, although there is some controversy as to which minimum concentration is sufficient to limit growth rate. Strickland (1965) claims that little evidence has been produced that rate limitation occurs in the marine environment. Usually a shortage of nitrogen is believed to be responsible for stopping the growth of phytoplankton populations. A shortage of phosphorus in a cell must be extreme before energy-related processes, such as enzymatic reactions, and photochemical processes in photosynthesis may be affected. The role of phosphorus is different from that of nitrogen because phosphorus is used primarily in the energy cycle of the cell.

Nitrogen: Phosphorus Ratio in the Sea

Harvey (1926) noted that the growth of phytoplankton in the English Channel resulted in the simultaneous depletion of both nitrogen and phosphorus. Furthermore, phosphorus and nitrogen are available in ocean water (N : P = 15 : 1) in very nearly the proportions usually required by phytoplankton (Redfield et al., 1963). The coincidence of phosphorus to nitrogen ratio in the sea and in phytoplankton requirements can be explained by the hypothesis that growth of phytoplankton cells, followed by sinking and decomposition, controls the N : P ratio in phytoplankton and seawater. A balance of nitrogen fixation and denitrification would fix the overall value of nitrogen, relative to phosphorus. Despite the probable correctness of this general hypothesis, numerous examples exist of phyto-planktonic species that deviate from the "typical" N : P uptake ratio (see Dugdale, 1976).

Silicon

Silicic acid is a constituent of seawater essential for the shells of diatoms. Depletion of silica results in the inhibition of cell division and eventually the suppression of metabolic activity of the cell (Werner, 1977). In natural waters depletion of silica can limit phytoplankton populations (e.g., Smayda, 1973) and may direct the course of subtropical succession toward phytoplankton lacking a siliceous test (Menzel et al., 1963).

Trace Substances

Such metals as iron, manganese, and zinc serve important functions in oxidase systems (iron is the cofactor in the oxygen evolution step of photosynthesis) and as cofactors for enzymes essential for plant growth (e.g., molybdenum, zinc, cobalt, copper, and vanadium). Iron has been shown to be a limiting factor to phytoplankton and a shortage lowers photosynthetic potential (Ryther and Guillard, 1959; Menzel and Ryther, 1961a). Chelators may be important in affecting the utilization of trace metals. Phytoplankton may synthesize and release chelating substances into the water to alter the availability of trace metals, such as iron.

Organic trace nutrients, particularly vitamins, may also be of great significance in the sea. Almost all marine phytoplankton species are auxotrophic and require cobalamine, thiamin, or biotin (Lewin and Lewin, 1960; Provasoli, 1958; Droop, 1968; Guillard, 1968). In mixed cultures, vitamin production and release by one species may stimulate the growth of another (Carlucci and Bowes, 1970), although most vitamin production is probably by bacteria. In the Sargasso Sea small diatoms requiring vitamin B_{12} increase in abundance relative to coccolithophores (that need only thiamin) at the time of year when the vitamin B_{12} concentration is great (Strickland, 1965).

Nutrient Cycling by Heterotrophs and Chemoautotrophs

Certain phytoplankters can grow heterotrophically on dissolved organic carbon sources—sugars, organic acids (e.g., lactic acid, amino acids, or alcohols) (Hellebust and Lewin, 1977). Generally many substances can be taken up by a given planktonic species but will not support its growth in darkness. Species that cannot grow in the dark may nevertheless accumulate one or another organic molecule. For a given species, certain organic molecules taken up may inhibit growth, although other quite similar compounds (e.g., a different amino acid) may be used heterotrophically. In some cases, light inhibits heterotrophic uptake (e.g., Hellebust, 1971).

Heterotrophic capabilities are widespread among the pennate diatoms and rare among centric diatoms. Pennate diatoms are predominately benthic, but heterotrophic pennate and centric diatoms are found in planktonic habitats with high concentrations of dissolved organic matter (Hellebust and Lewin, 1977). A few dinoflagellates can engulf particulate matter and phagocytosis may complement photosynthesis (see Bold and Wynne, 1978, p. 439).

The presence of heterotrophic uptake in phytoplankton suggests an overlap of nutrition with saprophytic bacteria. In shallow bays bacteria may also compete with phytoplankton for nutrients regenerated from the bottom.

Phytoplankton consume the majority of dissolved inorganic nutrients in well-lit surface waters. Phytoplankton are also responsible, especially nearshore, for heterotrophic consumption of dissolved organic nutrients. Bacteria, however, are probably the principal heterotrophic consumers in the water column; they efficiently utilize both dissolved and particulate material including chitin and cellulose.

Although ubiquitous, bacteria are most abundant in association with suspended particulate matter. In shallow basins during spring and summer they often coincide with concentrations of particulates at the base of the thermocline. In the open ocean they are similarly associated with the oxygen layer below the thermocline, which is itself associated with microbial consumption of organic matter. Resuspended particulates near the bottom also harbor large bacterial populations.

Bacteria in the water column are responsible for the rapid uptake of such organic substrates as carbohydrates, amino acids, and peptides. Therefore these dissolved substances can be converted to a particulate form available to consumers, such as protozoa and salps. Bacteria (and heterotrophic algae) attached to particulate organic matter may be a major source of food for copepods nearshore and in estuaries.

A wide variety of bacteria is specialized to utilize specific inorganic dissolved substrates to generate the energy (ATP) required for synthesis. Such chemoautotrophs oxidize substances like NH_4 (nitrifying bacteria) and H_2S (sulfur bacteria). These bacterial groups tend to be inviable or very inefficient in oxygenated waters. Consequently, chemosynthesis is largely restricted to anoxic or poorly oxygenated waters. As discussed in Chapter 12, these bacteria are active in anoxic pore waters of sediments.

Kinetics of Nutrient Uptake

Given that nutrients are usually superabundant at the beginning and very low in concentration toward the end of a phytoplankton bloom, we wish to determine the relationship between nutrient concentration and uptake rate. We shall define the uptake rate of a single or a group of phytoplankters as V (measured as amount taken up per cell per hour); we then define cell division rate as μ (number of cell doublings per cell per day). We assume for the moment that μ is linearly related to V; that is, increased nutrient uptake corresponds to increased cell division rate. We might expect, and much data suggest, that V is nutrient concentration dependent and linearly proportional to the concentration of a nutrient like ammonia. Because of the inherent limitation in nutrient uptake (e.g., a rate-limiting step, such as membrane transport of the nutrient), however, there may be a concentration above which uptake (V) no longer increases. So we expect a curvilinear relation between nutrient uptake V and nutrient concentration. Because of our assumption, we must also expect a similar relation between cell division rate μ and nutrient concentration.

    This relationship is analogous to the kinetics of enzyme-substrate interactions described by the Michaelis–Menton relationship (Caperon, 1967; Dugdale, 1967):

$$V = \frac{V_m\,S}{K_s + S}$$

where V is the rate of nutrient uptake as above, V_m is the maximum rate of uptake, K_s is the substrate concentration at which $V = V_m/2$, and S is the concentration of the nutrients.

    Usually μ is measured rather than V. Given our assumption that V is linearly related to μ,

$$\mu = \frac{\mu_m S}{K_s' + S}$$

where μ_m is the maximum cell doubling rate and K_s' is the nutrient concentration at which $\mu = \mu_m/2$. Figure 10-8 shows this relationship for data obtained by Eppley and Thomas (1969); an adequate fit to the model is found for an S versus μ plot and a s/μ versus μ plot (straight-line relationship). We assume that K_s should equal K_s'.

    Parameters derived from the model may be used to predict nutrient uptake differences among phytoplankton species living under different nutrient regimes. Figure 10-9 shows a testable hypothesis concerning a phytoplankton species living under low-nutrient concentrations and another living under high-nutrient concentrations. We presume that a species living under high-nutrient conditions would be capable of taking up nutrients at higher ambient concentrations than a second species adapted to low-nutrient conditions; hence μ_m is greater for the former species. We also assume that the low-nutrient species must be adapted to take up nutrients more efficiently at low concentrations relative to

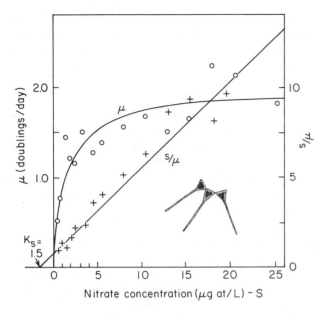

Figure 10-8   Growth rate (doublings per day) of *Asterionella japonica* (open circles) as a function of S, the concentration of nitrate. K_S is estimated at 1.5 by plotting nitrate concentration/doubling rate (S/μ) as a function of nitrate concentration (S—see crosses in diagram). (From Eppley and Thomas, 1969, with permission from the *Journal of Phycology*)

species adapted to much higher nutrient concentrations. So the value of K_s' should be greater for the latter species, as illustrated in Fig. 10-9.

MacIsaac and Dugdale (1969, 1972) show that K_s for coastal phytoplankton is usually greater than 1 μmole per liter for nitrate uptake whereas oceanic phytoplankton have K_s values of about 0.1 to 0.2 μmole. Clones of the same diatom species show high and low values of K_s, depending on whether the clones are isolated from nearshore or oceanic waters, respectively (Carpenter and Guillard, 1971). This fact indicates that the oceanic phytoplankton are more efficient at taking up nutrients at low-nutrient concen-

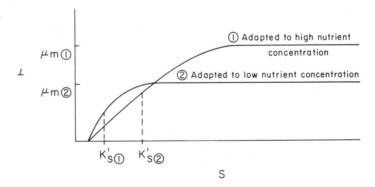

Figure 10-9   A theoretical representation of the differences in substrate uptake expected for a phytoplankton clone adapted to inshore high nutrient levels versus another clone adapted to offshore low nutrient levels. In this example, there is a lower threshold concentration for nutrient uptake.

tration. Thus they may be competitively superior to coastal forms in the low-nutrient concentrations of the open sea. Differences in K_s, linked with differences in the ability to photosynthesize under differing light conditions, might determine the sequence of phytoplankton species in spring–summer succession. However; data to support this hypothesis are scarce.

Differences in nutrient limitation among competing species may also contribute to coexistence. For example, species 1 might be superior at phosphate uptake while species 2 could be superior at silica uptake. So under conditions of varying nutrient concentration (spatial patchiness) different species might be favored. Under some conditions of intermediate phosphate and silica each species might be limited in population size by a different nutrient; coexistence would thus be possible. Titman (1976) demonstrates that two lake diatom species can coexist in culture under an intermediate ratio of silica/phosphate. One species excludes the other under high silica/phosphate conditions, while the latter wins under low silica/phosphate conditions.

It is a natural inference that nutrient uptake dynamics can be directly related to the growth of phytoplankton populations. We would assume that cell doubling rate μ is linearly proportional to nutrient uptake V. It is probably true in many cases. Yet phytoplankton species may take up nutrients without concomitant cell divison. Intracellular nutrient pools may also influence phytoplankton cell division (Droop, 1973). Thus a phytoplankton cell might take up nutrients at one time and employ them for growth sometime later. The dinoflagellate *Gonyaulax polyedra* absorbs nutrients at depth during the night and swims into the nutrient-depleted photic zone to photosynthesize during the daytime (Eppley et al., 1969).

Can interspecific competition for nutrients determine the success of some phytoplankton species relative to others? Major differences in K_s values—oceanic versus nearshore, for example—seem to indicate that this may be true on a grand scale. Still, many phytoplankton species seem to coexist with no competitive exclusion and similar nutrient preferences. This coexistence has been characterized as the "paradox of the plankton" (Hutchinson, 1961). Hulbert (1970) points out that an abundant species will rarely force a less abundant one to extinction because cell densities required for an individual phytoplankton cell's nutrient depletion zone to overlap with that of other cells would be much greater than typically observed in natural phytoplankton populations. Thus relative population densities of phytoplankton species shift in response to changes in light and to nutrient conditions. But it is unlikely that one species can drive others to complete extinction.

10-5 GRAZING BY THE ZOOPLANKTON

Grazing in the Sea

As discussed earlier in our hypothetical seasonal plankton cycle, zooplankton appear in great abundance in inshore waters after the peak of the phytoplankton bloom has passed. This abundance probably occurs (a) because of the time lag in reproduction of the zooplankton and (b) because of the gradual grazing away of the phytoplankton standing crop as the zooplankton population increases. Phytoplankton (e.g., diatom) population growth can be described as

$$C_2 = C_1 e^{rt}$$

where cell concentration C_2 is produced by a previous concentration C_1 growing after t time units with a rate r (see Chapter 3). With a grazing rate g, this equation becomes

$$C_2 = C_1 e^{(r-g)t}$$

If $g > r$, the zooplankton will graze the phytoplankton population to extinction.

Because the phytoplankton have a much shorter generation time than the zooplankton, phytoplankton extinction would be a destabilizing force in marine planktonic communities, causing strong zooplankton population oscillations. When the phytoplankters reach very low densities, however, zooplankton might not find them. A low-density refugium for phytoplankton allows a subsequent increase when zooplankton decrease in abundance.

Harvey and co-workers (1935) observed alterations of abundance of phytoplankton and zooplankton with an additional sudden drop in the phytoplankton population as the zooplankton first appeared in the nearshore waters of Plymouth, England. This finding suggests that the zooplankton have an important influence on the decline of phytoplankton. Phytoplankton in the Antarctic are grazed down by the zooplankton, resulting in large numbers of broken diatom tests being deposited on the sea floor (Hart, 1942).

Yet in some waters grazing apparently does not contribute to the decline of the phytoplankton. Tisbury Great Pond, a brackish enclosed pond, shows minimal effects of grazing by zooplankton on phytoplankton standing crop (Deevey, 1948). In Long Island Sound the phytoplankton decrease in late spring is attributed to the loss of nutrients to the stable water column rather than to any effect by the zooplankton (R. Conover, 1956). There is some evidence that a dense population of very small phytoplankton species develops in the summer and is never significantly grazed by the zooplankton.

The effectiveness of grazing probably varies not only from place to place but also from time to time with any given locality. On George's Bank phytoplankton and zooplankton abundances correlate positively with each other, both increasing in the early spring. In May, however, phytoplankton show a precipitous drop, accompanied by a very large rise in the zooplankton population. This situation probably indicates the effectiveness of grazing (Riley and Bumpus, 1946). It is possible that a threshold in zooplankton abundance is reached, above which the phytoplankton can be cropped faster than renewal is possible. Also, at this time it is possible for the water column to stabilize, thus reducing the return of nutrients to the upper parts of the water column and concomitantly prohibiting the phytoplankton population from renewing itself at the former rapid rate.

The literature tends to support the proposition that in inshore waters of great environmental variability long periods of time pass in which phytoplankton are not effectively grazed down or controlled by zooplankton grazing. Apparently in more open marine conditions zooplankton may be regarded as a consistently important controlling force throughout the year. Steemann Nielsen (1958) argues for this point of view; he points out the constancy in phytoplankton standing crop throughout the year in the tropics and suggests that zooplankton grazing is the stabilizing influence.

Modest zooplankton grazing on phytoplankton may have a positive effect on phytoplankton production. Using a freshwater phytoplanktivorous fish, *Notropis spilopterus*, Cooper (1973) shows that net primary productivity was directly related to grazing pressure up to a certain point and then inversely related. Animal excretion may stimulate phytoplankton growth with sufficiently low grazing pressure. With more intense grazing, the effect is canceled.

Porter (1976) has shown that colonies of the colonial freshwater planktonic green alga, *Sphaerocystis schroeteri*, are only partially disrupted and only a fraction of the cells are digested and assimilated by *Daphnia magna*, a natural predator. During passage through the gut the cells that remain viable take up nutrients (e.g., phosphorus) from algal remains and *Daphnia* metabolites. This nutrient supply stimulates *Sphaerocystis* carbon fixation and cell division and enhances algal growth. Thus nutrients regenerated by grazers may produce the summer bloom of gelatinous green algae *(S. schroeteri)* during the seasonal succession of late phytoplankton. In other words, grazing shifts the species composition of the phytoplankton community toward less digestible forms.

Zooplankton Feeding

Research on calanoid feeding behavior suggests that the stabilization of phytoplankton populations may be affected by phytoplankton cell concentration and cell-size differences. Mullin (1963) and his students and colleagues have elegantly investigated the feeding of *Calanus*, extending the work of Marshall and Orr (1955) and others. Frost (1972) investigated ingestion of diatoms in culture by the copepod *Calanus pacificus*, as a function of diatom concentration (cells ml^{-1}) and cell size (cell volume and cell carbon). He estimated the rate of ingestion of diatoms by copepods by comparing diatom population growth rates in grazed and ungrazed containers. Assuming exponential growth [$C_2 = C_1 \exp (r - g)t$] as described earlier, grazing rate g can be calculated. We can then measure I, the number of cells ingested per copepod per hour, as

$$I = \frac{\overline{C}Vg}{N} \quad \text{(cells copepod}^{-1} \text{ hr}^{-1})$$

where V is the volume of water in the experimental container, N the number of copepods in the container, and $\overline{C}$ the mean cell concentration during the course of the experiment in the grazed container.

For a given cell size, I increases linearly with cell concentration until a critical maximum, $\overline{C}_m$, where ingestion increases no further (Fig. 10-10). Below $\overline{C}_m$, the response is indicative of an animal that searches, encounters, and feeds on particles in direct proportion to their concentration. At the saturation level, ingestion increases no further. The maximum I implies a successively decreasing feeding efficiency above the upper corresponding value of $\overline{C}_m$. Feeding efficiency could be decreased by saturation of the filtering apparatus at the critical maximum concentration or fullness of the gut. Both would permit no more particles to enter. Therefore above the critical maximum the animal

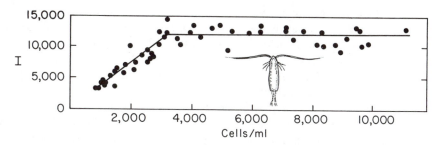

Figure 10-10   Ingestion rate, I, for the copepod *Calanus pacificus* as a function of cell concentration of the diatom *Thalassiosira fluviatilis*. (After Frost, 1972)

may be filtering the same amount of water or moving at the same speed but simply getting no more reward for its effort.

Frost's analysis neglects the possibility of two alternate thresholds. First, feeding could be reduced at a lower threshold because particles present cannot meet the energy requirements of searching or because the animals switch from a feeding mode to a "search" mode (Lam and Frost, 1976). Alternatively, the feeding rate could be reduced above an upper threshold because no more nutritive gain is accomplished above the critical maximum concentration. These two thresholds can be described with the feeding equation of Ivlev (1961) as

$$R = R_{max}[1 - e^{k(p_o - p)}]$$

where R is the size of the ration, R_{max} the maximum ration that a copepod will accept, p the concentration of prey (diatoms), and p_o the threshold prey concentration below which grazing ceases.

There is some argument as to whether a lower threshold exists, due to the high experimental variance of results at low food concentrations (see Mullin, 1963). Mullin and others (1975) analyze available data and conclude that the data do not justify choosing among various models of feeding. Lam and Frost (1976), however, predict, on energetic grounds, a lower threshold below which feeding behavior must change. At very low particle concentrations, feeding should be at a reduced rate. Above this level, feeding rate should increase with particle concentration until some maximum rate is achieved. Filtering rate should then decrease above a critical maximum cell concentration.

Cell size also influences feeding in copepods. Richman and Rogers (1969) found significantly higher feeding rates on two-celled chains than on single cells of *Ditylum brightwelli*. Frost (1972) examined cell ingestion rate, *I*, for different-sized species of diatoms and found that *I* increased with cell volume at cell concentrations below the critical maximum. From this we may conclude that *Calanus* handles larger diatom cells with greater effectiveness than small cells. When cell volume is expressed in terms of carbon/cell, the critical upper maximum is the same, independent of the size of the cell fed upon. Thus when the nutritive reward can no longer increase, further increases of ingestion with respect to carbon do not occur.

These studies ignore within-species differences, such as age, size, and previous nutritive experience, and also ignore between-species differences and the effects of quality of food. Mullin and Brooks (1970) investigated feeding by the planktonic copepods *Rhincalanus nasutus* and *Calanus helgolandicus* on diatoms. Adult *Rhincalanus* grew less rapidly than *Calanus* whereas *Calanus* nauplii did not grow at all. Feeding and growth for both species slowed down at 10°C relative to 15°C and rate of ingestion per unit body carbon decreased with increasing body carbon. Respiration per unit of body carbon decreased with increasing body carbon. Differences in food quality are possible between different foods with the same particle size. Invertebrate eggs and lecithotrophic larvae should have a higher carbon content per unit particle volume than phytoplankton. No mechanism, however, is as yet known by which animals distinguish between foods of the same particle size. Stoecker and colleagues (1981) have demonstrated that tintinnids (protozoa) prefer dinoflagellates to other algal cells of the same size. Detrital particles also serve as food if colonized by microorganisms (Heinle et al., 1977).

Assimilation efficiency may vary with type of food ingested and with particle-size concentration. Conover (1966) estimates assimilation efficiency of copepods by comparing composition of food with that of feces. Assimilation A in percent is

$$A = \frac{J - L}{(1 - L)J} \times 100$$

where J is the ash-free dry weight/total dry weight taken in food and L is the same ratio in feces. Ash weight must be considered because, in diatoms, ash content is high as a result of the silica shell. This equation is useful for carbon, but nitrogen and phosphorous are in such low concentration that small changes in ash content greatly change the calculated value of A. Values of A range from 50 to 95% and are relatively independent of temperature or cell concentration.

Particle-Size Selectivity

In natural waters copepods are presented with a great variety of particle sizes that continually change in relative abundance with season. Adaptation to the particle-size spectrum would maximize the efficiency of particle uptake. Poulet (1972, 1974) investigated the feeding of *Pseudocalanus minutus* on natural spectra of particles and found that copepods could shift their grazing from small to large particles to compensate for a reduction in abundance of small particles. Richman and others (1977) investigated ingestion of natural particles by copepods living in Chesapeake Bay and calculated ingestion over various natural distributions of particle size. They found that copepods fed selectively at particle-size abundance peaks but also generally preferred larger-sized particles. Because this shift could be observed by using natural particle distributions with differing size-frequency distributions, a rapid shift of selective feeding is suggested. In an experiment with differing densities of copepods, a shift from preferred large-sized particles to smaller particles was demonstrated.

The mechanisms of selective feeding are poorly understood at present. Rapid shifts of copepod feeding behavior seem to occur when particles of differing size are encountered (Conover, 1968). When small particles are grazed, adult *Pseudocalanus minutus* move their feeding appendages rapidly and feeding and swimming occur simultaneously. *P. minutus* swims slowly and makes several loops before its appendages stop moving. This pattern is stopped when large detrital particles or chains of cells are encountered. Chains of *Chaetoceras* sp. are seized, rolled with the mouth appendages, and consumed, starting at the end of the chain. Thus filter feeding and particle seizure are mutually exclusive processes, but switching from one mode to the other occurs in a short amount of time.

The hydrodynamics and exact nature of particle capture are something of a mystery. Poulet (1974) suggests that the evidence supporting selective feeding on larger cells (e.g., Frost, 1972) is simply a reflection of the greater experimental concentration of total cell volume of larger diatom cells as opposed to small cells. The selective feeding on large cells may therefore be due to a concentration effect rather than a cell-size effect per se.

The simplest hypothesis would explain feeding as a sieving process, governed by the spacing between the setules. The clearance of particles measured in feeding experiments should therefore correspond to intersetular distances; they do not. The feeding of the inshore copepod *Temora longicornis* cannot be rectified with the sieving hypothesis unless the measured intersetular distance is multiplied by 1.5 (Ninivaggi, 1979). The feeding appendage might operate as a leaky sieve (Boyd, 1976). Movement of the maxilli might cause the setules to spread apart. Furthermore, particles exactly the dimension of the intersetular spacing might slip through.

It is even possible to challenge the notion that copepod feeding appendages act like a sieve at all. The Reynolds number of a fluid can be used to estimate the ratio of inertial to viscous forces acting on particles or small objects moving in a fluid medium (Purcell, 1977). Calculations by Ninivaggi (1979) suggest that viscous flow dominates around a moving second maxilla of *Temora longicornis*. Under these circumstances particles might not flow freely through the setules; high-speed photography by Alcaraz and others (1981) suggests that this viscosity permits the feeding appendage to pluck particles out of the fluid medium. Still, the available feeding data cannot rule out the leaky-sieve hypothesis; we must also explain the evidence suggesting that the copepods may be able to track changes in the size-frequency distribution of particles in the water column. At this writing, the exact mechanism(s) of filtration remains to be firmly established.

The data suggest that copepods can track changes in the size spectrum of the phytoplankton and graze the most abundant size classes. Selective grazing can therefore favor the less abundant species and the less common size classes. Copepod grazing thus can affect the species composition of the phytoplankton and prevent one species from competitively excluding all others. If there is a threshold below which feeding is suppressed, then the phytoplankton always have a refuge, thereby preventing grazing to extinction. Recovery is possible because zooplankton will decrease in abundance and the lag time for zooplankton reproduction will allow the phytoplankton to recover if nutrients are available. Because the phytoplankton generation time is, in general, shorter than that of the zooplankton, the phytoplankton can recover quickly. This factor suppresses strong oscillations in both phytoplankton and zooplankton abundance.

10-6 DEFENSE AGAINST PREDATION

The composition of planktonic communities must be affected by differential susceptibility of planktonic species to predation. Yet few studies have demonstrated this process to be important in marine systems. In freshwater plankton, differential susceptibility is probably a common underlying explanation for the relative abundance of species (e.g., Brooks and Dodson, 1965; Porter, 1973; Zaret and Suffern, 1976).

The development of body spines and armature in many phytoplankton and zooplankton is probably an adaptation to increase the difficulty of capture and ingestion. Common diatoms, such as the genus *Chaetoceras,* have large projecting spines that increase the effective body size. Many planktonic crustacea are similarly armed with sometimes elaborate spination. The presence of spines may place the prey out of the size range of the predator or may make the prey difficult to handle and seize.

Bioluminescence is a common feature of many planktonic phytoplankton and zooplankton species. Dinoflagellates commonly luminesce under turbulent conditions. Calanoid copepods avoid luminescent dinoflagellates and favor grazing on nonluminescent forms (Esaias and Curl, 1972). As copepods feed or as they swim among swarms of luminescent dinoflagellates, the resultant microturbulence induces flashes of light. These flashes might endanger the copepods by making them visible to fish predators (Burkenroad, 1943). Many deep-sea fishes have opaque digestive tracts that might protect against the obvious signal that a luminescent prey in a transparent fish may create for the fish's potential predators (McAllister, 1961; Porter and Porter, 1979).

Many phytoplankton species, notably the blue green algae and dinoflagellates, are toxic to grazers and this toxicity undoubtedly influences the distribution of species of phytoplankton in grazed systems. Species of the golden-brown alga *Phaeocystis* produce large amounts of acrylic acid, an effective antibiotic capable of sterilizing the guts of consumers (Sieburth, 1960, 1961). It might inhibit microbially mediated digestion and could induce future avoidance of the alga. Schantz (1971) suggests that bioluminescence in some dinoflagellates may be an evolved "warning" system that has coevolved with the production of toxins and the presence of grazers.

10-7 SPATIAL STRUCTURE IN THE PLANKTON

Patchiness

A sample of plankton rarely comes from a homogeneously distributed population. Both phytoplankton and zooplankton have aggregated spatial distributions (see Chapter 3) at all scales of sampling. McAlice (1970) sampled horizontally at 6-m intervals and found all phytoplankton species to be significantly aggregated. Several species seemed jointly to increase and decrease in abundance. Figure 10-11 shows the patchy distribution of chlorophyll and zooplankton in the North Sea (Steele, 1974). Of note is the inverse relation between phytoplankton and zooplankton, due to grazing.

Chlorophyll (mg/m³)

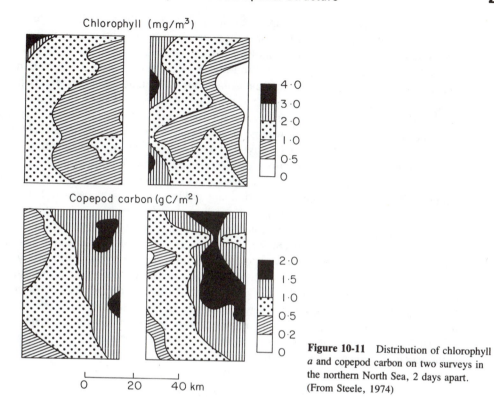

Copepod carbon (gC/m²)

0   20   40 km

Figure 10-11 Distribution of chlorophyll *a* and copepod carbon on two surveys in the northern North Sea, 2 days apart. (From Steele, 1974)

Horizontal patchiness is probably related to spatial variation in chemical, physical, and biological parameters (e.g., salinity, turbulence, and grazing). It seems reasonable that this variation is more frequent in nearshore, shelf, estuarine, and bay plankton. Open-ocean physical and chemical spatial variability and concomitant plankton variation should be less. Generally open-ocean phytoplankton is less abundant and more uniformly distributed than near-coast phytoplankton (Lorenzen, 1971; Venrick, 1972).

Processes promoting patchiness include (Parsons and Takahashi, 1973, p. 30)

1. Physical–chemical boundary conditions, including light, temperature, and salinity gradients.
2. Advective effects as in water transport, including small-scale variations due to turbulence.
3. Grazing.
4. Reproduction rates within the population.
5. Social behavior in populations of the same species.
6. Intraspecific interactions resulting in either attraction or repulsion between species.

Wind moving over the surface generates spatial structure over a wide range of scales. Langmuir circulation results from the creation of vortices caused by wind-driven water movement. These vortices are small in scale and result in small divergences and convergences of water. Phytoplankton cells therefore may be sorted by small currents; zooplankton may be trapped in an upward current while attempting to swim downward to avoid surface light (Stavn, 1971). Wind-driven water motion often arranges Langmuir circulation along linear tracks such that collections of plankton along transects parallel to the wind are much more variable than those taken perpendicular to the wind.

Turbulence is also important in generating medium scale (ca. $10^2 - 10^3$ m) patchiness of phytoplankton in shallow waters. Apparent between-station patchiness may simply reflect time-dependent changes in phytoplankton density at any one location. In the shallow waters of Nova Scotia, time-dependent variation of chlorophyll at a fixed point is consistent with a turbulence model using scales of 1 to 1000 m (Platt, 1972). Between-station differences in primary production are evident on a daily basis but not when sampling is integrated over longer periods of time.

A concentrated patch of phytoplankton cells must inevitably diffuse outward because of the transfer of wind and current forces into water energy of smaller scales that move water omnidirectionally from a given point. If phytoplankton growth is sufficiently rapid, a large phytoplankton patch can withstand a measurable degree of diffusion and still maintain its presence as a patch. The probability of survival of a patch should increase because population growth increases with the area of the patch, but the loss of phytoplankton increases at a slower rate—namely, with the perimeter of the patch. For a circular patch, an increase of patch size increases the area/perimeter ratio ($d/4$, where d is the patch diameter), allowing phytoplankton growth to overcome diffusive loss and a bloom to materialize. The patch will survive if

$$d > 4\left(\frac{k}{r}\right)^{1/2}$$

where r is the growth rate of the population and k is a constant estimating the diffusion (Kierstead and Slobodkin, 1953). If the patch diameter is less than d, then the patch will have diffused away before a bloom can occur. The implicit assumption is that the patch includes an area of water of fixed size surrounded by water in which phytoplankton cannot grow; the simplest case would be where the patch contains nutrients that may or may not support a phytoplankton growth rate sufficiently rapid to counteract diffusive loss.

Because there is an empirical relationship between the scale of a patch and k (Okubo, 1971), expected patch size can be calculated, given the known range of population growth rates for phytoplankton populations. For a doubling time of 1 to 10 days, minimum patch sizes of 2 to 50 km are predicted (Steele, 1976). Although data consistent with this prediction are available (e.g., Platt et al., 1970), the relationship between observed and predicted patch size requires more study.

Measurements of plankton spatial distribution suggest that patchiness is usually determined by an interaction of several factors. Using time-series analysis, Mackas and Boyd (1979) show that phytoplankton and zooplankton occur in patches at all conceivable spatial scales (fractions of 1 km to ca. 40 km); however, zooplankton tend to occur more frequently in small patches relative to phytoplankton. This may relate to zooplankton behavior. Furthermore, zooplankton patches tend to alternate with phytoplankton. Overgrazing by patches of zooplankton probably plays a significant role in these alternations (Harvey et al., 1935). Steele (1976) points out that we might expect copepod populations to graze phytoplankton patches and to homogenize spatial variation. Yet the emergence and decline of successive copepod generations in a spatially discontinuous manner might be accompanied by concomitant depression and explosion of phytoplankton populations.

Turbulence-generated patchiness may influence the availability of phytoplankton to zooplankton. Models predicting zooplankton abundance from phytoplankton production assume spatial homogeneity, giving an average ration to an average copepod. But turbulent processes may concentrate copepods in microvolumes where phytoplankton are not sufficient to support their energy requirements. Mullin and Brooks (1976) examined water samples pumped from various depths and simultaneously measured abundance of copepod developmental stages and phytoplankton carbon. Compared with laboratory measurements of energy requirements, as many as 41% of the individual 150 to 200 liter samples had phytoplankton standing stocks too small to satisfy the energy requirements of *Calanus pacificus* numbers collected in the same samples. At any given station, however, there was at least one depth where the requirement was being met. An overall positive correlation between phytoplankton carbon and copepod biomass was found. The duration of the patchiness is not known, but short-term mixing might make the impact of the patchiness negligible.

The vertical structure of phytoplankton during the season in a temperate–boreal inshore water column changes dramatically as the phytoplankton bloom develops and the water column stabilizes. Vertical turbulence in winter causes an initial uniform distribution of phytoplankton biomass with depth. The rate of productivity (in terms of photosynthesis per unit phytoplankton biomass—mg C/mg Chla/day) varies with depth, however. Phytoplankton are inhibited near the surface by very high illumination, and production has a maximum at some depth below the surface, where the light regime is optimal. The profile of phytoplankton biomass with depth changes over time because as the phytoplankton biomass increases near the subsurface photosynthetic maximum, the extinction coefficient of light increases due to absorption by pigments and scattering by phytoplankters. Therefore the self-shading effect reduces the daily compensation depth and the depth of maximum phytoplankton growth decreases. Eventually the standing stock of phytoplankton is concentrated near the surface. With the reduction of nutrients near the surface, phytoplankton standing stock decreases and primary productivity again peaks at a greater depth.

Sinking rates also greatly influence the vertical distribution of phytoplankton, a factor that is also important in regard to transfer of nutrients in the water column. Zooplankters, for example, form fecal pellets having high sinking rates. Fecal pellets of copepods sink approximately 36 to 376 m/day whereas living phytoplankters sink only 0 to 30 m/day. Table 10-1 shows variation in sinking rates of various planktonic organisms.

Deep scattering layers, which reflect sonar, consist of schools of fishes, euphausids, or other groups, and usually occur anywhere from 200 to several hundred meters deep. Many zooplankton groups live at different depths, depending on season. Behavioral

TABLE 10-1 SINKING RATES OF SOME COMPONENTS OF THE PLANKTON[a]

Group	Sinking rate (m/day)
Phytoplankton	
Living	0–30
Dead, intact cells	<1–510
Fragments	1500–26,000
Protozoans	
Foraminifera	30–4800
Radiolaria	ca. 350
Other zooplankton	
Chaetognaths	ca. 435
Copepods	36–720
Pteropods	760–2270
Salps	165–253
Fecal pellets	36–376
Fish eggs	215–400

[a]Modified from Parsons and Takahashi, 1973, with permission of Pergamon Press, Ltd.

responses of zooplankton to light and salinity gradients can play a major role in concentrating zooplankton. Zooplankton are often found at boundary layers, such as discontinuities in salinity.

Vertical Migration

One of the most interesting phenomena relating to the vertical distribution of zooplankton is vertical migration. Zooplankton groups descend to great depths during the day and ascend during the night. In the Clyde Sea, *Calanus* individuals migrate to 100-m depth during the day and move to the surface at nightfall (Raymont, 1963). Such a general description of vertical migrations masks their complexity. Marshall and Orr (1955) document in detail the seasonal vertical migration of the copepod genus *Calanus*. The story is complex indeed. In some areas *Calanus* shows no diurnal vertical migrations at all whereas in others the migration is of the classical day-down, night-up variety. Often differences are associated with season, sex, and molt stage. Vertical migrations of zooplankton are detected at depths of 1000 m (Waterman et al., 1939). Vinogradov (1968) suggests a ladder relationship whereby energy is transferred by vertical migrations from one major depth zone of zooplankton to the next deeper depth zone.

Hardy and Bainbridge (1954) developed the plankton wheel, a circular glass tube that is rotated, allowing zooplankton to swim vertically and continuously. The speed is adjusted until the animal's movement appears stationary. The measured speeds of ascent were remarkable. The genus *Calanus* moves at an upward rate of about 15 m an hour whereas the planktonic polychaete *Tomopteris* moves faster than 200 m per hour. *Calanus* was able to swim downward at 100 m per hour. Therefore the swimming speeds of these groups are certainly within the range of the postulated vertical migrations.

The factor that triggers vertical movement may be far different from the evolutionary forces that selected for vertical migrations. Diurnal changes in pH, oxygen, salinity, and temperature are probably too small to trigger migrations. Diurnal light variation seems to be the most likely explanation for the triggering mechanism. Because downward movements of zooplankton often precede dawn, it is conceivable that daily light periodicity sets an internal biological clock (Enright and Hamner, 1967; Esterly, 1917, 1919).

Because zooplankters actively seek discontinuity layers, are able to respond to pressure, and can detect different levels of light, a complex series of behavioral patterns may be involved in vertical diurnal migrations. The relative importance of these different factors may change in different groups. If light is important, one of the intriguing questions raised is how organisms detect diurnal light change at a depth of greater than 1000 m (Waterman et al., 1939), where the intensity is ca. 10^{-13} of the surface intensity in clear ocean water.

Adaptive explanations fall under the following categories.

1. Zooplankton are adversely affected by strong light and must get away from the surface during the day. They come back to feed on phytoplankton at night.

2. Because the upper layers of the sea often move at higher speeds than the waters at depth, the movement of plankton toward the surface allows animals to be carried greater distances while moving at night in the upper layers than during the day when they are at depth. The great movement of plankton in different directions and over great distances allows the zooplankton to feed on larger areas of phytoplankton (Hardy, 1956).

3. This hypothesis is similar to (2), but here the major effect is that it maximizes gene flow between populations and preserves genetic variability (David, 1961).

4. Migration downward during the day permits a recovery of the phytoplankton and thus gives a greater nutritive reward than continuous grazing (McLaren, 1963).

5. It is energetically advantageous to be in colder water during the day to respire at a lower rate and then come up at night to feed on the phytoplankton (Conover, 1968; McLaren, 1963).

6. To avoid predation by visual predators (e.g., fish, seabirds), zooplankton swim to depths beyond the illumination levels of predator detection. A nighttime return permits feeding on surface phytoplankton when fish and surface-feeding seabirds do not hunt prey (see McLaren, 1963).

The light intensity hypothesis may be excluded because of the great depths to which daily migrations extend. It is hard to believe that a copepod would go as deep as 100 m because light is harmful at the surface. Hypotheses (2), (3), and (4) could never work without the majority of the zooplankton initiating the behavior at once. How any accepted model of selection among individuals and evolution could accomplish this change is difficult to conceive. If all copepods left the surface, why wouldn't some "cheaters" remain at the surface to take advantage of the remaining phytoplankton? Hypotheses (2), (3), and (4) require a "cooperation" among the zooplankton to move together. The group evolution of cooperation seems unlikely in these groups.

We are thus left only with the predation and energy hypotheses to entertain seriously. The predation hypothesis is attractive, for it provides a measurable selective force that could favor the migration. If a gene for migration were present in the population, then strong-enough visual predation should select for diurnal migration, given that the animal must be in the surface waters to feed. In Gatun Lake, Panama, the calanoid copepod *Diaptomus gatunensis* undergoes vertical migrations in a water column with no thermal stratification. Although it can be a food for the planktivorous fish *Melaniris chagresi*, it is rarely found in gut contents, indicating the effectiveness of the vertical migration. In a temperate pond, the daytime (1300) depth distribution of a species of *Daphnia* lies below the depths in which the golden shiner *Notemigonus* can see to hunt prey (Zaret and Suffern, 1976). Some shortcomings of the predation hypothesis are (a) many zooplankton migrate to depths far deeper than needed to avoid predators, (b) many vertically migrating species bioluminesce at night, a behavior hardly conducive to invisibility, and (c) forms that are probably relatively invisible due to their transparency (e.g., ctenophores) also undergo vertical migrations.

The energy hypothesis would be credible if the energy conserved at depth surpassed that lost in swimming. The additional demand of migration on the energetic demands of zooplankton is believed to be trivial (Hutchinson, 1967; Vlymen, 1970); so a net energetic gain is conceivable. This energetic gain depends on a reduction of metabolic rate by a poikilotherm in colder waters at depth. The result should be more tropical vertical migrating species because the depth-dependent temperature differential is greater in lower latitudes. Such apparently is the case. Furthermore, copepod species are known to descend to deeper colder water after the seasonal phytoplankton bloom (e.g., Marshall and Orr, 1955). The migration of *Diaptomus* in a lake with less than 0.2°C diurnal variation, however, indicates that thermal variation with depth cannot be the only factor at work.

McLaren (1974) has calculated the potential advantage of vertical migration in thermally stratified water in terms of fecundity. Fecundity is increased by part-time residence in deeper, colder waters. Using data for growth and fecundity as a function of temperature for *Pseudocalanus minutus,* a calculated advantage occurs in seasonally

TABLE 10-2 SOME EFFECTS OF VERTICAL MIGRATION OF THE COPEPOD, *PSEUDOCALANUS MINUTUS*[a]

Parameter	Calculated values for	
	Nonmigrant	Migrant
Survivorship from Stage VI to hatch of first brood	0.8741	0.8741
Length of female (mm)	0.908	0.940
Number of female young per brood	7.35	8.32
Rate of increase per generation	1.000	1.033

[a]Based on calculations of McLaren, 1974.

interrupted growth in food-limited populations or for migration of later developmental stages when food is limiting and mortality is great in early developmental stages. Because juvenile mortality is large and earlier stages may show less pronounced migration than later stages (e.g., *Daphnia galatea mendotae;* Zaret and Suffern, 1976), there may indeed be a general energetic advantage to migration. Table 10-2 shows the calculated advantage for one case of migrating and nonmigrating *Pseudocalanus minutus.*

SUMMARY

1. The spring phytoplankton increase is a prominent feature of temperate–boreal waters. In general, diatoms bloom in spring and decrease in summer. A successional sequence of phytoplankton species usually occurs.

2. Nutrients are superabundant in winter, but light is limiting and turbulence carries a substantial proportion of the phytoplankton population to depths with low light. As the spring progresses, the establishment of the thermocline stabilizes the water column. This factor and the increasing photoperiod set off a phytoplankton bloom. The stability of the water column, in combination with sinking of dead cells, precludes the replenishment of the bloom from deeper waters with nutrients. The phytoplankton therefore decrease.

3. Light decreases exponentially with depth and exerts an obvious effect on photosynthesis. Phytoplankton cells can compensate for lowered light via an increase of photosynthetic pigments.

4. Nutrients are those constituents required by plants. Phosphorous and nitrogen are of particular importance to phytoplankton. Phytoplankton have evolved to different nutrient abundances; species living in offshore, nutrient-poor waters are relatively efficient at taking up nutrients. Species living in nutrient-rich areas are less efficient but can probably take up larger total amounts of nutrients.

5. The effect of zooplankton grazing in the sea is quite variable; it is probably substantial in offshore waters of low primary productivity. Often inshore phytoplankton show no evidence of control by zooplankton grazing. Some evidence suggests that copepods can shift grazing to specialize on differing particle sizes; overall, copepods seem to prefer larger cells. These effects might regulate the composition of the phytoplankton community.

6. Plankton are usually patchily distributed, primarily because of hydrography. Grazing and zooplankton behavior contribute to some extent as well. Diurnal vertical migrations are a fascinating instance of behavioral patchiness; zooplankton species migrate downward during the day, perhaps to avoid predators or to save on the lowered metabolic cost encumbered in the cooler deep waters.

11 food webs and productivity in the plankton

11-1 FOOD WEBS: DEFINITION, COMPLEXITIES, EFFICIENCY, AND DATA

The Food Chain Abstraction

The simplification required to abstract a natural food web into a food chain is illustrated in Fig. 9-2, a model of the North Sea. In the simplified food chain algae are eaten by copepods, which are, in turn, consumed by herring. But the complexities masked by this abstraction are enormous. So it is essential to remember some of the inherent assumptions in food chain analyses.

Simplified food chains assume that a single species or group of species at a single trophic level is consumed by a species (or several species) belonging to the next higher trophic level. The right-hand diagram of Fig. 9-2 shows that no simple abstraction of levels is readily apparent. Food chains usually only consider the adult stages of dominant species, making assignment to a few levels more practicable.

A second assumption in the construction of food chains is that a species maintains the same trophic position in the food chain throughout its life cycle. Yet such is clearly not the case in many instances. Growing animals may assume different feeding habits as they acquire the ability to attack larger prey. The herring *Clupea harengus* feeds on different food types as size increases. Cod switch from a diet of crustacea to herring at a length of ca. 50 cm. The changing diet of growing consumers can result in the exchange of the respective roles of predators and prey in the plankton. The adults of young prey species might feed on the juveniles of predator populations. These complexities suggest that food chains can be tenuous abstractions.

A last shortcoming of the food chain abstraction is the assumption of transfers from a given trophic level to the one immediately above. Consumers in many food webs can feed at several trophic levels below. Copepods can consume detrital particles, capture smaller zooplankton, and feed on phytoplankton. Some salps have a broad range of feeding, including phytoplankton and fish larvae.

Transfer Between Trophic Levels

If we accept the food chain abstraction, we can tabulate the losses from one trophic level to the next.

1. Some proportion of a given trophic level evades predation through escape, unpalatability, or unavailability. Phytoplankton with large spines or toxins might be avoided by zooplankton. A phytoplankton species might consist of cells too small to be ingested by zooplankton. Phytoplankton blooms might be too ephemeral for the appearance of a zooplankton population.

2. Some proportion of the food that is ingested is, nevertheless, not converted into growth. Ingested food may be balanced as follows:

$$I = E + R + G$$

where I is the amount ingested (measured in calories), E the amount egested, R the amount consumed in maintenance (respiration), and G the amount used in growth. G may be partitioned into somatic and reproductive growth.

Losses therefore occur when ingested food is egested with no assimilation by the consumer. In many cases, ingested diatoms are egested intact and viable (see Lopez and Levinton, 1978). In other cases, digestive enzymes may kill the prey, but no benefit in terms of assimilation is acquired. Even if assimilation occurs, some of the food is consumed in maintaining the metabolic activities of the consumer. This proportion of the food is effectively "burned away" and hence not available to the next trophic level. Some food goes toward indigestible supportive tissues that are similarly unavailable.

In terms of food chain dynamics, all losses can be summed up in the parameter known as *ecological efficiency (E)*, defined as

$$E = \frac{\text{amount of energy extracted from a trophic level}}{\text{amount of energy supplied to that trophic level}}$$

(Slobodkin, 1961). Although growth efficiencies (percentage of assimilated food used in growth) of copepods can be as high as 30 to 45% (Mullin and Brooks, 1970), the ecological transfer efficiencies are not nearly as high. Ecological efficiency of *Daphnia* populations is, on average, 8.5% and may be as much as 12 to 13% (Slobodkin, 1961). A literature survey revealed ecological efficiencies of natural commodities from 5.5 to 13.3%. Since Slobodkin's early estimates, later numbers have revealed a great variation—in some cases, significantly higher than 10% (Slobodkin, 1970). Ecological efficiency may thus be greater than the 10 to 12% often assumed. Higher-latitude planktonic systems probably have discontinuous production periods of higher efficiency than tropical systems (Gulland, 1972).

Errors in the estimation of transfer efficiencies at lower levels of the trophic web will often be magnified up the web, in attempts to estimate the standing stock or potential standing stock of fisheries. The magnification of production and estimates of ecological efficiency up the food web can be described as follows (Gulland, 1972):

$$P = BE^n$$

where P is the annual production, B the annual biomass at the primary trophic level, E the ecological efficiency, and n the number of trophic levels. A change from an E of 0.1 to 0.2 would result in an order-of-magnitude increase in our estimate of production at the fifth trophic level.

Using available data on production and ecological efficiency, Gulland (1972) estimated the potential annual catches of the more familiar types of commercial fish as being approximately a hundred million tons (Table 11-1). In the 1960s fish catches were increasing at a rate of some 7% per year. But Gulland points out that potential annual yields of the fishes now exploited do not greatly exceed the catches typically encountered today. Therefore the use of ecological efficiencies and estimates of potential productivity has direct use in fisheries biology. Cushing (1971) estimated the potential annual production in upwelling areas of the world as being about a hundred million tons. Both estimates indicate that the fisheries resources of the world ocean are limited and they will probably be fully exploited in coming decades (Cushing, 1975).

TABLE 11-1 POTENTIAL ANNUAL YIELDS OF MAJOR GROUPS OF COMMERCIALLY FISHED SPECIES COMPARED TO CATCHES IN 1970 (MILLION TONS)[a]

	Temperate areas				Tropical areas			
	North Atlantic	North Pacific	South Atlantic	South Pacific	Atlantic	Pacific	Indian	Total
Large pelagic (tunas and salmon)	trace	0.6	[b]	[b]	1.1	1.9	0.7	4.3
Demersal (e.g., cod, plaice)	10.6	3.5	4.9	0.8	2.9	12.4	7.4	42.5
Shoaling pelagic (herring, anchoveta, mackerals)	10.3	5.7	6.7	12.3	4.9	7.7	6.0	53.6
Crustaceans (shrimps, crabs, lobsters)	0.3	0.4	0.1	0.1	0.4	0.8	0.2	2.3
Total	21.2	10.2	11.7	13.2	8.9	22.8	14.3	102.7
1970 Catches	15.4	8.0	4.1	9.5	3.3	12.7	2.7	55.7

[a]From Gulland, 1972.
[b]Combined with corresponding tropical area.

Patterns of Food Chain Variation

The structure of food chains in different parts of the world oceans may contribute to their dynamics, community structure, and potential yield to fisheries. Ryther (1969) partitioned world oceanic food chains into three basic types: oceanic, continental shelf, and upwelled (Fig. 11-1). The open-ocean system has five trophic levels with a low primary production of about 50 g C/m^2/year. The coastal type includes three trophic levels; primary productivity is about 100 g C/m^2/year. The upwelling food chain type occurs in such areas as the Peru Current and the Antarctic and contains only two trophic levels. Because upwelling leads to a continuous supply of nutrients, production here is significantly higher: about 300 g C/m^2/year. Ryther assumes that ecological efficiency is highest in phytoplankton-herbivore transfers and lowest for secondary and tertiary carnivore transfers. Therefore he concluded that oceanic food chains have 10% overall efficiency, 15% for continental shelf food chains, and 20% for upwelling area food chains. Upwelling areas have the greatest potential fish production per year (Table 11-2).

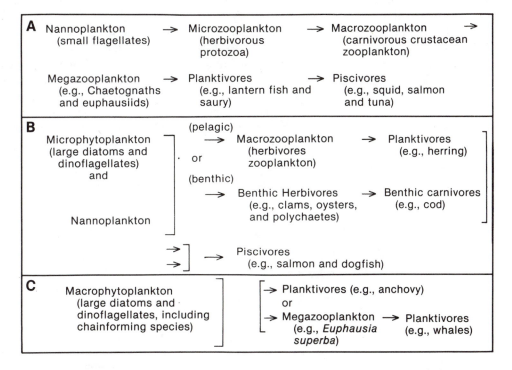

Figure 11-1 Types of oceanic food chains. (a) A gyre-center, nutrient-poor pelagic environment. (b) A coastal environment. (c) A coastal upwelling pelagic environment. (After Ryther, 1969, *Science,* vol. 166, pp. 72–76. Copyright 1969 by the American Association for the Advancement of Science.)

TABLE 11-2 CHARACTERISTICS OF OCEANIC, CONTINENTAL SHELF, AND UPWELLING FOOD CHAINS[a]

Food chain type	Primary productivity g C/m²/y	Number of trophic levels	Ecological efficiency	Potential fish production (mg Cm⁻²/y⁻¹)
Oceanic	50	5	10	0.5
Shelf	100	3	15	340
Upwelling	300	1-2	20	36,000

[a]After Ryther, 1969.

Figure 11-2 summarizes variations in the trophic organization of pelagic ecosystems (Landry, 1977). The physical environment exerts the major influence on the length of food chains in different oceanic environments. High-nutrient, turbulent (e.g., upwelling) environments favor the growth of large, chain-forming diatoms, which are consumed efficiently by larger herbivores. Large prey therefore favor the presence of larger fish and invertebrate predators. Rich nutrient supplies also favor dense phytoplankton populations. Larger consumers may thus utilize larger phytoplankton directly and dense patches of zooplankton may be exploited by consumers as large as blue whales (exploiting euphausids). The resultant shortening of the food chain increases the yield to the top trophic levels and thereby increases the potential of a commercial fishery (Ryther, 1969; Landry, 1977).

Areas of low nutrients and low turbulence (net downward loss through plankton sinking) select for smaller phytoplankton and planktonic bacteria. So primary consumers are small. Several trophic levels must be passed before the primary production is available to fish stocks and invertebrate carnivores are more abundant. Consequently, even though

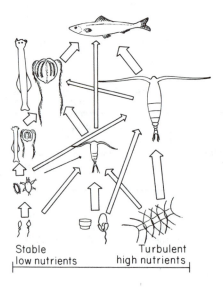

Stable low nutrients — Turbulent high nutrients

Figure 11-2 Variations in the trophic organization of pelagic ecosystems, ranging from stable, open-oceanic, low-nutrient environments to turbulent, nearshore, high-nutrient environments. (After Landry, *Helgoländer wissenschaffliche Meeresuntersuchungen*, vol. 30, pp. 8–17, 1977)

tropical oceans have about one-half the annual primary production of a rich temperate area, the potential tropical yield of fishes is orders of magnitude less (Table 11-2) because food is transferred over many more trophic levels (Ryther, 1969).

Environmental stability may also contribute to the relative complexity of oceanic food webs relative to upwelling and shelf webs. Lower amplitude of annual variation might permit the speciation-extinction balance to tip in favor of more species surviving. In more uncertain environments, the extinction rate might be greater, shifting the balance toward lower standing diversity.

Environmental stability might also favor the survival of complex multilevel food chains. Pimm and Lawton (1977) demonstrate that complex food chains are inherently unstable. Thus extrinsically imposed environmental instability (as in shelf environments) might limit the stability of food chains still further. An unstable environment might readily cause the extinction of one species at a given trophic level, causing strong fluctuations at other levels.

11-2 ESTIMATION OF PRIMARY PRODUCTIVITY

The estimation of primary production in the sea is of obvious importance in understanding the base of the food web and the potential productivity of the sea for humans. An accurate measurement of primary production in the sea would allow us to predict the potential amount of organic matter available for consumption by the zooplankton and, in turn, for consumption for commercially useful species. We must have an accurate and relatively simple field-oriented technique for estimating the amount of photosynthesis occurring in a given body of water.

Oxygen Evolution

The most common estimation of photosynthesis in the marine environment uses the fact that photosynthesis is accompanied by oxygen production. This technique depends on the accurate measurement of dissolved oxygen, usually in oxygen-tight biological oxygen demand (BOD) bottles that have been filtered of zooplankton with a 150 to 300 μ plankton net. An entrapped sample of algae can be incubated in a given light source for a specified number of hours, oxygen being measured before and after. Two techniques have been used to measure oxygen: (a) the Winkler method, a method based on chemical titration, and (b) the polarigraphic oxygen electrode, an electrode that measures an electrical current proportional to the concentration of oxygen in the water.

If we measure the change in oxygen concentration over time in a light bottle, we estimate the amount of oxygen evolved during photosynthesis, minus the amount of oxygen consumed in respiration by the algae and other associated organisms in the container. We can estimate the amount of oxygen consumed by the phytoplankton in respiration by using measurements of oxygen changes concurrently in transparent and light-tight bottles. Light-tight bottles are usually sealed in black tape. Oxygen decreases in the light-tight bottle due to respiration. So in order to get an estimate of the total amount of photosynthesis, we add the (negative) change in oxygen in the dark bottle to the (positive) change in oxygen in the light bottle.

Practically speaking, oxygen loss is primarily due to the respiration of bacteria and other microorganisms—because of the very low respiratory activity of the phytoplankton. Healthy algal cultures usually show approximately 10% respiration of the oxygen yield of maximum photosynthesis. So generally respiration is indirectly estimated from photosynthesis (see McAllister et al., 1964). Because of its difficulty of measurement, we are not really sure about the importance of the respiration of algae and its change with environmental conditions. Respiration of phytoplankton has been shown to be enhanced in the light (photorespiration).

The oxygen method assumes a fixed quotient of the molecules of oxygen liberated during photosynthesis, divided by the molecules of carbon dioxide assimilated during photosynthesis. This *photosynthetic quotient,* however, depends on the type of photosynthate manufactured. Theoretically the photosynthetic quotient must equal unity with a cell producing hexose sugars, 1.4 with a cell making only lipids, 1.05 if protein (with ammonia as a source of nitrogen) is manufactured, and as high as 1.6 if nitrate is the source and protein is the product manufactured (see Strickland, 1965). Similarly, in estimating and substracting respiration, we must know the *respiratory quotient:* the ratio of molecules of carbon dioxide liberated during respiration, divided by the molecules of oxygen assimilated during respiration. Thus

$$\text{Gross primary production (GPP)} = \frac{375(L - D)X}{PQ}$$

$$\text{Community respiration (CR)} = 375(I - D)RQX$$

$$\text{Net primary production} = \text{GPP} - \text{CR}$$

where I equals the initial oxygen content of the water added to the light and dark bottles, D equals oxygen in the dark bottle after a selected time period, L equals oxygen in the light bottle after a selected time period, X equals length (depth) of a 1-m square column of water, RQ is the respiratory quotient, PQ is the photosynthetic quotient, and 375 is a conversion factor from mg oxygen to mg carbon when the photosynthetic quotient equals one. After a specified amount of time, the amount of oxygen in the light bottle is $L = I + P - R$. In the dark bottle, however, because no photosynthesis occurs, the amount of oxygen in the bottle after that same specified time is $D = I - R$. Therefore the amount of oxygen in the light bottle, minus the amount of oxygen in the dark bottle, is $L - D = (I + P - R) - (I - R)$. This subtraction gives us P, the amount of oxygen due to photosynthetic processes (given the assumptions mentioned earlier).

Radiocarbon Technique

The radiocarbon technique takes advantage of radioactively (C-14) labeled bicarbonate taken up by phytoplankton during photosynthesis (Steeman Nielsen, 1952). In this method, the carbon is measured directly, not indirectly as in the oxygen method. Carbon uptake can also be calculated from pH changes in seawater. Nevertheless, pH change measures are only sensitive to high levels of primary productivity.

Steeman Nielsen (1952) developed a rapid and convenient shipboard technique for estimating photosynthesis, using the rate of labeled radioactive carbon uptake. A sample of seawater is innoculated with a small volume of radioactive bicarbonate solution and phytoplankton are allowed to photosynthesize for an hour. Afterward the plankton are filtered onto a membrane filter and the radioactivity of the phytoplankton is measured with a Geiger Counter or scintillation counter. Using this technique,

$$\text{GPP} = \frac{(R_s - R_b) \times W \times 1.05}{R \times N}$$

where W is the weight of bicarbonate in water (mg/C/m^3), N the number of hours during the experiment, R the counting rate expected from the entire C-14 added to the sample, R_s the counting rate of a light-bottle sample, R_b the counting rate of the dark-bottle sample, and 1.05 a constant allowing for a fractionation of C-14 relative to C-12. Because carbon dioxide is present in seawater mainly as bicarbonate, this technique probably estimates productivity quite accurately. In normal seawater the gaseous dissolved carbon dioxide concentration is a small fraction, usually less than 1% of the total carbon dioxide present. There is some argument as to the relative utilization of carbon dioxide and bicarbonate ions for photosynthesis by algae in the sea.

Radiocarbon generally gives lower estimates of photosynthesis than the oxygen evolution technique. Ryther and Yentsch (1958), for example, showed that radiocarbon estimates are lower than that of oxygen evolution in coastal waters of high productivity. Phytoplankton may excrete much of the photosynthate as it is being produced (Strickland, 1965; Fogg, 1975). Therefore the radiocarbon technique is probably a more accurate approximate measure of the formation of particulate organic matter. We should be aware that estimates of global productivity as given by the radiocarbon technique will be lower than estimates given by the oxygen evolution technique. Radiocarbon is the preferred technique in oligotrophic waters, where changes in oxygen are often too small to measure in a short period.

The greatest problem in field sampling and estimating of primary production is to obtain representative samples from a patchy environment and to take undamaged samples so that phytoplankton photosynthesis approximates field conditions (Strickland, 1965). In order to estimate the primary production of an area—in particular, to estimate the effects of respiration—we must have samples throughout a 24-hour cycle. Samples must be taken at different depths to estimate the variation in production with depth as discussed earlier.

In order to estimate primary production, three main techniques can be used. First, bottles can be suspended in the water column under close-to-natural conditions. This process has the disadvantage that it is necessary to return 24 hours later to the same open-water locality to retrieve bottles (or tie up the ship for a day). Secondly, bottles can be incubated on shipboard, using natural light, a method with the obvious disadvantage of unnatural light intensity and spectral differences at the surface. Finally, incubators with specified light sources and photoperiods can be used in the laboratory aboard ship; here we must be careful to incubate the samples at natural temperatures.

The following protocol, taken from Strickland, is a useful approach for estimating productivity via an incubator technique. Phytoplankton in a seawater sample are exposed to optimal light intensity (I_{max}) for the specified location, thus obtaining the maximum primary productivity (P_{max}) in that area. If the P versus I curve is known, plus the I curve as a function of water depth, then theoretically it is possible to calculate the productivity of a given area, integrating the amount of primary production as a function of depth. Laboratory measurements can be calibrated to individual situations by towing a duplicate surface sample, suspended by a buoy or trailing behind the sampling ship. Estimates of population densities through more detailed sampling can also be used to calibrate primary production estimates to a given locality.

11-3 GEOGRAPHIC DISTRIBUTION OF PRODUCTION

Latitudinal Variation

The spring diatom increase is a seasonal cycle that begins with unlimited nutrients, proceeds with nutrients used up by the phytoplankton concomitant with water column stabilization, and ends with the demise of the phytoplankton bloom because of utilizations of nutrients and grazing. Seasonal cycles in production vary with latitude (Cushing, 1959). The Antarctic shows a production cycle of long duration, a fact partially explained by the long period of continuous daylight. The long generation time and hence slow reaction in grazing intensity of the major grazers (Euphausids) also permit large standing stocks of phytoplankton. In contrast, the production cycle is much shorter in the Arctic. But all high-latitude production cycles do show one peak per year. In temperate waters there seem to be two peaks: one in the spring and one in the fall, separated by a low in the summer. The second peak is probably due to fall storms that overturn water from the bottom and recycle nutrients from the bottom to the surface. The autumn burst may also be due to a decline in the zooplankton. A peak of zooplankton follows the first peak of phytoplankton in both the Arctic and the temperate regions. In the tropics the amplitude of seasonal phytoplankton production is reduced. The production cycle and zooplankton variations in the Sargasso Sea near Bermuda show little variation over the year (Menzel and Ryther, 1960, 1961b). A peak in production occurs from December to late February. These data suggest that the seasonal cycle of phytoplankton and zooplankton production is more kurtotic in high than low latitudes. The marked seasonality of light intensity in high-altitude waters and the seasonal temperature variation in midlatitudes both contribute to strong production peaks. Figure 10-1 shows a diagrammatic representation of production cycles as a function of latitude (Cushing, 1959). Generation time increases and the number of generations of copepods decreases with increasing latitude. The time lag between the phytoplankton peak and a herbivore peak thus increases with increasing latitude (Cushing, 1959, 1975). The minimal lag time in the tropics may explain the general low standing stock of the phytoplankton and the poor development of oscillations between phytoplankton and zooplankton.

Regional Variation in Production

Continental shelf upwelling areas.   These regions are areas of high productivity because of a consistent wind parallel or at a slight angle to the coast. As a result of the Coriolis effect, surface waters are deflected at right angles to the wind (to the right in the Northern Hemisphere and to the left in the Southern Hemisphere). As the spring and summer season progresses, the axes of the trade winds shift, resulting in the movement of the principal locus of upwelling toward the pole.

The movement of water offshore results in replacement by cooler, nutrient-rich water originating from the bottom. Nutrients have originally been deposited on the bottom due to great amounts of sedimentation in these areas of very high productivity. The nutrient-rich water, when reaching the surface, sets off phytoplankton and zooplankton blooms of high amplitude. Consequently, a large amount of primary production is available for grazing and ultimately for fish production. Therefore the great fisheries of the world are located in these regions of high productivity. The California, Peru, Benguela, and Canary currents are all eastern boundary currents in the Atlantic or Pacific oceans.

Upwelling can occur from as deep as 200 m but usually occurs from 100 m depth or less. In the Peru coastal region, the Humboldt Current and the Peru Coastal Current cause upwelling from depths of about 100 m. In this region dense standing crops of phytoplankton and zooplankton occur, as well as huge populations of anchoveta. These fishes further serve as food for tuna, both species forming the base of major world fisheries. Anchoveta populations feed extensive populations of fish-eating birds whose guano production covers the famous guano islands found here and in other upwelling regions.

Upwelling in Antarctic seas is also a major source of large standing stocks of phytoplankton. In some areas upwelling is seasonal—for example, in the monsoon regions of southern Asia alternating periods of upwelling and low fertility occur throughout the year. Production in such an area can exceed 1.5 to 2 g $C/m^2/day$ for protracted periods (Strickland, 1965). Strong upwelling can take place adjacent to submarine ridges, as well as in areas of strong currents, such as the Faroe–Iceland ridge, where 2.5 g $C/m^2/day$ occurs.

Coastal areas.   Because of the shallowness of the water and the regeneration of nutrients from the bottom, waters close to shore are generally extremely productive realtive to the open sea. This high productivity is partially due to a large standing crop with accompanying high growth rates. The absence of grazers at the beginning of a phytoplankton bloom allows standing stocks to increase rapidly. In some of these regions upwelling contributes greatly to high productivity, as in the Grand Banks of Newfoundland and George's Bank.

Estuaries are normally rich sources of nutrients; however, particulate matter is abundant and the depth of active photosynthesis (euphotic zone) is correspondingly shallow. In contrast, outer-shelf surface waters are relatively nutrient poor; yet a reduced amount of resuspended particulates results in a deeper euphotic zone. Inner-shelf surface waters might be expected to have both moderate nutrient levels and euphotic depth. Figure 11-3 shows geographic variation in the euphotic depth, nutrient concentration and productivity, on a transect from the coast of Georgia to the edge of the continental shelf. The integrated result of intermediate nutrient concentration and euphotic depth is a maximum annual primary production in inner-shelf surface waters (Haines, 1979).

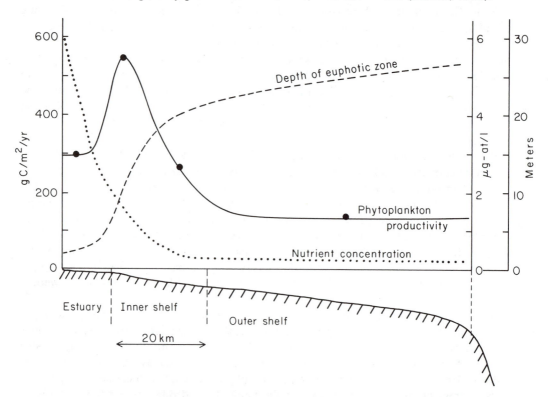

Figure 11-3 Geographic variation in the euphotic depth, nutrient concentration, and productivity, on a transect from the coast of Georgia to the edge of the continental shelf. (After E. Haines, "Interaction between Georgia salt marshes and coastal waters: a changing paradigm," in R. J. Livingston, ed., *Ecological Processes in Coastal and Marine Systems,* 1979, with permission of the Plenum Publishing Corporation)

Convergences and fronts.   Horizontal variation in hydrography has a significant effect on patterns of production. Surface convergences concentrate nutrients and plankton; seabirds and fishes also congregate at convergences to take advantage of the food supply. Such fronts may be temporary (as are intrusions of slope water moving onto the shelf) or long lived (as are some fronts off the shelf slope break near Nova Scotia and the southern Bering Sea). Fronts may be demarcated by temperature and salinity.

In the southern Bering Sea a series of fronts divides the pelagic realm into several distinctive trophic webs (Iverson et al., 1979). A middle shelf zone supports the production of large diatoms that are not effectively grazed by the small herbivores. Accordingly, diatoms sink to the bottom and subsidize large populations of benthic invertebrates and fishes. The outer-shelf primary production, however, is effectively grazed by the zooplankton. They, in turn, support large populations of pelagic seabirds, mammals, and fishes. Because the region is devoid of extensive lateral advection, these trophic realms maintain their integrity.

Barren central oceans and gyre centers.   Productivity is extremely low between latitudes 10 and 40 and within tropical gyres, such as the Sargasso Sea and the North Pacific gyre. The nutrient regime is poor but is compensated by the year-round growing season and the great euphotic depth. Therefore productivity over the year is about 50 g C/m^2, with an average rate of about 0.1 g C/m^2/day. An important feature of such areas is the permanence of the thermocline, prohibiting nutrients from being regenerated from deeper water. Near the Bermuda Islands the thermocline breaks down during the winter and early spring, resulting in upwelling and higher rates of productivity (0.9 g C/m^2/day) than surrounding deep-water areas.

Atlantic Ocean.   The Atlantic Ocean is by far the best-studied ocean in terms of measurements of primary production (Steeman Nielsen, 1975). The North Atlantic and the South Atlantic are both characterized by large circulating current systems, clockwise in the North Atlantic and counterclockwise in the South Atlantic Ocean. Eastern boundary currents, such as the Benguela Current off South Africa, produce a large upwelling system in the southern Atlantic. The Benguela Current runs along the west coasts of South Africa and then the South Equatorial Current crosses the southern Atlantic Ocean westerly toward South America, just south of the equator. Finally, the Brazil Current moves southerly along the coast of South America. Therefore the center of the South Atlantic at midlatitude constitutes a large anticyclonic counterclockwise eddy. A similar clockwise eddy is found in the North Atlantic, in the Sargasso Sea.

In the North Atlantic gyre the rate of gross production is generally below 0.1 g C/m^2/day. Near Bermuda, however, the production is somewhat higher. Waters in the open sea are generally of low productivity whereas in inshore waters production increases significantly. Adjacent seas, such as the North Sea and the Baltic, can have considerable productivity; the Mediterranean, on the other hand, has low productivity.

Pacific Ocean.   The Pacific Ocean is divided into the North Pacific and South Pacific and has a circulation pattern similar to that of the Atlantic. Because of upwelling, the Peru Current has a high rate of primary production, reaching values of more than 10 g C/m^2/day. Production is also rather high adjacent to continental areas, as off the coast of Washington and Oregon. Table 11-3 shows production levels in the Pacific. In general, Pacific productivity is higher than in the Atlantic, although such patterns as central ocean low productivity and shallow-water high productivity follow those observed in the Atlantic Ocean.

TABLE 11-3 ANNUAL PRIMARY PRODUCTION AREA, AND TOTAL YIELD OF DIFFERENT SURFACE WATER TYPES OF THE PACIFIC[a]

Water type	Area ($\times 10^3$ km^2)	Annual primary production (g C/m^{-2})	Total yield (g C/y^{-1})
Oligotrophic gyres	90,106	28	2.5×10^{15}
Transitional waters	33,358	49	1.6×10^{15}
Divergences near the equator and subpolar regions	31,319	91	2.9×10^{15}
Waters off the coasts	20,423	105	2.1×10^{15}
Neritic waters	244	237	5.8×10^{13}

[a]Data from Koblentz–Mishke, 1970.

Indian Ocean.   The rate of primary production in the central Indian Ocean is fairly similar to the South Atlantic and is characterized by low production rates (Steeman Nielsen, 1975). North of the equator, however, very high production rates are found off the coast of Somalia, in the Arabian Sea, and off the south coast of India. Winds shift seasonally because of the monsoon seasons. Production rates vary concomitantly with alternating periods of upwelling and influx of nutrient-poor surface water from offshore. The average rate of production is about 0.24 g C/m^2/day in the western part of the ocean and about 0.19 in the eastern part. The average is probably somewhat higher than the Pacific or Atlantic.

Antarctic.   One of the most productive areas of the world, the Antarctic supports a dense diatom flora and production is probably 1 g C/m^2/day during the phytoplankton growing period of about 100 days (Strickland, 1965). A slow upwelling of nutrient-laden water contributes to this high productivity. Considerable productivity can occur under sea ice (Bunt, 1963).

Arctic.   Production can be high in the Arctic in summer but year-round figures are low because of the short growing season and the diminution of productivity through ice cover. Optimally, levels of 1 g C/m^2/day may be reached.

Equator.   Upwelling a few degrees on either side of the equator results in fertile tropical open oceans in both the Pacific and the Atlantic. Productivity varies around 0.3 to 0.4 g C/m^2/day (Strickland, 1965).

World Productivity Estimates

The preceding discussion indicates the great complexity and geographic heterogeneity in rates of primary production. The situation is further complicated by seasonal variation and even smaller microgeographic patchiness in production. Nevertheless, a world estimate is necessary because of our need to know the amount of fish standing stock available

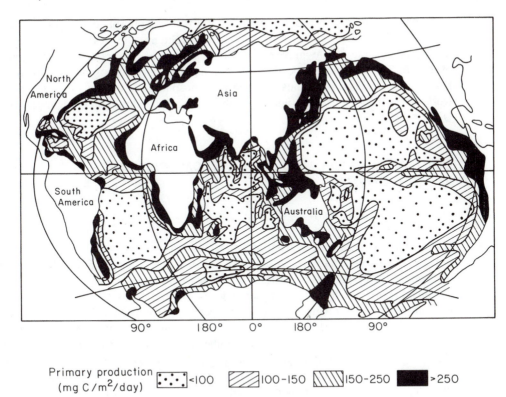

Primary production
(mg C/m²/day) [·.·.·] <100   [///] 100-150   [\\\\] 150-250   [███] >250

Figure 11-4 Distribution of primary production in the oceans. (After Koblentz-Mishke
et al., 1970)

to us. Figure 11-4 is a map showing the variations of world productivity (Koblentz–Mishke
et al., 1970).

Computation of global productivity varies with the analytical technique employed.
Measurements based on oxygen evolution give values of 100 to 200 g C/m²/year. In
contrast, the various estimates using radiocarbon techniques by Steeman Nielsen and his
colleagues indicate values of about 25 to 35 g C/m²/year. Ryther (1969) estimates that
a realistic world average is about 50 g C/m²/year.

Estimates of productivity are liable to have error bars of as much as 30% (Steeman
Nielsen, 1975). In understanding the availability of various fish stocks in the sea, an
estimate even within an order of magnitude would be useful for world planning. Using
the upper estimate of 200 g C/m²/year with a working surface area of the ocean of $3.5
\times 10^{14}$ sq m, the annual fixation of carbon in seawater will be 7×10^{16} g of carbon per
year. The lower estimate of 50 g C/m²/year would correspondingly give a total annual
fixation of 1.75×10^{16} g C/year. The upper level of 7×10^{16} g of carbon fixed would
require nitrogenous salts corresponding to 1.4×10^{16} g of nitrogen and 2×10^{15} tons
of phosphorus (Russell Hunter, 1970). These requirements are in excess of the total
amount of nitrogen and phosphorus dissolved in the ocean in any one time. In order to

achieve the estimated levels of productivity, nitrogen and phosphorus must turn over from plankton to the water column and back to the plankton again. Russell Hunter estimates that every year the nitrate–nitrogen content of the surface layers of the ocean must turn over from 5 to 50 times, depending on geography, whereas phosphorus must turn over from about 3 to 20 times. Thus nitrogen and phosphorus limitation to productivity on a world scale is obvious. Nitrogen is probably the usual limiting nutrient (Ryther and Dunstan, 1971).

11-4 QUANTITATIVE PREDICTION OF PLANKTON ABUNDANCE AND PRODUCTIVITY

Describing the System

Quantitative analysis ranges from statistical description and simulations of planktonic processes, using natural data, to deductive models predicting plankton abundance from our knowledge of processes, using simulation techniques (Steele, 1974). Riley (1946) attempted to predict phytoplankton standing stock from several key measured parameters. He measured the following quantities on George's Bank off the Maine coast: depth of water (D), temperature (t), mg-atoms phosphate per m^2 in the upper 30 m (P), mg-atoms nitrate per m^2 in the upper 30 m (N), number of grazing zooplankton in thousands/m^2 (Z), and plant pigments in thousands of Harvey units/m^2 (PP).

Using multiple regression analysis, he calculated a series of linear equations relating these six quantities. For each month he calculated a linear equation, such as the following:

$$September\ PP = -.011D - 23.8t + 5.02P + .371N + .26Z + 829$$

This equation shows, within the limits of error, the amount of variation in the phytoplankton crop that can be obtained by varying any or all the environmental factors at a given time. (Table 11-4 shows the means of George's Bank plankton and environmental

TABLE 11-4 ORIGINAL OBSERVATIONS MADE DURING 1939–1940 CRUISES ON GEORGE'S BANK[a]

	Sept.	Jan.	Mar.	Apr.	May	June
Water depth (m)	247	209	135	209	82	206
Temperature (upper 30 m)	15.2	4.6	2.6	3.8	5.1	9.7
mg-atoms phosphate m^{-2} (upper 30 m)	14.4	33.7	34.7	21.9	16.6	19.2
mg-atoms nitrate m^{-2} (upper 30 m)	153	209	172	129	285	155
Animals ($\times 10^3\ m^{-2}$)	135	14	24	32	106	103
Plant pigments (Harvey units $\times 10^3\ m^{-2}$)	560	118	828	2303	871	478

[a]Used by Riley (1946) to develop multiple correlation equations (mean values given).

factors over months.) If phosphate were increased one mg-atom, it would increase the calculated value for plant pigments (*PP*) by the amount of the phosphate constant (5.02). If the phosphate varied a "normal amount," as indicated by its standard deviation, the plant pigments would change ±6.5%. Using standard deviations of data in this way, Riley estimated the importance of a given factor in a given month.

Proportionality constants varied with season; and the effect of changing the mean value for each component by one standard deviation caused seasonal effects, indicating the *temporary* importance of each factor. Water depth, for example, played an important role in the inception of spring diatom flowering, the proportionality constant increasing strongly in March. The proportionality constant for temperature is negative because increased respiration uses up part of the store of energy that would otherwise be used in the production of new plant material. Phosphate and nitrate have positive proportionality constants. During the height of the diatom bloom the importance of the nutrient proportionality constants increased. Toward the end of the phytoplankton peak in May the proportionality constant for zooplankton increased, indicating that grazing was an important explanatory factor in phytoplankton abundance. Thus no one factor exercised complete control over phytoplankton abundance and shifts in dominance of factors occurred throughout the season. A multiple correlation equation was developed from the seasonwide cruise averages:

$$PP = -153t - 120P - 7.3N - 9.1Z + 6713$$

Using this equation, the average error difference between expected and observed phytoplankton standing stock was about 20% for all seasons (Fig. 11-5).

The shortcoming of this study is that one cannot measure all parameters and therefore

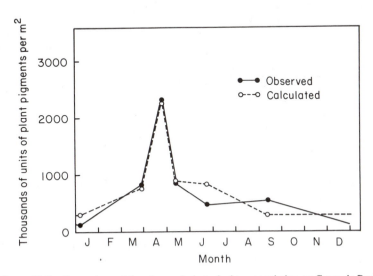

Figure 11-5 Comparison of the observed phytoplankton population on George's Bank with the calculated population as determined by a multiple correlation analysis of the relationship between phytoplankton and environmental factors. (From Riley, 1946)

the parameters explaining the data are restricted to one's biological intuition. Furthermore, these inferences only explain how much of the variance can be explained potentially by correlations between two different parameters. Correlations do not necessarily imply causality and many spurious correlations often lead to false conclusions.

Riley's original statistical technique has been amplified in later years by multivariate statistical analysis using nonlinear equations. Walsh (1971) demonstrated that most of the variation in phytoplankton abundance at the Antarctic convergence could be explained with the following parameters: water mass structure, silicate, light, turbulence, and an index of heterotrophic conditions. Walsh emphasized the correlative nature of the technique and the difficulty of identifying causative factors.

The Peru upwelling system has been intensively modeled, using computer-calculated partial differential equations relating state variables, such as nitrate, to a series of environmental components (e.g., water velocity, uptake rate of nitrate by phytoplankton). In this system upwelling brings nutrients to the surface as surface currents move offshore due to wind-stress interactions with the Coriolis effect. A longshore component moves water parallel to shore as well. The simulation model predicts nutrients, phytoplankton, and detritus as a function of water movement, light, grazers, and carnivores (Walsh, 1975). Figure 11-6 shows the effectiveness of the simulation model. The utility of the model lies in pinpointing important parameters and evaluating their quantitative contribution to state variables.

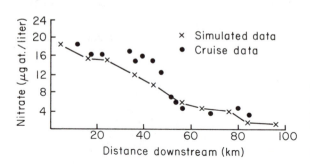

Figure 11-6 Comparisons of the observed and simulated nitrate and phytoplankton nitrogen over 0 to 10 m in the Peru upwelling ecosystem. (After Walsh, 1976, © Blackwell Scientific Publications, Limited)

Simple Deductive Models

A complementary approach is the prediction of processes through simple deductive models. These models sacrifice precise prediction for general insight into natural processes. Riley (1946) used the following equation in attempting to explain the rate of change of the phytoplankton population as a function of the difference in reaction rates between the processes of accumulation of energy by the phytoplankton population and processes of energy dissipation, such as grazing. Thus

$$\frac{dP}{dt} = P(Ph - R - G)$$

in which the rate of change of the phytoplankton population *(P)* is a function of the photosynthetic rate per unit of population *(Ph)*, the rate of phytoplankton respiration *(R)*, and the grazing rate of zooplankton *(G)*.

Each rate has its own mathematical function and must be accounted for in terms of environmental variation. For example, photosynthesis, in the absence of nutrient depletion, is proportional to illumination. So

$$Ph = pI$$

where *Ph* is the photosynthetic rate, *I* radiation in g $Cal/cm^2/min$ at the depth of the photosynthesizing plankton, and *p* a photosynthetic constant (taken to be 2.5). Because light varies exponentially with depth, the intensity of light, I_z, is

$$I_z = I_0 e^{-kz}$$

Therefore

$$p - hz = pI_0 e^{-kz}$$

where *k* is the extinction coefficient (defined as 1.7 divided by the depth of the Secchi disk reading). The mean photosynthetic rate in the euphotic zone can be found by integrating this equation from the surface to a finite depth where no photosynthesis effectively occurs and dividing by that depth.

$$\overline{Ph} = \frac{pI_0}{kz_1}(1 - e_1{}^{kz})$$

$\overline{Ph}$, as mentioned, is a function of nutrient abundance. To account for nutrient depletion, the mean photosynthetic rate is multiplied by a factor $(1 - N)$, where *N* is the reduction rate due to nutrient depletion. Vertical turbulence will also reduce the mean photosynthetic rate. Therefore the mean photosynthetic rate is multiplied by still another factor $(1 - V)$, in which *V* is the reduction in rate produced by water movements. This leads to a final equation for mean photosynthetic rate:

$$\overline{Ph} = \frac{pI_0}{kz_1}(1 - e_1{}^{kz})(1 - N)(1 - V)$$

Respiration rate increases with higher temperatures, being approximately doubled by a 10° increase in temperature. The temperature effect can thus be written

$$R_T = R_0 e^{rT}$$

in which R_T is the respiratory rate at any temperature *T*, R_0 the rate at 0°C, and *r* a constant expressing the rate of change of the respiratory rate with temperature. (This value is 0.069 when the respiratory rate is doubled by a 10° increase in temperature.)

Finally, grazing can be estimated as follows:

$$G = gZ$$

in which *G* is the rate of grazing, *g* the rate of reduction of phytoplankton by a unit quantity of animals, and *Z* the quantity of zooplankton in grams of carbon per square meter. The assumption is that a fixed proportion of the phytoplankton will be consumed in successive units of time. It is inaccurate because zooplankton do not consume phytoplankton above certain densities and may cease feeding at very low phytoplankton densities. Furthermore, there is probably a difference in grazing rate as a function of temperature. Riley uses 0.0075 as an average of the minimum values of G for September and January plankton populations. Thus

$$\frac{dP}{dt} = P\left[\frac{pI_0}{kz_1}(1 - e_1^{-kz})(1 - N)(1 - V) - R_0(e^{rT} - gZ)\right]$$

The rate of change of the population therefore depends on six variables: the incident solar radiation, transparency of the water, quantity of phosphate, depth of the mixed layer, surface temperature, and quantity of zooplankton. Using the integral form of this equation, Riley calculated predicted phytoplankton abundance. Predictions compare favorably with actual phytoplankton abundance as measured in seasonal samples on George's Bank. This deductive approach is far more satisfying in that it uses biologically determined processes and simplified mathematical approximations to predict phytoplankton abundance. A later attempt at this approach by Kremer and Nixon (1977), for Narragansett Bay, Rhode Island, is recommended to the interested reader.

Exploring Stability of the Planktonic System

It is impossible here to describe in detail all the predictive models for understanding the plankton. In a particularly interesting attempt, Steele (1974) attempted to analyze and simulate plankton processes in the well-studied North Sea. Steele simulates a natural ecosystem, employing extreme simplifications. "The simulation to be developed here is part of a process to discover how necessary it is to introduce relatively complex responses and describe the interaction between predators and their prey" (Steele 1974, p. 58). His simulation model links certain features (that he considered) typical of open marine environments in mid- to high-latitude waters.

The following considerations are necessary for the production of his model.

1. The vertical mixing of water through a stable thermocline is a process that transfers nutrients upward into the upper euphotic zone where plant production occurs.
2. A nutrient cycle is used, where supply occurs by vertical mixing and herbivore excretion, and nutrient consumption is by plants. Biologically the problem is the relation between the rate of nutrient uptake and the growth kinetics of the plant population as a function of nutrient concentration.
3. Grazing by copepods.
4. The growth rate of copepods.
5. The predation rate on the herbivores

Using a set of deductive equations, Steele simulates North Sea population dynamics, concentrating on the phytoplankton and the copepods. The model reveals great fluctuations in the cycles in plant and zooplankton carbon.

An interesting use of the model is to demonstrate the influence of certain factors on the stability of the entire system. Variations in the relationship between nutrient concentration and the rate of nutrient uptake, for example, do not have much significance for the general stability of productivity of the system. The system is greatly altered, however, when a threshold feeding response of the zooplankton was introduced. The system survives with a basically unstable predator–prey relationship, but stability is greatly improved when feeding ceases at a certain lower threshold. Stability is defined as the ability of the system to return to its original state after being subjected to some perturbation.

Landry (1976) critically evaluates Steele's model and has made several significant modifications in the component equations determining phytoplankton and zooplankton abundance. The Steele model allows the existence of only one cohort of zooplankton at any given time. Landry introduces several coexisting cohorts of copepods as grazers and finds that oscillations are greatly dampened and stability ensues after about 100 days or so. Multicohort reproduction provides for a more immediate herbivore response to change in phytoplankton concentration rather than the dramatic oscillations inherent in single coexisting cohort responses. Furthermore, predation on the zooplankton also enhanced stability in the community. Landry concludes therefore that feeding thresholds are not necessarily the only parameters that stabilize a community.

It is of interest that in the models, such as Riley's (1946), changes in environmental variables do not induce violent changes in the planktonic system. Responses of the phytoplankton to nutrient pulses are met with subsequent grazing pressure. Growth of grazer populations is checked by overgrazing of the phytoplankton, and selective shifts to abundant phytoplankton size classes (Richman et al., 1977) prevent exclusion of all phytoplankton species by one superior competitor for nutrients. In tropical oligotrophic regimes, nutrients are rapidly recycled in surface waters and changes at one trophic level are liable to be rapidly stabilized by responses at another. A pattern of stability emerges from an incredibly complex ecosystem.

Marine Ecology and Fisheries

Humankind has exploited a wide variety of estuarine, shelf, and open-ocean fish, invertebrate, and mammal species. Has our fishing effort exerted a significant effect on fish stocks? In many cases, the answer was "Maybe *too* significant." This effect was noticed in fish-catch statistics between and during major wars in the twentieth century. During World Wars I and II the total fishing effort declined; correspondingly, the catch per unit effort increased (see Cushing, 1975). In many fisheries, overfishing caused dramatic declines in stocks, as in stocks of herring in the North Sea (Cushing and Bridger, 1966) and in the Antarctic whale fishery (Schevill, 1974). These episodes have aroused the concern of commercial fishers, governments of major fishing nations, and conservationists. Such organizations as the International Whaling Commission have attempted to temper the political realities arising from international economic needs with the scientific advice necessary to manage the fishery successfully (see McHugh, 1974).

Sound management depends on (a) an accurate assessment of the population size of the fish stock and (b) an understanding of the dynamics of the population so that the numerical consequences of fishing can be estimated. The first problem is especially difficult; how can we estimate the population size of a species dispersed over broad areas of the ocean? The second requisite demands an expertise that depends on sound theory and accurate estimators of the parameters needed to use population dynamic models.

Because fish-catch statistics are often readily available, we have some hope of estimating the fish stock. It can be shown, formally, as follows (Cushing, 1975, p. 137):

$$N = R'e^{-z}$$

where N is the number of fishes remaining in the stock at the end of the first year; R' the number of recruits to a year class at the start of the year's fishing, and Z the instantaneous coefficient of mortality.

$$Z = F + M$$

where F is the instantaneous mortality rate from fishing and M is the instantaneous mortality rate from natural sources.

The number dying during the year is

$$R' - N = R' - R'e^{-z}$$
$$= R'(1 - e^{-z})$$

The catch in numbers, Y_n, or the proportion of deaths from fishing should be

$$Y_n = \left(\frac{F}{Z}\right)R'(1 - e^{-z})$$

Over λ years

$$Y_n = \left(\frac{F}{Z}\right)R'(1 - e^{-\lambda z})$$

λ is the number of years that a given year class lasts or the number of age groups present in the stock in any given year.

We can represent catch per unit effort of fishing as Y_n/F:

$$\frac{Y_n}{F} = \frac{R'}{Z(1 - e^{-z\lambda})}$$

The expression on the right represents the stock in terms of numbers of a year class throughout its existence (equal to numbers in an age spectrum in any single year). Catch per unit effort, Y_n/F, therefore correlates with standing stock.

Models of fish population dynamics permit management decisions to maximize the continuing yield to fishers. Cushing (1975) and Gulland (1974) discuss the principles of the theory of management in great detail; we briefly discuss several principles here.

Consider a population growing according to the logistic equation, discussed in Chapter 3:

$$\frac{dN}{dt} = rN\left(\frac{K - N}{K}\right)$$

Population growth is zero when $N = 0$ and $N = K$; maximal growth is achieved when $N = K/2$. If this model is appropriate, useful management decisions arise. First, as shown in Fig. 11-7, an appropriate rate of removal by fishing could maintain the stock at $K/2$; this represents the *maximum sustainable yield*. Secondly, removing less of this increment only allows the population to increase above $K/2$; dN/dt, or potential yield, correspondingly decreases. So an increase of fishing effort can actually increase the yield if the population level is near K and can be pruned down.

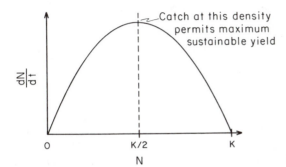

Figure caption within figure: Catch at this density permits maximum sustainable yield

Figure 11-7   Population growth as a function of population size, according to the logistic model.

Models estimating recruitment can be used to make predictions of fish population dynamics based on assumptions of density dependence. At low stock density, recruitment might be sufficient to increase the stock greatly. At high stock density, however, predation on juveniles and juvenile cannibalism are intense enough to diminish recruitment below that needed to replenish the stock. *Stock-recruitment* models can be devised to account for density dependence and to calculate parameters designed to estimate the extent of density-dependent effects on recruitment (Ricker, 1954; Cushing, 1975). Such models have been successfully applied to Pacific salmon, Arcto-Norwegian cod, and North Sea haddock (Cushing and Harris, 1973). Clearly these models can be used to predict the fishing intensity sufficient to reduce density-dependent mortality and to increase recruitment and hence fishing yields.

Life history theory (see Chapter 7, Section 7-1) can be readily applied to the study of fishery management and fish population biology. Fishery production is strongly related to the life history traits of the fish species to be exploited. For example, a species population with low reproductive effort, great longevity, and late age of first reproduction will be collapsed with relatively modest fishing pressure. In contrast, a species with early reproduction and high reproductive effort will naturally provide the greatest yields with even heavy fishing pressure. Species under selection for increased productivity (equivalent to *r*-selection) would therefore be best exploited with heavy fishing pressure. Species that are *k*-selected would be best managed more carefully; modest fishing pressure and tight monitoring of stocks would be recommended.

Commercial fishing can be a consistent source of mortality for certain age classes of fishes in a region. If adults are fished preferentially, then we have a classic case of selection for reproductive tactics that shift reproduction to earlier ages. This general effect has been observed in several fishes, where one can compare areas that are heavily fished with others that experience no substantial fishing pressure. The spiny dogfish, *Squalus acanthias,* consititutes a major fishery in Europe but is not fished to any great degree in coastal Atlantic waters of the United States. Correspondingly, greater population densities are found in the United States. As might be expected, spiny dogfish in Maine show later first reproduction, greater longevity, less reproductive effort in the first season, and reduced growth rate, relative to the European stocks (Woodhead, 1977). As some of these differences probably have a genetic basis, fisheries management can play a sophisticated role in shifting reproductive tactics toward more productivity and, therefore, more fish will be caught.

SUMMARY

1. A food web is an assemblage of feeding or trophic links within a community. We often simplify a complex web by reducing it to a food chain, with several trophic levels. Of interest is the efficiency with which energy is transferred from one trophic level to the next. Losses from one level to the next probably set limits on the number of levels possible in a community.

2. Nearshore areas and upwelling regions generally have fewer trophic levels than open-ocean oligotrophic (low nutrient level) pelagic communities. The fish harvest is less in the latter communities because of the substantial losses through more trophic levels.

3. Primary productivity can be estimated indirectly by measuring the oxygen evolved in photosynthesis. Incorporation of radioactively labeled carbon more directly estimates primary productivity.

4. Extensive geographic variation exists in the pattern and magnitude of primary productivity. Seasonal pulses in primary production diminish in amplitude with decreasing latitude. Continental shelf upwelling areas tend to be the most productive pelagic environments. Gyre centers tend to be the least productive.

5. Mathematical models enable us to judge whether our understanding of the pelagic

ecosystem function is sufficient to predict successfully parameters in the plankton. Simple equations describing response to light intensity, nutrient uptake, and similar features can be combined to predict measurable variables successfully, such as phytoplankton standing stock. Estimates can be made as to how important given parameters might be in explaining dependent variables (e.g., phytoplankton standing stock).

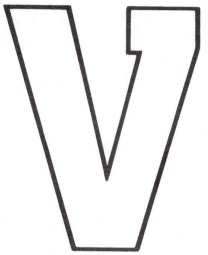

Substrata and Life Habits of Benthic Organisms

We inquire into the nature of bottom substrata and the parameters of greatest importance to benthic organisms. We discuss the life habits of benthic creatures and demonstrate how morphology adapts to environment. Finally, we examine the diversity of feeding adaptations in the sea.

12 the nature of benthic substrata

12-1 SUBSTRATUM CHARACTERISTICS

In this chapter we discuss and describe sedimentary characteristics that affect the distribution, functional morphology, and behavior of benthos. Before discussing bottom substrata, it would be useful to have a size classification of benthic animals to relate to grain sizes of soft sediments. *Macrofauna* (macrobenthos) include animals whose shortest dimension is greater than or equal to 0.5 mm. *Meiofauna* (meiobenthos) are less than 0.5 mm but greater than the *microfauna* (microbenthos), who are less than 0.1 mm (100 μ) in size.

Substratum type is the major controlling factor in the distribution of benthic species. Adaptations to rocky surfaces usually preclude optimal functioning in a watery mud. Intertidal organisms living on rock surfaces are more exposed to environmental changes at low tide than organisms living within the sediment. Adaptations to different substrata may thus determine morphology, mode of feeding, and physiological adaptations to changes in water temperature, salinity, and chemistry. Completely different suites of adaptive types are thus likely to occupy different substratum types.

We first distinguish between hard and soft substrata. Hard substrata are those that present a surface to which the organism must attach or into which it must bore. These substrata are composed of a single material, such as rock, hard skeletons (e.g., corals), wood, or recemented sedimentary grains (as in coral rubble lithified through solution and precipitation of calcium carbonate). Many species are well adapted to boring into calcareous rock and wood while others may even excavate crevices in weathered granite. Relatively impenetrable surfaces, such as fresh rock, however, often only support organisms attached to the surface. Surficial texture also influences the attachment success

235

of species. Rocks with grooves and crevices are more attractive to settling invertebrate larvae (see Chapter 8). Crevices often retain water at low tide and create refuges for organisms sensitive to desiccation.

Hard substrata also exert an important selective influence on the types of organisms that can survive on surfaces. Mobile organisms must possess an organ of locomotion that permits movement along the surface but at the same time resists dislodgment due to wave and current action. Bivalve mollusks normally relying on a probing foot in soft substrata must secrete attachment threads, nestle firmly in crevices, or cement to the substratum to prevent dislodgment. Mobile polychaetes, such as *Nereis,* can survive only when moving on the microtopography generated by other organisms like algae and mussels.

Soft substrata consists of sedimentary mineral grains, small organic particles derived from dead and decaying plants and animals (particulate organic detritus), and water. The definition of soft substrata must be related to the benthic organisms. In a sediment consisting of sand particles 2 mm in diameter, larger burrowing macrofauna push through sedimentary grains whose diameters are small relative to the body size of the burrower. But meiofaunal animals must move through spaces amongst packed sand grains larger than themselves; consequently, they can usually move only through narrow crevices. This situation has selected for vermiform body shape in many unrelated phyla of meiofauna (Fig. 12-1). To still smaller microorganisms (< 100 μ in size), sedimentary grains of 2 mm in diameter may serve as attachment sites (as in the case of smaller diatoms, bacteria, and protozoa). To microbenthos, the sediment is effectively a hard substratum.

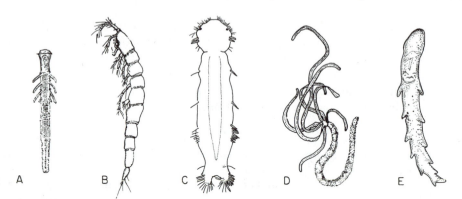

Figure 12-1 Examples of some meiofaunal organisms, showing convergence to a worm-like body form in different phyla: (a) polychaete, *Psammodrilus;* (b) harpacticoid copepod, *Cylindropsyllis;* (c) gastrotrich, *Dactylopodalia;* (d) hydroid, *Halammohydra;* (e) opisthobranch gastropod, *Pseudovermis.* (After Swedmark, 1964)

12-2 SOFT-SEDIMENT PARAMETERS

Grain Size

The size of sedimentary particles is of obvious importance in determining the distribution of marine benthic species; for instance, well-washed gravelly sediments will preclude species dependent on the ingestion of fine-grained organic particles. The diameter of sedimentary particles also indicates the current strength; stronger currents can transport larger particles. So areas in which the mean grain diameter is large can be expected to be areas of high current velocity. Sediments from areas with minimal current strength consist of very fine grained sediments.

Mechanical analysis of the size distribution of sediments can be performed through (a) individual measurement of sedimentary particles, (b) passing the sediment through a series of graded sieves, and (c) measuring the settling velocity of sedimentary grains (larger grains have a greater settling velocity than smaller sedimentary grains). Most methods of size distribution of particles in sediments employ the weight percentage of grains in each size class as opposed to the number of grains in that class. In general, size distribution data are graphically presented with a histogram—a block diagram giving the percentage of grains in each size class—or with a curve of the cumulative weight percentage of sediment as a function of size class (Fig. 12-2).

Sedimentary grain diameters are often expressed in a logarithmic scale, the phi (ϕ) scale. Grain diameter is expressed as its negative logarithm to base 2. This method of representation permits percentage differences between small particles to be conveniently presented alongside percentage differences between spectra of particles of very large size. Such geometric grade schemes—the Wentworth scale (see Table 12-1), for example— are well adapted to this description of sediments because they give equal significance to

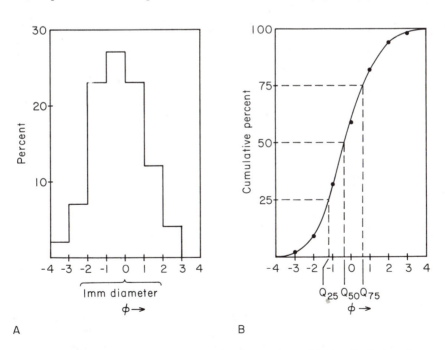

A   B

Figure 12-2 Graphical methods of presenting data on the particle size distribution of sediments: (a) histogram; (b) cumulative frequency curve, showing Q_{25}, Q_{50}, and Q_{75} (see text for explanation).

TABLE 12-1 WENTWORTH'S PARTICLE-SIZE CLASSIFICATION

Grade limits (diameter in millimeters)	Grade limits (diameter in phi units)	Name
>256	< −8	Boulder
256–128	−8– −7	Large cobble
128–64	−7– −6	Small cobble
64–32	−6– −5	Very large pebble
32–16	−5– −4	Large pebble
16–8	−4– −3	Medium pebble
8–4	−3– −2	Small pebble
4–2	−2– −1	Granule
2–1	−1–0	Very coarse sand
$1-\frac{1}{2}$	0–1	Coarse sand
$\frac{1}{2}-\frac{1}{4}$	1–2	Medium sand
$\frac{1}{4}-\frac{1}{8}$	2–3	Fine sand
$\frac{1}{8}-\frac{1}{16}$	3–4	Very fine sand
$\frac{1}{16}-\frac{1}{32}$	4–5	Coarse silt
$\frac{1}{32}-\frac{1}{64}$	5–6	Medium silt
$\frac{1}{64}-\frac{1}{128}$	6–7	Fine silt
$\frac{1}{128}-\frac{1}{256}$	7–8	Very fine silt
$\frac{1}{256}-\frac{1}{512}$	8–9	Coarse clay
$\frac{1}{512}-\frac{1}{1024}$	9–10	Medium clay
$\frac{1}{1024}-\frac{1}{2048}$	10–11	Fine clay

size ratios whether the ratios occur in gravel or clay. The difference of 1 cm in the size of a boulder is negligible whereas the difference as small as 1 μ in the size of a colloidal clay particle may be sufficient to double or halve it.

The cumulative weight percentage curve (Fig. 12-2) is of particular use because several parameters can be derived to characterize the sediment. Sediments can be characterized by quartiles; they are calculated by following the 25, 50, and 75% lines of the graph to the right, to their intersection with the cumulative curve. The values on the size scale (abscissa axis) at the intersection are the values of the quartiles (e.g., Q_{75} is the value for 75% cumulative weight). The second quartile is the median particle diameter.

Typically marine benthic ecologists are most concerned with distinguishing fine-grained sediments, usually finer than $1/16$ of a millimeter or 62 μ, from coarse, sandy sediments. The so-called *silt-clay* fraction is measured as the percentage (by weight) of sediments finer than 62 μ in diameter (Sanders, 1958). This fine-grained fraction of sediments correlates well with the abundance of fine-grained organic matter in the sediment, an important source of food and attachment substratum for the microfauna and microflora of sediments. The percent clay (particles < 4 μ) is a more sensitive indicator of those characteristics of the sediment that determine benthic animal abundance.

Sorting and Angularity

Sorting refers to the range of particle sizes in the sediment. A sediment is poorly sorted when most of the sediment is not in a narrow range of contiguous size classes; sediments are well sorted when the range of particle size variation is small. Two sediments may have the same median grain diameter but very different sorting.

Sorting can be expressed as the coefficient

$$S_0 = \frac{Q_{25}}{Q_{75}}$$

S_0 approaches unity as the sediment becomes perfectly sorted (all the sediment within one size class).

A well-sorted sediment indicates that water currents have worked extensively on the sediment. Poorly sorted sediments may result from proximity to the parent rock source or deposition by strongly variable currents in an area. Sedimentary grains that are angular usually are close to their parent rock source as well. Thus a current-worked sediment is usually well sorted and consists of grains of low angularity.

Biogenic Sorting

Burrowing organisms often generate strong vertical inhomogeneity in the sedimentary column. Typically, sediment-ingesting organisms consume preferentially small particles and transfer them to the sediment surface. As a result, biogenically reworked sediments often have a surface layer of fine particles, underlain by a layer of larger particles. The western Atlantic maldanid polychaete, *Clymenella torquata,* for example, resides in a tube head down. Particles less than 1 mm are ingested and defecated at the sediment surface (Fig. 12-3). Thus an initially poorly sorted sediment is sorted into a coarse fraction at depth and a fine fraction at the surface.

Vertical Distribution of Sediment Chemical Properties

Soft sediments are not chemically homogeneous with depth below the sediment–water interface. Sediments generally are reducing environments, devoid of oxygen, except for a surficial oxidized layer of a few centimeters or less. This zonation is generated as a balance between consumption of oxygen by benthic organisms and organic matter in

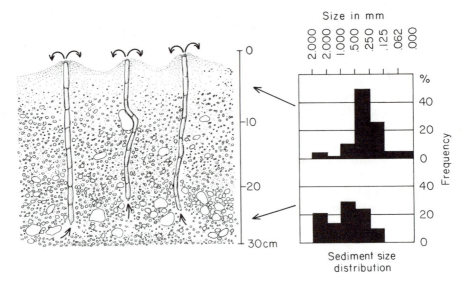

Figure 12-3   (a) Vertical reworking of intertidal sediments by the tubeworm, *Clymenella torquata*. (b) Change in the vertical distribution of particle sizes as a result of vertical reworking of the sediment by the tubeworm. (After Rhoads, 1967, with permission from the *Journal of Geology* and the University of Chicago Press)

sediments and pore water and the transport of oxygen from the overlying water into the sediment. In a sediment devoid of burrowing organisms, this latter transport is controlled by the diffusion of dissolved oxygen into the sediment pore water from the overlying water. The result would be an interface of a few millimeters depth, above which organisms with aerobic respiratory metabolism are found and below which anaerobic processes occur.

The redox potential discontinuity (RPD) is the depth below which oxidation processes are supplanted by reduction. The RPD can be ascertained by a platinum Eh electrode, recording the depth where Eh potential (oxidation) decreases from positive to negative. Below the RPD, H_2S can exist in the absence of oxygen and is generated by bacterial (e.g., *Desulfovibrio*) reduction of sulfate near the discontinuity. Vertical gradients can be visually detected as a surficial brown layer, where oxygen is present, overlying an anoxic black layer. Figure 12-4 shows variation of chemical properties across this interface.

A microbiota adapted to the anoxic zone below the RPD environment can decompose organic material through (a) fermentation, where some organics are used as hydrogen acceptors for the oxidation of other compounds, yielding end products, such as alcohols and fatty acids, or (b) dissolved sulfate, nitrate, carbonate, and water can be used as hydrogen acceptors by different bacteria, yielding reduced compounds like H_2S, NH_3, CH_4, and H_2 (Fenchel and Riedl, 1970). The mineralization of organic matter, although dependent on anaerobic processes, can be significant. In an experiment using sediments

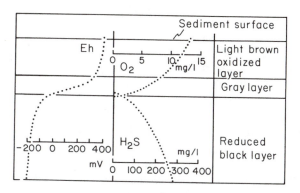

Figure 12-4 Schematic representation of the vertical distribution of some chemical properties of a typical marine bottom mud. (Modified after Fenchel and Riedl, 1970, from *Marine Biology*, vol. 7)

mixed with *Zostera* detritus, over half the oxidation of organic matter in the sediment could be related to the activity of sulfate-reducing bacteria (Jørgensen and Fenchel, 1974).

The conversion of organic matter below the RPD into bacterial biomass may be an important source of food for benthic invertebrates. Some macrofauna feed at depth below the RPD while others may take advantage of material transported to the sediment–water interface by deep feeders. Some meiofauna, such as nematodes, can tolerate long periods of anoxia and are bacterial feeders. They might be consumed by larger benthic animals.

Biotic Effects on Chemical Properties

The presence of burrowing organisms changes the transfer of oxygen into the sediment from a diffusion-dominated process to one dominated by advective exchange of oxygen with the overlying water. Intensive biogenic mixing and irrigation of the bottom accelerate the transfer of oxygen into a *biogenically reworked zone* of up to several centimeters in depth (Rhoads, 1974). Chemical processes influenced by burrowing activity are

1. Rate of exchange of dissolved or absorbed ions, compounds, and gases across the sediment–water interface.
2. Form of vertical gradients in Eh, pH, and pO_2; depth of the RPD.
3. Transfer of reduced compounds from below the RPD to the aerated surface sediment pore waters,
4. Cycling of carbon, nitrogen, sulfur, and phosphorus.

The formation of tubes within the sediment and the production of fecal pellets increase the area of contact between the aerated and anoxic zones across the RPD. Macrofauna irrigating tubes and burrows pump oxygen-rich water into the sediment and produce oxidized burrows (Teal and Kanwisher, 1961). A cross section taken of a sediment core will reveal burrow holes surrounded by a light brown halo of a few millimeters, surrounded, in turn, by black sediment. At the burrow walls, decomposition of organic matter and sulfide deposition are both active (Aller and Yingst, 1978). In invertebrate fecal pellets, the oxygen consumption balance may tip in favor of anaerobic processes, making a fecal pellet rich sediment a complex of oxidized and reduced microniches (Jørgensen, 1977). Sulfate-reducing bacteria and anoxic chemical processes may therefore occur in a complex geometric relationship with the oxidized part of the sediment. Diffusion of reduced end products of anaerobic decomposition into the aerobic zone is possible, permitting oxidation by aerobic chemoautotrophic bacteria.

Sediment feeding and the formation of fecal pellets further increase the heterogeneity of the sediment through concentration of substances relative to the average sediment in which the organism lives. Invertebrate fecal pellets often consist of particles finer than the average grain size of the sediment (Rhoads and Stanley, 1965; Hylleberg, 1975a). Furthermore, fine particles support a larger microbial community because of their higher surface area-to-volume ratio (Zobell, 1938); pellets are thus sites of enriched microbial growth. Feeding within the sediment by polychaetes often creates a pocket of aerated sediment near the worm. Many marine bacteria thrive poorly in the absence of oxygen (Zobell, 1946) and bacteria utilization is significantly increased in the presence of dissolved organic carbon (Otsuki and Hanya, 1972). Therefore irrigation by the worm effectively stimulates growth of readily digestible microorganisms (see later) and creates a "garden" (Hylleberg, 1975b). Feeding and fecal pellet production may also locally concentrate heavy metals, such as Pb and Co.

12-3 BIOLOGICAL INFLUENCES ON MASS PROPERTIES OF SEDIMENTS

Sedimentary properties are often strongly controlled by the presence or absence of large benthic populations. The presence of actively burrowing and feeding animals in densities of 10^3 to 10^5 m^{-2} has a profound influence on the distribution and abundance of marine benthic species. Earlier we discussed chemical properties and biogenic graded bedding. Benthic burrowing organisms exert several other important effects on sediments.

Effects on Grain Diameter

The feeding processes of some benthic animals result in the ingestion of sedimentary grains, their compaction into fecal pellets, and subsequent egestion back into the sedimentary environment. Some fecal pellets, such as those of crustacea, can last a long time whereas other fecal pellets—for instance, those of some bivalves—may disintegrate within a few days. Alternatively, sedimentary grains unsuitable for ingestion may be collected by a feeding organ and rejected at or near the mouth as pseudofeces. Pseudofeces may be bound in mucus and ejected onto the sediment–water interface.

The feeding activities of a dense population of deposit-feeding animals convert the sediment from a largely fine-grained sediment consisting mainly of silt and clay-sized particles into a sediment of fecal pellets that is usually much larger in size. As long as extensive deposit-feeding activity exists, a fecal pellet-rich sediment will consist of larger sedimentary grains (i.e., pellets), with pore space occupied by water. Typical muddy bottom sediment taken from a depth of 20 m in Buzzards Bay, Massachusetts, shows a completely different size-frequency distribution of grain diameter when disaggregated in an electric blender and then sieved, as opposed to being wet sieved immediately without any prior disaggregation of the sediments (Rhoads, 1967). Pellets are produced principally in the upper 2 to 5 cm of the sediment and may or may not be reingested by deposit-feeding organisms.

Sediment Water and Oxygen Content

The combination of fecal pellet production and constant lateral burrowing by deposit-feeding organisms increases the sediment's water and oxygen content. Without pelletization and burrowing, the sediment would probably be stable, reducing transport of oxygen into the sediment. The zone of fecal pellets produced near the sediment–water interface has been termed the flocculant zone (Sanders, 1960) and can often be easily observed in a grab sample of bottom as a watery light brown layer of sediment over a dark-black, more compact layer of sediment. As noted, the light brown color indicates the presence of dissolved oxygen in pore waters. A fecal pellet-rich sediment would be expected to have more organisms not requiring communication to the sediment-water interface for oxygen by means of siphons or open burrows and tubes.

Mud bottoms that are actively burrowed by great densities of deposit-feeding organisms generally have higher water contents than those with low densities. Typically the near-surface fecal pellet zone is greater than 60% water and can be as high as 80% in bottoms with large amounts of biogenic reworking of sediments (Levinton, 1977; Rhoads, 1970; Rhoads and Young, 1970). As discussed in Chapter 18, this high-water content has a strong negative influence on the presence of suspension feeders in biogenically reworked muddy bottoms.

TABLE 12-2 BURROWING SPEEDS OF THE NUCULID BIVALVE, *NUCULA PROXIMA,* IN SEDIMENTS THAT HAVE OR HAVE NOT BEEN BIOGENICALLY REWORKED[a,b]

Sediment type	Number of burrowing thrusts (+ s.e.)	Time until burial (sec ± s.e.)	Number of observations
ca. 20% silt-clay, reworked	6.0 ± .08	25.7 ± 1.19	7
ca. 20% silt-clay, unreworked	8.2 ± .23	32.6 ± 1.55	28
ca. 50% silt-clay, reworked	5.7 ± .06	20.7 ± 2.89	9
ca. 50% silt-clay, unreworked	7.6 ± .24	33.1 ± 1.99	30

[a]After Levinton, 1977. Reprinted from *Ecology of Marine Benthos,* by Bruce C. Coull, editor, by permission of University of South Carolina Press, copyright 1977.

[b]In both sediment types, it takes more time and burrowing thrusts to penetrate the unreworked sediment ($p < .05$); there is no significant difference, however, when different reworked, or unreworked, sediments are compared.

The increase of water content alters the mass properties of the sediment and greatly reduces the effort needed for burrowing organisms to penetrate the substratum. Sediments of high-water content are *thixotropic*—that is, effort required to move through the sediment decreases with increased velocity through the sediment. With decreased water content, plastic deformation of sediments occurs and required burrowing effort increases. Table 12-2 shows that burrowing speed of deposit-feeding bivalves is more rapid in bioturbated sediment (Levinton, 1977).

Water content of sediments can be determined by rapid freezing of cores carefully collected and transported to the surface by divers. Frozen cores are sliced perpendicular to the core's axis and segments are weighed wet and dry. Sediment cross sections may also be examined with a sediment–water interface camera, allowing pictures of significant interfaces and biogenic structures. Using this camera, the flocculant zone can readily be seen as an oxidized layer of pellets (Fig. 12-5).

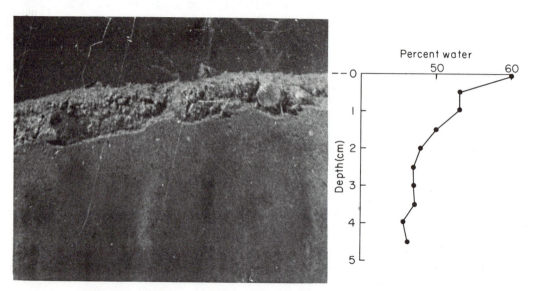

Figure 12-5 A sediment–water interface photograph of a typical shallow-water marine mud. The percent water content as a function of depth below the sediment–water interface is shown to the right. Note the thin layer of pelletized sediment near the interface. (Compiled from Levinton, 1977)

Properties of the Sediment–Water Interface and Turbidity of Overlying Water

The erodability of sediments by bottom currents is affected by water content. Muds with high clay content that contain greater than 50% water are potentially unstable in the presence of weak bottom currents. The granular surface produced by burrowers makes the sediment–water interface hydrodynamically "rough"; this factor increases the probability of converting laminar to turbulent flow near the bottom. Large amounts of sediment over watery mud buttoms are therefore resuspended into the overlying water. The great majority of sediment settling toward the bottom has thus originated from sediment eroded from the bottom (Young, 1971).

Because the sediment that is resuspended tends to be fine grained in nature and fine-grained sediments tend to have higher organic contents, this resuspension mechanism is an important means of recycling organic matter into the overlying water, perhaps importing nutrients via bacteria back to the bottom. Thus the activities of burrowing organisms influence not only the sediment in which they live but also many properties of the overlying water. Rhoads (1973) has suggested that the highly turbid overlying water could perhaps be used as a source of food for oyster culture.

Microtopography

Topographic relief is minimal on soft bottoms but is increased to a degree by biogenic reworking of sediments. Mounds or cones may be produced through the accumulation of fecal material at the anal sediment openings of such deposit feeders as polychaetes and holothurians. Organisms requiring constant maintenance of burrows may cast sediment on the surface in a mound, as in the borrowing shrimp *Callianassa*. The fecal mounds produced by the holothurian *Molpadia oolitica* (Fig. 18-9) in Cape Cod Bay, Massachusetts, permit colonization by several species of suspension-feeding, tube-swelling polychaetes not found on the level mud bottom (Rhoads and Young, 1971).

Sediment Reworking and Biogenic Sedimentation

The ingestion rate of individual deposit feeders, combined with population density, usually results in the reworking of bottom sediments many times per year. Because ingestion of sediment results in digestion of some of the organic matter and changes in sediment grain size, dense populations of deposit feeders can be said to regulate sediment characteristics completely. Normal populations of the deposit-feeding bivalve *Nucula annulata* rework the permanent annually deposited sediment 1 to 5 times per day (Young, 1971). As a result, the sediment is often completely or nearly pelletized (Moore, 1931; Rhoads, 1967; Levinton et al., 1977; Levinton and Lopez, 1978).

Suspension feeding from the water column also changes the character of the sediment by enrichment with fecal pellets derived from water-borne material. Compaction and formation of pellets increase their mass and hence settling velocity. In the Waddenzee, populations of the suspension-feeding cockle *Cardium edule* biodeposit 100,000 dry weight metric tons of suspended matter per year (Verwey, 1952). Because fecal pellets have a high sinking rate, planktonic feeders also supply suspended matter to the bottom at a dramatic rate (Moore, 1931).

12-4 PARTICULATE ORGANIC MATTER IN SEDIMENTS

Benthic Particulate Detritus and Decomposition

We define particulate detritus as particulate matter derived from nonpredatory losses in the food web. Benthic particulate detritus is usually derived from the breakdown of bodies of benthic organisms, deposited plankton organisms (excluding mineralized skeletons), decomposed planktonic organisms that have settled from the overlying water, and matter in various states of decomposition derived from organisms originally living in freshwater or on land and so on. Characterization of both qualitative and quantitative aspects of organic matter in sediments is important because this material is probably an important source of nutrients for bacteria and other microorganisms that live in sediments. So it is the base of a detrital food chain that interacts strongly with deposit feeders living in the bottom.

Entry of Organic Matter into Detritus Pathways

Primary productivity of coastal environments is very high, providing a great deal of potential food for primary consumers. In the pelagic realm, 60 to 90% of this production is consumed by herbivores; the rest decomposes and releases nutrients to the water column or sinks to the bottom. Much of the primary production of benthic seaweeds and kelp forests, sea grasses, mangroves, and marsh grasses, however, is seldom consumed to any significant degree by herbivores. The large majority of benthic primary production (500 to 1000 g C/m^{-2}/yr^{-1}) enters the food chain as dead material or organic detritus (Mann, 1976). The consumption of detritus by deposit feeders and carrion feeders and its rate of decomposition are thus of great interest in the economy of sedimentary biomes.

Leaves of the salt-marsh grass *Spartina* may die on the stalk and be immobile for long periods of time or are rapidly chopped off by winter ice or storms (see Lopez et al., 1977). The common North Atlantic kelp *Laminaria longicruris* elaborates a ribbon-shaped blade of up to several meters in length, growing at the base while the tip is simultaneously eroded. Copious amounts of detritus are thus deposited on the sediment (Mann, 1972). During the winter, when laminarian growth is active, decomposition is confined to the tips. But in the summer, when growth is slowed and conditions poor, senescent algal tissues are attacked over the entire blade and plant tissues are probably attacked and disrupted by bacteria more specifically adapted to metabolism of such substances as alginate (Mann, 1972, 1976). *Laminaria* decays rapidly and is available in great amounts to benthic ecosystems. In contrast, eel grass, *Zostera marina,* decays much more slowly, with only 35% loss of dry weight in the laboratory, after 100 days at 20°C (Harrison and Mann, 1975). Coastal sediments in eel grass beds typically have accumulations of eel grass fragments, perhaps providing a reserve of organic matter for benthic consumers. Thus a bottom adjacent to an eel grass bed receives the majority of eel grass production as slowly decomposing detritus. In St. Margaret's Bay, Nova Scotia, about 15% of the seaweed production reaches the sediment as deposited organic detritus (Mann, 1976). In Georgia estuarine creeks over 90% of the detritus is of *Spartina* origin (Odum and de la Cruz, 1967).

Decomposition of Detritus

Leaves of marine vascular plants and algal fronds undergo three types of breakdown. *Leaching* of soluble organics occurs immediately on death of the leaf, resulting in a rapid transfer of soluble organic matter to seawater. This soluble material, probably a complex including soluble proteins, amino acids, and carbohydrates, is attacked rapidly by bacteria and quickly mineralized (Fenchel, 1977). *Bacterial attack* also may occur on the less soluble substrates of the particulate detritus. Finally, *mechanical fragmentation* due to wave action, feeding, and movement of animals and ice reduces the size of particles and exposes new areas for bacterial attack.

Aerobic decomposition of detritus is usually accompanied by changes in the carbon:nitrogen ratio. Although carbohydrate—and hence carbon—is lost, as detritus is initially leached and attacked by bacteria, colonization by the microbial saprophytic community enriches the nitrogen content. Thus the C:N ratio decreases over time (Fig. 12-6), making aged detritus a suitable food for grazing organisms (Odum and de la Cruz, 1967; Harrison and Mann, 1975; Mann, 1976).

As noted, anaerobic decomposition can be at least as important as aerobic processes.

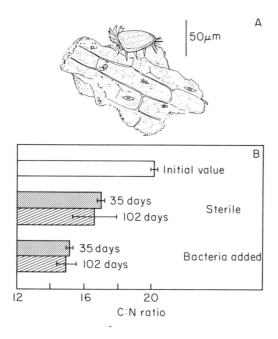

Figure 12-6 (a) A piece of turtle grass (*Thalassia testudinum*) detritus and its microbiota. (Redrawn from Fenchel, 1972) (b) Change in the C:N ratio in particulate organic matter, with and without bacteria. (Modified after Harrison and Mann, 1975, © Blackwell Scientific Publications, Ltd.)

Bacterial Abundance, Grain Size, and Particulate Organic Detritus

Most bacteria in sediments are found attached to the surfaces of particles. Because the surface area-to-volume ratio of a sphere decreases with increasing particle diameter, a surface-limited bacterial population would be expected to be more abundant in a unit volume of fine-grained sediment relative to a unit volume of coarse-grained sediments. As a general rule, fine sediments do harbor richer bacterial populations than coarse sediments (Zobell, 1938; Dale, 1974). It is an interesting detail, however, that attached microbial organisms occupy less than 10% of the surface area of particles (Fenchel, 1972). The increase of microbial abundance with decreasing grain size may be related to nutrient absorption onto particle surfaces or to the greater proportion of particulate organic matter found among small particles. In experimental vessels, both the presence of particles and nutrient enrichment can increase microbial abundance (Jannasch and Pritchard, 1972).

Earlier we noted that particulate organic matter tends to be more abundant in fine-grained sediment and may influence the abundance of decomposing microorganisms. Fine-grained sediment contains a variety of organic particles that vary in their degree of resistance to microbial attack. In a study of eel grass decomposition, fine particles ($<$ 10 μ) were shown to lose more organic matter (due to decomposition) than larger particles (Harrison and Mann, 1975). Thus sediments composed of finer particles may have larger microbiotas due to the presence of fine organic particles. Dale (1974) investigated bacterial abundance as a function of organic carbon and nitrogen and of grain size. The statistical technique of partial correlation coefficients demonstrated a significant correlation of bacterial abundance with both carbon and nitrogen, independent of grain diameter.

Colonization of Detritus by Microbial Organisms and Their Removal by Detritus Feeders

Although there is some evidence for initial colonization by fungi, Morrison and co-workers (1977) and Lopez and colleagues (1977) find that detritus is initially colonized by bacteria with later colonization by fungi, algae, ciliates, and flagellates. The muramic acid:ATP ratio is a useful indicator of this succession, for muramic acid is only found in the cell walls of bacteria and blue green algae (Salton, 1960) and ATP is a measure of microbial biomass on litter (Ausmus, 1973). When oak or pine leaf fragments are kept in seawater, the ratio decreases over 5 weeks. This decrease reflects a succession from bacterial domination (blue greens were absent) to a reduction of bacterial importance (Morrison et al., 1977). The later colonization by ciliates and flagellates probably increases grazing on bacteria and causes a reduction of bacterial standing crop on detritus (Barsdate et al., 1974; Harrison and Mann, 1975). Bacterial (and fungal) degradation of organic detritus is therefore accompanied by bacterial grazing by ciliates, flagellates, and larger grazers (Fig. 12-6) while the entire microbial community is grazed by macrofaunal detritus eaters.

248

Detritus feeders seem to assimilate organic detritus poorly but efficiently digest the microbial community living on the detritus (Hargrave, 1970a; Fenchel, 1970). Although the detritus-consuming amphipod *Parhyalella whelpleyi* mechanically reduces the size of turtle grass *(Thalassia testudinum)* detritus during feeding, only the microorganisms are digested as the particles pass through the gut (Fenchel, 1970). Fecal material is then recolonized by microorganisms with rapid bacterial recovery in several days or less (Hargrave, 1976; Levinton and Lopez, 1978).

Grazing on the microbial community stimulates microbial growth and detrital mineralization. Oxygen consumption of microbes on detritus (corrected for microbe-free consumption) increases in grazed relative to ungrazed detritus (Hargrave, 1970b; Fenchel, 1970). Barsdate and others (1974) measured turnover of phosphorus in a bacteria-detritus-protozoan system and found that protozoan grazing on bacteria increased phosphorus uptake by a factor of nearly eight relative to ungrazed systems. This observation is important because particulate detritus itself is poor in such essential nutrients as nitrogen and phosphorus. Turnover and exchange rate of nutrients (Fig. 12-7) is a rate-limiting step in detrital breakdown (Fenchel, 1972). The stripping of bacteria and other microorganisms during macrofaunal grazing may favor faster-growing forms or may shift microbial succession toward a state with higher bacterial biomass and activity. Recolonization of detritus usually involves an increase to a maximum and then a decrease of microbial oxygen consumption (Hargrave, 1976). Grazing may keep bacterial growth in this active state, enhancing detrital mineralization.

Other possible mechanisms of microbial stimulation by grazing include (a) exploitation of bacteria limited by space, thus increasing bacterial productivity and nutrient requirements from detritus; (b) stirring by macrofauna, providing a local increase of oxygen and stimulation of aerobic bacteria; and (c) excretion of nitrogen and phosphorus by grazers stimulating microbial growth.

Lopez and co-workers (1977) investigated the effect of grazing by the amphipod

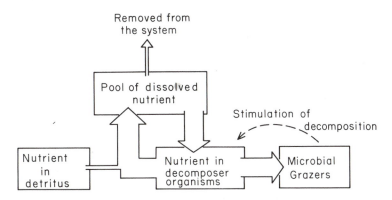

Figure 12-7 Schematic representation of the nutrient cycling of a decomposer system showing the relationship between nutrient cycling, mineralization of the organic substrate, and grazing. (Modified after Barsdate et al., 1974)

Orchestia grillus on the microbial community and decomposition of *Spartina* litter (salt marsh grass). Grazing increased the rate of colonization of the microbial community (as estimated by ATP) and hastened the removal of nitrogen from the detritus. The positive correlation between macrofaunal abundance and bacterial abundance in the sediment (Fenchel, 1972; Tunnicliffe and Risk, 1977) may reflect the stimulation of microbial growth by grazers.

SUMMARY

1. Substratum type determines the morphology, dominance patterns, feeding, and interactions of benthic species. We can distinguish between hard and soft substrata. Soft substrata can be characterized by grain size distribution, percentage of fine particles, and the variance in particle-size class abundances (sorting).

2. Soft sediments vary extensively in chemistry of pore waters, water content, and the nature of the sediment–water interface. Soft sediments are generally anoxic at some distance below the sediment–water interface.

   Burrowing organisms increase the oxygen content of the surface sediment, increase the water content, and thereby increase the fluidity of the sediment

3. Large amounts of particulate organic matter enter sediments because a large proportion of sea grass and seaweed production is never consumed by herbivores. Organic particulates also include material derived from dead plankton. Organic particulates decompose through mechanical fragmentation, leaching of dissolved substances, and microbial attack. Invertebrates ingest particulate organic matter and accelerate decomposition through direct digestion, fragmentation, and stimulation of microbial populations.

13 life habits and adaptations of benthic organisms

13-1 INTRODUCTION

Benthic animals and plants must successfully attach to or penetrate the substratum while maintaining an orientation that permits feeding and respiration. The nature of the substratum presents the fundamental selective force in determining the life habits and morphology of benthic organisms. Current strength controls the nature of the substratum and also presents a force to which attached and crawling species must adapt. Currents also bring planktonic food and organic detritus to the bottom and deliver planktonic larvae to their settling sites. In this chapter we discuss adaptations of benthic organisms to life on and within the substratum.

13-2 EPIBENTHOS

Epibenthic noncolonial (Fig. 13-1) organisms are often permanently attached to the bottom by holdfasts (stalked barnacles, seaweeds), roots (stalked crinoids, sea grasses), and basally cemented structures (oysters, serpulid polychaetes, acorn barnacles). Bushlike and vine-shaped colonial forms (e.g., ectoprocts, hydrozoa) also permanently attach with holdfasts. Sheetlike colonial forms, such as encrusting ectoprocts (e.g., *Membranipora*), colonial tunicates, and the hydrocoral *Allopora*, cement to the surface. But boring sponges like *Cliona* maintain a sheet of living tissue on the surface while boring into the calcareous hard substrate. Figure 13-2 illustrates some epibenthic colonial forms.

Some epibenthic forms remain attached to the bottom for a while but relocate frequently. Anemones remain attached to the surface by a basal disk but can move

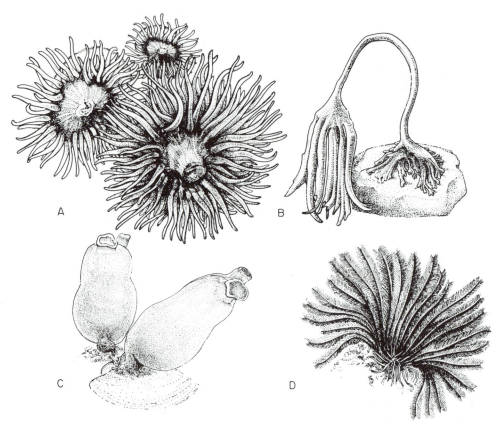

Figure 13-1 Some epibenthic solitary species. (a) Anemone, *Anthopleura elegantissima;* (b) seaweed, *Laminaria;* (c) tunicate, *Halocynthia;* (d) feather star (unstalked crinoid), *Pontio*.

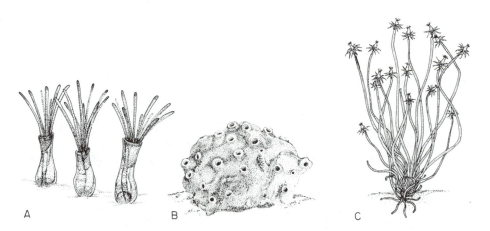

Figure 13-2 Some epibenthic colonial species. (a) Closeup of individuals of the bryozoan *Membranipora;* (b) the sponge *Halichondria;* (c) the colonial hydroid, *Tubularia*.

laterally. Mussels remain attached to the bottom by means of byssal threads but can move by sequentially severing and producing new threads. Feather stars (unstalked crinoids) hold fast to the bottom by means of specially adapted cirri. Movement by using the mobile cirri and arms is also possible, however. When currents are swift, the large Caribbean feather star *Nemaster grandis* can rapidly retreat to a crevice (Meyer, 1973).

Water turbulence strongly influences the morphology and taxonomic composition of epibenthic biotas. Areas of strong wave energy usually harbor epibenthos with robust skeletons and supportive tissue. In quieter waters, more fragile forms are common. Storms and breaking waves often cause mechanical damage and even complete dislodgment of epibenthic seaweeds and invertebrates (Dayton, 1971; Koehl and Wainright, 1977).

Epibenthic species may have one of several adaptations to wave energy.

1. Adoption of a short and squat profile to minimize exposure to shear stress (e.g., some anemones). In general, water current velocity is strongly diminished near the bottom. This boundary effect greatly reduces the stress suffered by a squat organism relative to one protruding into the "mainstream" current.

2. Hiding in holes or behind protrusions to avoid the full force of the wave or current.

3. Presence of stout and rigid support structures (e.g., the robust California mussel *Mytilus californianus* has thick byssal threads).

4. Presence of mechanical structures that permit extensibility or "flopping over" to minimize shear stress (see Wainright and Koehl, 1976; Koehl and Wainright, 1977).

On the West Coast of temperate North America the anemone *Metridium senile* is found in relatively sheltered subtidal habitats whereas the anemone *Anthopleura xanthogrammica* occupies the wave-swept intertidal zone. Although *M. senile* occurs in calm waters, its tall, relatively slender form exposes it to the full velocity of the current. Because currents are, as a rule, strongly diminished near the water-substratum interface, *A. xanthogrammica* suffers a strongly reduced shear relative to the mainstream current.

Nevertheless, *A. xanthogrammica* individuals still experience current velocities that are an order of magnitude greater than those encountered by *M. senile*. The adaptation of *A. xanthogrammica* to current shear stress can be related to the stress generated in a beam. Tensile stresses in a bending beam increase with length and decrease with beam diameter. Flow-induced stress in the tall, slim *M. senile* can be shown to be an order of magnitude greater than in *A. xanthogrammica* (see Koehl, 1976). The short, stout *A. xanthogrammica* feeds upright on mussels while *M. senile* individuals bend over in the current and suspension feed via oral disks.

Tall, sessile seaweeds often encounter rapid currents. Koehl and Wainright (1977) studied the structure and mechanical function of the stipe of giant kelp, *Nereocystis luetkaena*. This kelp forms dense forests just off wave-beaten shores from Alaska to California. Gas-filled floats support the fronds and waves or tidal currents continually subject the stipes to tensile forces. Although the strength of the stipe is not particularly great, its crossed-helical array of cellulose fibers embedded in a matrix permits the stipe to be very extensible. Thus *Nereocystis* resists water flow by being flexible and extensible.

Corals show variation in skeletal strength that may be related to the water energy of the habitat (Chamberlain, 1978). The strength of the erect-branching elkhorn coral *Acropora palamata* is much greater than that of the massive coral *Montastrea annularis*. This difference may be related to the exposure of the branching form to strong "mainstream" current velocities. Massive colonies are better suited to resist strong currents because of their colony geometry; they do not subject many individual branches to tensile stress and maintain a low profile, away from the mainstream current.

Attached epibenthic species also adapt to different water velocity regimes through changes in orientation. As discussed, moving water generates tensile stresses within individuals and colonies protruding into the current. Proper orientation may minimize stress on the individual or colony (Koehl and Wainright, 1977; Chamberlain, 1978). But water currents are also a source of food for suspension feeding and organisms orient to increase the probability of food capture (Meyer, 1973; Warner and Woodley, 1975; Wainright and Dillon, 1969).

Many invertebrates have a planar structure that is kept at, or grown into, an attitude that is at right angles to the prevailing current. In suspension-feeding Ophiuroids (brittle stars) and comatulid crinoids (feather stars) a series of tentacles are arrayed in a plane to capture particles effectively. Feather stars found in directional currents have arms lined with pinnules that are themselves arranged in a plane. In the brittle star *Ophiothrix fragilis* tube feet arise from either side of the tentacle and are also arranged in a two-row plane (Fig. 13-3). Food particles are captured by the tube feet and compacted into a mucous-clad bolus that is passed down the arm (Warner and Woodley, 1975).

Feather star species found in crevices generally experience multidirectional currents. Unlike species in unidirectional currents, these latter species normally do not arrange their arms in a plane. Furthermore, pinnules are arranged in four rows at approximate right angles, which maximizes food capture from the several possible directions (Meyer,

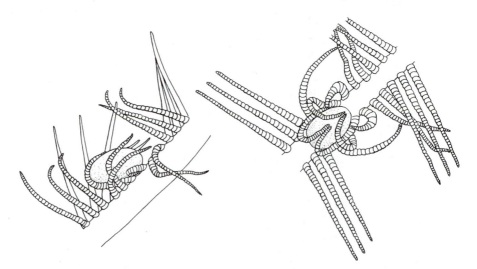

Figure 13-3 The bolus-collecting wave of a suspension-feeding ophiuroid, showing the planar arrangement of the pinnules. (From Warner and Woodley, 1975)

1973). In multidirectional currents, *Ophiothrix fragilis* (an ophiuroid) deflects alternate tube feet aborally to form two more filtering rows (four-way orientation as in feather stars).

Sea fans of the genus *Gorgonia* commonly show a preferred orientation with the fan normal to the current. Unlike feather stars, these organisms are attached and so must continuously grow in the preferred orientation. Wainwright and Dillon (1969) found that the degree of orientation increased with the size of the sea fan colony. This situation can be explained by a change in orientation as the sea fan grows. It is not clear why younger and smaller sea fans do not grow in as precise a preferred orientation as larger colonies. Perhaps smaller sea fans experience more irregular currents near the bottom while other sea fans are more exposed to the unidirectional mainstream current.

Mobile epibenthic forms (Fig. 13-4) may crawl over the bottom, using (a) pedal musculatory waves, as in snails and flatworms, (b) legs or struts for walking and hopping, as in crabs and the fish *Benthosaurus*, or (c) tube feet in the case of echinoderms (urchins, sea stars, sea cucumbers).

A   B

C   D

Figure 13-4 Some mobile epibenthic forms. (a) The gastropod, *Conus;* (b) the benthic fish, *Benthosaurus;* (c) the asteroid starfish, *Asterias;* (d) the nudibranch, *Tridachia*.

The distinction between mobile epibenthic forms and swimmers is arbitrary because swimmers generally spend some time at rest on the bottom. The killifish *Fundulus heteroclitus* is capable of rapid swimming but can be found foraging for small crustacea and detritus on New England salt-marsh bottoms in summer. Swimming scallops normally rest on the bottom or barely cover themselves with a thin layer of sediment while being partially buried. When a predator comes near (e.g., a starfish) and attempts to seize the scallop, a swimming response is accomplished by rapid flapping of the valves and ejection of water through openings formed by the mantle lobes near the auricles (Fig. 13-5). The sea hare, *Notarchus*, also uses a form of jet propulsion made possible by a sac formed by two parapodia. This movement allows escape and supplements normal crawling with pedal waves (Martin, 1966). Benthic squids and *Octopus* use a compressible mantle cavity in combination with a nozzlelike opening. *Octopus* uses this jet propulsion to augment movement accomplished with sucker-lined arms.

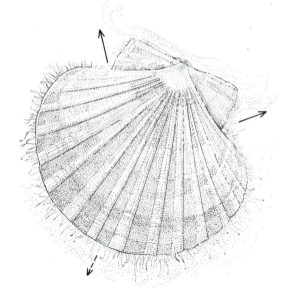

Figure 13-5 Swimming of the scallop, *Pecten maximus*. Dashed arrow indicates swimming direction; solid arrows show water propulsion.

Undulatory swimming and crawling are a common means of movement, the tube-shaped body being thrown into sinuisoidal waves, usually passing from head to tail. The movement requires a skeleton, either hydraulic (nematodes, polychaetes) or an axial bar (fish vertebral column). Sinusoidal waves produce a component of thrust tangential to the body surface and a normal component [Fig. 13-6(a)] that exerts the forward push, against either the bottom or the water (Trueman, 1975). Swimming polychaetes, such as *Nereis*, adapt the sinusoidal wave to permit parapodia exposed to the water to exert a power stroke and those protected to perform a recovery stroke (Clark, 1964) [Fig. 13-6(b)]. Nepthyid polychaetes use swimming to thrust into the sediment.

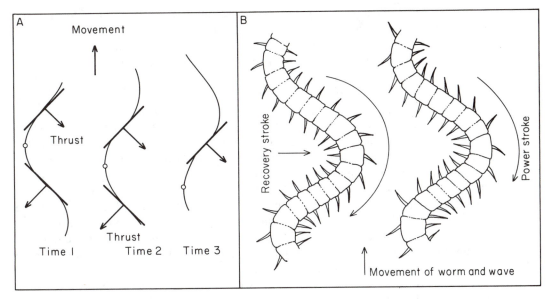

Figure 13-6 (a) Diagram illustrating the forces generated in undulatory swimming in water; circle indicates direction of muscular contraction. (b) Undulatory swimming, as aided by parapodial movements in the polychaete *Nepthys*. (Modified after Trueman, 1975)

13-4 INFAUNA

In order to penetrate soft sediments, infauna must exert a forward thrust within the sediment while maintaining temporary points against which a force can be exerted. Soft-bodied burrowing animals penetrate the sediment by forming (a) a *penetration anchor* [Fig. 13-7(a)], permitting extension of the body forward with the aid of circular or

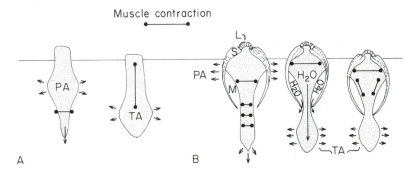

Figure 13-7 (a) The burrowing of a soft-bodied animal, showing the formation of a penetration anchor *PA*, and dilation of a distal region, forming a terminal anchor *TA*. (b) Burrowing of a generalized bivalve mollusk: S = shell; M = mantle cavity; L = flexible ligament; lines ending in filled circles indicate muscle contraction. (After Trueman, 1975)

transverse muscles, and (b) a *terminal anchor*, permitting the contraction of longitudinal or retractor muscles to pull the animal into the sediment (Trueman and Ansell, 1969). In the lugworm *Arenicola* the proboscis is everted and penetrates the sediment surface with a series of thrusting and probing actions. When in the sediment, the penetration anchor is formed by flanging of the anterior segments and protrusion of chaetae, which allow the proboscis to extend forward. The terminal anchor is then formed by dilation of the anterior segments, permitting the rear of the worm to be pulled into the sediment (Trueman, 1975). *Arenicola* shows reduced segmentation. In polychaetes with better-developed intersegmental walls, burrowing is more in the form of waves of contraction, where the anchor location passes along the length of the worm during locomotion.

In bivalve mollusks penetration is accomplished through (a) probing of the sediment surface, (b) thrusting of the foot in the sediment, (c) righting of the shell, (d) siphon closure, dilation of the foot, and (e) drawing of the animal downward. A typical bivalve is enclosed in a bivalve shell whose ligament and hinge system springs open the shell unless held closed by the adductors. When the shell has entered the substratum, a penetration anchor is formed by keeping the valves slightly apart. Hydraulic pressure thrusts the bulbous foot downward and a terminal anchor is formed by contraction of the valves, thus dilating the foot. Siphon closure ensures that water does not pass out of the animal during adduction of the shells. Pedal retractor muscles then draw the shell downward after the foot [Fig. 13-7(b)].

The speed of burrowing is related to the cross-sectional area of the valves thrusting in the sediment and the degree of ornamentation. Small, streamlined bivalves with cylindrical, bladelike or disklike shapes tend to be rapid burrowers. Increased ornament generally decreases burrowing speed (Stanley, 1970). Many bivalves squirt water into the sediment to facilitate burrowing, but watery, biogenically reworked sediment allows easier penetration than unburrowed sediment (Levinton, 1977). Some bivalves (e.g., *Divaricella*) have chevron-shaped sculpture that actually facilitates burrowing when the shell is rocked as penetration occurs. Alternately, each subparallel set of grooves acts as a penetration anchor, facilitating stronger gripping of the sediment by the shell (Stanley, 1970).

Many infaunal animals secrete tubes or reside in permanent burrows. Worms of the family Maldanidae construct vertical tubes of sediment bound by mucus. Particle gathering for tube construction is usually selective. The ice cream cone worm, *Pectinaria*, constructs a cone-shaped enclosure of closely fitted, even-sized fine sand grains [Fig. 13-8(a)]. The worm can move the tube laterally. An elaborate U-shaped tube is constructed by the parchment worm, *Chaetopterus*. The worm is so delicate that adults cannot live in the sediment without the tube.

Permanent open burrows are excavated by many crustacea, including fiddler crabs *(Uca)* and the burrowing shrimps *Upagebia* and *Callianassa*. Unlike soft-bodied creatures, crustacea use a lever-muscle system to burrow into the substratum. Intertidal fiddler crabs frequently retreat to burrows to escape predators and desiccation. *Callianassa major,* an Atlantic intertidal and shallow subtidal form, excavates elaborate anastamosing tunnels as much as 1 m deep and several meters wide. *Upagebia* excavates and maintains a Y-shaped burrow. Somewhat less distinct and permanent burrows are maintained by a wide

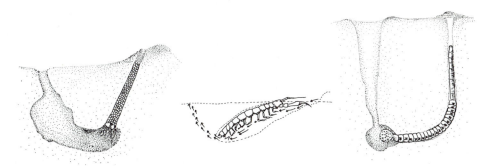

Figure 13-8 Living positions of some burrowing invertebrates. (a) The burrowing polychaete, *Pectinaria;* (b) the amphipod, *Corophium;* (c) the polychaete, *Abarenicola.*

variety of infaunal animals for feeding and probably escape from predators. The Atlantic
razor clam *Ensis directus* keeps open a tubular burrow, down which it can rapidly escape
predatory attack (Stanley, 1970). The burrowing amphipod *Corophium* excavates a temporary burrow [Fig. 13-8(b)] with (a) its walking and swimming legs passing sand to the
surface and (b) by beating of the pleopods that propel the animal into the sand. The body
is flexed and reversed as the antennae protrude from the main burrow opening. A secondary
burrow is excavated, resulting in a horseshoe-shaped burrow.

The lugworms *Arenicola* and *Abarenicola* are common residents of intertidal flats.
They maintain a U-shaped burrow [Fig. 13-8(c)] consisting of a mucous-lined gallery
and tail shaft and a head shaft formed by consumption of sand by the worm at the gallery
entrance. Periodically the worm backs up the tail shaft to egest the sand castings characteristic of *Arenicola* flats. Fishes, such as some gobies and the sand eel *Ammodytes*,
maintain burrows in the sediment as protection from predators.

Many soft-sediment-dwelling infauna maintain no burrows or remain stationary for
short periods of time and burrow laterally. These animals (e.g., protobranch bivalves like
Nucula, Yoldia; polychaetes like *Nephthys incisa*) are responsible for the fluid, pelletal
nature of subtidal fine-grained sediments.

Interstitial forms (most are meiofaunal in size) live among sedimentary grains and
are therefore restricted in size and morphology in order to negotiate crevices successfully.
In order to move, many species have evolved a vermiform shape (Fig. 12-1). Harpacticoid
copepods move through interstices with a wriggling movement unlike any calanoid copepods. The hydroid *Halammohydra* has evolved a wormlike body with a reduction of
tentacle number from 36 to 7 in connection with the change in shape (Swedmark, 1964).
A large number of interstitial forms has evolved means of temporary adhesion to sand
particles. Gastrotrichs employ adhesive tubes to attach temporarily but glide along grain
surfaces by means of cilia. Nematodes may attach to grains with the aid of caudal adhesive
glands. The restriction of interstitial space to optimally adapted forms results in strong
preferential movement of given interstitial species into sediments of their preferred grain
size (Boaden, 1962). In general the size of interstitial spaces has selected for body sizes
less than 2 to 3 mm in a wide variety of phyla.

Tidal beaches lose water as the tide recedes except where interstices retain water due to capillary action. As a result, gravels and coarse-grained sands can only support interstitial forms that are able to live in the moist film adsorbed to grain surfaces at low tide. The vertical distribution of interstitial forms in sandy beaches thus may coincide with the upper limit of the water table at low tide (McIntyre, 1969). Consequently, interstitial forms may be found at successively greater minimum depths below the sediment surface as we go landward from the low tide mark. Sediments retaining water at low tide also dampen the variation in temperature and salinity and allow interstitial forms to survive.

In addition, grain size controls the depth of penetration of oxygen, for fine-grained sediments often contain finely dispersed organic detritus colonized by oxygen-consuming microorganisms. So oxygen is absent only a few centimeters below the sediment–water interface. This fact restricts all meiofaunal forms but some nematodes to the upper couple of centimeters. Compaction makes fine-grained sediments difficult to penetrate and further restricts interstitial forms at depth. In sands, interstitial forms can be commonly found 0.5 to 1.0 m below the sediment surface.

13-5 BORING ANIMALS

Boring holes into hard substrata is accomplished through mechanical abrasion and chemical weakening of the substratum. The boring sponge *Cliona* and the boring turbellarians *Stylochus* and *Pseudostylochus* efficiently employ chemical attack. The latter two species effectively bore and attack oysters and barnacles. Most other well-known boring forms use mechanical and sometimes chemical means, but the exact nature of chemical attack is usually obscure. Boring has been adapted for predation, as in the drilling turbellarians and snails. Seven superfamilies of bivalve mollusks have independently evolved the boring habit to varying degrees.

Rock borers, such as *Zirphaea crispata* and the shipworm *Teredo navalis*, use the valves as abrasive tools. In a typical dimyarian bivalve, the valves open and rotate about an axis parallel to a line going through the adductors of one valve. This trend is altered slightly in deep-burrowing forms, such as *Mya*, where the hinge is reduced and some movement of the valves about a dorso-ventral axis is present to permit accommodation of siphons as they are withdrawn (Nair and Ansell, 1968a). In the rock borer *Zirphaea* the ligament is lost, the anterior part of the shell has a pedal gape and is strongly sculptured, siphons are elongated and fused, and muscles and hinge line are highly modified to rock the shells about a dorso-ventral axis and scrape the rock when the adductor or muscles contract (Purchon, 1955; Nair and Ansell, 1968b). The foot forms an anchor and draws the shells down against the borehole walls to facilitate abrasion.

In the shipworm *Teredo* (Fig. 13-9), valves are articulated at dorsal and ventral points and shell movement is exclusively about the dorso-ventral axis. The foot grips the end of the hole while the ridged valves rock and scrape away wood. The mantle is elongated into a vermiform tube terminating at the borehole opening and attaching to the wood with two calcareous plates. *Teredo* can also digest the wood with the aid of cellulase. Wholesale destruction of wood pilings is often possible.

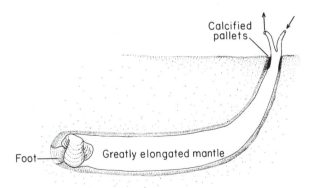

Calcified pallets

Foot

Greatly elongated mantle

Figure 13-9 Cross section showing the living position of the boring bivalve, *Teredo*, within its wooden substratum. (After Trueman, 1975, *The Locomotion of Soft-Bodied Animals*, with permission of Edward Arnold (Publisher) Ltd)

Borings are excavated by sea urchins and limpets in rocks, particularly those softened by weathering. The owl limpet, *Lottea*, leaves a recognizable scar while the temperate Pacific urchin *Strongylocentrotus purpuratus* often excavates circular crevices. Smaller animals, such as the polychaete *Polydora*, are effective borers of calcareous rock, as are many species of algae. This situation causes a great deal of erosion of scleractinian coral skeletons.

13-6 BENTHIC FREE-LIVING COMMENSALS AND MUTUALISTS

Commensal species are those that depend on another species but return no benefit to the latter. It is often difficult to distinguish between those commensal relationships in which the dependent species is specifically adapted to this protection or acquires it by accident. Barnacles, for example, may be attached to leafy algae and escape predation from starfish incapable of climbing the algal frond. This relationship is clearly not an evolved one if the majority of the barnacle population lives elsewhere and the settlement on algae is fortuitous. But mussel larvae attach for a few days to filamentous algae; a number of species are suitable as hosts. The commensal relationship is temporary but clearly evolved, for the initial settlement is necessary for early survival (see Chapter 8). A more special case is where the commensal species maintains a relationship with only one host species; proboscidactylid hydroids are specialized for settlement and life on tube-building Sabellid polychaetes (Campbell, 1968).

Commensal relationships usually involve (a) protection against predators or (b) provision of a suitable substratum or crevice. In both cases, the dependent species may also feed on material inedible by the host or food that escapes the host's feeding apparatus. Species of the crab *Pinnixa* live in polychaete tubes and feed on material entering the tube. *Pinnixa chaetopterans* is a commensal of the parchment worm *Chaetopterus*. Crabs enter the tube as juveniles and grow too large to depart as adults. Burrows of the echiurid worm *Urechis caupo* harbor a gobiid fish, a polynoid polychaete, and a pinnotherid crab as commensals (Fig. 13-10). The polychaete feeds on some of the mucous bag maintained by *Urechis* for suspension feeding (Fisher and MacGinitie, 1928).

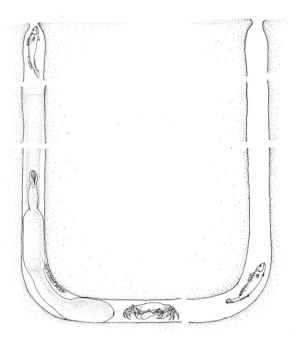

Figure 13-10   Part of the burrow of the Echiuroid worm, *Urechis caupo,* showing the following commensals: the goby *Clevelandia,* the polynoid polychaete, *Harmothoe,* and the pinnotherid crab, *Scleroplax.* (After Fischer and MacGinitie, 1928)

Commensals living on the surfaces (epibionts) or around other species are also common, as in the case of *Proboscidactyla*. Comatulid crinoids often harbor cryptically colored shrimp among their arms. The polychaete *Aphrodite* often has byssally attached bivalves on its body surface.

In most cases of mutualism in free-living benthos, one mutualist protects the other against predation. Coelenterates are involved in many such relationships. Large coral reef anemones, such as *Actinia*, harbor small pomacentrid fishes (*Amphiprion* or *Dascyllus*) among their tentacles. The fish approaches the anemone, brushes against it, and acquires some mucus. With later approaches, the anemone fails to detect the fish as a prey. Eventually the fish lives among tentacles and is in no danger of being stung by anemone nematocysts (Le Danois, 1959). The fish may attract other fishes, which are consumed by the anemone. Many crabs use anemones for protection while the anemone uses the crab as a substratum. The Hawaiian crab *Melia tessellata* detaches small anemones from the bottom, maintains them on the claws for defense, and obtains food from among the anemone's tentacles. The anemone *Adamsia* wraps itself around the shell occupied by the hermit crab *Eupagurus prideauxi*. As the crab increases in size, the growth of the anemone increases the capacity of the chamber within which the crab must live (Cott, 1940).

Symbioses seem most elaborately developed in the tropics. In particular, coral reefs harbor many of the most fascinating mutualisms in the sea; they are discussed in Chapter 21, Section 21-2.

13-7 PARASITES

Parasites decrease the fitness of species with which the former is associated by a variety of nutrient-gathering and shelter-establishing functions. A discussion of parasites and their host relationships is beyond the scope of this book, but many benthic parasites exert important effects on benthic populations. We can distinguish between ectoparasites and endoparasites. Ectoparasites live on or are embedded into body walls, gills, and other structures and range in morphology from mobile organisms indistinguishable from related nonparasitic forms to those with highly modified morphology suited to a parsitic mode of existence. Members of the opisthobranch gastropod family Pyramidellidae use a pumping pharynx, chitinous jaws, and piercing stylets to feed on blood of bivalve mollusks and polychaetes. Pyramidellids reported as ectoparasites, however, may also be found free living in the mud (Sanders, 1960). Members of the genus *Stylifer* live embedded in the body walls of echinoderms, lack a radula, and have a vestigial foot and a proboscis modified for sucking tissue fluids.

Endoparasites live within host tissues and have a highly modified morphology consonant with life within body cavities and food uptake and absorption of fluids. Also typical of these groups are complex life cycles with a different host for each stage.

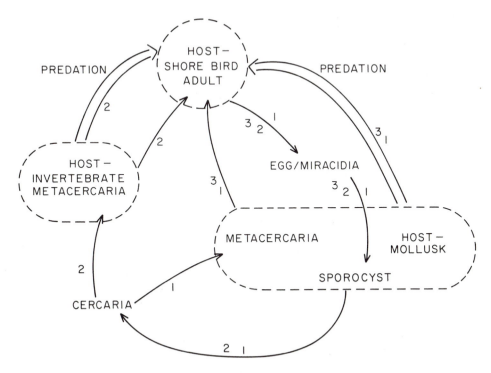

Figure 13-11 Variations (pathways 1,2, or 3 are possible) in the life cycle of digenetic trematodes living in coastal habitats of California. (Courtesy S. Obrebski)

Although all parasites are of obvious potential harm by feeding on tissues, many directly affect reproductive functions. Rhizocephalan copepods, such as *Sacculina*, feed on fats and thus eliminate reserves that would to to egg production in crabs. The infestation also causes atrophy of gonads (parasitic castration) and affects secondary sexual characteristics. Infection of *Hydrobia ulvae* by larval trematodes may cause gigantism and shell abnormalities (Rothschild, 1936).

More complex life cycles are found in Epicaridian isopods where the intermediate host is a copepod while the final host is a crustacean, such as *Upogebia*. Digenetic trematodes may have one or two intermediate invertebrate hosts and a final vertebrate host. S. Obrebski, however, has found a remarkable amount of variation within this group. Figure 13-11 shows the possible variations in life cycle. The vertebrate acquires the parasite by preying on the invertebrate intermediate or through direct infection by a free-swimming cercaria stage (as in the debilitating *Schistosoma* in humans).

13-8 ANTIPREDATOR AND ANTIFOULING DEVICES

The diversity and ubiquity of predators impose a constant chance of extinction of prey populations. Any genetic variant capable of discouraging a predator would therefore be favored under conditions of heavy predation. Many benthic organisms have evolved a wide variety of antipredator adaptations. Although presently a matter of speculation, it is possible that many predators have coevolved with their prey and have therefore improved modes of prey capture and consumption (Vermeij, 1978).

Benthic organisms similarly face the danger of being fouled by other organisms settling as invertebrate larval or algal dispersal stages from the plankton. Sessile individuals or colonies face the danger of being smothered by other sessile organisms seeking some space. This danger has been met with the evolution of antifouling adaptations.

Mechanical Adaptations

Many invertebrates have spines or thickened skeletons to discourage predators. The tropical urchin *Diadema* has long slender spines with reverse barbs. Although the spines retard many carnivores, triggerfish can expel a jet of water that upends the urchin and exposes the relatively unarmed oral side. Many epifaunal hard-surface-dwelling mollusks bear ridges and spines that discourage predators. Spines greatly increase the difficulty of fishes in attempting to crush tropical gastropods (Palmer, 1979). Similarly, gastropods living in areas of heavy crab predation often have thickened shells, a thickened outer lip, a low spine, and small or narrow apertures to discourage intruders (Zipser and Vermeij, 1978). Many tropical crabs have large master claws with molariform teeth that easily crush or puncture fragile-shelled forms.

Chemical Defense

A great number of plants and sessile and mobile invertebrates produce or sequester a wide spectrum of toxic substances. These substances are used to discourage predation, kill potential fouling organisms, or perhaps prevent competitors for space from intruding on the living space of sessile species. Some species exude noxious inks into the water to discourage predators (e.g., *Aplysia*). It is not yet clear how these substances first appeared in evolution. Whittaker and Feeny (1971) suggest that many organic toxic compounds found in plants (e.g., terpenes, alkaloids) originated as metabolic end products or wastes that were subsequently selected for increased production to discourage predation. The implication is that toxic compounds have the singular purpose of retarding predators, fouling organisms, parasites, or pathogens. But Seigler and Price (1976) point out that many so-called defense compounds have rapid turnover times in cells, so they may serve other metabolic functions.

Acid secretion is perhaps the best-known case of defense in benthic organisms. Acid serves as both an antipredator and an antifouling device. Gastropods, tunicates, and the seaweed *Desmarestia* all use acid to discourage predators (Thompson, 1969; Stoecker, 1978; Irvine, 1973). In gastropods, acid is produced by epidermal acid gland cells and acid discharge occurs by cell rupture and follows mechanical disturbance. In some species, subepidermal acid glands produce acid that is expelled through pores. Fishes invariably refuse to eat gastropods with acid secretions (Thompson, 1960). Many tunicates secrete acid and are similarly repulsive to fishes (Stoecker, 1980).

Both marine plants and animals have a broad spectrum of toxic organic compounds that retard predation. Halogenated organic compounds are synthesized by many red algal species and often deter herbivores and inhibit bacterial attack (Fenical, 1975). Saponins (triterpene glycosides) are commonly produced by invertebrates and discourage predators and fouling organisms (Bakus and Green, 1974; Patterson et al., 1978). Fishes that are force fed pieces of saponin-bearing tropical holothurians invariably die (Bakus and Green, 1974). Tropical sponges are similarly toxic and are avoided by fishes. The sea hare *Aplysia californica* is capable of sequestering halogenated compounds in the digestive gland (Stallard and Faulkner, 1974). In some cases, symbiotic organisms have evolved mechanisms to sequester or detoxify the poisons.

Despite the wide occurrence of toxic organic compounds, some symbiotic, predatory, and fouling species apparently are not harmed by toxins and are successful at sequestering or otherwise detoxifying such substances as saponins. The polynoid polychaete *Arctonoe fragilis*, for instance, maintains a symbiotic relationship with the starfish *Evasterias troschelii* in Puget Sound. The polychaete is immune to the saponin synthesized by the starfish. Yet the closely related *A. pulchra* is deleteriously affected by *Evasterias* saponin. Therefore *A. fragilis* has evolved tolerance to a posion concomitant with the development of its symbiotic relationship with the starfish.

Many tunicates concentrate transition elements from seawater. The ascidian *Phallusia nigra* concentrates vanadium and transports it to the tunic surface, where it is found in concentrations of 1000 ppm dry weight! Stoecker (1978) showed that agar disks with vanadium kill settling larvae and agar pellets with vanadium discourage feeding by fishes and crabs. In the natural habitat (hard substrata in the American tropics) this species has a conspicuous black color, is unfouled, and is not touched by grazing fishes. *P. nigra* is also acidic and probably synthesizes toxic organic compounds. Such obnoxious beasts are common in the tropics.

Color and Vulnerable Structures

The evolution of toxic chemicals seems intimately involved with the color and morphological structures of benthic organisms. Many poisonous organisms have conspicuous color patterns in contrast to relatively inconspicuous nonpoisonous prey (Thompson, 1960; Russell, 1966). The poisonous *Phallusia nigra* is a conspicuous black tunicate (Stoecker, 1978). Many toxic tropical species are bright red or yellow. Similarly, the Panamanian tunicate *Rhopalea birkelandii* is acidic and a bright electric blue. Some mollusks bear tentacular structures that are conspicuous and almost appear to be temptations for fishes. Species of the bivalve *Lima* may have mantle tentacles that can be autotomized and are poisonous. Fishes bite such tentacles, spit them out, and then leave the animal unharmed. Unfortunately, there is no good evidence to my knowledge that the association of color with poison leads to avoidance learning in fishes.

Transparency seems common in planktonic and benthic animals that are vulnerable to predators. Many zooplankton are nearly transparent; reduction of pigmentation seems correlated with the presence of visual predators (Zaret, 1972). Scallops and fishes are usually pale colored on the ventral surface; this may serve as effective camouflage against predators looking upward for prey.

Escape Behavior

Many mobile animal species have evolved escape behaviors when confronted by predators. In Puget Sound the voracious starfish *Pycnopodia* elicits violent escape responses in sea cucumbers (violent undulation through the water) and some anemones (detachment and swimming; see Margolin, 1976). Many scallops rapidly swim away from starfish when the latter attempt to grip the valves. In some cases, escape responses seem only to follow direct contact (e.g., *Pycnopodia→Parastichopus*) whereas in others a water-soluble, diffusible chemical exuded by the predator (e.g., Saponins from starfish) can induce escape behavior in the prey.

Geographic Patterns of Chemical and Mechanical Defense

Current evidence suggests that the evolution of chemical and mechanical defenses seems best developed in the tropics. Bakus and Green (1974) found that the percentage of toxic sponges and holothurians is inversely related to latitude and can reach 100% in diverse coral reefs. Mechanical adaptations of gastropods to resist crushing similarly increase toward the tropics (Palmer, 1979). Vermeij (1977) showed that the arms race has progressed to a greater extent in Pacific coral reefs relative to the Caribbean. Pacific snails (Guam) and their crab predators are both more robust than their Caribbean counterparts.

It is not totally clear why these patterns exist. Vermeij (1977) argues that the differences represent fundamentally different evolutionary responses in different biogeographic realms. One possibility is that tropical, high-diversity realms are sites of more intense predation, a situation that would select for a greater frequency of poisonous and armored species. But as discussed in Chapter 16, predation can be devastating in low-diversity areas where few poisonous species are found. It is also possible that the environmental stability of high-diversity environments (see Chapters 5 and 19) permits selection for investment in defensive mechanical structures and chemical defense. High-stress, variable low-diversity environments may select for investment in reproduction. This hypothesis assumes a tradeoff between those adaptations that confer resistance to predators and those that permit rapid growth and devotion of energy to reproductive output.

SUMMARY

1. Benthic, nonparasitic species can be divided into epibenthic, infaunal, interstitial, boring, swimming, and commensal-mutualistic forms. We briefly discuss parasites.

2. Epibenthic species attach temporarily or permanently to the substratum surface. Water turbulence strongly influences morphology and behavior. In swift currents, epibenthic creatures must do one of the following:
   (a) adopt a short, squat profile;
   (b) hide in crevices;
   (c) have stout holdfasts and stalks; or
   (d) be flexible enough to bed with the current. Unidirectional currents permit an epibenthic creature to develop an optimal orientation that minimizes stress on delicate structures and yet maximizes the possibility of capturing suspended food.

3. Burrowers must displace sedimentary grains and obtain a purchase within the sediment. Soft-bodied burrowers first form a turgid penetration anchor. When it is formed, a terminal anchor is made by expanding a structure within the sediment. Once this purchase is obtained, the burrower can drag the rest of the body into the sediment. Hard-bodied crustacea use a lever-muscle system to dig into the sediment.

4. Boring invertebrates use a combination of mechanical and chemical means to enter the substratum. Such bivalves as shipworms scrape wood with the valves and digest the wood with the aid of cellulase.

5. Commensals and mutualists both derive benefit from another species. Commensals return no benefit to their host, however, whereas mutualists do. Such relationships are beneficial as
   (a) protection against predators or
   (b) provision of a suitable substratum or crevice.

6. A wide variety of parasites infect marine species. Ectoparasites live attached to the body while endoparasites live within the body. Parasites may have complex life cycles, with stages in more than one host species. Endoparasites are often strongly modified for existence within the body of the host. Parasites negatively affect their host by causing lesions, consuming tissues, or consuming fats that might otherwise be available as an energy reserve.

7. Benthic species have evolved various antipredator and antifouling devices. Spines in urchins and mollusks, thickened shells in mollusks, and other mechanical devices foil or discourage predation. Many species produce obnoxious chemicals, such as acid (seaweeds, gastropods), and halogenated organic compounds (seaweeds, sea whips). The percentage of such adaptations in the benthos increases with decreasing latitude. Mobile invertebrates also have evolved escape responses to elude mobile predators.

14

the diversity
of feeding adaptations
of benthic animals

A classification and description of feeding are essential in characterizing the ecological space occupied by a species. Such a task is complicated by the different types of feeding found in a single species (e.g., suspension feeding and deposit feeding by the bivalve *Macoma*, deposit feeding and scavenging by the mud snail *Nassarius*) and by the difficulty of classifying foods. (Is eroded and resettling detritus consumed by suspension feeders or detritus feeders?) These problems always arise if we do not acknowledge the flexibility of feeding habits and the difficulty of reconciling food source and mode of obtaining food in a feeding classification. Here we consider modes of obtaining food:

1. Suspension feeders
2. Deposit feeders
3. Herbivores
4. Carnivores
5. Scavengers.

14-1 SUSPENSION FEEDING

Particle Capture

We define suspension feeding as the collection of particles from the water column. Although modes of capture vary (see later), there is a great deal of convergence among many unrelated groups in particle-collecting ability because of a similar food source and similar hydrodynamic and mechanical constraints in collecting particles (Jørgensen, 1966).

269

Although we intuitively think of particle capture as a sieving process (as in calanoid copepods; see Chapter 10), actually, there is a range of possible mechanisms of particle interception by a biological filter (Rubenstein and Koehl, 1977). Sieve capture may, in fact, be rather rare as a particle-capture method.

Figure 14-1 illustrates the following different possible methods of particle capture:

1. Sieving
2. Direct interception
3. Inertial impaction
4. Motile particle deposition
5. Gravitational deposition.

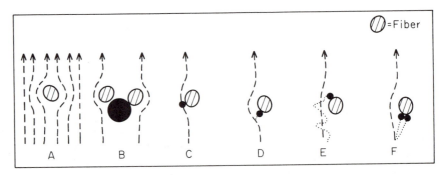

Figure 14-1 Possible means of capture of suspended particles by fibers. (a) Streamlines of flow around fibers; (b) Sieving; (c) Direct interception; (d) Inertial impaction; (e) Motile-particle deposition; (f) Gravitational deposition. (Modified from Rubinstein and Koehl, 1977)

Sieving involves trapping particles with diameters greater than interfiber distances. Direct interception occurs when a particle following hydrodynamic streamlines comes within a distance of one particle radius. Inertial impaction takes place when a particle deviates from the streamlines due to the particle's own inertia. Motile particle deposition on fibers occurs because small particles move randomly about relative to streamlines and may come within one particle radius of the fiber. Finally, particles denser than the fluid sink as they move and may deviate from streamline into the fiber.

Note that simple sieving is by no means the only possible mode of particle capture by a series of fibers. Suspension feeders may capture particle sizes of diameters much smaller than the distance between two fibers. On the other hand, fibers might spread apart as a particle is trapped and lose the particle. So two adjacent fibers might lose a particle somewhat greater than the distance between the two fibers.

Rubenstein and Koehl (1977) report calculations of the probable relative importance of different modes of particle capture other than sieving. For a variety of suspension feeders (echinoderm larvae, rotifers, brittle stars, some sea anemones), direct interceptions are the hydrodynamically most likely mode of particle capture by fibers. La Barbera (1978) investigated suspension feeding by the Pacific brittle star *Ophiopholis aculeata* and found that the animal captured particles much smaller than the average distance between the tube feet. These data suggest that suspension feeding by benthic invertebrates may involve processes other than simple sieving.

Some of the more common modes of suspension feeding are described next. In many cases, a suspension-feeding organ may be likened to a series of fibers, as discussed earlier. The analogy, however, does not always hold.

Modes of Suspension Feeding

Mucous-bag and mucous-sheet suspension feeding.   Particles are directed toward, and collected on, a mucous sheet. The polychaete *Chaetopterus* resides in a U-shaped tube and creates a current that is passed across a sheet of mucus stretched between a pair of specialized parapodia (Fig. 14-2). The sheet fills up with particles and is rolled into a bag that is passed to the mouth. This procedure provides an efficient means for trapping all particles that enter the tube (MacGinitie, 1939a).

Ascidians use a mucous sheet to trap particles as small as 1 µ (MacGinitie, 1939b). Water enters an inhalent siphon, across a branchial basket studded with pores. Mucous sheets across the basket's inner surface trap particles and cilia-aided tracts move mucous strings toward the esophagus. Without the mucous sheets, the pores would trap only larger phytoplankton (Werner and Werner, 1954); feeding, however, is essentially nonselective when mucus is employed.

Ciliary-mucous mechanisms.   In this type of feeding, mucus is also used to entrap suspended particles. But various combinations of cilia arranged in rows permit selectivity in particle feeding by excluding particles that are too large and, to some extent, ones that are too small. Such feeding has been adopted by a wide variety of benthic suspension-feeding groups, including bivalve mollusks, gastropods, suspension-feeding polychaetes like serpulids and sabellids, brachiopods, and bryozoa.

The polychaete families Serpulidae and Sabellidae (Fig. 14-2) feed with a crown of gills containing filaments equipped with cilia that drive water through the gill crown, transporting particulate matter toward the mouth (Nicol, 1930; Jørgensen, 1966). The filaments forming the branchial crown are arranged at regular intervals while the crown is fully extended during feeding. Water currents are produced by cilia arranged on the pinnules, resulting in eddies that move freshwater over the frontal ciliary tracts. Particles collect on the tract and are transported toward the base and, finally, the gill folds. Large particles are rejected whereas medium-sized particles are passed between gill folds and are eventually transported to a storage area for particles to be used in tube formation. The finest particles pass straight down between the folds and wind up in the basal grooves, which carry fine particles directly to the mouth. High concentrations of particles provoke the production of mucus, which is used to entrap the dense particles and transport them to rejection tracts.

Within the bivalve mollusks, many ctenidial (feeding gill) types have evolved to

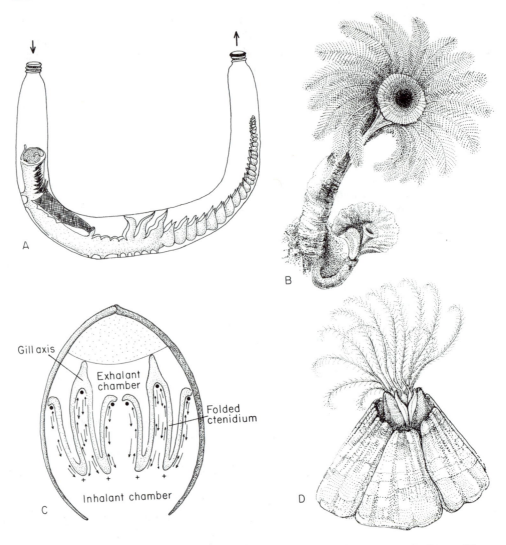

Gill axis

Exhalant
chamber

Folded
ctenidium

Inhalant chamber

A

B

C

D

Figure 14-2 Some suspension-feeding invertebrates. (a) The polychaete *Chaetopterus;* (b) serpulid polycheate *Serpula;* (c) cross section showing ctenidium (gill) of bivalve mollusk; arrows and crosses denote ciliated tracts transporting particles; (d) acorn barnacle *Balanus.*

deal with suspended matter (see Fig. 14-2). The many adaptations for suspension feeding in bivalves cannot be discussed in detail here; we refer the reader to the work of D. Atkins (summarized excellently in Jørgensen, 1966) for details of ciliary anatomy and to Jørgensen (1966) for a summary of different bivalve feeding types. Suspension-feeding bivalves utilize the mantle, gills, and mouth palps to collect, sort, and reject particles.

Setose suspension feeding.   Currents created by the movements of pairs of thoracic limbs are drawn through a series of filters formed by limbs equipped with setae. This type of feeding is limited to marine arthropods: acorn and stalked barnacles, amphipods, and other small crustacea. Crisp and Southward (1961) have studied the method of feeding in acorn barnacles (Fig. 14-2). Six pairs of cirri on the head and thorax beat and create a water current; the mouth and mouth parts are located between the bases of the first and second cirri. The cirri are extended and withdrawn rapidly. During this withdrawal motion-suspended matter is caught on the setae of the cirri and then scraped off by specially adapted setae on cirri closer to the head. Particles are passed to the mouth.

Tentacle-tube feet suspension feeders.   A tentacular structure (see Fig. 13-3) entraps food. Mucus may be employed to facilitate the capture of the particle, which is usually large—on the order of a zooplankter. Stalkless comatulid crinoids use tube feet in conjunction with ciliary grooves that carry particles toward the mouth. This kind of feeding is found in crinoids, some brittle stars, and many types of zooplankton-feeding anemones and corals. In the crinoid *Antedon difida,* pinnules bear tube feet in groups on each side of the food groove. Tube feet beat toward the food groove when plankton are added to the water. Particles are entangled in mucus secreted by tube feet and then carried to the food groove by shorter tube feet. In the food groove, ciliary currents carry these particles trapped in mucus toward the mouth. Zooplankton are a major part of the diet (Meyer, 1973).

Spionid polychaetes feed with the aid of two elongate tentacles clad with mucus. When bottom currents are sufficiently strong to saltate particles along a mud flat, spionids deploy the tentacles in an erect helical coil; particles are entrapped with the aid of the sticky mucus. When water currents are slight, the tentacles are used for deposit feeding (Taghon et al., 1980).

Other examples of tentacle feeders are anemones and corals that use various types of stinging, entrapping, and mucus-laden nematocysts to entrap prey. It is difficult to define where suspension feeding on small zooplankton, such as in the genus *Metridium,* ends and carnivory begins—as in those anemones like the Pacific anemone *Anthopleura xanthogrammica* that feed on large prey, such as mussels.

Sponges.   Sponges are a large homogeneous group of suspension feeders adapted to feeding on very small particles. Sponges can trap particles of diameters as small as 0.2 μ (Jørgensen, 1966, pp. 3 and 4). Sponge morphology varies greatly, but basically particles pass through the pores of the surface epithelium and eventually reach flagellated chambers that provide the driving force for water movement through the sponge. Only the smallest particles can pass through the prosopyles, larger particles having been effectively filtered out before reaching the flagellated chamber. Flagellated cells collect particles on a structure known as the collar; particles are then transferred to phagocytic archaeocyte cells where digestion occurs. Exhalent water is then expelled into an excurrent canal. A group of excurrent canals are usually organized into still larger canals that eventually reach the exhalent chamber or osculum. There may be more than one osculum on a sponge, but its function is to expel water rapidly away from the sponge so that only freshwater will reach the feeding surfaces.

Passive suspension feeding.   Suspension feeders may take advantage of velocity gradients above the bottom to facilitate feeding currents. With laminar flow, as discussed, water velocity decreases as the bottom is approached. A vase-shaped sponge might take advantage of this gradient as follows. The relatively stronger current passing over the top of the vase (excurrent opening) creates a pressure differential such that water is drawn into the incurrent openings on the outside of the vase, closer to the substratum. Therefore the flow through a vase-shaped sponge may be facilitated by flagellated cells and a passive flow mechanism (Vogel and Bretz, 1972). Many encrusting sponges and ectoprocts have regularly spaced chimmeylike structures which probably are deployed to facilitate passive expulsion of waste water.

Food of Suspension Feeders

Selectivity and feeding rate.   Although mucous-bag suspension feeding seems unspecialized with regard to particle size, most other types select for a narrow range of diameters. In crustacea, intersetal distance may determine (see Chapter 10, Section 10-5) the upper and perhaps the lower limit of ingested particle sizes. Ciliary-mucous feeding employs a series of ciliary tracts to select and reject particles. Apparently particle retention efficiency drops at small particle diameters whereas large particles are rejected from tracts leading to the mouth.

Rates of suspension feeding can be measured in two ways: (a) successive sampling of particles in chambers with suspension feeders and (b) measurement of particle density near the inflow and outflow of a suspension feeder. Particle number can be estimated with direct counts of diluted volumetric samples of the chamber or by use of a counting device, such as a Coulter counter. This instrument detects changes in conductivity in a small volume of electrolyte when a particle enters and displaces some of the electrolyte. The change in conductivity is proportional to the particle's volume.

Particle concentration affects feeding rate. The rate of particle ingestion increases with particle concentration to a plateau above which ingestion rate no longer increases (see Chapter 10, Section 10-5, for details). Below this plateau, digestion efficiency per particle (e.g., per diatom) may drop off, for the rate or particle transport through the gut is too fast for efficient digestion. At very high concentrations of particles, suspension-feeding organs are usually easily clogged and feeding ceases or is inhibited. Thus particle size and concentration both have optimal ranges for suspension feeders.

Nutritive sources.   The quality of suspension-feeding food is difficult to determine because of the diverse nature of particles ingested by most suspension feeders. Selection is usually for particle size rather than particle type if we exclude coelenterates feeding on small zooplankton. Suspension-feeding food includes the following possible sources:

1. Phytoplankton, ranging from diatoms a few hundred microns across to flagellates only a few microns in diameter.
2. Bacteria suspended in seawater, $< 5\ \mu$ in diameter.
3. Bacteria and other microorganisms living on inorganic particles and particulate organic detritus delivered by water currents.
4. Bacteria and other microorganisms living on particles resuspended from the bottom (Rhoads, 1973).

Clearly groups like sponges are adapted to feeding on particle sizes corresponding to bacteria. But the majority of ciliary-mucous modes of suspension feeding seem adapted to size ranges that include common phytoplankton. Probably the great majority of benthic suspension feeders living outside estuarine areas or zones adjacent to seaweed or seagrass beds feed principally on phytoplankton. Jørgensen (1966) discusses a series of studies demonstrating that phytoplankton diets are capable of supporting vigorous growth in a great variety of suspension-feeding groups. Bacteria seem insufficiently abundant to be important to macrofaunal suspension feeders.

Nutritional differences exist among species of phytoplankton. Walne (1963) demonstrated that veligers of the British oyster *Ostrea edulis* grew successfully on only 14 of 19 algal species tested. Some evidence suggests that diets of mixed algal cultures produce better growth than monospecific algal cultures. Digestion efficiencies can be extremely high, surpassing 90% (Marshall and Orr, 1955). Algae with tough and thick cellulose walls are generally less digestable than naked phytoplanktonic algae (Jørgensen, 1966).

Dissolved organic matter.   The role of dissolved organic matter in the nutrition of suspension feeders has been a subject of enduring controversy (Jørgensen, 1976). Typical oceanic values of DOC range from 0.4 to 1.0 mg C liter^{-1} but may reach 8 mg C liter^{-1} inshore. Pelagic algae and benthic seaweeds excrete copious quantities of DOC.

Stephens and coworkers have demonstrated that nearly all invertebrate phyla are capable of absorbing dissolved organics. Amino acids and sugars are taken up by 12 invertebrate phyla, but not by arthropods (Stephens and Schinske, 1957; 1961; Stephens, 1964; Stephens, 1975). In some cases, excretion of carbon occurs at rates comparable to uptake (Johannes et al., 1969; Webb et al., 1971). Polychaetes and bivalves, however, can be shown to be net absorbers of dissolved amino acids (Stephens, 1975; Jørgensen 1976).

Using measurements of uptake rates, ambient concentrations, and metabolic requirements, Jorgensen (1976) estimated the role that dissolved organic matter in seawater might play in the energy budgets of marine animals (Table 14-1). The polychaetes *Nereis* and *Capitella,* inhabiting sediments rich in organic matter, seem able nearly to match their energy requirements. But this source is probably minuscule relative to microbial and other particulate organic matter available in sediments. In contrast, epifaunal and planktonic species can only cover a small fraction of their requirements from normal seawater.

TABLE 14-1   THE ROLE OF DISSOLVED AMINO ACIDS IN THE ENERGY BUDGETS
OF SOME AQUATIC ORGANISMS[a]

Species	Type	Percent oxygen requirements accounted by uptake
Aurelia aurita polyps	epifaunal coelenterate	10
Tetrahymena pyriformis	protozoan	0.5–20
Mytilus edulis	epifaunal mussel	9–30
Strongylocentrotus purpuratus embryo	sea urchin	15–40
Capitella capitata	infaunal polychaete	60–90
Nereis diversicolor	infaunal polychaete	80

[a]After Jørgensen, 1976.

Particulate organic matter.   Another important source of possible suspended food is organic detritus. Several studies show that the detritus itself has little or no nutritive value for suspension-feeding organisms and will not maintain them in the laboratory. Detritus with attached bacteria and algae living on surfaces, however, may be of great importance, particularly over seagrass beds and in estuaries where microorganism-rich organic detritus is abundant.

Specialization.   Several questions about suspension feeding remain, one being the ability of suspension feeders to specialize on given food sources relative to the spectrum of suspended matter in the water column. At present, available evidence suggests that suspension feeders by and large are only limited by the particle-size spectrum within which they can feed. Qualitative specialization on bacteria coincides with feeding specialization on small particles in sponges. A number of studies suggest that suspension-feeding gut contents usually mirror the spectrum of particles in the surrounding water. This result is expected because benthic suspension feeders attached to a substratum can only feed on whatever the overlying water presents to them. The great fluctuation in concentrations of various types of particles favors a generalized mode of feeding so that any type of food would be potentially available for ingestion and subsequent digestion. Even in phytoplankton-rich water the great variety of phytoplankton species and phytoplankton succession causes variation in food quality and quantity. So we would not expect to see qualitative specialization among suspension feeders except where their anatomical peculiarities preclude feeding on given particle sizes.

Because of the great fluctuations of quality and quantity of suspended food in the overlying water, we might also suspect that competition for food among suspension-feeding species would rarely result in niche subdivision explained primarily by food specialization. This would lead to the prediction that sympatric species of suspension feeders would not feed on different but contiguous subsamples of the available size spectrum of suspended food in the water. Unfortunately, data supporting or falsifying this prediction are not available in any complete way. A study by Ryland (1975) on sympatric species of bryozoa, however, does tend to support this point of view. Sympatric bryozoan species have rather surprisingly similar lophophore diameters, suggesting a similar means of exploitation of food. Sympatric species of bryozoa have thus converged on feeding exactly the same way; this is hardly indicative of competitive exclusion leading to morphological niche subdivision. More impressive is the degree of convergence among unrelated suspension-feeding groups as opposed to divergence. Jørgensen (1959) summarized data on feeding rates of sponges, bivalve mollusks, and ascidians. With a few exceptions, all fed at rates of the same approximate order of magnitude. Turpaeva (1954) suggests the possibility of stratification of suspension feeders above the bottom. There may be some basis for this claim, for hydrodynamic properties of water within a few millimeters and several centimeters above the interface may be quite different and may support different types of food. Within a few millimeters of the bottom, food might be resuspended from the bottom due to turbulent interaction of currents with the bottom.

14-2 DEPOSIT FEEDING

We define deposit feeding as feeding on particles in soft sediments, usually several at a time or in short succession, on or in the bottom by either appressing or aiming a feeding organ against or toward the substratum. This definition obscures the details of size, origin, and qualitative differences among particles, which is justifiable because many macrofaunal deposit feeders consume a wide variety of particles, and (as discussed later) effective comparisons can be made between species that are specialized to different sediment types. We avoid an exact definition of the term particle. Ingesting particles several at a time or several in short succession implies the exclusion of macroalgal feeders.

Modes of Deposit Feeding

Swallowers.   Swallowers ingest sediment many particles at a time with little particle size or qualitative selectivity. In some cases, a mouth is applied directly to the sediment. Polychaetes like *Arenicola*, snails like *Ilyanassa obsoleta,* spantangoid urchins, and sipunculids feed in this way. The lugworm *Arenicola marina* resides in a tube, as described earlier, and feeds through swallowing with its protrusible muscular proboscis (see Fig. 13-8c for a related form). Because the sediment usually contains 2% organic matter or less, the organism processes large amounts of nutritionally valueless material in order to obtain food. Consider the possibility of eating a couple of ounces of peas diluted in about 6 pounds of sand.

Tentacle feeders.   This subdivision includes those organisms ingesting sediments that do not directly apply the mouth to the sediment–water interface or within the sediment during burrowing but use a tentacular structure to gather detrital particles and transport them to the mouth. They include sea cucumbers, nuculoid bivalve mollusks, and some tube-dwelling polychaetes that gather detritus by means of tentacles at the sediment–water interface (Fig. 14-3).

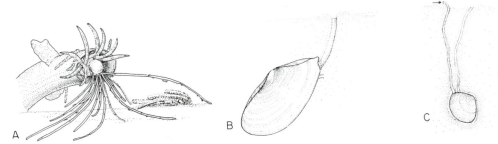

Figure 14-3   Deposit-feeding invertebrates. (a) The polychaete *Hobsonia;* (b) the nuculanid bivalve *Yoldia;* (c) tellinacean bivalve *Macoma*. (Drawing of *Hobsonia* copied from original by P. A. Jumars)

Surface siphon feeders.   Unlike the first two types of deposit feeding, this feeding is restricted to bivalve mollusks of the superfamily Tellinacea. The siphons are separate, and the inhalent siphon is adapted to ingesting sediments like a vacuum cleaner (Fig. 14-3). Deposit feeding is not common to all the Tellinacea, as originally suggested by Yonge (1950), but is restricted to a few genera, including *Tellina, Macoma, Scrobicularia,* and *Abra.* Suspension-feeding tellinaceans usually have protective siphonal tubercles or tentacles that keep out large particles typically ingested by deposit feeders (Pohlo, 1969). The behavior of the inhalent siphon varies greatly, from darting motions with frequent contacts with the sediment–water interface to very slow and deliberate extensions of the siphon with subsequent withdrawal toward the siphon hole. This mode of feeding is well suited for consumption of microalgae living on or near the sediment–water interface.

Setose deposit feeding.   A number of benthic crustacea move appendages equipped with setae to trap sedimentary particles. The amphipod *Corophium volutator* and the cephalocarids are good examples. *Corophium* resides in a tube or burrow and scrapes surface detritus in its burrow with its gnathopods (Fig. 13-8). The animal retreats in the burrow and draws in particles from this material in a current created by the pleopods. Long plumose setae on the first and second gnathopods first form a basket that traps particles and then a combing structure that rejects unsuitable material and transports suitable material to the mouth via other appendages (Meadows and Reid, 1966).

Other modes.   A number of species of marine benthic invertebrates primarily adapted to suspension feeding seems to ingest the same food as that ingested by deposit feeders. This is particularly true of those suspension-feeding organisms that live near large sources of current-resuspended organic detritus, such as within estuaries or adjacent to salt marshes. The ribbed salt-marsh mussel, *Guekenzia demissa,* is primarily adapted to ingesting suspended matter in seawater, but this suspended matter consists of great amounts of finely divided organic detritus derived from marsh grass, *Spartina.*

The Food of Deposit Feeders

Sediments are complex mixtures of inert mineral grains, particulate organic matter, and microorganisms. Current evidence suggests that deposit feeders efficiently digest and assimilate the microbial organisms living among and attached to sedimentary particles. Generally, however, they are far less efficient with particulate organic matter normally found in sediments. Particulate organic matter usually consists of such compounds as structural carbohydrates (derived from cord grass—*Spartina;* eelgrass—*Zostera;* turtle grass—Thalassia, etc.) that are too refractory to be broken down by the digestive enzymes available to most deposit feeders (e.g., Newell, 1965; Hargrave, 1970a; Fenchel, 1970). Some particulate organic matter derives from nitrogen-rich seaweeds and may be directly utilized to some degree (Tenore et al., 1979). Deposition of dead diatoms and copepod fecal pellets during the spring diatom bloom might supply digestible particulate organic matter to deposit feeders living in shallow-water muddy bottoms.

If microbial organisms provide most food for deposit feeders, then particulate organic matter must be converted by microbes before it is nutritionally useful. Nitrogen nutrition strongly depends on microbial production. Most particulate organic matter would be nitrogen deficient without a microbial community (Newell, 1965; Fenchel, 1970). As organic matter ages, the establishment of a rich microbial community decreases the overall carbon/nitrogen ratio of the organic fraction of the sediment.

Even if the percentage assimilation of carbon from nonliving organic matter is low, the role of POM in deposit feeders nutrition may be significant. If the carbon in sediments is 99% nonliving, then even a 100% assimilation efficiency will yield only 1% return when sediment is ingested. But a mere 1.01% assimilation efficiency on nonliving carbon will gain the same amount of nutrition! This problem is significant because current techniques preclude the measurement of assimilation of just a few percent. Furthermore, estimates of microbial standing crop, based on the ATP content of sediments suggest that there is less than 3% living carbon in sediments (Ferguson and Murdoch, 1975). There is even some question as to whether microbial carbon would be sufficient to satisfy the energy requirements of deposit feeders (Tunnicliffe and Risk, 1977; Cammen et al., 1978).

Feeding Rate

Feeding rate is normally estimated as egestion rate of fecal material, an accurate measure of ingestion rate (Hargrave, 1972). Because the assimilated microbial fraction contributes little to the weight loss of the sedimentary material passing the gut, this method is justified. Four factors can significantly affect feeding rate: (a) food quality, (b) degree of starvation, (c) fraction of the sediment available for feeding, and (d) population density. Deposit feeders generally do not feed or feed very slowly on sediments devoid of microbial colonists. *Hydrobia totteni* will feed on glass beads only when they have been incubated with a microflora for a few days (Lopez and Levinton, 1978). The polychaete *Pectinaria gouldii* can adjust its sediment ingestion rate to the nutritive value of the sediment (Gordon,

1966). Calow (1975) measured egestion in two freshwater gastropods and found a decrease in egestion rate when the snails were starved. This decrease was explained as an attempt by the snail to slow passage of food through the gut and thereby increase digestive efficiency. Yet Hargrave (1970b) found no relation between assimilation efficiency and egestion rate in the amphipod *Hyalella*.

Despite potential variation in feeding rates, some generalizations can be made across a wide variety of deposit-feeding organisms. Sediment ingestion rate, as estimated by surface area of fecal pellets, shows a strong correlation to animal dry weight for a wide variety of unrelated taxonomic groups (Hargrave, 1972). Furthermore, the rate of ingestion of organic matter is also closely related to body size across a wide spectrum of deposit-feeding species; bivalves are excluded (Cammen et al., 1978). The ingestion rate of organic matter is also relatively independent of the concentration of organic matter in the food. Thus it seems that deposit feeders have evolved to feed at rates consonant with the procurement of a sufficient ration from the sediment.

Coprophagy and the Nutritive Value of Fecal Pellets

Invertebrate fecal material is abundant in marine bottoms and is believed to play an important nutritive role in benthic communities (Moore, 1931; Frankenberg et al., 1967). Earlier we discussed the significant capacity for biodeposition of both zooplankton and benthic pellets. Relative to carbon production, pellets from the shrimp *Callianassa* can provide a carbon source up to ca. 30% of pelagic diatoms (Ragotzkie, 1959) and 10% of benthic diatoms (Pomeroy, 1959) in Georgia salt marshes.

When no food is available, some deposit feeders have been observed to ingest their own pellets (*Hydrobia ulvae*, Newell, 1965; *Palaemonetes pugio*, Johannes and Satomi, 1966). Doubly digested fecal material was avoided by the shrimp *P. pugio* until it had aged and presumably been recolonized by microorganisms.

Coprophagy would seem a poor approach to feeding, given that all extractable food has been digested as the material passes the gut. In *P. pugio* two passes through the gut render the detritus food temporarily without value. Lopez and Levinton (1978) fed *H. ventrosa* sediment particles < 10 μ in diameter and measured a digestion of about 50% with respect to diatoms. Whole pellets were not reingested unless they were disrupted into constituent sedimentary particles. Although *H. ventrosa* fed on them, assimilation was negligible despite a diatom density of 5×10^5 cells/mg^{-1} dry weight sediment. Avoidance of fecal pellet reingestion may therefore be an adaptation to allow recolonization of pellets by the microflora. Recolonization by bacteria takes less than 3 days (Hargrave, 1976; Newell, 1965), but complete diatom recolonization might take 10 days or more (Fenchel and Kofoed, 1976; Levinton and Lopez, 1978). The inability of *H. ventrosa* to gain immediate nutrition from its own pellets does not mean that other species might not efficiently extract the remaining microbiota.

The Particle-Size, Body-Size Paradigm

One of the potentially most important observations made about deposit feeders is the positive correlation between feeding on larger particles and increased body size. This capacity might result in the evolution of body-size differences among coexisting species to relax competition for given particle-size classes (Fenchel, 1975a,b).

Although several studies document a particle-size versus body-size relationship (e.g., Fenchel, 1975a,b; Whitlatch, 1974), few have elucidated the mechanism that permits differential particle-size selection. Hughes (1973) investigated feeding by the tellinacean bivalve *Abra tenuis* on sand grains. The diameter of the inhalent siphon was linearly related to shell length. Correspondingly, inhalent siphon diameter was linearly related to particle diameter (Fig. 14-4). No further sorting was accomplished in the mantle cavity or on the palps and the animals could not distinguish among grains of identical size but different specific gravities. A deposit-feeding ampharetid polychaete can distinguish between particles of differing specific gravities; the capacity for selectivity increases with increasing body size (Self and Jumars, 1978).

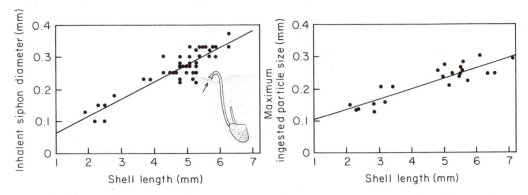

Figure 14-4 (a) Shell length versus siphon diameter in the deposit-feeding tellinacean bivalve, *Abra tenuis*. (b) Particle size ingested versus siphon diameter. (After Hughes, 1973, with permission of the Zoological Society of London)

The possible importance of food quality was investigated by Whitlatch (1974) in the deposit-feeding ice cream cone worm *Pectinaria gouldii*. The polychaete is common in New England soft sediments and is responsible for considerable sediment turnover (Gordon, 1966). Worms showed an approximately linear relation between body length and mean particle diameter. Positive selection, however, was shown for fecal material, floc aggregates, and mineral grains encrusted with organic matter. Through selective feeding, worms concentrated an average of 32.7% possible organic matter found in the sediment to an average of 42.7% in gut contents.

Although a relationship between body size and particle size seems general for many deposit feeders, strong variation in particle ingestion is to be expected in natural sediments. If only fine particles are available in the sediment, then large bodied creatures will, of necessity, likely be ingesting the same sized particles as small creatures. Similarly, if small creatures are in sediments of large grain size, a switch of feeding mode might be possible. For example, mud snails of the genus *Hydrobia* are active swallowers of sedimentary grains. Snails can scrape the microflora from the surface of large grains as well (Lopez and Levinton, 1978).

Is there an optimal particle size for deposit feeders? The relationship of body size to ingested particle size raises the question of whether there is an optimal particle size for deposit feeders. The knowledge of the optimum would predict baseline of body size from which we would expect evolutionary divergence under competitive pressure.

Unfortunately, such a calculation is not simple for two alternative particles could be imagined. First, feeding on inorganic sand grains "coated" with microflora is common in deposit feeding. Because the surface area/volume ratio of a sphere decreases with increasing particle diameter, a unit volume of large particles will have less surface area and hence less attached microflora than a unit volume of smaller particles (see Levinton, 1980; Taghon et al., 1978). So it would be advantageous to consume volumes of small particles. What if the edible morsels consist of cells of microalgae, however? Then it would make sense to eat larger particles, especially if the energy invested in feeding is proportional to the number of particles seized and ingested. Because most sediments are a mixture of whole edible morsels and indigestible particles with surface-bound microbes, it is difficult to make a sweeping prediction.

14-3 HERBIVORE BROWSERS

Herbivore browsers graze algae or marine grasses from the surface of hard substrata or directly consume plant material. Many herbivore browsers are capable of consuming animals as well as plants. The grazing urchin *Arbacia punctulata* usually feeds on macroalgae and microalgae living on rock and wood piling surfaces. At Beaufort, North Carolina, however, where macroalgae are rare, epifauna are the major diet of this species (Karlson, 1978). The mud snail *Hydrobia* often deposit feeds on fine particles. Nevertheless, it can readily feed by scraping microalgae and smaller macroalgae off rock and sand grain surfaces.

Scrapers and Chompers

Gastropods and chitons. Chitons and many gastropods use a ribbon of chitinous teeth, or radula, located in the posterior of a muscular buccal mass. Muscular action everts the buccal mass and presses the radula in contact with the substratum (Fig. 14-5a). The movement of the subradular membrane over a cartilaginous portion of the buccal mass causes the teeth to become erect and scrape food from the surface. As the teeth point posteriorly, the effective stroke occurs as the radular belt moves backward. The radula and the buccal mass are then retracted and food trapped on the teeth is thus delivered to the buccal cavity.

Arrangements of radular teeth and the musculature drawing the radular belt vary and may be highly specialized. In the macroalgal and microalgal grazing periwinkle *Littorina littorea*, the radular median (rachidian) teeth and lateral teeth are well developed and scrape food off the substratum or rasp the surface of such algae as *Ulva*. In the

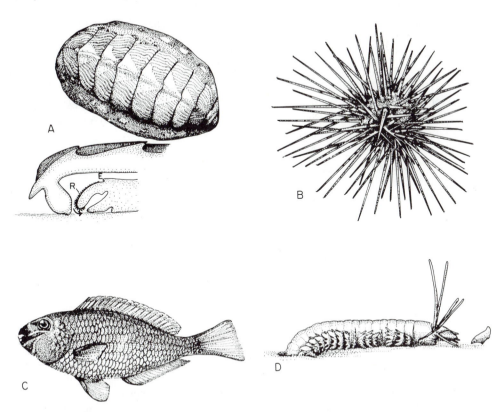

Figure 14-5   Benthic herbivores. (a) Chiton *Tonicella;* inset shows anterior sagittal cross section and illustrates action of radula (arrow) in scraping food from substratum (R = radular tooth belt, E = esophagus); (b) sea urchin *Arbacia;* (c) a scarid fish (parrotfish); (d) nereid polycheate *Nereis vexillosa*. (Copied from original by K. Fauchald)

abalone, *Haliotis,* lateral teeth brush food off of the substratum, but other teeth merely transfer particles into the buccal cavity (Fretter and Graham, 1962). This radular structure is too weak to feed on anything more than algal films on rock surfaces. In the limpets *Patella*, extensive muscular adaptability permits drawing of the entire radular belt across the substratum, causing an effective rasping action, without the conveyor beltlike movement described. A crude jaw system is present in some herbivorous forms (e.g., the opisthobranch *Aplysia*), permitting pieces of seaweed to be seized while rasping takes place.

Urchins.   Urchins possess an *Aristotle's lantern*, a complex of teeth and ligaments operated by muscular action and capable of powerful rasping and tearing of macroalgae. Food is conveyed to the mouth by tube feet. The urchin *Tripneustes ventricosus* is capable of consuming turtle grass, *Thalassia,* whereas the genera *Arbacia* and *Strongylocentratus* are active consumers of seaweeds on intertidal and shallow subtidal rocks (Fig. 14-5b).

Fishes.   Many fishes scrape algae off the surfaces of rocks and coral skeletons. Members of the family Scaridae (parrot fishes) have jaw teeth fused into plates capable of cutting material from the surface of coral skeletons. Parrot fishes are active eroders of coral reef material (Fig. 14-5c). Acanthurids (surgeon fishes) are also active coral reef algal grazers (see Chapter 20).

Crustacea.   Many crustacea consume algae by triturating fronds with mouth parts. Such isopods as *Ligia* cut up seaweeds with the mandibles. Gammarids like *Orchestia* feed on algae and algae fragments by tearing them off with maxillipeds.

Polychaeta.   Many polychaetes consume fleshy algae; species of the genus *Nereis* can tear material with chitinous teeth (Fig. 14-5d). Some species attach pieces of sea lettuce (*Ulva*) to their tubes and maintain algal gardens.

More Unusual Forms of Herbivory

Wood feeders.   Wood-boring invertebrates have the ability to digest wood or depend on the marine microbiota living in wood. The wood-piling bivalves *Teredo* and *Bankia* scrape wood particles and use the digestive enzyme cellulase to attack the cellulose. The wood-boring isopod *Limnoria* is similarly capable of hydrolyzing cellulose (Ray, 1959). *Limnoria* also requires wood-boring fungi for nutrition and will not survive in wood stripped of fungi. Fungi are the major source of nitrogen (Schaefer and Lane, 1957).

Cellulose feeders.   Like wood, cellulose per se is digested by few organisms. Cellulose is typically beyond the digestive capacities of most invertebrates and fishes. However, the green turtle, *Chelonia mydas,* can digest cellulose derived from sea grasses (e.g., *Thalassia*) with the aid of postgastric fermentation, analogous to the digestion of ruminants (e.g., cows). Symbiotic bacteria and protozoa facilitate digestion (Fenchel et al., 1979).

Symbiotic autotrophs.   Many marine heterotrophs maintain autotrophic algal symbionts and garden them as a source of nutrition (Smith et al., 1969). Two means of harvesting are used: (a) absorption by the animal of a photosynthate, such as carbohydrate, and (b) actual digestion of the algal symbiont. In almost all known cases, the symbiont is obtained by ingestion and is transported to a specified site (Table 14-2). The opisthobranch gastropod *Elysia viridis* uses a modified radula and pumping pharynx to puncture the alga *Codium tomentosum.* It obtains and transports chloroplasts to cells of the hepatic tubules. Zooxanthellae are derivatives of dinoflagellates (see Chapter 20) and may be obtained by ingestion. In the giant clam *Tridacna,* zooxanthellae are maintained in blood

TABLE 14-2 SOME EXAMPLES OF SYMBIOTIC ASSOCIATIONS BETWEEN ALGAE
(AUTOTROPHS) AND INVERTEBRATES (HETEROTROPHS)

Autotroph	Heterotroph	Location of autotroph	Mode of infection
Dinoflagellate, diatoms green algae	foraminifera	cytoplasm (usually)	?
Platymonas convoluta	Convoluta roscoffensis (flatworm)	intracelluar, subepidermal	ingestion
Codium chloroplasts	Elysia viridis	hepatic tubule cells	ingestion
Zooxanthellae	scleractinian corals	endoderm, intracellular	ingestion (?)
	gorgonians	endoderm, intracellular	ingestion (?)
	anemones	endoderm, intracellular	ingestion (?)
	Cassiopeia (benthic scyphozoan)	mesoglea	?
	Tridacna (bivalve mollusk)	blood sinuses	?

sinuses and the animal is highly specialized to expose the sinuses to the light. Senescent zooxanthellae are removed and eventually intracellularly digested by amoebocyte lysosomes (Fankboner, 1971).

14-4 CARNIVORES AND SCAVENGERS

Carnivores seize and capture animal prey (Fig. 14-6). Many are scavengers when live food is not abundant. Because of the diverse means of prey capture, we treat important groups by taxon.

Great variation exists in the prey specialization of carnivores. This fact is effectively illustrated by feeding preferences of asteroid starfishes. Some like the North Pacific *Mediaster aequalis* feed on a wide variety of animal prey but may be herbivores, deposit feeders, or suspension feeders as well (Mauzey et al., 1968). In contrast, Birkeland (1974) found that the sea star *Hippasteria spinosa* consumed only the sea pen *Ptilosarchus*.

Prey handling and capture time may influence patterns of carnivory. Generally the size of invertebrate prey increases with the size of the predator (Kohn and Nybakken, 1975; Menge and Menge, 1974; Birkeland, 1974). As a predator grows, its prey-size

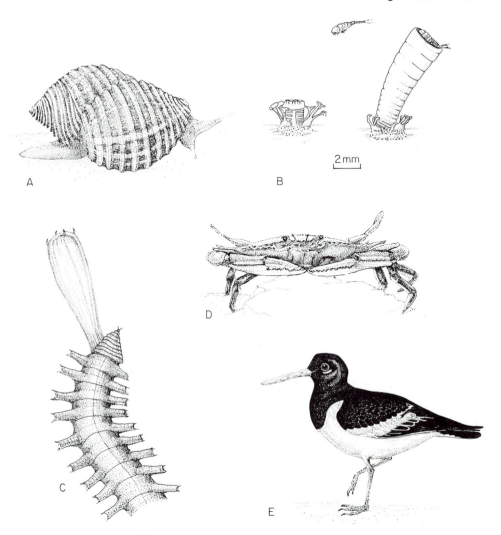

Figure 14-6   Marine benthic carnivores. (a) Drilling gastropod *Thais;* (b) bivalve mollusk *Cuspidaria,* trapping a small crustacean; (c) polychaete *Glycera;* (d) decapod crab *Callinectes sapidus;* (e) oyster catcher *Haematopus ostralegus.*

preference may shift it to different, larger prey species (e.g., Kohn and Nybakken, 1975). But capture time and consumption time also increase with prey size, yielding a size above which diminishing returns are gained. The advantage in consuming large prey must therefore lie in a proportionately greater food return relative to the increase in time needed to consume the prey. There is evidence for selective consumption of larger invertebrate prey than expected from a random sampling of the population (Connell, 1961b; Birkeland,

1974). But prey may reach "size refuges" where size is too great for the predators' capturing capability (e.g., Paine, 1976). See Chapter 3 for a discussion of optimal foraging theories.

Protozoa

Sarcodina, such as foraminifera, are omnivorous and trap prey on sticky mucous-laden pseudopodia. Engulfment by pseudopodia is often accompanied by secretion of proteolytic enzymes, allowing feeding on other protozoa, algae, and even small interstitial metazoa. Many interstitial ciliates are capable of engulfing relatively large prey with an anterior, distensible mouth (e.g., the freshwater *Didinium*). Suctorian ciliates are attached to particles and have tentacles capable of trapping and paralyzing small prey.

Coelenterates

Most anemones, scleractinian corals, and other benthic coelenterates are carnivorous, trapping prey on tentacles armed with nematocysts specialized for puncturing, ensnaring, or trapping with mucus. Contact with tentacles and extracts of the prey usually triggers nematocyst firing and drawing of tentacles toward the oral opening (see Pantin and Pantin, 1943). Prey are passed into the oral opening. Scleractinian corals generally consume small prey, such as zooplankton, whereas larger anemones like *Anthopleura xantho-grammica* may consume and efficiently digest bivalves up to 30 cm long (see Fig. 13-1a).

Nemerteans and Turbellaria

Nemertines are exclusively predators and seize prey with an eversible proboscis sometimes armed with stylets (Hoplonemertea) capable of piercing prey. The proboscis coils around the prey and immobilizes the victim with glandular secretions or an injected poison (Hyman, 1951). Nemertean worms range from only a few millimeters to the many meters long *(Cerebratulus lacteus)* and feed principally on polychaetes and carrion.

Free-living Turbellaria (flatworms) catch small prey, such as small crustacea, nematodes, and protozoa, with a muscular pharynx (Straarup, 1970). Flatworms like *Stylochus* and *Pseudostylacus* bore into oysters and barnacles and feed on the soft bodies contained within the test (Yonge, 1963). *Cycloporos* feeds by sucking zooids of colonial tunicates.

Gastropods

Many mesogastropods and neogastropods are specialized predators, using adaptations of the buccal apparatus and radula for seizing and swallowing prey. The prosobranch families (Fig. 14-6a) Muricidae (*Urosalpinx, Murex*), Naticidae (*Polinices*), and Thaiidae (*Thais*) drill holes in calcareous exoskeletons of mollusks and barnacles. *Urosalpinx cinerea* is a predator, devastating oyster and barnacle beds, that drills a hold by alternating radular rasping and secretions of an accessory boring organ. After the circular hole is completed, the proboscis is inserted through the hole and the soft tissue is consumed. The composition of the secretion is unknown, but it apparently loosens crystals of calcite, enabling the mechanical drilling action of the radula (Carricker, 1961). Drilling gastropods, such as *Urosalpinx* and *Thais,* orient to prey "upstream" and apparently are capable of olfactory detection (Morgan, 1972). Feeding is often restricted to one prey type after "ingestive conditioning" has occurred (Wood, 1968).

Some snails like *Buccinum* and *Busycon* grip bivalve mollusks with their feet and wedge the valves apart with the edge of the shell. *Busycon* often abrades the valve margins with the edge of its own shell and then inserts a long thin proboscis. The genus *Conus* has a highly movable proboscis and long, barbed, radular teeth. The proboscis is shot out and one or a few barbed teeth stab the prey and convey a poison through a tooth groove (Kohn, 1956, 1959). The virulence of the poison and the speed of attack permit different species to specialize on mollusks, mobile polychaetes, and even fishes (see Chapter 21). Prey are then consumed whole (Kohn, 1959).

Many gastropods, such as wenteltraps (*Epitonium*) and nudibranchs, actively feed on epifauna like sponges and coelenterates (e.g., Robertson, 1963). Some nudibranchs consume nematocysts of their coelenterate prey, transfer them undisturbed to sacs in the cerata, and use them for defense against predators (Graham, 1938).

Bivalve Mollusks

Members of the Septibranchia can consume copepods, ostracods, and other small prey with a protrusible siphon rapidly aimed at the prey through the hydraulic action of a pumping septum (Reid and Reid, 1974). The septum, a modified ctenidium, divides the mantle cavity into upper and lower chambers and is perforated with closable pores. A sudden movement of the septum expels water out the exhalent siphon and draws water into the distensible inhalent siphon. For larger prey (Fig. 14-6b), the inhalent siphon is constricted by a sphincter at the tip after a successful capture. Prey are transferred to the mouth and triturated in a muscular, chitin-lined gizzard stomach (Yonge, 1928). Stomach contents of *Cuspidaria obsea* include bivalves and gastropods 1 to 2 mm in length, gastropod radulae, and ostracods up to 3 mm long (Reid and Reid, 1974).

Polychaetes

Many polychaetes are voracious carnivores. *Glycera* (Glyceridae) has a long tubular proboscis armed with four hook-shaped teeth (Fig. 14-6c). *G. americana* moves through the sediment, seizes, and feeds on other polychaetes. The European *G. alba* lives in a burrow and feeds on a broad spectrum of prey living at the surface, including polychaetes and crustacea (Ockelmann and Vahl, 1970). Members of the genus *Nephtys* have a proboscis armed with a pair of large chitinous jaws. Although *Nephtys incisa* is apparently a deposit feeder in New England waters (Sanders, 1956), other species appear to be carnivores, feeding on polychaetes (Clark, 1962). Orientation to prey is probably aided by olfaction (as in *Nereis;* Copeland and Wieman, 1924).

Crustacea and Other Arthropods

Carnivorous crustacea usually spend some time as scavengers as well. Crabs are effective predators, tearing and seizing prey with their claws. The blue crab *Callinectes* (Fig. 14-6d) is a devastating predator on small bivalve mollusks, such as *Mulinia* (Virnstein, 1977). The green crab *Carcinus maenas* is believed to be a major factor in population size of the steamer clam, *Mya arenaria* (Ropes, 1968). Stomatopods kill prey with a movable "finger" on the second pair of thoracic appendages. A rapid extension and retraction of the finger can slice the prey (e.g., shrimp) in half (MacGinitie and MacGinitie, 1949). The class Pycnogonida feed on hydroids, bryozoa, sponges, and other epifauna by ripping off pieces with the chelicerae. Vermeij (1977) has demonstrated a regional correlation between the size of the "master" (larger) claw of crabs and the degree to which molluscan prey have spines, thickened shells, and narrow apertures. Such a relationship suggests a long period of coevolved interaction between predators and prey.

Asteroid Starfish

Almost all phyla have representatives that are consumed by these effective carnivores. The classic prey–predator interaction is the feeding of a starfish, such as *Asterias forbesi,* on the scallop *Argyropecten irradians.* The starfish wraps its arms around the clam with the mouth pressed against the gape. Tube feet attach to the valves and exert pressure to open the valves enough for the starfish to evert its stomach between the valves and extracellularly digest the prey. A remarkable symbiosis between a sponge living on valve surfaces and a scallop interferes with the adhesive ability of asteroid tube feet and allows the scallop to escape its grasp (Bloom, 1975). The difficulty of opening given prey bivalves influences the starfish's preferences (Kim, 1969).

Many starfishes ingest prey whole and feed on a broad variety of prey. The starfish *Luidia sarsi* uses its arms to leap onto ophiuroid prey. If the prey escapes before the sea star "lands," another leap is attempted. Leaping behavior can be induced by soaking cotton with extracts of brittle stars. When captured, the starfish buries itself with the prey in the mud, drawing the ophiuroid toward the mouth with tube feet and then further seizing the prey with an eversible stomach (Fenchel, 1965). This process takes up to half an hour.

The European sea star *Astropecten irregularis* moves along the sediment surface and detects prey in the sediment beneath with great precision. It feeds on shallow infauna, such as small bivalves, by pushing them into the mouth with the aid of suckerless tube feet surrounding the mouth. This species does not seem to evert its stomach and digest food outside the body (Christensen, 1970).

Fishes

Many benthic fishes feed actively on benthic invertebrates and other fishes. Flatfishes, such as the plaice *Pleuronectes platessa* in the British Isles and the winter flounder *Pseudopleuronectes americanus* on the Atlantic Coast of North America, feed on polychaetes, small crustacea, small bivalves, and bivalve siphons (Levinton, 1971; Bregnballe, 1961; Trevallion, 1971). Many fishes, such as killifishes (*Fundulus*), move into shallow waters in summer and consume large numbers of small crustacea and polychaetes. Pipefishes and sea horses feed on copepods, mysids, and other small prey by sucking victims into a tubular snout. The angler fish *Lophius piscatorius* lies camouflaged on the bottom and displays a dangling lure composed of a moveble, modified dorsal spine. Fishes are lured within striking range and are suddenly attacked and swallowed.

A large number of fishes and rays are armed with jaw parts adapted to seize and crush shells of mollusks. Spiny puffers (*Diodon*) take a shell in their mouth and position it on the jaw-crushing plates by a series of "inhaling and exhaling" motions. The shell is crushed repeatedly and the body, plus a small amount of shell, is ingested after as much of the shell as possible is spat out. Spines on the shell increase the effective size, thus discouraging the predator limited by prey size (Palmer, 1979).

Birds and Other Vertebrates

Shore birds feed extensively on invertebrates living on tidal flats, intertidal rocks, and shallow subtidal bottoms. Sandpipers and plovers feed on surface-dwelling small invertebrates by turning over stones or picking off small macrofauna (crustacea, polychaetes) washed out of the sediment surface by the tide. Curlews, willets, and phalaropes employ a long thin bill for probing in soft substrata. Gulls feed on polychaetes, urchins, crustacea, and bivalves at low tide. Bivalves and urchins are often dropped on the rocks from the air to smash the shell. The oyster catcher, *Haematopus ostralegus* (Fig. 14-6e), feeds on polychaetes and shellfish in the intertidal zone (Heppleston, 1971). They feed on mussels in two ways. The bird hammers a hole with its chisel-shaped bill into mussels exposed by the tide, but drives its bill into the gape and severs the adductor mussel of those prey covered by water. Oyster catcher diet and feeding behavior vary in different tidal flats, depending on prey availability, differences in shell strength of the prey, and firmness of prey attachment to the substratum (Norton–Griffiths, 1967).

Nearshore mammals, such as sea otters, can be significant predators on benthic communities. Otters (*Enhydra*) pry urchins and abalone from shallow subtidal bottoms and feed at the surface by smashing prey against rocks resting on the chest.

Scavengers

Many herbivorous and carnivorous forms are scavengers as well. Urchins and crabs feed actively on decaying carrion. The normally deposit-feeding mud snail *Ilyanassa obsoleta* feeds actively on decaying animals and algae and secretes the complement of hydrolytic enzymes necessary to break down the principal constituents of algae. But its radular morphology is that of a typical gastropod carnivore (Brown, 1969). Thus feeding in this species is very generalized. Orientation to dissolved extracts of carrion results in large concentrations of snails around the source.

SUMMARY

1. Benthic animals can be divided into the following feeding types:
    (a) suspension feeders,
    (b) deposit feeders,
    (c) herbivores,
    (d) carnivores,
    (e) scavengers.
2. Suspension feeders largely depend on phytoplankton for food. A wide variety of particle-capturing adaptations has evolved to capture particles ranging from bacteria to smaller zooplankton. Although particles may be captured in sievelike devices (e.g., setules of barnacles), particles may directly impact on sticky tentacle structures (e.g., crinoids). Most suspension feeders have elaborate particle-sorting structures and only ingest a specified range of particle sizes; little qualitative selection is evident. Dissolved organic matter is taken up by suspension feeders but is unlikely to be a major nutritional source.
3. Deposit feeders ingest sedimentary grains but only efficiently digest the microbial organisms living attached to particles. Particulate organic matter is usually too refractory to be digestible; nitrogen-rich seaweeds, however, may provide nutrition in certain instances.
4. Herbivores display a large number of adaptations to scrape algae from rocks and to chomp macroalgae. Numerous mouth parts have been evolved by different phyla for this purpose. A few feed on wood. Many species harbor autotrophic algal symbionts that provide nutrition for their hosts.
5. Carnivores range widely in the degree of specialization for prey. Some asteroid starfishes are relatively unspecialized whereas others are confined to only a few prey species. Some carnivores are restricted to general classes of prey, such as naticid gastropods feeding on mollusks. Some carnivores actively pursue mobile prey and have acute senses of sight and olfaction (e.g., fishes); many others literally stumble on their prey while moving across a substratum, sometimes moving in the direction of a chemical signal (e.g., gastropods, asteroid starfishes).

VI Coastal and Benthic Habitats

We introduce several important coastal and benthic habitats and examine the factors important to the organization of the associated biota. Our coverage is not encyclopedic; generally we emphasize only those aspects of a particular habitat type that are of special interest or useful in understanding more general principles of marine ecology. For example, we examine in detail interspecies interactions in the intertidal zone; work in this habitat provides a paradigm for such research elsewhere. In contrast, we have devoted a great deal of attention to the coral reef habitat per se; the fascination of coral reefs and the exciting research now underway calls for such attention.

15 the intertidal zone: adaptations and trophic relations

15-1 ESTABLISHMENT AND MAINTENANCE OF VERTICAL ZONATION

The intertidal zone is the best-understood and most examined natural marine habitat type. A strong microgeographic gradient of environmental change creates an ideal situation for the study of adaptations to heat, desiccation, and submergence time. Easy access permits field experimentation and manipulation of populations. The latter advantage explains our relatively complete knowledge of interspecies interactions.

An important feature of the intertidal zone is *vertical zonation,* the occurrence of dominant organisms in distinct bands whose respective upper and lower limits coincide with specific horizontal levels relative to a tidal datum, such as mean low water. In temperate and boreal rocky shores, a nearly universal zonation appears throughout the world. From highest to lowest the zones are (a) a black lichen or myxophycean zone above the highest normal extent of the tide; (b) a periwinkle (littorine gastropod) zone; (c) a barnacle-dominated zone; and (d) a zone dominated by different species, depending on locality. On North American shores, mussels (e.g., *Mytilus edulis*) dominate below the barnacle zone.

The boundaries between zones often coincide with specific tidal levels. For example, the upper limit of the seaweed *Laminaria* in sheltered waters coincides with extreme low-water spring tides. Due to wave-exposure variation, such upper limits often vary dramatically from place to place. Wave splash can extend livable sites meters above that encountered in sheltered waters (Fig. 15-1). Furthermore, interlocality variation in interspecies interactions can completely change the dominant species. Therefore tidal data, such as mean low water, fail to predict the exact pattern of zonation more often than they succeed.

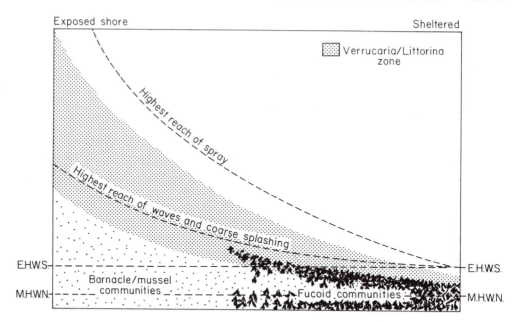

Figure 15-1   The effect of wave exposure in broadening biotic zones of the rocky intertidal of the British Isles. (After Lewis, 1961)

The establishment and maintenance of zonation are controlled by the following parameters:

1. Selective larval settlement.
2. Behavioral patterns, keeping mobile organisms at a certain level of the shore.
3. Physiological tolerance of intertidal organisms.
4. Wave action and tidal range.
5. Intraspecific and interspecific competition.
6. Predation and algal grazing that is tidal height dependent.

We discussed larval settlement preference in Chapter 7. Selective settlement permits pelagic invertebrate larvae and various algal dispersal stages to find suitable living sites on the shore. Mobile organisms, however, must be adapted to maintain a fixed position relative to water level or a given tidal height. They can do so through (a) the maintenance of a home territory or (b) a series of fixed behavioral reactions to directional stimuli that keep the animal at a given station on shore.

Territoriality and maintenance of a home base keep mobile animals in fixed favorable positions on the shore. Many species of limpets remain in one spot at the time of low tide and graze on algae in the vicinity of this spot when covered by water. Scars due to adhesion to the rocky surface mark this home base. The Pacific Coast owl limpet, *Lottia gigantea,* maintains a territory around its home scar and actively prevents other limpets from entering its territory (Stimson, 1973). Other grazing limpets and snails stay in wet, cool cracks in rocks or amongst the byssal threads of the mussel bed during low tide. At high tide they make short feeding excursions to graze on algae on open rock surfaces near their low tide shelters.

The response to light and gravity of the high-intertidal periwinkle, *Littorina neritoides*, illustrates the utility of behavior in maintaining tidal position (Fraenkel, 1927). When submerged in seawater, animals are negatively geotactic. They are negatively phototactic when immersed and right side up but positvely phototactic when immersed and upside down. When the animal is moist but not submerged, it has no light response. Figure 15-2 illustrates how these responses interact to lead *Littorina neritoides* to its typical highest intertidal splash zone habit after a hypothetical dislodgement by waves or a feeding excursion to lower down in the intertidal.

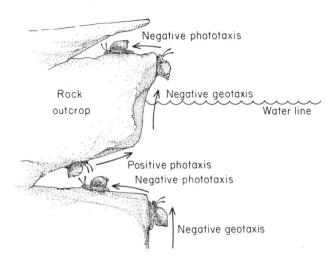

Figure 15-2 The contribution of behavioral responses of the gastropod, *Littorina neritoides,* to the regulation of its vertical position on rocky shores. (After Newell, 1979)

Light-compass reactions appear to regulate shore movements of many invertebrate species inhabiting featureless mud flats or horizontal rock benches. Periodic reversals of response to light cause mobile intertidal invertebrates to make looped tracks, which maintain a general fixed position on the shore but permit exploitation of resources in the immediate vicinity (Brafield and Newell, 1961). The deposit-feeding bivalve *Macoma balthica* is found on mud flats, usually at about midtide level. It rapidly clears adjacent surface mud of edible material and so would be at an advantage to move to a new area (Brafield and Newell, 1961). When exposed at low tide for 1 hour, the animals are photopositive, moving toward the sun. But after 5 hours of exposure, this response reverses, leaving a loop-shaped track on the mud surface. This response permits feeding excursions without leaving the optimal shore level (Brafield and Newell, 1961). The periwinkle *Littorina littorea* shows a similar pattern when living on horizontal intertidal rock benches (Newell, 1958a,b). A reversal of light response occurs after several hours of air exposure. The structure of the littorine eye permits orientation to polarized light (Charles, 1961).

15-2 ADAPTATIONS TO STRESS IN THE SOFT BOTTOM AND ROCKY INTERTIDAL

Heat Stress and Desiccation

The advent of low tide means an immediate change of temperature from sea to air for all intertidal organisms not residing in tidal pools. Even residents of tidal pools, however, experience strong temperature change. Temperature changes in near-surface interstitial spaces as the water table drops on a sandy or muddy beach. Thus intertidal organisms must have morphological, physiological, and behavioral adaptations to resist or compensate for rapid changes in temperature. The lack of seawater contact results in drying out or desiccation of essential membranes and body walls. Loss of moisture is hastened if the air is dry and hot.

Temperature shock can

1. Affect metabolic and biochemical processes, such as enzyme function and oxygen demands.
2. Retard cellular activities, such as ciliary motion.
3. Inhibit behavioral activities, such as feeding and protection against predators.
4. Inhibit reproductive behavior, such as egglaying and copulation (see Chapter 2).

Because of differences in air exposure periods at different intertidal levels, organisms high in the intertidal zone and hence exposed for a greater proportion of the day should be subjected to more heat stress than those lower in the intertidal zone. Species living in the upper intertidal should be more resistant to thermal shock than those living closer to mean low water. This expected hierarchy has been found for lethal tolerance and for measures of cellular activity, such as epithelial ciliary movement. Fraenkel (1966) examined the lethal tolerance limits of a series of Japanese intertidal gastropods living at various heights and found that the high-intertidal species have higher thermal survival limits than low-intertidal species. Similarly, Ushakov (1968) examined cell thermostability of three neritid gastropods living on rocky shores in southeast Asia and found that the cell survival times of high-intertidal species were greater than those of low-intertidal species subjected to heat shock (Fig. 15-3). Presumably the species inhabiting the high intertidal have evolved enzymatic and physiological mechanisms that deal more efficiently with thermal stress than those of the low intertidal zone.

Behavioral reactions to thermal stress may occur as well. In order to avoid excessive temperatures, motile organisms, such as crabs, fishes, and snails, may migrate below the level of low water at the time of low tide. Still other species can retire to moist and cool

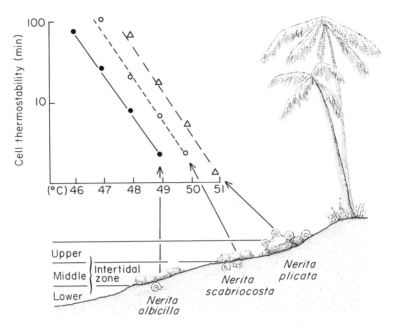

Figure 15-3   Survival of ciliated epithelial cells to high temperature in three species of the intertidal gastropod genus, *Nerita,* and the relation of thermal tolerance to position on the shore. (From Ushakov, 1968)

crevices, seaweed beds, or the byssal threads of mussels. Several species of invertebrates, such as the West Coast nereid polychaete, *Nereis vexilosa,* depend on such refuges to avoid heat and desiccation stress. Seaweeds and mussel beds retain moisture and harbor large biotas of otherwise desiccation-prone invertebrates. In winter the western Atlantic mud-flat snail *Illyanassa obsoleta* moves downshore and lives subtidally to avoid the danger of freezing.

Dark substrata, such as volcanic sands and dark volcanic rocks, absorb heat and increase in temperature more rapidly than light-colored substrata (e.g., sandstones). Wholly different snail faunas may exist on dark- and light-colored rocky substrata in the tropical Caribbean (Vermeij, 1971a,b). Basal areas of tropical high-intertidal neritid gastropods and limpets are reduced to lower conduction of heat from the rock (Vermeij, 1971a,b, 1972b, 1973).

Organisms living in the sand flats and mud flats face the same thermal problems as those living on hard substrata. But the presence of soft sediments permits deep burrowing to avoid the effects of heat and desiccation near the sediment–water interface (Fig. 15-4). Deep-burrowing invertebrates are more common in intertidal soft bottoms than in subtidal sediments (Rhoads, 1967). In sand flats of Barnstable Harbor, Massachusetts, polychaetes, bivalve mollusks, the burrowing sea cucumber *Leptosynapta,* gastropods, sipunculids, the acorn worm *Saccoglossus,* and others burrow deeply in the sand

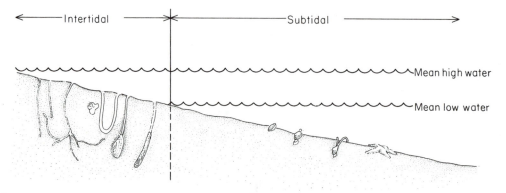

Figure 15-4 Generalized cross section of the bottom of nearshore soft sediments. The depth of burrowing is deeper in the intertidal where exposure to desiccation and temperature variation at the sediment's surface is greatest. (From Rhoads, *Discovery* (Yale Peabody Mus. Nat. Hist.) 2(1), pp. 19–22, 1966)

to avoid desiccation and temperature shock. A casual inspection of the beach surface on a hot day will reveal the selective advantage of such a strategy. Deeper than about 25 cm, sand flats are relatively temperature and dryness stable. Deep-burrowing clams, such as the razor clam, *Ensis directus,* and the soft-shell clam, *Mya arenaria,* have gapes that might permit water loss. In contrast, the shallow-burrowing hard-shell clam, *Mercenaria mercenaria,* can completely seal off its shell.

Water loss is an important problem for intertidal organisms and interacts with temperature stress. Problems of water loss are most severe in the upper intertidal zone, where air exposure is the longest. Mechanisms of losing heat to compensate for temperature stress may involve water loss. Intertidal organisms like snails do not display the temperature of a blackbody but are several degrees cooler. This relative coolness is accomplished through evaporative water loss, concomitant with circulation of body tissue fluids.

Those species with hard parts capable of sealing off the external environment (e.g., acorn barnacles, mussels, and limpets) resist water loss more effectively than soft-bodied organisms that cannot seal themselves off (e.g., anemones, polychaetes, and ascidians). By closing a shell, barnacles and mussels retain a parcel of water that prevents desiccation for long periods. It is possible to keep the mussel *Mytilus edulis* for days in a dry 40°C incubator with no mortality. As might be expected, smaller mussels die preferentially because of their greater surface area per unit volume. Organisms like anemones can behaviorally react through contraction of the body, thus reducing the surface area available for evaporation. The dichotomy in protection against water loss is reflected in the predominance of soft-bodied, water-unretentive animals under rocks, in crevices, and in tidal pools. Animals with protection against water loss, such as acorn barnacles, survive on sunlit rocky faces.

Intertidal algae face particularly severe desiccation problems. Dense populations of mid-intertidal and high-intertidal algae often "burn out" as the summer proceeds. In particular, filmy forms like *Ulva* and *Enteromorpha* are not too resistant to heat combined with desiccation. In Flax Pond, New York, *Ulva* occur in dense populations on mud flats in winter but begin to look unhealthy and die off in late spring and early summer. *Ulva* can competitively dominate other algae in tidal pools but generally burn out and lose in competition to tougher algae on open rock faces in summer (Lubchenco, 1978).

The effect of air exposure and desiccation on intertidal macroalgae has not been extensively studied. The photosynthetic rate in air of the brown algae, *Fucus distichus,* was six times that found under submergence, at similar light and temperature (Johnson et al., 1974). Yet other studies have shown higher photosynthesis under submerged conditions (see Brinkhuis et al., 1976). In the intertidal salt-marsh fucoid, *Ascophyllum nodosum,* photosynthesis in air changes little over a broad range of temperature, light, and disiccation. Photosynthesis is inhibited when greater than 50% water loss occurs (Brinkhuis et al., 1976). Therefore intertidal exposure cannot be simply regarded as a period of extreme environmental stress to which algae must survive before photosynthesizing under more favorable submerged conditions. Growth rate of *A. nodosum* is greater on the lower parts of the shore, however. Furthermore, growth form differs with intertidal height, especially when the protective canopy of *Spartina* is missing in the winter and spring. Lower intertidal forms are larger, have broader and flatter fronds, and show more dichotomous branching (Brinkhuis, 1976). High-intertidal forms grow similarly to low-intertidal forms when the former are transplanted to the lower shore (Brinkhuis and Jones, 1976).

Strong diurnal changes in temperature (due to solar heating), pH and oxygen (as a balance between photosynthesis and respiration), and salinity (due to variations in rainfall and evaporation) make tide pools a highly fluctuating environment. Residents of high-intertidal pools are tolerant of strong fluctuations of temperature and salinity. The rock pool ostracod *Heterocypris salinus* is tolerant of environmental change and is negatively phototactic, confining it to the bottoms of rock pools during the day. This behavior allows the animal to avoid the strong fluctuations in the surface waters of the pool. At night *Heterocypris* is found near the surface (Ganning, 1967). The harpacticoid copepod genus *Tigriopis* contains intertidal pool species very tolerant of wide salinity fluctuations.

Dissolved Oxygen and Gas Exchange

With the advent of low tide, intertidal animals requiring oxygen must respire from air or from a small, contained supply of oxygenated water (as in barnacles and mussels) that will soon be depleted of oxygen. Oxygen may be taken directly from air at low tide. The ribbed marsh mussel, *Geukenzia demissus,* maintains its valves open and takes oxygen from air at low tide (Lent, 1969). Although some evaporative cooling advantage is demonstrable, it does not contribute to increased survival under conditions of heat shock. Air gaping is therefore an adaptation to aerial respiration in the high-intertidal environment, where mussels are exposed much of the time. The California mussel, *Mytilus californianus,* also air gapes under moist conditions (Bayne, 1976).

Many intertidal aerobic organisms do not have the option of air breathing. When all the oxygen is depleted from a reserve, such as water trapped within a barnacle shell, anaerobic forms of respiration must be adopted to employ carbohydrate and other reserves as energy sources. One way of dealing with the lack of oxygen is to reduce behavioral and metabolic activity at low tide. In bivalves, heartbeat slows down when the valves are closed at the time of low tide, and polychaetes reduce activity. Some other problems arise, such as the accumulation of acidic end products related to anaerobic metabolism. Low tide, however, is of such short duration that it is not a period of major accumulation of poisonous substances.

A mechanism that can be used to deal with low oxygen at low tide is the respiratory pigment, such as hemoglobin, hemocyanin, and chlorocruorin. They are present in a wide variety of marine intertidal invertebrates. Pigments capable of binding and releasing dissolved oxygen may be used for oxygen storage. Investigations of the lugworm *Arenicola marina* and the sipunculid *Sipunculus nudus* suggest that the oxygen storage capacity of such pigments in the blood is sufficient for the whole period of low tide. Pigments may store oxygen between successive periods of irrigation of burrows (in the case of polychaetes) or during intermittent shortages of oxygen.

Reduced Feeding Times

Because the high intertidal is exposed to air for a greater proportion of the time than the low intertidal, animals dependent on water or water currents for feeding get greater rations with decreasing level in the tide zone. Several lines of evidence point to the period of emersion as a food-limiting parameter. Browsing periwinkles *(Littorina littorea)* cease feeding when the tide recedes on hot, dry days. Higher intertidal periwinkles feed for a shorter time per day than lower tide individuals. Upper intertidal forms, however, compensate for the limitation by feeding more rapidly (Newell et al., 1971). Size gradients observed from low to high tide in organisms like mussels and barnacles suggest a relation between immersion time, ration, and growth rate. Smaller mussels are usually found in the higher intertidal (Fig. 15-5). Harger (1970) experimentally determined growth differences in the intertidal mussels *M. californianus* and *M. edulis* and found that both species grew more slowly in the high intertidal than the low intertidal. Because gametic output is proportional to body weight in *M. californianus,* the reproductive value of a mussel should decrease with increasing tidal height. Barnacles have been observed to produce fewer eggs higher in the intertidal zone(Barnes and Barnes, 1968). Presumably there is a tradeoff in the use of glycogen for egg production and for energy reserves in metabolic processes.

Carnivores seem particularly inhibited by long emersion times in the intertidal zone. In fact, one important source of zonation in the intertidal is the vertical tidal level above which carnivores cannot feed, as discussed later. The asteroid *Pisaster ochraceus,* a resident of West Coast rocky regimes, prefers the mussel *M. californianus* but only has a limited time to prey on the higher mussel bed. A feeding trip high in the intertidal, involving the loosening and retrieval of a mussel, runs the risk of desiccation. Starfish are only capable of reaching approximately midtide level to prey on mussels. Conse-

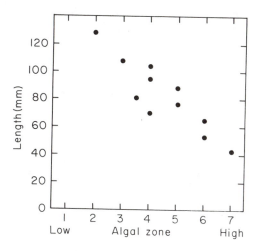

Figure 15-5 Relation between algal zone (indicating intertidal height) and mean shell length in samples of *Mytilus californianus* from Tatoosh Island, Washington. (After Levinton and Fundiller, 1975)

quently, there is usually a boundary between the mussel bed above and an area below that is bare of mussels (Paine, 1974).

Wave Shock

The impact of waves and the material they carry (sand, logs, etc.) is important in selecting morphological adaptations to intertidal life. We need only visit an intertidal area on a stormy day to witness the tremendous energy that is expended and concentrated on intertidal habitats. As waves travel toward the shoreline, the rotating motion of parcels of water begins to affect the bottom and interacts with the shore (See Chapter 1). If the slope is gradual, as in a sandy area, waves affect the bottom when the depth of water is approximately one-half the wave length. At this time the crest of the wave steepens until it falls forward, breaking and dissipating wave energy. The impact of this energy has profound effects on both the soft-bottom and rocky intertidal zones.

Because of seasonal differences in storm intensity, the slope of wave-exposed sandy beaches changes throughout the year. In winter, storm activity usually produces a steep sloping beach whereas in summer lower energy conditions allow sand deposition and result in a shallower slope (Fig. 1-18). Because of wave and current action, sediments constantly shift, providing a rigorous environment for sandy beach organisms. Almost no large algal species and few macroinvertebrates can tolerate the intense wave energy of this environment. The barely subtidal surf clam *Spisula solidissima* is large and thick shelled and is capable of rapid burrowing when uncovered by wave action. The intertidal sand crab *Emerita,* a resident of exposed beaches, is capable of rapidly reburying itself within the sand after being uncovered by waves or currents.

The surf clam, *Donax denticulatus,* lives in the saturated wash zone of West Indian sandy beaches and can only maintain its infaunal life position in sand saturated with seawater that is free of wave action. The upper and lower boundaries between which *Donax* can burrow are set by the opposing factors of the extent of water saturation on the upper part of the shore and the frequency of wave-generated turbulence on the lower end. This zone is not static but moves up and down the beach with the tide (Trueman, 1971, 1975).

On the rising tide, *Donax* emerge from the sediment as a response to the increasing instability of the sand–water mixture. It emerges by pushing downward with the foot and is carried shoreward by the incoming wave, reburrowing on deposition at the tidal level with sand of the appropriate type. As the tide falls, *Donax* again emerge from the sand and are carried down the beach to moister sand. Ansell and Trueman (1974) determined the energy cost of such migrating behavior and found that it took 6.4 cal day^{-1} to maintain position in the wash zone, whereas only 0.87 cal day^{-1} was expended in migratory behavior. Survival in the wash zone entails frequent reburial and is apparently much more costly than migration.

Breaking waves can damage rocky shore organisms in the following ways (Jones and Demetropoulos, 1968):

1. *Abrasion*. Particles in suspension or floating debris (e.g., logs) scrape delicate structures. Water turbulence may whip seaweeds and hydroids against rocks and cause similar damage.
2. *Pressure*. The hydrostatic pressure excreted by a breaking wave could crush or damage delicate and compressible structures, such as gas-filled bladders of seaweeds. Most intertidal organisms are liquid filled and hence relatively incompressible.
3. *Drag*. Water exerts a directional force against intertidal organisms that may rip apart support structures (as in seaweeds; see Chapter 13) or dislodge holdfasts from their points of attachment (e.g., byssal threads of mussels).

Wave action modifies the extent of the tide zone and vertical zonation (Fig. 15-1). The pronounced wave action of an open coast expands the occurrence of an intertidal biota up a rock cliff (Evans, 1947; Lewis, 1964). The upper extent of the intertidal, however, usually coincides with the extent of highest high tide level in quiet coves or bays with little wave action. Protected areas may also more dramatically experience the effects of desiccation relative to continuously moistened wave-exposed coasts. Thus the outer coast of Washington has a rich intertidal flora extending several meters above mean low water whereas the algal flora is poorly developed because of desiccation in rocky areas of the protected San Juan Islands, Washington. Some organisms on exposed coasts live continuously out of reach of normal vertical tidal motion but can survive on the moisture delivered by wave spray. Barnacles like the outer Pacific Coast *Balanus cariosus* can feed instantaneously after being immersed by sea spray.

Demographic Aspects of the Intertidal Gradient

The vertical intertidal gradient strongly influences the demography of mobile organisms. The effects of desiccation above and wave shock and predation below generate strong selective forces on the behavior of intertidal animals. The behavior of *Littorina neritoides*, as discussed previously, is a case in point. An elaborate series of cues permits the snail to keep at a specified vertical zone in the rocky intertidal.

These adaptations force a potentially dense population to occupy a rather narrow band in the intertidal. In Oregon populations of the limpet *Acmaea digitalis,* the band can change seasonally in vertical extent; snails ascend in winter to avoid potentially higher chances of mortality (Frank, 1965). In Washington populations of the intertidal snail *Tegula funebralis,* juveniles recruit to the upper intertidal, presumably to avoid the great abundance of predators in the lower intertidal. They then migrate downward with increasing age (Paine, 1971a). As noted, the downward migration relates to the greater time available for feeding while submerged or moist.

Populations of the limpet *Acmaea scabra* in the high and low intertidal of Bodega Head, California, point out the influence of habitat and population dynamics on resource limitation (Sutherland, 1970). Despite the considerably lessened feeding time in the high zone, individuals obtain larger sizes than those in the low zone. In contrast to *Tegula,* recruitment to the low zone is continuous, dense, and results in a population capable of outstripping the renewable algal resource. In contrast, recruitment to the high zone is sparse and sporadic; algae are usually abundant and not as limiting to growth. Thus high-zone individuals grow faster despite the occupation of an ecologically more marginal zone and the attendant dangers of catastrophic mortality due to desiccation (as shown by Frank, 1965).

15-3 FOOD WEBS AND NUTRIENT CYCLING

The process of enumerating the sources and fates of energy is complex even in experimental environments (Slobodkin, 1970). The intertidal is difficult for several reasons. First, much of the energy is imported from the outside—often in an unpredictable fashion. Large amounts of seawater drain through sandy beaches, making plankton, phosphates, nitrates, and other dissolved nutrients available to the infauna (McIntyre et al., 1970). Some water is filtered by suspension-feeding organisms while nutrients are removed by bacteria, benthic diatoms, blue green algae, and other photosynthesizers. The same problem holds for the rocky intertidal. Organic matter is imported as drift algae and consumed by herbivores. In some cases, these herbivores have evolved means to capture this algae. The extensible aboral tube feet of sea urchins can remove entrapped floating debris from spines. Clearly this type of import is not constant over time and hence difficult to assess. Suspension feeders also import food into the rocky intertidal zone.

An equally important problem in studies of the intertidal is the potential loss of energy to adjacent environments. Longshore currents can erode animals and plants from a beach and transport them subtidally or to another beach. Shifting sands in wave-swept beaches also provide a rigorous environment for primary production, which may be as little as 5 g C m^{-2} year $^{-1}$ (Steele and Baird, 1968). Wave energy and log damage can remove a great deal of the energy fixed within a given trophic level. Significant losses also occur through leaking of dissolved organic matter fixed in intertidal algae to moving water (Mann, 1976). Such loss can amount to over a third of gross productivity. Both the rocky intertidal and sand-mud flat intertidal zones are therefore open systems, with a quantitatively important exchange with other environments.

Temperate Sand and Mud-Flat Food Webs

The probably substantial import and loss of dissolved and particulate matter make the study of intertidal food webs especially difficult. One might expect that quiet depositing shores are relatively closed systems, as compared to open exposed sandy beaches. Some evidence, however, supports the often restricted nature of nearshore circulation and limited exchange with offshore waters. Suspension feeders residing in sandy beaches probably remove much of the phytoplankton from the overlying water. But mineralization of invertebrate excretions and particulate organic matter by microorganisms may result in the loss of significant amounts of nitrogen from the bottom to the overlying water. This loss may be the primary nutrient source for near-beach surf phytoplankton blooms on the western U.S. coast (Lewin et al., 1975). As beach-dwelling suspension feeders graze on these blooms, extensive recycling in this system is evident.

In quieter sand and mud flats, both microorganisms and particulate organic matter play an important role in the food web. Algae and bacteria living among, or attached to, sedimentary grains constitute the primary source of nutrition for the large numbers of deposit feeders living in the flats. Below the surface oxidized zone, bacteria adapted to anoxic conditions (see Chapter 12, Section 12-4) convert substantial amounts of particulate organic matter to usable food for deep-feeding deposit feeders. Substantial proportions of the macroalgal production of mud flats are consumed by herbivorous gastropods and polychaetes. Some tube-building polychaetes like *Diopatra* maintain clumps of algae on the openings of their tubes as gardens (Woodin, 1978). Most of the macroalgal production, however, probably decomposes and enters detritus pathways. This is certainly true for relatively indigestable grasses, such as the marsh grass *Spartina*. It is possible that some of the more nitrogen-rich macroalgae are consumed and digested by the meiofauna and macrofauna. Most, however, is probably converted first to microbial food before being consumed by deposit feeders.

The interaction of the herbivore and detritus-consuming pathways can be illustrated in a study of the fate of the green algae *Ulva, Monostroma,* and *Enteromorpha* in a protected embayment (Hylleberg, 1977). False Bay, San Juan Island, Washington, is a circular-shaped, protected intertidal sand flat of ca. 10^6 m^2. Large amounts of green algae grow in the bay and are uprooted and moved about the bay by storms. Pieces of algae ranging from several thousand cm^2 to 100 μm^2 are found in sedimentary columns, in pools, and at the strandline.

Hylleberg estimated that 34 tons dry weight of green algae disappeared from the bay in three fall months of 1973. After this disappearance, about 0.5×10^6 fragments m^{-2} (5 g dry wt. m^{-2}) were found in the sediment. After storms, algae were ripped apart by turbulence and much material was transported to the strandline. Rolling logs crushed the algae while other clumps fermented on the beach. Killed and stranded algae returned to the bay lost 10% of their organic matter as dissolved material.

The polychaete *Nereis brandti* is probably the only major herbivore to consume the live green algae. The worm draws fragments of the algae down into the sediment and assimilates about 70% of the organic matter ingested. Fragments of algae not torn loose by the jaws remain in the sediment to decompose. Hylleberg's estimate of ca. 10^6 *N. brandti* in the bay, with a population ingestion rate of 0.7 g d^{-1} m^{-2} dry weight of algae, leads to the conclusion that only 10% of the loss of 34 tons of algae was channeled through the worm population. The fate of the rest is diagrammed in Fig. 15-6. Animals feeding on the microbial community decomposing the detritus clearly provide a major pathway. Some algal material must have been lost to the strandline and some may have been exported from the bay.

The large amounts of particulate organic matter entering intertidal sand flats and mud flats (and shallow subtidal systems) have been estimated for shoals in New England, as shown in Table 15-1. The primary source is decomposing eelgrass *(Zostera marina)*. The fact that organic matter does not accumulate in the sediments suggests that decomposition of the detritus by bacteria, enhanced by bacterial grazing of interstitial forms and macrofauna, results in the transfer of nutrients borne in the detritus to higher trophic levels. The great fish and shellfish productivity of shoals of Rhode Island probably results from the transfer (Marshall, 1970). The transfer is complex because macrofauna consume

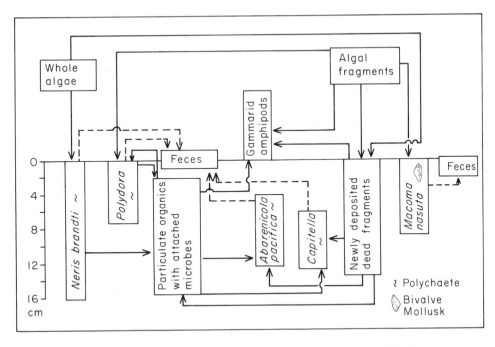

Figure 15-6 Sources and pathways of particulate organic matter on depositing shores. Particulates also cycle, through feces, back to deposit feeders. (Modified after Hylleberg, 1977)

TABLE 15-1   ESTIMATES OF ORGANIC CARBON CONTRIBUTIONS TO ESTUARINE
SHOALS IN SOUTHERN NEW ENGLAND[a]

Source	g $C/m^{-2}/y^{-1}$
Macroflora, including *Zostera* and epiflora	125
Benthic microflora	90
Phytoplankton deposition	50
Allochthonous matter	0–10
Conversion of dissolved organic matter	?
Total	265–275 + ?

[a]From Marshall, 1970.

interstitial forms but also compete with them for bacteria and small diatoms attached to organic detrital particles. Moreover, suspension-feeding macrofauna may play a major role in incorporation of organic matter into the sediment, making it available for the microbial and meiofaunal communities. Populations of the oyster *Crassostrea virginica* can deposit 2400 kg hectare^{-1} week^{-1} of feces and pseudofeces. All types of algal cells from the surrounding water are represented (Haven and Morales-Alamo, 1966). In the Waddenzee, populations of the cockle *Cardium edule* biodeposit 100,000 tons dry wt/yr of suspended matter (Verwey, 1952).

   Populations of meiofauna and small macrofauna may respond to short-term influxes of organic matter. Buildups of deposited organic matter from spring deposition of planktonic algae, feces of zooplankton (Moore, 1931), and other sources are accompanied by spring and early summer buildup of benthic populations (see Schneider, 1977). These populations probably subsist on the microbial community living on detritus. The accumulation of organic matter may be met by an increase in demand by the microbial community. Unfortunately, balance estimates, using measurements of organic input and oxygen consumption of sediment cores, have had confusing results (McIntyre et al., 1970; Marshall, 1970; Weiser and Kanwisher, 1961).

Constructing the Trophic Web

In the previous section we primarily discussed the paths through which particulate organic matter flows in sand flats. A significant amount of carnivory, however, is liable to occur as well. Most conspicuous are the effects of large mobile predators on macroinvertebrates (as discussed in Section 16-2). The common Atlantic nereid polychaete *Nereis virens*, for example, is a major predator of amphipods and small polychaetes. It is possible to study such predation through traditional experimental approaches, as discussed earlier.

   Establishing links between consumers and meiofauna or larval stages is much more difficult. Gut contents usually are an amorphous mass (mess!) that cannot easily be differentiated to species except where hard parts of prey items are involved. As a result, more elaborate techniques are required to work out the extent of the sand-flat food web. One such technique is the use of antigen–antibody reactions employed in immunology. Extracts of invertebrate species are injected into rabbits and the specific antibodies so generated are used to assay for the presence of different prey species in predator gut contents. Using this technique, Feller and others (1979) have shown that many juvenile and larval stages of common tidal flat invertebrates are ingested by species normally regarded as deposit feeders. This fact may imply that deposit feeders in mud flats strongly influence the degree of recruitment of larval stages and thereby affect community composition.

Salt-Marsh Food Web

Salt marshes are protected shore areas where salt-resistant plants form the majority of the primary production and help stabilize and promote accretion of bottom substrate. In the tropics various species of mangroves form emergent islands with large trees whereas grasses of the genus *Spartina* dominate many temperate salt marshes. Marshes are best developed on coastlines where sea level is rising relative to the shore (Ranwell, 1972).

Although marsh topography is initially dominated by the topography of the shore, plant growth and sediment accretion processes shape *Spartina* meadows of low-relief, adjacent intertidal mud flats and anastamosing channels (Fig. 15-7). There is usually a fundamental distinction between *low* and *high* marshes, bounded at about the level of mean high water (Chapman, 1960). In New England salt marshes it is the boundary between the grasses *S. patens* (high marsh) and *S. alterniflora* (Chapman, 1940). Figure 15-8 shows a schematic cross section of a typical New England salt marsh.

Spartina marshes are of great interest as intertidal food webs because of the seasonal cycle of grass production and detrital decomposition. In spring and summer marshes are highly productive, although little of the *Spartina* is grazed directly by marine consumers (see Burkholder, 1956). In Sapelo Island, Georgia, insects like the grasshopper *Orchelimum fidicinium* consume a small fraction of the living *Spartina* (Odum and Smalley, 1959). The resistance to grazing relates to the relatively low nitrogen content of *Spartina*, a characteristic of plants with C-4-type photosynthesis. Nitrogen fertilization increases the nitrogen content of *Spartina* and attracts herbivores.

In spring and summer, but mainly in late fall, *Spartina* leaves die and begin decomposing on the stalk. Water turbulence, floating debris, and ice (in northern marshes) dislodge decaying leaves and transport them to mud flats, the strandline, and other locations around the marsh. Thus the great majority of the *Spartina* production enters the food web as detritus (Odum and de la Cruz, 1967). Particles from over a meter to a few hundred microns are recognizable in sediments and on the sediment surface. Benthic algae and macroalgae also play an important role in transfer to consumers. In particular, the aufwuchs community is a complex of hundreds of species of bacteria, fungi, and microalgae attached to *Spartina*, eelgrass (*Zostera*), macroalgae, and the sediment surface. They are grazed on by such interstitial forms as protozoa, foraminifera, nematodes, and

Figure 15-7 The Spartina salt marsh at West Meadow Creek, north shore of Long Island, New York. Upper: West Meadow Creek and marsh; lower left: an eroding marsh bank, with *Spartina alterniflora;* lower right: closeup of eroding bank showing individuals of the ribbed mussel, *Geukenzia demissa*.

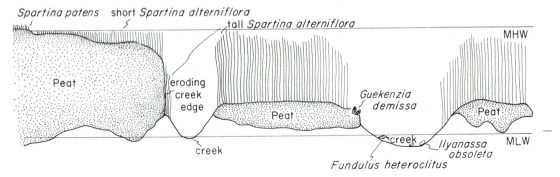

Figure 15-8 Subhabitats of the *Spartina* salt marsh ecosystem. The tall form of *S. alterniflora* is associated with the high nutrient supply of flowing creeks. (After Redfield, 1967)

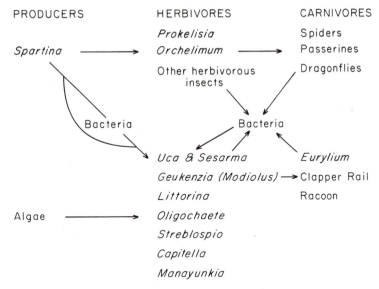

Figure 15-9 Food web of a *Georgia Spartina* salt marsh (After Teal, 1962 copyright 1962, the Ecological Society of America.)

harpacticoid copepods (see Lee et al., 1975). The nutritional requirements of consumers like the harpacticoid copepod *Nitocra typica* are apparently complex and different algal species have varying advantages, depending on season and reproduction (Lee et al., 1975). Phytoplankton also play a major role in supporting suspension-feeding species in marsh flats.

Teal (1962) measured the rates and determined the pathways through which energy flows in a Georgia salt marsh (Fig. 15-9, Table 15-2). Production at all levels of the food web is converted to kilocalories or energy units. Almost half the net primary production was exported to adjacent estuarine waters in the form of organic detritus. Detritus could conceivably play a major role in the economy of harvestable benthic species, such as crabs and shrimp. Table 15-2 shows an energy budget for the salt marsh.

TABLE 15-2 SUMMARY OF A SALT-MARSH ENERGY BUDGET FROM A *SPARTINA* MARSH IN GEORGIA[a]

Sources and losses	Estimated values
Input as light	600,000 kcal/m^2/y
Loss in photosynthesis	563,620 or 93.9%
Gross production	36,380 or 6.1% of light
Producer respiration	28,175 or 77% of gross production
Net production	8,205 kcal/m^2/y
Bacterial respiration	3,890 or 47% of net production
Primary consumer respiration	596 or 7% of net production
Secondary consumer respiration	48 or 0.6% of net production
Total energy dissipation by consumers	4,534 or 55% of net production
Export	3,671 or 45% of net production

[a]After Teal, 1962.

A comparative study of energy flow in the herbivorous grasshopper *Orchelimum fidicinium* and the detritivore *Littorina irrorata* gives insight on some of the effects of detritus on food webs. Odum and Smalley (1959) used seasonal sampling, respirometry, and calorimetry to compare production in *Spartina* and energy flow (defined here as somatic growth m^{-2} d^{-1} plus respiration m^{-2} d^{-1} in calorific equivalents) in the two consumers. In the grasshopper, high energy flow was confined to the relatively brief period of high *Spartina* production. High energy flow in *Littorina,* however, occurred over a broader period of the year because of the longer availability of detritus (Fig. 15-10). Thus the introduction of large amounts of organic matter into the detritus pathway has a buffering effect and provides a more stable source for consumers.

The dramatic seasonal production and decay of *Spartina* suggest extensive cycling of elements important in the economy of nearshore marine habitats. So it is of interest to ask whether any of this production and nutrient cycling has an impact on nearshore marine productivity or on exploitable fisheries. We discuss this subject in Chapter 17.

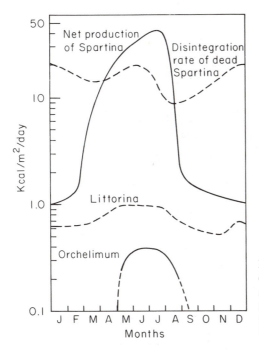

Figure 15-10 The annual pattern of energy flow in the deposit-feeding gastropod, *Littorina*, and the herbivorous grasshopper, *Orchelimum*, in relation to potential food sources. (From Odum and Smalley, 1959)

Rocky Intertidal Food Web

Paine (1966) examined three rocky intertidal food webs, focusing on trophic levels above the herbivores. Figure 15-11 (a), (b), and (c) shows the differences encountered in rocky intertidal habitats in Washington (outer coast), the northern Gulf of California, and the Pacific coast of Costa Rica. The Washington example shows the channeling of energy principally through one keystone species, the starfish *Pisaster ochraceus*. Predators lower down in the web, such as *Thais spp.*, were more trophically specialized. This generalization holds for the subtropical web as well. In the subtropical web, however, a greater number of predator species lessens the importance of *Heliaster* as a keystone species. The tropical subweb is of interest because of its simplicity. Despite the broad nutritional base of *Thais*, which includes cannibalism, there was little overlap with the other major predator, the gastropod *Acanthina*.

The herbivore part of the rocky intertidal food web consists principally of urchins, chitons, and gastropods. Rocks may be covered with anything from a film of diatoms to fleshy algae; a wide variety of herbivore adaptation is therefore to be expected. The quantitative importance of herbivores is immense, for they can completely clear off rocky areas of algae. The periwinkle *Littorina littorea* can rapidly deplete populations of leafy algae, such as *Ulva* and *Enteromorpha*.

The pattern of herbivory is influenced by the spatial heterogeneity and desiccation

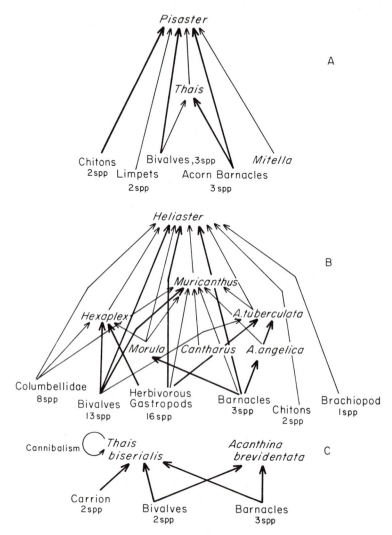

Figure 15-11 Feeding relationships and pathways in components of the food web on rocky shores of the Pacific coast of North America. (a) The subweb dominated by the starfish, *Pisaster*, on the wave-exposed coast of Washington; (b) The subweb dominated by the starfish, *Heliaster*, in the northern Gulf of California; (c) the rocky shore food web in Costa Rica. Major pathways in the food web are indicated by thickened arrows. (Modified after Paine, 1966)

stress of the rocky intertidal. Limpets often live within mussel beds for protection at low tide and then graze adjacent to the bed at the time of high tide. This process leaves a halo of grazed rock around mussels beds and in open areas within the mussel bed (Seed, 1969). Limpets and littorines are often confined to moist cracks at low tide and therefore also may have a limited cruising range, which similarly results in halos. Refuges for algae are therefore created where herbivore access is restricted. This restriction may be amplified by microtopographic barriers (Petraitis, 1979).

Many seaweeds are mechanically resistant to herbivores and harbor a variety of toxins that discourage grazing. Many red and brown algae contain organic allelochemics (e.g., quinones) that are toxic to invertebrates (see Chapter 13, Section 13-8). Others are tough and resistant to grazing (e.g., the tarlike encrusting red alga *Petrocelis middendorfi*). Littler and Littler (1980) have shown that mechanically and chemically resistant seaweeds tend to dominate later stages of algal succession.

SUMMARY

1. Vertical zonation is the most prominent feature of intertidal communities. Zone boundaries vary substantially, depending on exposure to waves and the presence of given important competitors or predators.

2. Zonation may be established through selective settlement of larvae or differential survival of sessile species. Mobile species have elaborate adaptations to maintain a given vertical position on the shore.

3. Living high in the intertidal has the following disadvantages:
   (a) heat stress and desiccation;
   (b) reduced access to dissolved oxygen;
   (c) reduced feeding times.

4. To some degree, intertidal habitats may be open systems. Particulate organic matter may be lost from a given area; intertidal suspension feeders may import phytoplankton from the adjacent sea. Large amounts of particulate organic matter enters relatively quiet sand-flat and mud-flat systems. In the rocky intertidal, however, herbivores consume algae attached to rocks and may themselves be consumed by a wide variety of carnivores.

16 the intertidal zone: community interactions

16-1 COMPETITION WITHIN AND BETWEEN SPECIES

Competitive interactions between species frequently determine the distribution and abundance of intertidal species. As will be shown later, predation and disturbance strongly modulate the outcome of competition. Spatial variation in community structure can be explained by differences in the relative impact of competition, predation, and disturbance. Strong physical gradients, such as high → low intertidal or wave-exposed → protected sites, often influence the degree of disturbance or abundance of predators. These influences, in turn, cause gradients in community structure. Intertidal species, however, are typically confined to physiologically unfavorable refuges because of predation and competition.

Dense settlements by pelagic dispersal stages are a major source of competition for space on rocky coasts. In Long Island Sound cyprids of the acorn barnacle *Balanus balanoides* carpet rock surfaces each spring. As individuals of the dense population grow, space becomes a severely limiting resource. In the lowest part of the intertidal, larvae of the mussel *Mytilus edulis* settle in late May and June, smothering the young barnacles that may have previously settled on bare space. This sequence of events reveals the importance of space as a limiting resource and suggests that experimentation by manipulation of field populations is a potentially effective technique for the study of the mechanisms and trajectories of intra- and interspecific competition. The following studies describe the importance and mechanisms of competition in intertidal systems.

Barnacles and Seaweeds: The Effects of Competition and Desiccation on Zonation

The acorn barnacles *Chthamalus stellatus* and *Balanus balanoides* are common on rocky shores of the temperate and boreal Atlantic. *C. stellatus* is a southern species, ranging from the Mediterranean to the Shetland Islands, Scotland, whereas the northerly *B. balanoides* ranges from the Arctic to southern Spain. In areas of overlap, *C. stellatus* adults are restricted to the upper intertidal. Adult *C. stellatus,* however, extend down to mean low water in the south and in northern sites where *B. balanoides* is absent (Connell, 1961a,b). In Cumbrae, on the Firth of Clyde, *C. stellatus* cyprid larvae settle lower in the intertidal than surviving adults. The obvious implication is that *B. balanoides* is a superior competitor to *C. stellatus* in the lower intertidal zone.

To investigate competition, Connell (1961a,b) transplanted rocks with newly settled *C. stellatus* to all levels in the intertidal. Some rocks were caged to avoid predation by the drilling snail *Thais lapillus*. Later in June when *Balanus* larvae had stopped settling, some rocks were cleaned of newly settled *B. balanoides* cyprids. On these rocks *C. stellatus* survival significantly exceeded undisturbed control sites, although survival was still somewhat less at levels lower than which the species was found locally.

Greater tolerance to desiccation allowed *C. stellatus* to outsurvive *B. balanoides* in the higher intertidal. In high intertidal areas settled only by *C. stellatus,* individual growth rates were slow and mortality resultant from intraspecific competition was rare. At lower tidal levels, interspecific competition with *Balanus* was a major cause of mortality for *Chthamalus*. Rapidly growing *Balanus* overgrew and undercut *Chthamalus* individuals, thereby resulting in the latter's elimination. Intraspecific competition also caused high mortality among *Balanus* individuals. Connell's transplanted rocks showed clearly that *Chthamalus* had markedly better survival when kept clear of settling *Balanus*. This effect has a seasonal component; periods when *Balanus* are growing rapidly (spring and summer) are periods of high mortality of *C. stellatus*. Those *Chthamalus* individuals that survive interspecific competition are smaller and have fewer young.

Space competition is intense and results in zonation with sharp boundaries due to the competitive success of *Balanus* relative to *Chthamalus*. The survival of *C. stellatus* in a high-intertidal refugium is related to its greater tolerance to desiccation. So it might be concluded that, in space-limiting systems, the lower limit of an animal species vertical zone may be determined by competition with a superior competitor whereas the same species' upper vertical limit is determined by its susceptibility to physiological shock, such as thermal and desiccation stress. Competitive failure is manifested in lowered survival of the inferior competitor and diminution of reproduction. Figure 16-1 summarizes the effect of principal limiting factors.

The interactive results of desiccation and competition/predation exert similar effects on competitive interactions among seaweed species. On New England rocky shores, the middle intertidal zone is dominated by the brown seaweed *Fucus vesiculosus* while the red seaweed *Chondrus crispus* (Irish Moss) dominates the lower intertidal. Experimental

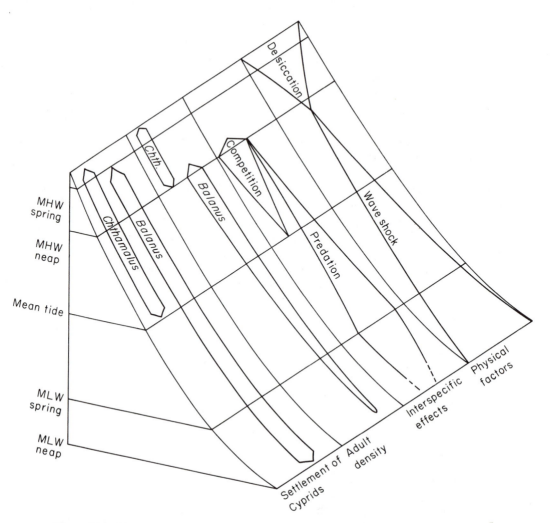

Figure 16-1 Distribution on rocky shores of Scotland of adult and newly settled larvae of the barnacles *Balanus balanoides* and *Chthamalus stellatus*. The relative effects of principal limiting factors are indicated by the width of the bars. (Modified after Connell, 1961b. Copyright 1961 by the Ecological Society of America.)

removal of the red seaweed results in the establishment of dense stands of *F. vesiculosus* in the lower intertidal—in the absence of grazing gastropods. When gastropods are present, the brown seaweed can still colonize the lower zone. These results demonstrate that interspecific competition typically restricts *F. vesiculosus* to the upper zone. Individuals growing in the lower zone show more vigorous growth (Lubchenco, 1980). Desiccation determines the upper limit of Irish Moss, which explains why *Fucus* can exist in the upper zone.

Mussels

Harger (1968, 1970, 1972) examined competitive interactions between the bay mussel *Mytilus edulis* and the California mussel *Mytilus californianus*. Both mussels occur in the rocky intertidal from Baja California to Alaska. *M. californianus* occurs mainly in wave-exposed, outer-coast environments, however, whereas *M. edulis* is most commonly found in protected embayments, such as San Francisco Bay and Puget Sound. An example of this transition is to be found in the Pacific Northwest of the United States. On the outer coast of Washington, in areas like Cape Flattery and Tatoosh Island, *M. californianus* is the predominant mussel. On Tatoosh Island it is most common, with *edulis* occupying a narrow band at the top of the main mussel bed. Along the Strait of Juan de Fuca, where exposed environments alternate with protected embayments and estuaries, the frequency of localities with the bay mussel as a dominant increases near the northern limit of Puget Sound (Levinton and Suchanek, 1978). Within Puget Sound, *Mytilus edulis* is dominant, correlating with the protected nature of this basin.

Mytilus edulis is smaller in size and has thinner byssal threads than *M. californianus*. Byssal threads are structurally and biochemically identical in both species, only thicker in the California mussel. Furthermore, *M. californianus* has a robust wave-resistant shell. Finally, *M. edulis* is a very mobile species, always attempting to crawl upward; *M. californianus* does not move nearly as much. In both species, adult locomotion is achieved through severing and production of byssal threads, permitting the animal to move in a given direction.

Harger (1968) investigated interspecific interactions at a locality near Santa Barbara, California, where the two species co-occurred naturally. He collected clusters of two-species assemblages of mussels and painted clusters to track individuals that had been on the outside at the beginning of the experiment. Individuals of *M. edulis* that had been in the center of a mussel clump soon reached the outside. Clusters that might have started as homogeneous mixtures of *edulis* and *californianus* now had *M. edulis* on the outside and *M. californianus* on the inside. Natural clumps show a similar pattern of occurrence. Growth was slower in experimental clumps of *M. californianus* surrounded by *M. edulis* than in clumps of pure *M. californianus* (Harger, 1972).

The activity of *M. edulis* is an adaptation to escape burial by sedimentation in its own preferred quiet-water environment. In bays and estuaries mud settles continually and may bury mussels with sediment. In addition, the copious production of mussel feces and pseudofeces may also smother a mussel bed. So in a quiet-water environment the relative ability to move upward and out of sediment would confer a competitive advantage for *M. edulis* relative to *M. californianus*. But this behavior is inappropriate for outer-coast environments because the relatively thin byssal threads are insufficient to prevent it from being swept away when it has reached the outside of a clump of mussels. It is reasonable to ask whether this behavioral pattern of crawling upward evolved in response to competition with *M. californianus*. Apparently not, for *M. edulis* shows the same behavior on the East Coast of the United States, where *M. californianus* is absent. Thus the general adaptation for life in calm bays allows *M. edulis* to occupy successfully a subset of potential mussel habitats on the West Coast of North America.

The regular occurrence of *M. edulis,* above the zone of *M. californianus* on exposed coasts, complicates this picture (Levinton and Suchanek, 1978). *M. edulis* also occurs within the *M. californianus* zone when the latter are dislodged by wave shock or predators. Its high larval recruitment rate permits it to colonize newly opened space. Predation by the snail *Thais* (Fig. 16-2) and ensuing competition from *M. californianus,* however, result in its elimination (Suchanek, 1978). It probably can occur above the *M. californianus* zone because of its greater tolerance to desiccation and heat. Thus *M. edulis* appears to have a broader niche than *M. californianus,* occupying bays, estuaries, and exposed coasts. *M. californianus* may be a superior competitor and more resistant to predation in the lower intertidal of exposed coast, thus explaining its dominance. In summary, the outer-coast vertical distribution of *M. edulis* and *M. californianus* may be explained in a way similar to Connell's (1961a,b) deductions for barnacles.

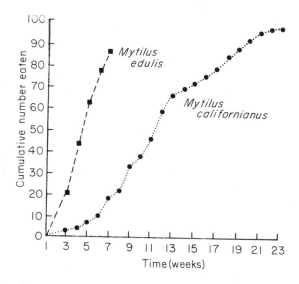

Figure 16-2 Rates of predation by the gastropod, *Thais emarginata,* on a mixed species clump of the mussels *Mytilus californianus* (closed circle) and *M. edulis* (closed squares). (From Harger, 1972)

Mussel–Barnacle Competition

In Long Island Sound *Mytilus edulis* occupies a distinct zone below a band of barnacles, *Balanus balanoides*. Competition between these two species maintains the sharpness of the interspecies zone boundary. The experimental removal of predators was used by Paine (1966) to enable competition to become a dominant process. Dense populations of predators reduce space-keeping organisms to densities at which intraspecific and interspecific competition cannot be a major determinant of species occurrence. Working in a rocky intertidal habitat in Mukkaw Bay (outer coast of Washington), Paine removed individuals of the starfish *Pisaster ochraceus*, a carnivore capable of devastating mussel and barnacle populations. Following removal of *Pisaster,* the acorn barnacle *Balanus glandula* settled successfully throughout much of the intertidal and occupied 60 to 80% of the space by September of that year. By the following June, *Balanus* were overgrown by small *Mytilus* and the bivalve *Mitella.* Eventually *Mytilus* took over the available space, smothering the previously settled barnacles (Paine, 1974). Higher in the intertidal zone of Atlantic boreal and temperate regimes, the relative tolerance to emersion allows the barnacle *Balanus balanoides* to occupy a refuge from competition with *Mytilus edulis.* Small *Mytilus* are prone to death through desiccation and heat.

Polychaete Competitors in Mud Flats

Unfortunately, competitive interactions between species in soft-sediment intertidal flats have attracted less experimental research. Yet several observations suggest that interspecific competition plays an important role in structuring communities. In an investigation of a sand-flat community in Barnstable Harbor, Massachusetts, Sanders and others (1962) observed that the bivalves *Gemma gemma* and *Mya arenaria* tended not to co occur. This non-co occurrence may be due to an inhibitory effect that *Gemma* exerts on *Mya. Gemma* adults may swallow *Mya* larvae, thus precluding successful settlement. Nonoverlap has also been shown for the tube-building polychaete *Clymenella torquata* and *Mya arenaria* and between the amphipod *Pontoporeia* and the bivalve *Macoma* in mud flats of San Francisco Bay (Vassallo, 1969). McIntyre and Eleftheriou (1968) found that the normally upper subtidal sand-dwelling bivalve *Tellina fabula* penetrates the intertidal zone when its congener *Tellina tenuis* is absent from its typical intertidal habitat. These observations are circumstantial evidence that competition is important in soft-bottom intertidal habitats.

Woodin (1974) examined polychaete abundance patterns in a mud flat on San Juan Island, Washington, and found that the abundance of common species could not be related to such physical factors as temperature-depth profiles, oxygen, and sediment type. An ingenious experiment revealed the intensity of interspecific competition. Several of the species are tube builders, whose larvae will not pass through a screen coated with diatoms and placed above a portion of intertidal mud. In contrast, settling larvae of the burrowing species *Armandia brevis* settle on the diatom-coated screen and burrow through to the mud below. Controlled experiments of the influence of tube-building polychaetes on burrowing polychaetes could thus be performed. The manipulation of tube-builder abundance showed that the burrowing species responded to an increase of space through increased settling success and higher densities (Table 16-1). Because the insertion of glass tubes seems to have the same effect as the presence of tube-building polychaetes, space is probably the limiting resource in this instance. Laboratory experiments confirm that the volume of sediment available for burrowing in *Armandia brevis* limits population density.

Woodin (1976) suggested that dense populations of soft-sediment invertebrates might be able to resist invasion by newly settling larvae. A successful year class of a suspension-feeding bivalve might thus be able to resist the settlement of future year classes of larvae, perhaps by ingestion or entrapment of pseudofeces. Although this may

TABLE 16-1 LARVAL RECRUITMENT OF TUBE-BUILDING AND ONE BURROWING
POLYCHAETE SPECIES TO CAGED AND UNCAGED SEDIMENTS OF A TIDAL FLAT
ON SAN JUAN ISLAND, WASHINGTON; AVERAGE NUMBER OF RECRUITS
PER CAGE IS GIVEN[a]

	Lumbrinereis inflata	Tube-building species		Total tube dwellers	Burrower *Armandia brevis*
		Axiothella rubrocincta	*Platynereis bicanaliculata*		
Without cage	130	140	335	606	50
With cage	119	134	36[b]	289[b]	143

[a]After Woodin, 1974.

[b]In these cases, there was a significant negative correlation between tube dwellers and burrowers ($p < .05$).

be true in some cases, an experiment performed by Williams (1979) yielded discouraging results. Larval settlement and survival of the clam, *Tapes japonica,* were monitored on intertidal flats with varying adult densities. Although adults depressed settlement, the effect was small even at unusually high adult density. Thus adult larval interactions may not be dramatically important in the dominance of soft-sediment populations.

Competition Between Sand-Flat Deposit Feeders

The coexistence of many ecologically similar species is a paradox, given the expectation that two species requiring the same resources cannot coexist. Competition for attachment space in the rocky intertidal results in the replacement of species by their competitive superiors. Superiority is determined by the ability to exploit a common limiting resource and to survive more successfully under given conditions of physiological stress.

In sand flats and mud flats many species of deposit-feeding benthic invertebrates seem to coexist over long periods of time. Because all species require ingestion of sediment particles for their ration, the co-occurrence of many deposit-feeding species suggests a lack of competitive exclusion or displacement or a subtle means of resource subdivision. It seems unlikely that much specialization with regard to food type occurs. Deposit feeders seem to assimilate a variety of microbial foods equally well (Kofoed, 1975; Hargrave, 1970a). Lee and co-workers (1975) have shown some differential nutritional value when different microalgal species were presented to an harpacticoid copepod.

Whitlatch (1981) examined the niche relations of deposit-feeding polychaetes of Barnstable Harbor, Massachusetts, and measured living position below the sediment–water interface and size sprectrum of ingested particles. Species were found to live at different levels beneath the sediment–water interface or to feed on different particle sizes when coexisting at the same level. Stratification of living positions has also been shown for subtidal deposit feeders (Levinton and Bambach, 1975; Levinton, 1977) and for infaunal suspension-feeding bivalves (Peterson, 1977). Whitlatch was also able to demonstrate a positive correlation between species richness and particle-size diversity. The interaction of particle-size spectrum and living position parallels the results of Schoener (1968), who showed that species of the lizard *Anolis* with dietary overlap tended to diverge in their use of microhabitats.

In Chapter 1 we suggested that ecological expansion, when competition is relaxed, is evidence of the influence of competition on niche breadth. Fenchel (1975a) examined the distribution of the mud snails *H. ulvae* and *H. ventrosa* in the Limfjorden, Denmark. Although *H. ventrosa* has an optimum of distribution in lower salinites (Fig. 16-3), there are broad regions of sympatry. Transitions of dominance from one species to the other occur at different salinites in different fjords. Examining about 100 localities, Fenchel (1975b) found that size-frequency distributions of single-species (allopatric) populations of *H. ventrosa* were identical to single-species populations of *H. ulvae*. When sympatric, however, the mean size of *H. ulvae* was 1.5 times that of *H. ventrosa* (Fig. 16-4). Fenchel interpreted this difference as character displacement (see Chapter 4): Two closely related species diverge morphologically when sympatric but remain nearly identical when allopatric (Brown and Wilson, 1956). The basis of the differentiation is a localized evolutionary event that relaxes competition and permits coexistence between the two species. Because the Limfjorden was not open to marine circulation until 1825, such an evolutionary event could have occurred in ca. 150 years (or 150 *Hydrobia* generations) or less. This type of evolutionary event must have independently occurred in many isolated bays in the Limnfjord.

The basis of the inferred character displacement is hypothesized to be particle-size specialization. Using HNO_3-treated gut contents, Fenchel (1975b) determined a log-linear relationship between median particle diameter and shell length. Both species of *Hydrobia* show the same relation (Fig. 16-5). Therefore a divergence in size would be sufficient to cause divergence in the spectrum of mineral grains and large diatoms ingested (Fig. 16-5).

One of the values of Fenchel's (1975a,b) work is that it relates exploitation of resources along one niche dimension—particle size—to a readily measured morphological

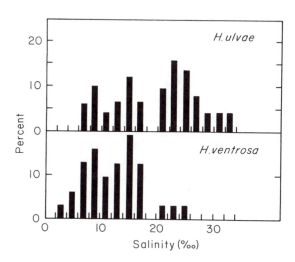

Figure 16-3 Distribution of two species of the mud snail *Hydrobia* in the Limnfjord, Denmark. Note that although *H. ulvae* tends to occur in higher salinity than *H. ventrosa*, there is broad overlap in habitat. (After Fenchel, 1975a from *Oecologia*, vol. 20)

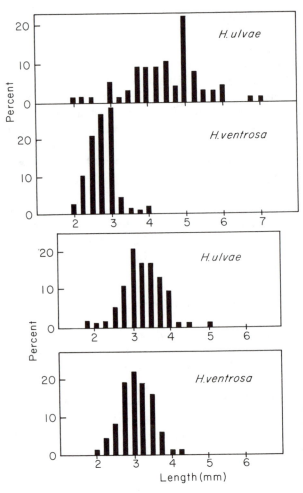

Figure 16-4 Length frequency distributions of the shells of *H. ulvae* and of *H. ventrosa* from a locality where the species coexist and from two localities where they occur singly. (After Fenchel, 1975b from *Oecologia*, vol. 20)

character, shell length. Menge and Menge (1974) investigated competition between the asteroid starfish *Pisaster* and *Leptasterias*. Competition had been demonstrated through experimental manipulations of field populations (Menge, 1972). In coexisting populations of the two species, dietary overlap was minimal, but separation was achieved in a complex way, based on a combination of prey size and prey species (Menge and Menge, 1974). Therefore simple expectations of niche separation between competitors along one niche axis (e.g., prey species) cannot always be expected. Because body length in bivalves is correlated with substrate, living depth below the sediment–water interface, and particle size ingested (Stanley, 1970; Hughes, 1973), an evolutionary change in body length can cause ecological change along three niche dimensions. As mentioned in Chapter 4, movement along two niche dimensions a small amount is ecologically equivalent, in terms

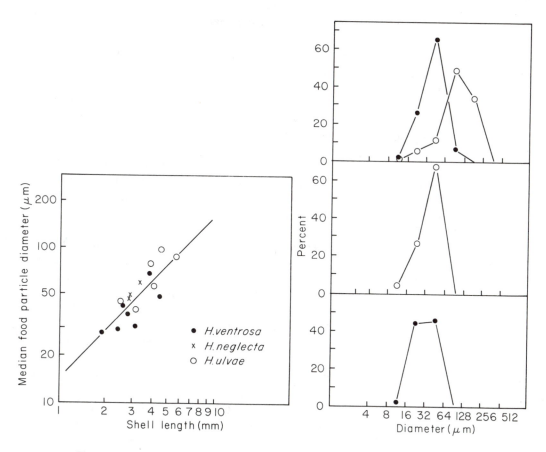

Figure 16-5   The relationship between body size and the median diameter of ingested food particles. The distribution of particle sizes in snails living singly and as a two-species combination are shown to the right. (After Fenchel, 1975b from *Oecologia,* vol. 20)

of competitive interaction, to a larger divergence along one niche dimension. Such complexities have so far been unexplored.

16-2 *PREDATION, NATURAL DISTURBANCE, AND FUGITIVE SPECIES*

If resources are not limiting and competition is depressed, then many species requiring the same resources may coexist. The relaxing of such pruning processes as predation and habitat disturbance tends to intensify competition with the subsequent elimination of a number of species. Interspecific competition therefore lowers the variety of species that co-occur within a habitat.

Predation: Experiments Employing Predator Removal

Predation refers to any process in which one living organism captures and consumes another, including the consumption of plants by herbivores. A predator capable of eating all the prey species below it in a food web ameliorates competition and delays local extinctions due to competitive superiority. Species that prevent resource monopolization by consuming prey species at lower trophic levels are *keystone species* (Paine, 1966). Keystone species may be carnivores, or herbivores, in which case the competing prey species are plants.

Pisaster ochraceus, the top carnivore of rocky intertidal food webs of the Pacific Coast of the United States, consumes barnacles, snails, and chitons but prefers the competitive dominant, *Mytilus californianus* (Landenberger, 1968). Removal of this predator therefore results in accelerated dominance by *Mytilus californianus* (Paine, 1966, 1974). A removal of the New Zealand top carnivore *Stichaster australis* for 9 months results in an extension of the vertical distribution of the mussel *Perna canaliculus* to 40% of the available intertidal range. Species richness of space keepers decreases from 20 to 14 (Paine, 1971b).

Keystone species interactions have held for herbivore-algae systems as well. In this case, however, the herbivore's grazing activities prevented a competitively superior algal species from outcompeting its inferiors. Lubchenco (1978) examined the role of periwinkle grazing *(Littorina littorea)* on rocky pools and exposed rocks in Massachusetts. The elimination of periwinkles from a pool resulted in the sudden appearance and complete dominance of a formerly rare algae, *Ulva*. This alga was relatively easy to graze, and snail grazing was sufficient to prevent dominance of *Ulva*. Paine and Vadas (1969b) removed the urchin *Stongylocentrotus purpuratus* from tidal pools at Mukkaw Bay, Washington, and monitored algal species in the following months. After variable periods of time, the brown algae *Hedophyllum sessile* dominated in the intertidal and a species of *Laminaria* dominated subtidally. Neither of the eventual competitive dominants was initially present in the experimental pools before the urchins were removed.

Predation pressure may also influence competitive interactions between intertidal seaweeds and invertebrates. The low intertidal of New England tends to be dominated by the mussel *Mytilus edulis* on headlands, but the seaweed *Chondrus crispus* occupies most space in protected embayments. Yet the wave exposure gradient itself is not the proximate explanation for the microgeographic difference in dominance. At protected sites the thaid gastropod *Thais lapillus* and the starfishes *Asterias forbesi* and *A. vulgaris* prey heavily on *Mytilus edulis*. Predator-removal experiments demonstrate that, in the absence of predation, *M. edulis* is a superior competitor over *Chondrus* in protected embayments. *Mytilus edulis* dominates in natural headland habitats because predators are excluded as a result of intense wave shock (Lubchenco and Menge, 1978).

Soft Bottoms and Predator Removal Experiments

Studies done in the rocky intertidal suggest the following prediction. When a keystone predator is removed from an area that formerly experienced high levels of predation, the

326

total biomass of prey should increase. Following this increase, a competitively superior species (e.g., mussel) displaces the other competing prey species. Although this paradigm has been repeatedly confirmed in hard-substratum intertidal systems, results are more complex in soft bottoms. Using cages, several studies demonstrate that biomass does indeed increase after predators are removed (e.g., literature summarized in Peterson, 1979). Nevertheless, the community is not simplified to one or a small number of competitive dominants. Instead species richness usually increases along with biomass and fails to simplify after a long period.

The failure to observe species reduction is probably due to the nature of competitive interactions in soft sediments compared to the rocky intertidal. In the rocky intertidal, a few species compete for two-dimensional space; often one species is capable of forming a blanket of individuals and smothering all other species. In soft sediments, such dramatic interference is probably less common. The usual effects of competition seem to be on somatic growth rates and emigration (e.g., Peterson, 1977). Furthermore, species are not competing along a two-dimensional space; one species can escape competition from another by living at different levels below the sediment–water interface (e.g., Levinton and Bambach, 1975).

It is not nearly as simple to use cages in the study of the effects of predation on soft-sediment biotas. Cages significantly affect the local hydrodynamic regime and cause the accumulation of soft sediments within the cage. Thus an increase in biomass and species richness cannot be reliably ascribed to the exclusion of predators; it may simply be due to the increased supply of fine sediments. In some cases, cages are actually the sites of recruitment of predators within the cages themselves. So cages must be used with caution in soft-sediment habitats.

Predator Influxes

High levels of predation can devastate a community. This process seems to be a major factor in the structure of intertidal sand and mud flats of the East Coast of the United States. On the North Shore of Long Island, the horseshoe crab, *Limulus polyphemus*, comes into the intertidal to breed in the spring and summer. During the mating season feeding and burrowing also occur. The combination of physical disturbance and predation on almost all members of the infaunal benthic community results in patchy and unrepeatable species distributions in protected intertidal flats.

Seasonal influxes of predators severely alter the character of intertidal sand-flat and mud-flat benthic communities of the northeastern coast of the United States. The major predators feeding on epifaunal and infaunal deposit-feeding benthos in New England salt-marsh pools are fishes of the genus *Fundulus* (killifish) (Valiella et al., 1977) and the shrimp genera *Palaemonetes* and *Crangon* (Nixon and Oviatt, 1973). In Rhode Island salt-marsh pools, killifishes arrive in large schools of juveniles in July. Shrimp also appear in the middle-to-late summer in great abundance. Colonization experiments of azoic mud trays in Sippewisset Marsh, Massachusetts, show that juvenile recruitment and growth increase the benthic population from May until July. But the advent of predator populations in July and August is accompanied by significant reductions of deposit-feeder density

(Schneider, 1977). The evenness component of diversity increases at this time because abundant species experience greater losses than less common species.

A similar decline is effected by shorebirds on the sand flats and mud flats of Plymouth Bay, Massachusetts. A summer migration of several seabirds that forage on benthos results in massive declines over the summer. A caging experiment conducted by Schneider (1978) showed that caged areas of mud flat showed no decline whereas a significant decline occurred in uncaged treatments. Birds tended to take common species disproportionately frequently and took comparatively smaller percentages of rarer species. The result was diminishing dominance by a single benthic species.

Goss–Custard (1977) showed that the redshank, *Tringa totanus*, could eliminate 16% of the tidal flat population of the amphipod *Corophium*. Feeding rate was shown to be independent of prey density. But populations of the oyster catcher, *Haematopus ostralegus*, could remove the majority of cockles, *Cardium edule*, from flats of southern Wales. The oyster catcher can also control populations of the mussel *Mytilus edulis* in North Atlantic mussel beds.

An indirect, although important effect of predation is its influence on the reproductive capacity of the prey population, lowering the prey's ability to produce young during the next reproductive period. Such an instance can be found in a predation system involving the plaice *Pleuronectes platessa*, which preys on the intertidal and shallow subtidal tellinacean bivalve *Tellina tenuis*, an infaunal bivalve commonly found in intertidal sand flats of British waters. Its siphons are separate and extend into the overlying water and are thus exposed to visually hunting plaice. Siphons can constitute over one-half the stomach contents of plaice (Trevallion et al., 1970). Although siphons can be regenerated after being snipped off, the time involved for regeneration is time lost in feeding and gonad production. Because of this negative effect, predation intensity has been related to occasional reproductive failures of *Tellina* populations (Trevallion et al., 1970). Above a critical rate of siphon removal, not enough energy is obtained to be able to allocate reserves to reproduction. Plaice probably shift to other prey species, however, as the density of *Tellina* decreases. This shift allows *Tellina* populations to recover from intense siphon cropping.

Vertical Zonation of Predators

Earlier we mentioned that predators do not seem to blanket the rocky intertidal zone. Searching, prey capture, and consumption can only be accomplished while the predators are moist. Predation is thus concentrated in the lowest portions of the rocky intertidal, which is covered by water for a greater proportion of the tidal cycle. Both starfish and drilling gastropods (e.g., *Thais*) fail to affect upper intertidal populations of mussels significantly (Seed, 1969; Paine, 1974). The limitation of predators to the lower intertidal often results in a sharp line, below which there are no abundant sessile invertebrates (Fig. 16-6). On the coasts of Yorkshire, United Kingdom, the starfish, *Asterias rubens*, and the gastropod, *Thais lapillus*, eliminate most lower intertidal mussels. In contrast, upper intertidal mortality is minimal (Seed, 1969). On exposed coasts of Washington state, the asteroid, *Pisaster ochraceus*, actively maintains a sharp line above which abundant mus-

Figure 16-6   Predation line on rocky shores of England. Below this line (adjacent to fissure), the gastropod *Thais lapillus,* and the starfish, *Asterias rubens* (not pictured), can clear the rocks of their prey, the mussel, *Mytilus edulis*. (Photograph courtesy R. Seed)

sels are found. If starfish are removed experimentally from a site, the mussel bed gradually extends downward (Paine, 1974). At the time of high tide, starfish move to the base of the mussel bed and dislodge a mussel; they then retreat to moist lower intertidal crevices with their prey. In the process, more mussels are dislodged and fall into subtidal beds of the anemone, *Anthopleura xanthogrammica* (Dayton, 1973a).

Prey Quality and Predator Selectivity

As noted earlier and in Chapter 13, many types of selectivity are exhibited during the process of predation. This selectivity can have important ramifications on community structure. The preference of *Pisaster ochraceus* for mussels (Landenberger, 1968), for instance, tends to reduce interspecific competition and allow competing species to coexist. An important aspect of selective feeding by intertidal carnivores is "ingestive conditioning" (Wood, 1968), where a carnivore feeds preferentially and often exclusively on the species that is most abundant. For example, if the oyster drill *Urosalpinx cinerea*, a drilling gastropod found in New England rocky regimes, is removed from a bed of barnacles, it will preferentially feed on barnacles in the laboratory. Fixing on a certain prey type allows the predator to develop a stereotyped series of behavioral activities that most efficiently use time to kill or develop a search image for the prey. Not overexploiting a rare prey species allows that prey to build up in numbers until it is an abundant food source. Switching is therefore expected to stabilize communities and reduce extinction. Predator switching to different prey is probably controlled by the quality and availability of prey species.

Differential quality or availability of victims may affect the rate of predation and hence the composition of the community. Although the calorific value of intertidal algae seems to show no apparent relationship to herbivore preference (Paine and Vadas, 1969a), seaweeds and diatoms that are, respectively, soft or poorly attached to the substratum are often consumed at greater rates by herbivores, such as periwinkles, than those composed of tough materials or firmly cemented to the substratum (Lubchenco, 1978). Many seaweeds characteristic of late succession contain various poisonous compounds, such as terpenes, that discourage grazing (see Chapter 13, Section 13-8).

Herbivory may explain the seasonal pattern of abundance of stages of heteromorphic seaweeds. As discussed in Chapter 7, many seaweeds display alternations of life stages of strongly differing morphologies. Seaweeds of widely differing taxonomic affinities show an alternation from an upright rapidly growing stage to a crustose, sometimes shell-boring stage. In both the rocky shores of New England and Oregon, the upright stage dominates in winter while the crust dominates in summer. Lubchenco and Cubit (1980) cleared away herbivores from rocky sites in summer; the upright stages immediately colonized, demonstrating that grazing, and not the summer climate, usually restricted the upright stage to winter dominance, when grazing was minimal. These results suggest that the heteromorphic life cycle may be adapted for resistance to grazing (crustose stage) and rapid population growth (upright stage) in the absence of grazing. This combination maximizes population growth should the opportunity arise (i.e., relaxed grazing) but permits the population to survive periods of intense predation. Although this hypothesis suggests that the species has evolved this life cycle in response to grazing, some skepticism is justified. The hypothesis fails to explain the occurrence of isomorphous life cycles of seaweeds especially prone to predation. The sea lettuces *Ulva* and *Enteromorpha* are particularly delectable to herbivores and yet have identical gametophyte and sporophyte stages. So although herbivory may explain the seasonal patterns of distribution of such heteromorphs as *Porphyra* and *Petalonia*, it seems difficult to prove that the heteromorphy evolved as a response to grazing in the first place.

Size may also strongly influence the availability of victims to predators. On West Coast rocky shores the length of prey mussels is positively related to the length of their predator *Pisaster ochraceus* (Paine, 1976). Paine interprets this relationship as reflecting a size limit above which the starfishes cannot successfully attack and open a mussel. Thus mussels could presumably outgrow their predators and obtain a size refuge. It is also possible that Paine's interpretation is incorrect and that starfishes are actually avoiding large mussels, not because they cannot subdue them but because it is uneconomical to spend the long time required when more easily subdued prey are available. This optimal foraging argument is consistent with the results obtained for crabs attacking mussels, as discussed in Chapter 4. Therefore the correlation found by Paine is ambiguous. The reality of size refuges, however, is still a likely force in these communities.

Natural Disturbances and Fugitive Species

Because the wave-exposed sandy and rocky intertidal zone is a habitat continually subjected to wave shock, waves may be a significant structuring force in the community. This action is especially apparent on sandy beaches below mean high tide, where wave energy can rapidly erode and transport sand parallel to the shoreline. Rapid removal of sand can unbury organisms so frequently that their major life styles involve adaptations to reburial, as discussed earlier.

The intensity of wave shock can exceed the survival capacities of many intertidal organisms. Floating logs are a devastating force in removing sessile epifaunal organisms from the rocky intertidal. Anyone watching a log resting next to a mussel bed in rough water can observe the crushing and dislodgement of mussels in the area of impact. Most damage in the intertidal, however, results from wave shock in winter storms. Dayton (1971) examined the effects of wave damage in a gradient on rocky shores of Washington. Wave shock and log damage initiated gaps in mussel beds. Continued prying by waves at this weak point stripped away mussels and left bare rock of primary space. Wave exposure also negatively affected common barnacles. The survival of nails implanted in rocks at each study site showed that barnacles survived best where nails were least dislodged.

Wave dislodgement is important in the sequential colonization of the rocky intertidal, for settling mussel larvae usually require *secondary space* or areas colonized by other organisms, such as barnacles or filamentous algae. Therefore organisms that settle readily on *primary,* or unoccupied, space will settle—some algae and barnacles, for instance. So the effects of wave shock and log damage may be qualitatively different than those caused by a predator that removes prey but does not completely eliminate the secondary spatial structure of the substratum.

With the constant availability of new space, whether primary or secondary, we might expect a group of species adapted to colonizing such areas. We might expect these species to be the ones that could not compete with eventual dominants and so take advantage of such physical disturbances. Good examples of such species in the rocky intertidal are barnacles, which can colonize primary space but are eventually outcompeted by mussels (e.g., *Mytilus californianus* winning out over *Balanus glandula*). Species adapted to colonizing newly opened ecological space were termed fugitive species by Hutchinson (1957). A good example of such a species is the palm seaweed *Postelsia palmaeformis* (Fig. 16-7), a resident of exposed outer-coast environments on the Pacific Coast of the United States. *Postelsia* occurs in the high intertidal and successfully colonizes areas of the mussel bed that have been ripped apart through log damage with subsequent further wave shock damage (Dayton, 1973a). Although the exact mechanism is not clear, the mussel *Mytilus californianus* must have some detrimental effect on sporophytes. Dayton cleared areas of mussels and showed that sporophytes colonized these areas with no concurrent colonization of control sites containing mussels. Sporophyte colonization distance, however, was only about 3 m from the edge of an existing *Postelsia* patch. Short-distance dispersal may therefore serve the purpose of maintaining a successful foothold on the opened space. Dayton concludes that colonization of a newly opened patch of primary space is an uncertain event and probably occurs through floating over long distances of a sporogenic plant.

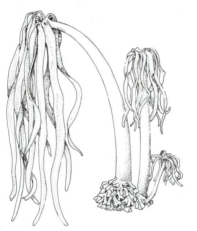

Figure 16-7   The sea palm, *Postelsia palmaeformis*. This brown alga colonizes wave-swept shores that have recently been denuded of the competitive dominant *Mytilus californianus*. (Reprinted from *Between Pacific Tides*, Fourth Edition, by Edward F. Ricketts and Jack Calvin. Revised by Joel Hedgpeth, with permission of the publishers, Stanford University Press. Copyright © 1968, by the Board of Trustees of the Leland Stanford Junior University.)

Recolonization of soft-bottom intertidal flats could be followed after a potent oil spill occurred near West Falmouth, Massachusetts, in the fall of 1969. The entire marine benthos was wiped out in adjoining intertidal areas and provided a somewhat unnatural experimental environment within which to view the effects of disturbance and subsequent responses. Grassle and Grassle (1974) examined the sequence of events that followed the elimination of the benthos, with particular regard to the polychaete fauna. Experimental colonization trays of frozen and then thawed mud were placed in an intertidal station to examine the sequence of colonization. A small polychaete, *Capitella capitata*, normally rather uncommon in the intertidal fauna, increased rapidly to spectacular densities (over 10^6 m^{-2} in some cases). This species is actually a complex of several sibling species (Grassle and Grassle, 1976) and is well known as a pollution indicator (Reish, 1970).

In adjacent settling trays, one *Capitella* population could be crashing while in another tray one might be in a phase of rapid increase.

Grassle and Grassle erected a hierarchy of opportunistic species, where opportunistic species are identified as those with rapid population increase, soon to be followed by high mortality. Figure 16-8 shows a colonization sequence for the polychaetes of the West Falmouth experiments. It is not clear whether this sequence implies a dependence on conditioning of the environment by previous opportunists or whether it is simply a hierarchy of species' respective abilities to respond to an opened environment; the latter

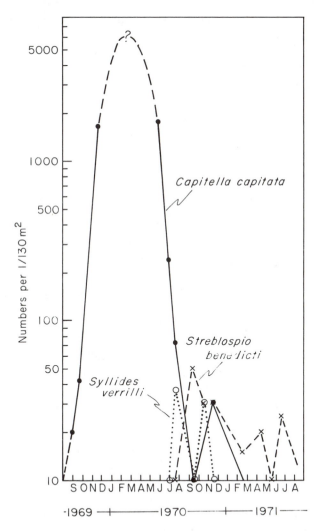

Figure 16-8 Colonization of polychaete species in experimental mud trays adjacent to a salt marsh following a major disturbance. (Modified after Grassle and Grassle, 1974)

seems most likely. Similar to the case of the algae *Postelsia palmaeformis* discussed, *Capitella* seems adapted to invading a local area and increasing rapidly within that area.

When an intertidal flat is undisturbed, dense populations of opportunistic annelids are usually rare (Sanders et al., 1965). What permits the buildup of great population density? The resident fauna might be capable of resisting invasion. Levinton and Stewart (1981) placed ten individuals of the oligochaete *Paranais litoralis* in Petri dishes with varying density of the deposit-feeding snails *Hydrobia totteni* and *Ilyanassa obsoleta*. In both cases, *P. litoralis* population buildup was depressed with increasing snail density. Concurrent experiments show that increasing densities of snails severely reduced the standing crop of microalgae (Levinton and Bianchi, 1981); therefore low standing crop of food probably contributed to depressed worm population growth. It is also possible that snails reduced the survival of newly hatched larvae; this hypothesis was not fully explored, however. It is likely that the longer life span of the snails permits them to outlast the shorter-lived oligochaetes; thus snails are usually the predominant species on mud flats because they have survived from the previous winter and now resist the invasion of colonizing worms through resource depression and perhaps direct interference. The resistance is not always completely effective, for annelid populations build up every summer in mud flats of Long Island Sound.

Disturbance and Spatial Heterogeneity

Hydrodynamic and biological disturbance of intertidal sand and mud flats both tend to be patchy in distribution. Localized sediment excavations by horseshoe crabs disrupt polychaete burrows and increase the fluidity of the sediment. Tidal currents generate ripple marks that strongly affect microscale spatial variation of sediment and current strength. Flats are usually covered with ephemeral channels that retain water for a greater proportion of the tidal cycle. These factors suggest that tidal flats are spatially heterogeneous in both sediments and current regime; such heterogeneity must affect organism distribution.

A possible source of spatial heterogeneity is the distribution of the benthic creatures themselves. Some tidal flats are forests of polychaete tubes projecting above the surface. Tubes may alter current patterns on a microscale and might influence the distribution of settling larvae. A dense stand of tubes might also interfere with the movement and burrowing activities of otherwise potential agents of biological disturbance, such as crabs and naticid gastropods. Tube concentrations may thus serve as a spatial refuge for associated free-living infauna (Woodin, 1978).

To assess the effect of tubes, Eckman (1979) planted sewing needles at regular intervals in a mud flat. Using autocorrelation analysis, it was possible to demonstrate that the abundances of several invertebrates showed spatial periodicity coincident with the needle implants. In other localities, spatial periodicities of distances as small as 2 to 3 cm could be established. A common periodicity of 15 cm was consistent for several species and persisted over time. The 15-cm periodicity may be explained by a transient set of ripples that existed at the time of settlement.

The local hydrodynamic disruption of current patterns induced by ripples or tubes might attract benthic organisms in several ways. Fine particles might accumulate in the troughs of ripples; deposit feeders might thus be attracted. Tubes might have attacked algae for herbivores or serve as an initial settling site that is relatively free of infaunal predators.

Eckman's results suggest that caution must be used in interpreting the results of field manipulative experiments using cages in soft sediments. Cages disrupt currents and are usually sites of fine-sediment accumulation. So the increased success of benthos in a cage may not be due to decreased predation; instead it may simply be due to the presence of fine sediments and their positive influence on some deposit-feeding or burrowing organisms. The reader is invited to consider the discussion of Woodin's (1974) experiments (discussed earlier) in this light.

16-3 INTERTIDAL COMMUNITY STRUCTURE: SYNTHESIS

Several points emerge from the previous discussion, giving some general insights into the structure of intertidal communities and marine communities in general. First, several processes operate simultaneously to control population size and species composition in a given locality: (a) colonization mainly through pelagic dispersal stages, (b) intraspecific and interspecific competition, (c) predation, and (d) natural disturbances. The relative importance of these four factors often varies from place to place and from time to time in a specific locality. A kilometer stretch of intertidal is often a complex of patches in various states of change after a biological or physical disturbance. The competitive dominant is frequently suppressed, allowing competitively inferior but early colonizing species to invade the patch (e.g., *Mytilus edulis* invading newly opened patches in a *Mytilus californianus* bed).

The unpredictable nature of intertidal community structure is partly related to differences in recruitment of larvae and other pelagic dispersal stages of algae and invertebrates. There are often good and bad years in larval recruitment, a fact that partly determines whether a given species requiring space will be a successful competitor. The high recruitment rate of mussels is an important factor in their ability to displace barnacle competitors (Paine, 1966). But mussel recruitment is sparse in many years. Moreover, the reasons why mussel larvae are plentiful in the plankton may be independent of the resident fauna of a given intertidal locality. Poor years for phytoplankton are also poor years for larvae of the acorn barnacle *Balanus balanoides* in British waters (Barnes, 1956). These years must be the same years in which *Chthamalus* may be able to extend farther down into the intertidal because of a lapse of competitive pressure from *B. balanoides*. The fact that we see good zonation that seems repeatable for general groups of organisms throughout the temperate and boreal world suggests, however, that recruitment is predictable enough to result in repeatable competitive hierarchies and zonation.

General changes in the physical environment also have great impact on the structure of intertidal communities. *Chthamalus stellatus* is basically a tropical species whereas *Balanus balanoides* is boreal–Arctic in distribution. In the area of geographic overlap in Scotland, these two species interact as discussed. But the direction and success of this competitive interaction are often a function of climatic changes that move temperature isotherms toward the north or south. When the waters of western Great Britain get warmer, *Balanus* is in a less favorable position due to its thermal adaptations and its competitive success, relative to *Chthamalus,* lessens accordingly. Thus biogeographic interactions with the physical environment have important influences on species composition.

These generalizations make the intertidal seem a stressful environment whose species structure is mainly governed by the impingement of a harsh physical environment on a biota fighting for its very existence against the elements (wave shock, desiccation, heat). In wave-swept sandy beaches, areas of freezing and ice cover, and in some other intertidal biomes, this is true. But in most intertidal environments, both hard and soft substratum, biological interactions are important in determining the nature and abundance of intertidal species. Paine (1974) discusses this problem in the framework of a model proposed by Sanders (1968). Sanders emphasizes the great physical stresses imposed by the intertidal zones as opposed to the relatively benign nature of the subtidal tropics or deep sea, where the environment changes less (at least with respect to classical oceanographic parameters, such as temperature and salinity). Sanders concludes that intertidal populations change and species composition should therefore be controlled by physical forces, with biological interactions, such as competition and predation, being of relatively little impact. Paine (1969b) describes the fauna of the rocky intertidal zone of the West Coast of North America, noting the homogeneity in species composition from Baja California to Alaska. There are striking similarities in major space keepers, predators, and the attendant general hierarchies or competitors and predation. The various elements of competitive superiority and predator preference perhaps suggest that this system has evolved into a balance that is dynamic, although stable (Paine, 1974, p. 118).

> A species has evolved, *M. californianus,* that can competitively hold a spatial resource against all other species; a predator, *Pisaster,* has evolved that consumes this prey preferentially, and in so doing predictably renews the limiting resource, making it available to a host of other species. It is precisely because the dynamic interaction is predictable that the rocky intertidal is biologically rich and vibrant community, one showing few signs of the lack of organization and absence of important or visible processes that one would associate with an ecologically stringent environment continually disrupted by events of unpredictable timing, position or magnitude.

Paine excludes the highest intertidal zone from this general characterization, where adaptation and response to desiccation become of major importance.

Perhaps this view is overstated, especially the point that the biological processes of predation and competition necessarily lead to some stability. The work of Dayton (1971) suggests a balance between evolved or preexisting hierarchies of competitive ability and the influence of predation and natural disasters. The spatial variation of the interaction between these factors causes the characteristics of any square meter of the intertidal to be unpredictable. We can generally state what *may* be in a given rocky shore, but the species composition of a given spot may vary, depending on recruitment, competitors, predators already present, and the chance possibility of wave damage. Even recruitment will differ if the space available is primary or secondary. Moreover, the presence of resident grazing animals will often disrupt successful recruitment. The bivalve *Gemma gemma* may swallow larvae and keep other bivalves from usurping its position on sand flats (Sanders et al., 1962). Furthermore, limpet grazing often bulldozes newly settled cyprid larvae from rocks, causing high mortality (Connell, 1961a,b; Dayton, 1971). The intertidal is thus best thought of as a dynamic mosaic rather than a biologically well-tuned superorganism.

SUMMARY

1. Vertical gradients in physical parameters set the framework for interspecies interactions. Dominant sessile species occupy specific zones. Usually they are limited above by their ability to tolerate desiccation and limited below by better competitors. Interference competition is the rule on rocky shores. Superior competitors bulldoze, smother, and undercut less effective competitors. There is some evidence for the role of exploitation competition in mud-flat deposit feeders. Body-size differences permit species to coexist by exploiting different grain sizes.

2. Predators alter the course of competitive interactions by pruning potential competitive dominants. Increased drying in the upper intertidal tends to confine predatory activities to the lower part of the shore. In some cases, predation is so intense as to create a line below which prey are virtually absent.

3. Wave shock tends to dislodge common species and provide open space for colonization. Intertidal habitats can be visualized as a mosaic of patches differing in the time since a major disturbance. Intertidal community structure can be regarded as an interaction between competition, disturbance, and predation.

17 estuaries

17-1 PHYSICAL AND CHEMICAL STRUCTURE

Introduction

Estuaries are the unstable interfaces across which the freshwater drainage of the terrestrial world communicates with the open sea. As such, they are highly variable in physical, chemical, and biological properties. This variability and the extremely low salinity have strong effects on both the composition and the dynamics of the biota. Seasonal and sometimes unpredictable changes in weather patterns exert particularly strong effects on estuaries because of their incomplete coupling with circulation in the open sea. Restricted exchange allows rapid changes in salinity, temperature, nutrients, and sediment load.

An estuary is a semienclosed coastal body of water that has a free connection with the open sea and within which seawater is measurably diluted with freshwater derived from land drainage (Pritchard, 1967). In contrast, *estuarine realms* are large coastal water regions that have geographic continuity and that are bounded landward by a stretch of coastline with freshwater inputs but bounded oceanward by a salinity front. For example, the northeast Atlantic estuarine realm is bounded landward by the eastern Atlantic coastline ranging north from Cape Hatteras and is bounded seaward by the Gulf Stream and Arctic Sea. Within this region, a large assemblage of rivers feeds freshwater into the sea, creating a surface mass of brackish water that overlies a deep-water mass of open-ocean water (McHugh, 1967). Finally, *brackish seas* are large water bodies (thousands of square kilometers) in which tidal stirring and seaward flow of freshwater do not exert enough of a mixing effect to prevent the body of water from having its own internal circulation pattern, aside from simple seaward flow. The Baltic and Black seas are good examples of brackish seas.

Origins of Estuaries

The origin of an estuary is controlled by changes in sea level via glaciations (lowering of sea level) and deglaciations. Therefore all present-day estuaries have postdated the last glacial maximum when sea level was approximately 100 m lower than today. Most estuaries on the Atlantic Coast of North America were probably established in the last 3000 to 5000 years. Since that time sinking of the coastline under the burden of the water column may have resulted in the final configuration of estuaries. Estuaries are probably far more common today than they were during the glacial maximum (Schubel and Hirschberg, 1978).

The specific morphology of estuarine basins may originate from (a) the drowning of river valleys, (b) the formation of barrier beach bars that enclose a shallow embayment of water, (c) the drowning of glacially carved valleys (fjords), and (d) the formation of a basin via tectonic activity (i.e., an active earth's crust). On the East Coast of the United States most estuaries are drowned river valleys (e.g., the Hudson River, which is also partially glacially excised). Long Island Sound, which might be regarded as a brackish sea, was a freshwater lake until a postglacial rise in sea level converted it into an estuarine system.

Estuaries are geologically ephemeral and can rapidly disappear with a relatively small change in the world's glacial ice budget. A modest (e.g., 10-m) drop in sea level could uncover most of Chesapeake Bay; a similar rise in sea level would probably drown and eliminate many small estuaries. Deposition of sediment and infilling by salt marshes inexorably proceed to fill in an estuarine system. So even estuaries enjoying a relatively long period of stable sea level are probably doomed to extinction. The deposition of a delta seaward of a river can eventually seal off the potential interaction between the freshwater river and the sea. Atlantic and Gulf coast estuaries probably have a life span of no more than 10,000 years.

Brackish seas can also be geologically recent phenomena. The origin of the Baltic Sea is complex but includes a transition from a series of freshwater-lake stages to a brackish sea following the most recent rise in sea level. In contrast, the Black and Caspian seas are ancient and harbor biotas that are in part a vestige of a formerly rich marine biota.

Estuarine Circulation Patterns

The upstream–downstream physical structure of estuaries varies in response to the net interaction of freshwater flow, friction, and tidal mixing. In a *highly stratified estuary* (Fig. 17-1), freshwater flows downstream over a deeper layer of higher salinity ocean water. In the absence of friction, seawater extends upriver to mean sea level. Frictional forces between the freshwater lense and the underlying seawater, however, tend to transfer

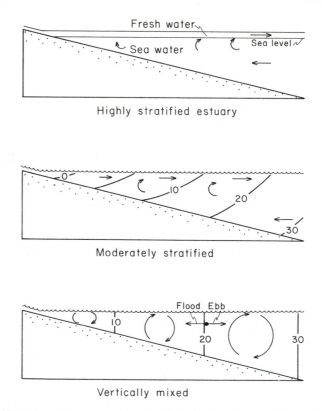

Figure 17-1 Salinity structure variation in estuaries. Lines of equal salinity are indicated for moderately stratified and vertically mixed estuaries.

denser seawater into the freshwater layer above. The transfer increases the transport seaward of surface water and requires a slow compensating up-estuary flow in the lower seawater layer (Pritchard, 1967). Highly stratified estuaries only exist where river flow strongly dominates over tidal motion, as in the Mississippi River.

With moderate tidal motion, mixing occurs at all depths and vertical exchange occurs downward as well as upward. This circulation generates a *moderately stratified estuary* (Fig. 17-1). The two-way vertical mixing causes the salinity of both the upper and lower layers to increase seaward. At any point in the estuary, however, the deep-layer salinity exceeds that of the surface. Vertical exchange tends to bring a greater volume of water to the surface layer than in the highly stratified estuary. Thus seaward surface flow increases concomitantly with an increased and compensating up-estuary flow in the deeper layer. Chesapeake Bay and the Savannah River estuary are good examples of tidally stratified estuaries.

Vigorous tidal mixing tends to homogenize the vertical salinity gradient and results in a *vertically homogeneous estuary* (Fig. 17-1). Because of a strong tidal control, the salinity at any point in the estuary changes radically, depending on the state of the tide.

At low tide, the salinity is dominated by downstream river flow whereas at high tide the inrush of seawater determines the salinity. Only estuaries of very small extent permit such domination by tidal motion. The Pocasset River estuary in Massachusetts demonstrates such tidally related salinity fluctuations (Sanders et al., 1965).

Freshwater Pulses and Salinity Structures

Changes in freshwater input can strongly influence the downstream–upstream distribution of salt. Seasonal variation in rainfall and spring snow melts both increase freshwater drainage and shift downstream lines demarking constant surface salinity (isohalines). During dry seasons isohalines shift upstream. The area drained by the estuarine system, relative to the volume of the estuary, determines the extent to which pulses in freshwater flow can alter the salinity structure of the estuary. In North Carolina estuaries, such as the Newport River, the diminutive nature of the estuary permits pulses of freshwater to cause enormous fluctuations in salinity over scales of a few days. In contrast, far larger fluctuations of freshwater input would be required to alter strongly the downstream salinity structure of a large estuarine system, such as Chesapeake Bay. Tropical storms and hurricanes can dump enough freshwater to change the salinity structure of even Chesapeake Bay enormously. In 1972 tropical storm Agnes caused tremendous increases in river discharge and record decreases in salinity.

Sedimentary Processes in Estuaries

Sedimentary transport in estuaries is regulated by the combined effects of estuarine flow and tidal action. Particles derived from land drainage are entrained in the upper layer of low salinity and transported seaward. In the lower reaches of the estuary, tidal forces resuspend sedimentary particles from the bottom. Particles also tend to be mixed downward from the surface layer due to the turbulence. Because of the net up-estuary flow in the deeper layer, particles may be transported up-estuary. The erosion caused by bottom currents tends to transport fine particles up-estuary as well. Estuaries thus serve as a sink for fine sediments; estuarine bottoms are typically dominated by silt and clay except in shallow water where bottom currents are swift.

The effects of tidal erosion of bottom sediments are related to both the tidal and the seasonal cycle. With every flood tide, particles are saltated from the bottom into the water column. This effect is sufficient in summer to generate a layer of turbid water several meters thick above the bottom (Rhoads et al., 1975). In winter biogenic activity is slight and the sediment tends to be more cohesive. Because of the cold temperature, bacterial film on fine sediments is not consumed by the meiofauna; this factor reduces the erodability of the bottom. In summer active bioturbation and pelletization of the sediment and consumption of the bacterial film tend to present a hydrodynamically rough surface to the overlying water. The critical velocity required to erode the bottom is thus decreased. The resultant erosion tends to increase turbidity in the water column and to increase the rates of exchange of nutrients with the overlying water (Rhoads et al., 1978; Yingst and Rhoads, 1978).

Estuarine sedimentation is strongly related to seasonal patterns in freshwater input. Spring pulses of freshwater deliver the majority of the annual fluvial input of suspended sediment to Chesapeake Bay. This deposited material is reworked during the rest of the year by tidal currents and wind-generated turbulence (Schubel, 1972). Infrequent storms may contribute disproportionately to sedimentation. Tropical storm Agnes generated floods that delivered more than 30 times the typical annual sediment input to the Susquehanna River–Chesapeake Bay estuary (Schubel, 1974). This and another flood period in 1936 account for at least half of all the sediment deposited in the upper Chesapeake Bay since 1900 (Schubel and Hirschberg, 1978).

17-2 SPATIAL PATTERNS OF THE ESTUARINE BIOTA

Estuarine Transitions

Salinity changes in estuaries present significant physiological challenges to marine organisms. As discussed in Chapter 2, changes in salinity require adjustments both in ionic composition of cellular fluids and in the total concentration of dissolved materials. The pattern of species richness in estuaries reflects responses and adaptations to selective forces related to salinity gradients.

It would be useful to focus on those areas of the estuary in which natural selection is likely to be intense. There are two such transitions. The mouth and lower reaches of the estuary constitute the first area where organisms must adjust physiologically to lowered salinity. In a small, tidally mixed estuary, the salinity gradient is liable to be unpredictable and rapidly changing. Salinity, for example, can change from fresh to completely marine in a diminutive estuary, such as the Pocasset River, Massachusetts, over one tidal cycle. In larger estuaries, such as Chesapeake Bay, however, estuarine salinity gradients are far more gradual and are affected by tidal motion to a far lesser degree. Seasonal changes in freshwater input may change salinity at any one point in a large estuary throughout the year.

If salinity is tidally regulated, benthic organisms may experience fresh and saltwater in a single tidal cycle. Such *daily estuaries* present more of a physiological challenge because of the time required for acclimation. *Seasonal estuaries* show seasonal shifts up- and down-estuary, but the rate of salinity change at any point is slow, thus permitting acclimation. Because of the stratified nature of seasonal estuaries, subtidal bottom marine organisms may penetrate the estuary farther than pelagic forms. Infaunal species experience less salinity variation over a tidal cycle in a daily estuary than epifaunal species because of the buffering effect conferred by sediment pore waters that reach slowly to the overlying water.

The second major estuarine transition is the *critical salinity* (Khlebovich, 1968; Kinne, 1971). This region encompasses an approximate salinity range of 5 to 8 ‰ and marks a pronounced minimum of benthic invertebrate species richness (Remane and Schlieper, 1958). A relatively rich fauna of bivalve mollusks and other invertebrates reside in freshwater. Freshwater species decrease, however, in numbers at a maximum salinity of 5 to 8 ‰. Estuarine marine bivalves are also rare at this salinity but increase steadily in species richness with increasing salinity. Khlebovich (1968) maintains that this critical salinity range is the threshold above which ion regulation is not necessary in most marine organisms.

The critical salinity marks a zone where Ca/Na and K/Na increase dramatically with decreasing salinity. The critical salinity also marks an empirically elucidated break in physiological adaptation, at least in bivalve mollusks (Gainey and Greenberg, 1977). Marine bivalves can regulate cell volume in higher salinities by changes in cellular free amino acids but are generally incapable of the ion regulation necessary at and below the critical salinity. Freshwater species have lost the capacity for extensive volume regulation but can regulate ionic concentration and maintain a hyperosmotic state in freshwater.

Biotic Responses to the Estuarine Gradient

Many biological variables correlate closely to the estuarine gradient. In some cases (e.g., physiological response to salinity), the role of salinity is direct. In others (e.g., niche expansion), however, the role may be indirect (i.e., by diminishing the number of competing species that can survive the lowered salinity). Consequently, the following effects are, to varying degrees, the result of both direct and indirect effects of salinity fluctuations.

Gradients in species richness.   As noted, species richness generally diminishes steadily up-estuary and reaches a minimum at the critical salinity. Species richness then increases again in freshwater (Fig. 17-2). The steady decrease in the estuary must be related to the steady reduction of species capable of extensive cell volume regulation. Certain groups, such as echinoderms, brachiopods, and nuculid bivalves, are notorious for being uniformly absent in low salinities. Other groups (e.g., bivalve mollusks, po-

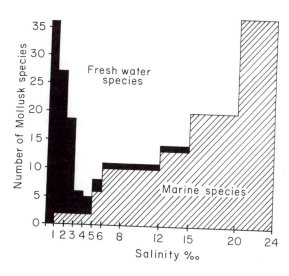

Figure 17-2   Species richness along the estuarine gradient of the Randersfjord, Denmark. (After Remane and Schlieper, 1971)

lychaetes) simply show a quantitative diminution in the number of species able to tolerate the stress.

Physiology.   As discussed earlier, volume regulation is required as we pass into those reaches of the estuary where seawater is measurably diluted with freshwater. At the critical salinity, regulation of the concentration of specific ions (e.g., sodium, potassium) is necessary as well.

Morphology.   Pronounced differences in morphology can be demonstrated with distance up an estuary. In many cases, the maximum size of bivalve mollusks decreases with decreasing salinity (Fig. 17-3; Remane and Schlieper, 1971). This effect, however, may simply be related to the negative role of decreased salinity in growth. Changes in meristic (countable) characters, such as numbers of fin rays in fishes, have also been observed. This finding suggests that genetic differentiation tracks the salinity gradient.

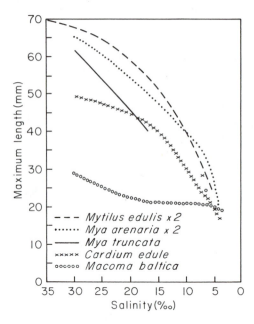

Figure 17-3   Reduction in maximum length of some bivalve mollusks along a salinity gradient. (After Remane and Schlieper, 1971)

Niche breadth.   The diminution of species richness in brackish water is accompanied by niche expansion of those species capable of invading the estuary or surviving in a brackish sea. The deposit-feeding tellinacean bivalve *Macoma balthica* is almost invariably intertidal in open marine North Atlantic (and East Pacific) mud flats. But in the Chesapeake Bay estuary and in the brackish Baltic Sea it maintains large subtidal populations in a variety of sediments. The gastropod *Hydrobia* is generally found in soft sediments except in the inner Baltic, where it replaces salinity-sensitive littorine gastropods on rocks and algae. These expansions suggest that competition for resources has restricted the ecological range of these species in more diverse environments.

A related phenomenon to niche expansion is the presence of enormous populations of a relatively few species. This situation is probably due to the relative lack of competitors and the nutrient enrichment of estuaries. Large fluctuating populations of phytoplankton, invertebrates, and fishes dominate the estuary. Chesapeake Bay harbors enormous populations of oysters *(Crassostrea virginica)* and blue crabs *(Callinectes sapidus)*. The bay is literally carpeted with both species at certain times of the year. The density of blue crabs is sufficient to devastate the benthic biota; benthic habitats are completely governed by crab predation (Virnstein, 1977). An obvious result of this dominance pattern is the status of estuaries as rich invertebrate fisheries. The Great South Bay of Long Island, New York, supports an enormous clam *(Mercenaria mercenaria)* fishery, the largest fishery in New York State waters.

Conservation of pelagic species.   As noted, estuarine flow results in a net transport of surface water to the open sea. This situation would dilute the larval populations of estuarine species unless some mechanism existed to conserve larvae. In many species, larvae are adapted to stay near the bottom during ebb tide and rise up into the water column during the flood. The result is to counteract the seaward dilution effect (see Chapter 8 for further discussion on larval loss). Because phytoplankton lack such adaptations, the presence of a unique estuarine flora is inversely related to the tidal exchange rate with the coastal waters.

Genetic divergence.   Extensive evidence exists for the presence of spatial changes in gene frequencies at the mouths of estuaries and brackish water seas. An excellent example is the entrance to the Baltic Sea. Sharp geographic differentiation has been found at polymorphic genetic loci coding for enzymes (see Chapter 6), for the eel pout, *Zoarces viviparus,* and the mussel, *Mytilus edulis,* between the North Sea and the Baltic Sea. For both species, distinct morphological differences have been recorded as well. Thus sharp differentiation has been found over a short geographic space in a species with viviparous reproduction *(Z. viviparus)* and one with dispersal via relatively long-lived pelagic larvae *(M. edulis)*.

Probably some form of natural selection enhances dynamically the differentiation observed (Levinton and Lassen, 1978). Although some estuaries may be somewhat hydrodynamically isolated from open marine waters, many estuaries have openings with extensive two-way exchange. So dynamic selection probably reinforces the genetic differentiation between estuarine and open marine populations (Levinton, 1980b). Selection may act on variation at individual loci, but it is probable that estuarine populations have diverged broadly into genetically distinct races relative to their open marine conspecifics. Consequently, fine-scale adaptation to local estuarine conditions has resulted.

17-3 ESTUARIES AS NURSERIES

As noted, estuaries harbor large populations of relatively few species. The lack of competitors and the high input of nutrients both contribute to conditions supporting dense populations of commercially exploited species. Estuaries are particularly important as nurseries for juveniles of many fish species. On the Atlantic Coast of the United States, the majority of commercially exploited fish species utilize estuaries as juvenile feeding grounds.

Many fish species spawn offshore, but juveniles drift to estuaries to spend one or a few years of their early life. The Atlantic menhaden, *Brevoortia tyrannus,* for instance, is a commercially important planktivorous fish in the eastern middle Atlantic seaboard of the United States. Adults spawn offshore, but larvae and juveniles live and feed in such estuaries as Chesapeake Bay. Larvae probably move passively toward shore in currents generated by Ekman circulation. Residual currents may move larvae of Pacific estuaries inshore as well.

Fishes face problems in remaining within the estuary. The seaward estuarine flow requires either that tidal exchange is slow or that retention mechanisms be developed to prevent flushing from the estuary. Small estuaries with extensive tidal flushing cannot support large nurseries of juvenile fishes; any retention adaptations would probably be insufficient to counteract the tidal exchange with the open sea. Larger estuaries with longer flushing times can support fisheries because of the reduced loss of larvae to sea. Within large estuarine systems, such as Chesapeake Bay, tributaries like the St. Mary's River have relatively low tidal exchange rates with the rest of the estuary. The reduced exchange tends to trap larvae and may be the reason for heavy larval sets of the oyster *Crassostrea virginica.* To counteract the estuarine flow that does exist, estuarine fish and invertebrate larvae have been observed to keep to the bottom during the ebbing tide and actively swim with the flood. Menhaden are more easily netted during flood tides, indicating their adaptations for retention within the estuarine system. Larvae prefer water of low turbulence and usually reside in tidal creeks and quiet shallow water.

Because of abundant food, low predation pressure, and the lack of many truly estuarine endemic fish species (McHugh, 1967), estuaries are ideal feeding grounds. They are among the most productive of marine environments; food abundance fluctuates strongly over time and space, however. The abundance and fluctuating nature of the food supply have favored the following characteristics of the feeding ecology of juveniles of common marine fishes in Atlantic estuaries, such as the spot *(Leiostomus xanthurus)* and the croaker *(Micropogonias undulatus)*:

1. Flexibility of feeding habits in time and space.
2. Omnivory.
3. Sharing a common pool of food resources among species.
4. Exploitation of food chains at different levels by the same species.
5. Ontogenetic changes in diet with rapid growth.
6. Short food chains based on detritus-algal feeders (Miller and Dunn, 1980).

Thus the most successful estuarine juvenile fishes have marked niche breadth; a wide variety of foods and habitat types is exploited.

Nutrient Inputs

Estuarine waters owe their productivity to the extraordinary amounts of nutrients that enter the estuary and to extensive recycling of nutrients between the overlying water and the biologically active bottom sediments. Inputs can be summarized as follows.

River input. Freshwater drainage delivers large amounts of nutrients in dissolved and particulate form. In the Pamlico River of North Carolina, phosphorous is always in abundant supply and never limiting to phytoplankton growth (Hobbie et al., 1975). Pulses of freshwater input often deliver large amounts of dissolved nutrients. In small estuaries, nutrients derived from washouts following rains may be the source of brief phytoplankton blooms in creeks. Particulate organic matter is also transported in the estuary, both continuously and episodically. Infrequent storms can deposit many centimeters of sediment and particulate organic matter in estuaries. The microbial attack of particulate organic matter subsequently facilitates the introduction of nutrient elements into the estuarine system.

Oceanic import. As discussed earlier, a net landward flow at depth compensates for the net seaward estuarine flow of surface water. Dissolved nutrients arrive with the deep-water flow into estuaries. Vertical mixture with the surface water permits the phytoplankton to use this nutrient source; benthic algae and sea grasses may also benefit. Landward transport of nutrients in deeper waters may be partially responsible for pulses of production in nearshore and estuarine waters.

Regeneration from the bottom. Bottom sediments consist of an active microbial community whose ability to decompose particulate organic matter permits the continuous recycling of nutrients between the bottom and the overlying water. Large amounts of particulate organic matter settle to the bottom of estuaries, but most nitrogen and phosphorous is mineralized and returned to the water column via diffusive and advective processes (see Chapter 12, Section 12-4 for a discussion of decomposition). Nutrient fluxes across the sediment water interface are important to primary production in summer when water-column-dissolved nutrients would otherwise be low (Boynton et al., 1980; Nixon et al., 1976). Such a mechanism could only work in an estuary sufficiently shallow for effective vertical mixing. In this case, resuspension would also be an effective means of nutrient exchange between the bottom and the overlying water.

Primary Production and Food Webs

The extensive variation of size, nutrient input, and tidal exchange among estuaries suggests a large variation in the pattern of connections between primary and secondary production. Most studies of nutrients and production of estuaries derive from a series of classic investigations of the dynamics of food webs in Georgia salt-marsh estuaries (Teal, 1962; Odum and de la Cruz, 1967). Most of the primary production is believed to derive from the cordgrass *Spartina alterniflora;* little of this production, however, is directly transferred to the herbivores. Most cordgrass escapes herbivory and circulates through detritus pathways. Many estuaries in the southeastern United States are dominated by suspended detritus and many of the secondary consumers ingest detritus and digest and assimilate the attached microbiota (e.g., Teal, 1962; Odum and Heald; 1975). Detritivorous fishes like mullets are major consumers in such estuaries.

Although detritus is undoubtedly a major factor in estuarine food webs, algal production and grazing of phytoplankton are now known to be of considerable importance as well (Haines, 1979). The stable carbon isotopic composition of suspended particulates in Georgia estuaries suggests its derivation from algal production rather than from cordgrass. Much of the particulate organic matter generated within Georgia salt-marsh estuaries may be trapped in marsh sediments and decomposed in situ. Salt-marsh creeks may therefore be a haven for abundant detritivorous invertebrates and juvenile fishes. We must conclude that marsh estuarine food webs are obviously complex.

If we consider other estuarine systems, among-estuary variation in dominance by primary producer groups is also obvious. Phytoplankton dominate the primary production of large estuarine basins having restricted interchange with the open ocean and having small marsh peripheries relative to basin size (e.g., Long Island Sound; Bedford Basin, Nova Scotia). Nutrient inputs from sewage outfalls tend to favor phytoplankton as well. Seaweeds, however, dominate the primary production of some open basins with extensive connection to the ocean (e.g., the kelp *Laminaria* in St. Margaret's Bay, Nova Scotia). Nutrient-rich water derived from the ocean, combined with swift within-bay currents, provides favorable conditions for luxuriant kelp forests (Mann, 1975). Small basins with quiet waters tend to be dominated by marsh grass, mangrove, and seagrass (e.g., *Zostera marina*) production at the basin peripheries. Detritus pathways dominate food webs in such basins because herbivores consume little of this production directly. Georgia salt-marsh estuaries probably fit best in this last category.

The Outwelling Hypothesis

E. P. Odum hypothesized in 1968 that outwelling of dissolved nutrients and particulate organic matter from estuaries provided a principal nutrient source for shelf productivity (E. P. Odum, 1980). The magnitude of detrital production alone within salt-marsh estuaries (Odum and de la Cruz 1967) suggested that estuaries subsidized the shelf. Coastal zone fronts of primary production in Georgia waters were often cited as a manifestation of outwelling from marsh estuaries. This idea was not too new. At the beginning of the century the Danish marine biologist Johannes Petersen suggested that the enormous production of seagrass detritus must have great importance in the North Sea food web leading to fishes. Yet the great eelgrass epidemic (see Chapter 5, Section 5-4) did not result in a collapse of fisheries, as might have been expected.

An emerging model of estuary-shelf interaction must minimize the importance of outwelling of particulate organic matter to nearshore shelf primary productivity. *Spartina* is quickly decomposed to a detritus of low N/C ratio and hence of poor nutritive quality. A marsh-derived epiphytic microflora, however, might be exported, riding piggyback on *Spartina* detritus. Unfortunately, available data militate against the significant export of particulates. As cited earlier, Georgia estuarine and nearshore shelf particulates are derived mainly from phytoplankton and not from cordgrass. A detailed study of exchange between a small salt marsh (Flax Pond, New York) and Long Island Sound suggests no net export of particulate carbon (Woodwell et al., 1977). It is possible, however, that basins with broad openings have the tidal exchange necessary to permit large-scale export of particulate organic matter (W. E. Odum et al., 1979).

Outflow of dissolved nutrients is more promising as a source of estuarine subsidy for shelf primary productivity. The high phytoplankton productivity near Georgia coasts is probably due to three major nutrient sources: (a) outflow of estuarine water, (b) inflow from intrusions of nutrient-rich water from the shelf margins, and (c) regeneration of nutrients between the shelf bottom and the water column. Haines (1975) estimated that estuaries probably provide less than 5% of the total dissolved nitrogen for nearshore shelf production. This overall contribution, however, might be spatially and temporally discontinuous; very nearshore pulses of primary production might be subsidized disproportionately by estuarine outwelling.

As discussed in Chapter 10, nitrogen is a critical limiting nutrient to marine ecosystems. The role of salt-marsh estuaries as exporters of nutrients to shelf environments is therefore estimated best through an understanding of the nitrogen budget of a salt marsh. The sources and losses of nitrogen in salt marshes include:

1. Groundwater flow and precipitation. Dissolved nitrate and dissolved organic nitrogen dominate.

2. Nitrogen fixation. Gaseous nitrogen is fixed as nitrate by bacteria associated with *Spartina* roots, bacteria living in sediment, and by blue green bacteria living in mats (e.g., *Oscillatoria*) on the sediment surface.

3. Tidal exchange. Salt marshes are usually connected to coastal waters via a small number of tidal creek channels. Marshes might import (or export) dissolved inorganic nitrogen, dissolved organic nitrogen, or nitrogen in particulate form.

4. Denitrification. Denitrifying bacteria reverse the direction of nitrogen fixation and convert dissolved inorganic nitrogen to gaseous nitrogen.

5. Accretion in sediment. Some nitrogen is buried in the process of sedimentation.

Table 17-1 shows a budget measured for Great Sippewissett salt marsh (Cape Cod, Massachusetts). The marsh is old, connected to adjacent Buzzards Bay through a single channel, and flooded twice daily with a semiequal high tide (Valiela and Teal, 1979). Given potential measurement errors, the small (11%) discrepancy between inputs and losses is probably inconsequential.

The balance suggests an ecosystem in approximate steady state. It is not clear whether this balance is fortuitous or due to a series of feedback processes that buffer nitrogen exchange. Examples of such feedbacks might be accelerated denitrification with

TABLE 17-1 NITROGEN BUDGET FOR GREAT SIPPEWISSETT SALT MARSH, A NORTH TEMPERATE *SPARTINA* MARSH LOCATED NEAR WOODS HOLE, MASSACHUSETTS[a,b]

	Input	Output	Net exchange
Precipitation	380		380
Groundwater flow	6,120		6,120
N_2 fixation	3,280		3,280
Tidal water exchange (total)	26,200	31,600	−5,350
$\quad$ NO_3-N	390	1,210	
$\quad$ NO_2-N	150	170	
$\quad$ NH_4-N	2,620	3,540	
$\quad$ Dissolved organic N	16,300	18,500	
$\quad$ Particulate N	6,740	8,200	
Denitrification		6,940	−6,940
Sedimentation		1,295	−1,295
Volitilization of NH_3		17	−17
Deposition of bird feces	9		9
Shellfish harvest		9	−9
Total	35,990	39,860	−3,870

[a]After Valiella and Teal, 1979, by permission from *Nature,* vol. 280, pp. 652–656, © 1979, Macmillan Journals, Ltd.

[b]Losses from the marsh to adjacent Buzzards Bay are shown as negative numbers in the net exchange column.

increased nitrate input or decelerated nitrogen fixation with increased ammonium input. Fertilizer-enhanced *Spartina* production argues against strong buffering. The dissolved inorganic nitrogen requirements of the plants in a typical unfertilized marsh are larger than the import, suggesting that nitrogen fixation, nitrification, and recycling within the marsh are all necessary to sustain production.

Most nitrogen exported through the ebbing tide is in the form of particulate ammonium and molecular nitrogen (see Haines, 1979 for a similar study of nitrogen pools in Georgia salt-marsh coastal waters). But the export of dissolved inorganic nitrogen from marshlands around the overall region of Buzzards Bay (assuming rates comparable to those measured for the Great Sippewisset marsh) could entirely replace the nitrate content of the bay in 100 days (Valiella et al., 1978). These results suggest an extensive interaction among uplands, marshes, and coastal waters with respect to nitrogen. In contrast, there is no net outflow of dissolved inorganic nitrogen from Flax Pond, a small salt-marsh system emptying into Long Island Sound (Woodwell et al., 1979). This marsh, however, exports significant amounts of ammonium-N in summer; seasonal subsidy is thus possible. Woodwell and co-workers (1979) discuss in detail the complexities behind any sweeping characterization of marshes as outwelling systems. Seasonal variation in nutrient cycling, regional variation in groundwater and precipitation inputs, variation in river input, and differences in the extent of salt marshes themselves all combine to suggest that strong geographic variation is likely in the extent of nitrogen outwelling.

Estuaries fertilized by sewage apparently do outwell and subsidize production of shelf waters. This is most apparent in the Hudson Estuary–New York Bight system (Malone, 1977). The lower Hudson is a partially mixed estuary with enormous inputs of domestic wastes. Within the estuary, primary production is not limited by nutrient supply. Most nutrients are transported to the inner shelf, where they support high levels of phytoplankton and zooplankton production (Malone and Chervin, 1979). In the winter and spring the estuary imports chain-forming diatoms via the salt wedge; in summer, however, most of the phytoplankton standing stock (nannoplankton) is explained by in situ primary production. Overall, however, the inner shelf serves as a tertiary sewage treatment plant, converting domestic wastes into secondary production.

SUMMARY

1. An estuary is a semienclosed coastal body of water that has a free connection with the open sea and within which seawater is measurably diluted with freshwater. The freshwater derives from land drainage and tends to float as a low-density surface layer over denser seawater. Tidal mixing can reduce or obliterate this stratification.

2. The decrease in salinity in an estuary precludes those species incapable of volume regulation (see Chapter 2). A critical salinity of 5 to 8 ‰ precludes species incapable of extensive regulation of specific inorganic ion concentrations.

3. The following biotic responses to the estuarine gradient are important:
   (a) a reduction of species richness with decreasing salinity;
   (b) changes in physiological adaptations;
   (c) changes in body size and genetically determined morphological features;
   (d) an increase in niche breadth as competitors disappear in the decreased salinity;
   (e) adaptations to avoid transport in the surface layer to the open sea;
   (f) genetic divergence from open marine conspecifics.

4. Many fish species spawn offshore but spend some period feeding in estuaries. Estuaries are therefore crucial to many species as nurseries.

5. A great controversy exists as to whether estuaries outwell large amounts of particulate or dissolved nutrients to the shelf.

18 the subtidal benthos: sampling, substratum control, and community interactions

18-1 METHODS OF SAMPLING THE SUBTIDAL BENTHOS

The collection of benthic samples from a remote shipboard location is a primary source of bias in the analysis of living marine benthic communities. Sample location is often imprecise and samplers usually differ strongly in collecting efficiency depending upon the substratum type encountered. A good benthic sampler should recover a sample over (a) a precise area and (b) a uniform depth below the sediment–water interface. Where benthic animals are sparse, large samples are required. Samplers operating at great depths must have closing devices to prevent elutriation of organisms from the sampler, as the latter is raised through the water column. Finally, samplers must hit the bottom slowly in order to prevent a "bow-wave" (created by the sampling device) from eroding sediment and animals from the bottom before sampling (McIntyre, 1971).

Gravity and spring-loaded grabs are designed to sample a precise area of bottom with two or more sharp indigging sections. As the Peterson grab hits the bottom and the supporting wire has some slack, the hook whose support depends on the wire's tension releases and allows a chain to pull the two sections closed (Fig. 18-1a). If the wire is suddenly slackened upon lowering, the device will fire prematurely. The Van Veen grab avoids this problem by having longer arms attached to the digging sections. The most efficient version is the Smith–McIntyre grab, a heavy spring-loaded device that digs efficiently in both sands and muds (Holme and McIntyre, 1971).

Shipboard-deployed gravity corers take a sample of uniform depth and specified area. Small-diameter (< 10 cm) cylindrical devices, such as the Phleger corer, are useful for meiofauna, sediment, and microbial samples. A weight drives the corer into the bottom. The box corer is a rectangular gravity corer that is guided into the bottom by a

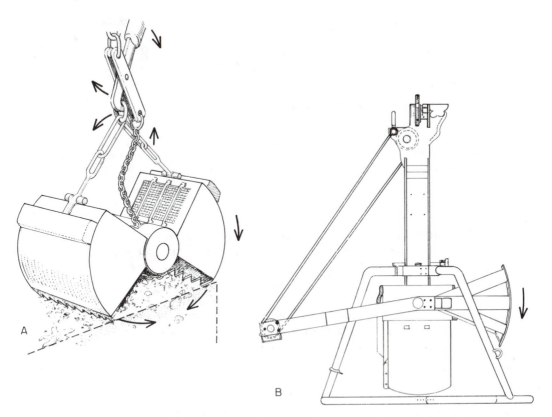

Figure 18-1 Some benthic sampling devices. (a) The Peterson grab taking a sample from the seabed. (From McIntyre, 1971, after Hardy, 1959) (b) A box core developed to take samples from the deep-sea bottom. (After Hessler and Jumars, 1974)

movable plunger mounted on a frame. A spade is released when the frame hits the bottom and digs into the sediment and closes the bottom of the corer as the frame is lifted by a wire (Fig. 18-1b). This device has been used effectively in sampling the deep-sea benthos (Hessler and Jumars, 1974).

Dredges are heavy metal frames with cutting edges designed to move within the sediment; an attached burlap or chain bag collects the sediment behind. Sleds are special versions with runners that permit the device to dig only to depths of a few centimeters. These devices are not quantitative, but permit large amounts of material to be collected; this is very useful in depauperate areas, such as the deep sea. Sanders (1958) designed a quantitative anchor dredge with a control plane that constrains the dredge to bite to a depth of approximately 7 cm. The sediment is collected in an attached burlap bag (Fig. 18-2); the volume of sediment collected divided by the depth (7 cm) is the area sampled.

Direct observation, via scuba diving, photography, and submersible vehicles, has been instrumental in our current understanding of the biology of subtidal communities and the deep sea. Scuba diving is effective only to depths of 40 m. However, it permits

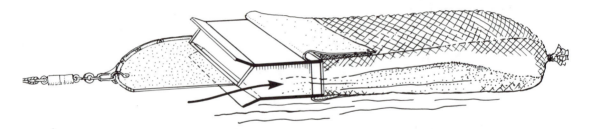

Figure 18-2 A deep-sea anchor dredge, used by Sanders to sample shallow and deep-sea bottoms. (From Sanders et al., *Deep-Sea Research*, vol. 12, pp. 845–867, 1965)

experimental manipulations, close-range visual observations, and precise sample location. Submersibles and submarine photography have been important in the study of the deep sea. An especially good example is the "monster camera" developed by Isaacs (e.g., Hessler et al., 1972) that was designed to observe organisms in the deep sea that attack a large bait can filled with fishes. This camera has discovered organisms otherwise not observable through standard trolling or benthic sampling procedures. Deep submersibles, such as the "Alvin," now perform excursions routinely and allow investigators to perform direct manipulative experiments on deep-sea benthic communities.

18-2 *SUBSTRATUM TYPE, LIFE HABIT, AND ABUNDANCE OF BENTHIC ORGANISMS*

Because most of the ocean bottom consists of soft sediments, we concentrate on the properties of soft sediments that affect this distribution and abundance of principal marine benthic adaptive types. The best framework within which to study is the dichotomy between the occurrence of suspension feeders and deposit feeders. As discussed in Chapter 14, suspension feeders feed on water-suspended particles whereas deposit feeders ingest sedimentary particles, usually assimilating part of the microbial community. It has been known since the time of Petersen (1918) that deposit feeders are most commonly found in muds whereas suspension feeders are generally found in greatest abundance in clean, well-sorted sands (this applies to noncarbonate sediments). These two major feeding groups tend not to co-occur, especially in muds (Rhoads and Young, 1970).

A classic paper by H. Sanders (1958) attempted to explain the distribution and abundance of marine benthic organisms in Buzzards Bay, an arm of sea enclosed by the New England mainland and Cape Cod, Massachusetts. He collected a large series of samples and measured the grain size of sediments concurrently with the abundance of the different dominant species and their feeding types. Deposit-feeder abundance posi-

tively correlated with percent silt-clay of the sediment. In particular, the clay fraction
(particles $< 5\ \mu$ in diameter) seemed to be the best predictor of deposit-feeder abundance.
Sanders inferred that the organic fraction of the sediment correlated positively with the
fine fraction of the sediment and provided more food for deposit feeders, thus causing
this relation. Microbial standing stock also increases with the increased particle surface
area, per unit volume, in finer grained sediments.

The remaining question is why suspension feeders do not occur in mud. Sanders
suggested that currents over mud bottoms brought less phytoplankton to the suspension-
feeding populations. With the unavailability of food, suspension-feeding abundance thus
decreases in muddy bottoms. Several studies on the effect of suspended fine particles on
suspension feeders, however, suggest an inhibition of an inhibition of growth and feeding.
Soft-shell clams (*Mya arenaria*) buried at various depths suffered higher mortality in mud
than in sand (Glude, 1954). Growth of the hard-shell clam *Mercenaria mercenaria* and
the oyster *Crassostrea virginica* suffers in high-turbidity relative to low-turbidity environments (Pratt and Campbell, 1956; Loosanoff and Engle, 1947). Peddicord (1977)
measured the condition index, CI [(dry soft tissue weight in g/shell cavity volume in cm^3)
$\times$ 100], of the estuarine clam *Rangia cuneata* over a year in the James River in sand
and mud. CI was greater in sand populations [Fig. 18-3(a)]. Transplants of boxes of
sediments and animals between environments show that a water column property, not the
sediment itself, influences CI. Over the mud bottom, suspended solids are more concentrated than over the sand bottom [Fig. 18-3(b)].

Rhoads and Young (1970) documented the nonoverlap between deposit feeders and
suspension feeders in subtidal New England bottoms. The presence of deposit feeders
in mud seemed to be the important correlative factor predicting the absence of suspension

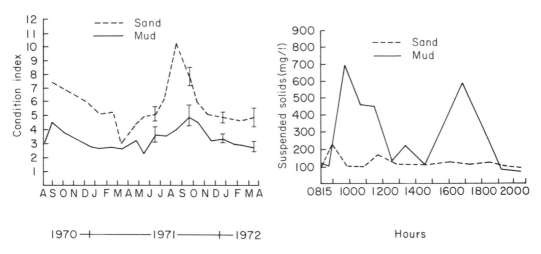

Figure 18-3   (a) Condition index of the calm *Rangia cuneata* in sand and mud (bars indicate 95%
confidence). (b) Suspended solids in water column over sand and mud bottoms. (After Peddicord,
1977, reprinted from *Marine Biology*, vol. 39)

feeders. A series of measurements of sediment properties (Rhoads and Young, 1970) demonstrated that deposit feeders alter the sediment and render it unstable and unsuitable for suspension feeders. Burrowing and fecal pellet formation of deposit feeders greatly increase the water content of muddy-bottom sediments. To demonstrate, short cores were carefully taken by divers from several different types of bottoms, including those with abundant deposit feeders and those where deposit feeders were relatively rare or absent. Cores were frozen and then extruded and weighed, wet and dry, to obtain the percentage of water content. As shown by Fig. 18-4, the water content of sediments is much higher in bottoms where deposit feeders are abundant. The effect is to make the sediment very fluid, particularly at the sediment–water interface. This fluidity can be documented photographically with the sediment–water interface camera (Rhoads and Young, 1970), which is essentially a prism of plexiglass filled with freshwater, with an underwater camera at the apex. The granular surface zone (Fig. 12-5) corresponds to the near-surface water content maximum. A suspension-feeding juvenile would thus lack a stable living position. The inability to orient the suspension-feeding organ in an unstable fluid medium increases the probability of clogging with silt in fluid muds.

The fluid and unstable nature of the substratum results in its relative erodability by weak bottom currents. Rhoads and Young (1970) demonstrated this situation by placing an oscillating vane over muddy sediment with and without large abundance of deposit-feeding bivalves (*Nucula proxima*). Bottoms that were granulated through fecal pellet formation by deposit-feeding organisms were more easily eroded than those where deposit feeders did not disrupt the sediment. Near-bottom maxima of turbidity therefore further

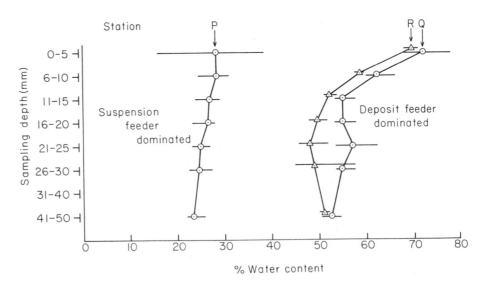

Figure 18-4 Water content of cores from soft bottoms in Buzzards Bay, Massachusetts, showing the difference in surface-water content between bottoms dominated by suspension feeders and burrowing deposit feeders. (From Rhoads and Young, 1970)

exascerbated the problems of suspension-feeding organisms that require the passage of relatively clear water across a suspension-feeding organ.

The interaction of weak bottom currents with granular bottoms by deposit-feeding activities is enhanced during the spring and summer when the thermocline develops. Bottom waters are stabilized through temperature gradients, trapping large amounts of suspended matter at or near the sediment–water interface (Rhoads, 1973). In Buzzards Bay, Massachusetts, current-suspended material is trapped near the sediment–water interface over silt-clay rich bottoms (Fig. 18-5). Rhoads and Panella (1970) transplanted individuals of the suspension-feeding bivalve *Mercenaria mercenaria* to trays at the sediment–water interface of a muddy bottom dominated by deposit feeders and to a tray approximately 1 m above the bottom. Growth was inhibited in bivalves living at the sediment–water interface. Rhoads (1973) placed oysters (*Crassostrea virginica*) in nets at various distances above the muddy sediment–water interface. Growth in bivalves above the sediment–water interface (approximately 1 m) was similar to the growth rates en-

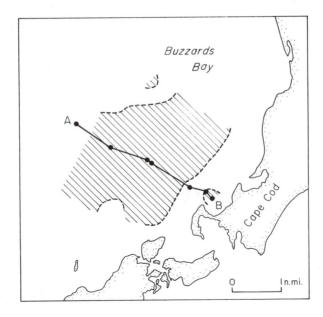

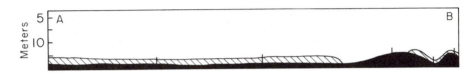

Figure 18-5 Bathymetric distribution of ≤ 90% transmissivity water (crossed-hatched areas) over muddy bottom of Buzzards Bay, Massachusetts. (After Rhoads, *American Journal of Science*, vol. 273, pp. 1–22, 1973)

countered in oysters placed in their normal barely subtidal habitat. Resuspension of detritus from the bottom provides a source of food for oysters and a potential for mariculture of oysters. This situation further indicates that it is unlikely that suspension feeders are rare in muds dominated by deposit feeders because of a paucity of food source.

In summary, the deposit-feeding trophic group exerts an adverse effect upon the suspension-feeding trophic group. When deposit feeders dominate muds, we expect alterations of the sediment to result in a watery fluid sediment–water interface that is inhibitory to the successful survival of most suspension-feeding groups.

The intensive burrowing activities of deposit-feeding organisms—leading to the production of large amounts of suspended matter near the sediment–water interface–raise interesting questions concerning the recycling of nutrients between the bottom and the overlying water. Bacteria, for example are important colonizers of organic detritus (Fenchel, 1969) and probably obtain a significant amount of their nutrition from dissolved organic matter (especially nitrogen). When organic detritus is recycled into the overlying water, there is probably a significant amount of import of nitrogen back to the bottom when the detritus finally returns. This situation suggests that the deposit-feeding system is probably coupled with the overlying water. Thus the deposit-feeding system can be

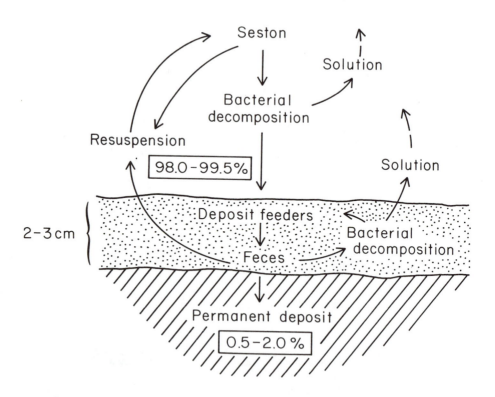

Figure 18-6   The movement of particulate organic matter in shallow-water environments. (After Young, 1971)

described (Fig. 18-6) as a system in which there is interdependence between detritus, bacteria, and deposit feeders and the overlying water. As suggested in Chapter 17, influxes of particulate nitrogen from salt-marsh systems may enhance deposit-feeder production, in contrast to most subtidal suspension-feeding systems, where suspension feeders must import matter from the overlying water but essentially exert no regulation on their food source or its abundance (Levinton, 1972).

18-3 COMMUNITY STRUCTURE

Community Delineation

Most subtidal benthic studies have relied upon remote shipboard sampling. A great deal of subtidal benthic ecological research has therefore been devoted to identifying and characterizing communities from sets of samples. The word community has nearly as many definitions as investigators, but most definitions fall into one of two categories: (a) all those organisms coexisting in a single place (Pielou, 1969) and (b) repetitive associations of species (Fager, 1957). The latter concept is essentially statistical and depends on the belief that a repetitive association implies an interacting group of species that respond to similar biotic and physical conditions. Petersen (1913) collected grab samples in Danish waters and classified them as to species content. This early establishment of repetitive associations of marine benthic species set a pattern that has been followed to the present day by most marine benthic ecologists. The procedure is described in the following paragraphs.

A set of samples is collected and a data matrix is constructed, consisting of n samples from which m total species are collected. Either a species' abundance or its presence in the sample may be registered. Then a new matrix of resemblances among the samples is constructed by calculating a similarity measure between all possible pairwise sample combinations. The new matrix shows the similarity in species content between any two samples. If species presence alone is used, a similarity measure, such as Jaccard's coefficient (see Chapter 6), can be used. The relative abundance of species in a sample is more appropriate, for the presence of a rare species everywhere will be equivalent to strong fluctuations in a second species over all samples, in a presence or absence measure. A relative abundance similarity measure was employed in Sanders' (1960) study of the soft-bottom community in Buzzards Bay, Massachusetts. The percentage composition of the various species in each sample was determined; all possible pairs were compared with regard to their faunal content. An index of affinity can be calculated as a measure of the percentage of the fauna common to a pair of samples. It is obtained by summing the smaller percentages of those species present in both samples (Table 18-1).

Having calculated a series of indices of affinity, samples of benthic animals can then be grouped by various techniques of cluster analysis (see Sneath and Sokal, 1974). With this approach, a tree of similarity grouping sample stations can be constructed. From the data matrix we can also calculate the affinities between species, based on occurrence among all the samples; this step permits the delineation of communities of

TABLE 18-1 A COMPARISON OF THE AVERAGE INDEX OF FAUNAL AFFINITY
AMONG STATIONS TAKEN AT ONE SUBTIDAL LOCATION IN BUZZARDS BAY,
MASSACHUSETTS (SEDIMENT WAS MAINLY SILT-CLAY, MACROFAUNA DOMINATED
BY DEPOSIT-FEEDING INVERTEBRATES), CONTRASTED WITH INDICES FROM OTHER
STUDIES[a]

Comparison	Index of affinity
Macrofauna at the subtidal location	69.3
Meiofauna at the subtidal location	55.7
Macrofauna of many localities throughout Buzzards Bay	21.4
Meiofauna of many localities throughout Buzzards Bay	27.2

[a]After Sanders, 1960.

related organisms. Neither approach presupposes that the co-occurrence of species is due
to anything more than statistical artifact. Co-occurrence may be the result of (a) a similar
response to some physical or environmental factor or (b) a series of among-species
interactions that results in an integrated community.

After calculating the degree of similarity between stations, a station ranking derived
from degree of similarity of pairs of stations can be compared with rankings of stations
compared in order of increasing or decreasing values of environmental parameters, such
as salinity or substratum. This procedure may allow us to ascertain the presence of causal
relationships between the physical environment and species distributions. Such an approach was taken by Nichols (1970), who examined benthic polychaete assemblages as
they relate to sediment type near Port Madison, Washington. Three replicate samples
selected at each station along a transect of water depth were used for the analysis.
Polychaete distributions most closely related to sediment type, as found in most other
studies.

Niche Delineation

Although various multivariate techniques organizing samples or species into communities
are useful, we would like to understand the interactions of individual species with the
environment and thereby determine which other potential overlaps in resource exploitation
occur between species within a community. Hutchinson (1957) defined the niche as an
environmental hyperspace consisting of several dimensions. A species' niche may be
described as a hypervolume in this hyperspace (see Chapter 2).

Green (1971) attempts a multivariate statistical approach to the Hutchinsonian niche
in his study of bivalve mollusks living in lakes of central Canada. He uses multiple
discriminate analysis to identify significant and independent ecological factors that separate species distributions. Multiple discriminate analysis starts with a data set consisting
of n measurements on M parameters, each of the n measurements being associated with
an individual belonging to one of g groups or species. The analysis reduces the data set

to *n* measurements on *k* new parameters that are linearly independent of each other and additive functions of the original *M* parameters. The data matrix is therefore transformed into *g* sets of points in a *k*-dimensional space. The technique assumes that species abundance is related to environmental parameters in a linear additive manner.

In this approach, species occurrence can be plotted on graphs by using these new *k* dimensions. Techniques described by Green allow us to determine which measured parameters (e.g., sediment organic content and sediment particle size) figure most prominently in that given dimension. Then the degree of overlap between species (i.e., the degree of potential competitive interaction) can be calculated unambiguously. Figure 18-7 shows that there is a significant degree of overlap of freshwater bivalve mollusks with regard to sediment organic content, sediment particle size, and depth. This approach is not significantly different from qualitative estimates of resource overlap in rocky intertidal organisms competing for space. The problem with subtidal soft-bottom benthic com-

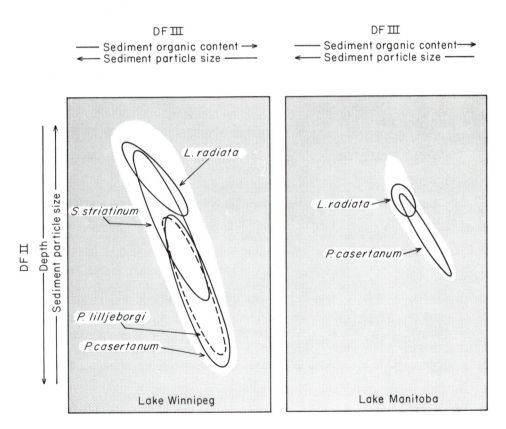

Figure 18-7 Multivariate approach to the Hutchinsonian niche. Discriminant function analysis shows overlap in niche parameters for freshwater bivalves in Lake Winnipeg and Lake Manitoba, Canada. (From Green, 1971. Copyright 1971, the Ecological Society of America.)

munities, however, is that (a) no single factor seems to be important in controlling benthic distributions and (b) it is difficult to observe population changes in subtidal benthic communities but not so difficult to sample them in a static manner. This technique, as do all multivariate techniques of sample organization, depends on the assumption that correlation implies causality.

With many variables acting together to determine the distribution and abundance of benthic animals and plants, we would like to distinguish among the variables. Doing so would involve a set of samples where density is measured, along with a number of environmental variables. Buzas (1969) examined environmental variables and densities of several species of foraminifera in the Choptank River, Maryland, a part of the Chesapeake estuary. Three stations were examined and sampled for foraminifera and environmental variables monthly. A multiple-regression analysis of variance model was used to examine relationships. Parameters for environmental variables (temperature, salinity, oxygen, chlorophyll *a, b,* and *c*), station differences, overall periodic differences, and the interaction of station and periodic differences were statistically compared, by species, with several more restricted models. He found significant periodicity and significant correlations with the whole set of environmental variables. Nevertheless, there were so many intercorrelations between different individual environmental variables that it was not possible to distinguish among them. That is a common problem in correlation studies.

18-4 SPATIAL STRUCTURE

Small Spatial Scales: Behavioral Interactions

Behavior affects the spatial distribution of mobile organisms on a small areal scale. Random patterns suggest an independence among individuals foraging over a uniform environment. Levinton (1972b) investigated spatial distribution of the burrowing subtidal deposit-feeding bivalve *Nucula annulata* in natural populations and in laboratory-maintained trays of mud. Over an area of ca. 0.25 m², *N. annulata* was randomly distributed in field populations with a tendency toward aggregation. Laboratory populations were analyzed by using x-radiography of sediment trays (Fig. 18-8). Nearest-neighbor analysis showed that the average distance to nearest neighbor did not deviate significantly from random. The deposit-feeding mud snail *Hydrobia ventrosa* shows similar random movement. So strong behavioral interactions leading to even spacing among mobile deposit feeders seem missing. Some deposit feeders, however, require a fixed living position and foraging area in which to feed. Some tellinacean bivalves are thus territorial and maintain uniform spacing.

Behavior can aggregate populations of a species when (a) a preferred resource is patchily distributed and (b) populations swarm for mating. Mobile feeders, such as herbivore browsers, can locate food through olfactory stimuli, clustering them around algal stands. At seasonal breeding periods, many benthic invertebrates form breeding swarms.

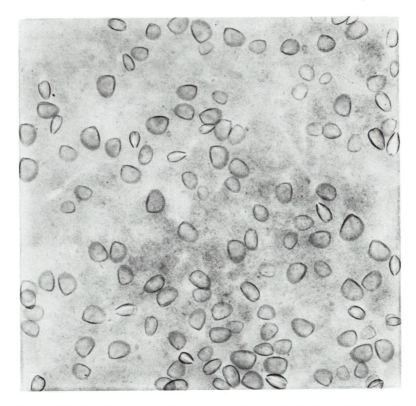

Figure 18-8   X-radiograph of a dispersion experiment with the Buzzards Bay mud-bottom deposit-feeding bivalve, *Nucula annulata*. (From Levinton, 1970b, *The Biological Bulletin*, vol. 143, pp. 175–183)

Medium Scale Spatial Variation

Gage and Geekie (1973) evaluated the spatial distribution of benthic species collected in samples from Scottish sea lochs (Table 18-2). They found a high degree of aggregation of all species. An aggregated spatial distribution pattern of samples taken over more than 10 m^2 is common in subtidal soft-bottom communities. Medium-scale patchiness may be explained by (a) nonrandom settlement of pelagic larvae, (b) patchy distribution of sediment variation, influencing survival, and (c) negative or positive interactions between resident benthic species and immigrating larval stages or adults.

Patchy larval settlement and juvenile migration may explain much of the spatial heterogeneity of marine benthic communities. Buchanan (1967) examined spatial patterns of several infaunal echinoderms in British waters. Echinoderm populations usually were distributed among large-scale patches, each occupying several square kilometers. These patches overlapped broadly and probably arose from the original aggregated settlement of pelagic larvae. Because of the low mortality rates of some ophiuroids and echinoids,

TABLE 18-2 SPATIAL PATTERN OF THE BENTHOS IN SCOTTISH SEA LOCHS.
VARIANCE/MEAN RATIOS FOR ABUNDANCE DATA OF SETS OF SAMPLES,
ORGANIZED BY ECOLOGICAL GROUPINGS. WHEN THE VARIANCE/MEAN RATIO
EXCEEDS A CRITICAL VALUE, THE ORGANISMS ARE AGGREGATED TO A
STATISTICALLY SIGNIFICANT DEGREE ($p < .05$).[a]

Species grouping		Percentage of ratios indicating significant aggregation	Total number of ratios considered
Sediment position	Epifauna	47.5	153
	Infauna	52.5	264
Feeding type	Deposit feeder Suspension feeder	34.9	191
	Suspension feeder	42.0	91
	Scavenger or carnivore	23.1	65
Mobility	Motile	44.4	239
	Sedentary or sessile	44.9	248
Polychaetes only	Errant	27.0	74
	Sedentary	61.5	152

[a]After Gage and Geekie, reprinted from *Marine Biology*, volume 19.

patches may have persisted for as long as 10 to 15 years. This pattern is general, although patch survival may vary. When subtidal muddy bottoms are disturbed in New England waters, larval swarms of the bivalve *Mulinia lateralis* often settle in dense patches (Levinton, 1970). Patch extinction is on the order of 1 year or less. Subtidal populations of the sea pen *Ptilosarcus gurneyi* in Puget Sound consist of hundreds of square meters of bottom with individuals of similar size and, by inference, belonging to the same year class. Sharp discontinuities of size range are often found in diving transects of sea pen bottoms. Birkeland (1974) followed repeated aggregated recruitments by *Ptilosarchus* forming overlapping patches of 20 to 200 m in length along an isobath. Recruitment is unpredictable and patchy and a mosaic of different age classes eventually builds up.

The alternation of sandy and muddy substrata may create a habitat patchiness that results in either selective settling by larvae or differential survival of different benthic invertebrate species. In Chapter 8 we discussed the problems faced by larval swarms failing to locate suitable substrata for settlement. Benthic organisms preferentially adapted to sandy substrata do not survive after settlement in muds (Thorson, 1950). Segerstråle (1962) demonstrated reduced survival of larvae of the suspension-feeding bivalve *Mytilus edulis* on muddy substrata. Sedimentary spatial heterogeneity may be created by the benthic organisms themselves. The fecal cones of *Molpadia oolitica* (Fig. 18-9) provide a relatively stable surface for settlement and growth of suspension-feeding bivalves,

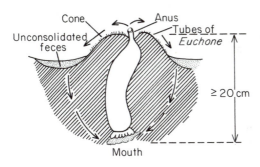

Figure 18-9 Cross section of the sediment showing the feeding position, surface cone, and overall microtopography generated by the burrowing sea cucumber, *Molpadia oolitica,* in Cape Cod Bay, Massachusetts. (From Young and Rhoads, 1971, reprinted from *Marine Biology,* volume 11)

amphipods, and polychaetes, where muds with high water content would normally preclude these ecotypes (Rhoads and Young, 1971). Therefore habitat heterogeneity in the muddy substrata of Cape Cod Bay, Massachusetts, was determined principally by biological mechanisms.

Woodin (1976) suggests that, in infaunal populations, interactions between resident species and newly colonizing species result in the preclusion of newly settling larvae (Table 18-3). Suspension-feeding bivalves may filter larvae. This type of interaction has been previously suggested by Segerstråle (1962). Tube-building polychaetes also may reduce the ability of larvae to penetrate the sediment surface and establish burrows of their own (Woodin, 1974). Such interactions should result in strong dominance by given

TABLE 18-3 FUNCTIONAL GROUPS OF INFAUNA AND TYPES OF INTERACTIONS[a]

Function group	Ingests or disturbs by its feeding activities surface or near-surface larvae	Filters larvae from water	Alters sediment	Larval type at settlement	Predicted dense co-occurring forms
Deposit-feeding bivalve	yes	no	destabilizes	surface or burrowing	burrowing polychaetes
Suspension-feeding bivalve	no	yes	much less than deposit feeders	surface	none
Tube-building forms	yes	depends on feeding type	stabilizes by attracting algal mat; destabilizes by increasing near-bottom turbulence; reduces space below surface, increases settling surface due to tubes	surface	epifaunal bivalves and tube epifauna

[a]After Woodin, 1976.

year classes of successfully settling invertebrate populations. Thus agonistic interactions of successful adult populations with newly settling larvae may also be a source of spatial heterogeneity in subtidal benthic communities.

Significance of Patchiness

The large-scale patchiness of soft-bottom subtidal environments probably indicates the transient nature of many subtidal species assemblages. Species composition in an area may be strongly influenced by the availability of larvae to given localities or by which species currently occupies a bottom, precluding larval settlement by new species or future larval swarms of the same species. The "year class effect," or strong fluctuations of larval settlers in different years, further exaggerates the transience of subtidal level-bottom community structure. The subtidal soft bottom may be envisaged as a target for often widely varying colonizing species with pelagic larvae. Certain sediments may preclude the presence of major adaptive types (e.g., deposit feeders from clean sands), but a number of species of a given adaptive type (e.g., deposit feeders) may settle in a given substratum (e.g., mud). If these species require the same resources, interspecific competition may result in exclusion or resource specialization. But transience may preclude the evolution of specialization that might eventually minimize interspecific competition. The pelagic dispersal mechanism itself may unpredictably recombine competing species, further precluding the evolution of niche subdivision.

Broadscale Variation on the Continental Shelf

Investigations of benthic communities of the continental shelf suggest that water depth strongly influences both species richness and temporal stability of the resident biotas. In all probability, this correlation is related to the decreasing thermal variation (Fig. 18-10) and reduced effects of storms in deeper habitats. This reduction is reflected in greater

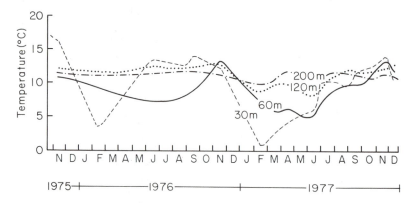

Figure 18-10 Seasonal variation in bottom water temperatures at four depths on the continental shelf and upper slope off New Jersey. (Courtesy D. F. Boesch)

species richness and more predictable species assemblages in deeper shelf environments (Boesch, 1979). Distinct depth zonation has been observed both on the continental shelves of northern Europe and the east coast of the United States (Glemarec, 1973; Boesch, 1979). Boesch (1979) suggests that the transition from benthic assemblages of shallow character to those of deep character is essentially gradual. This can be demonstrated with a plot of water depth versus a multivariate statistical measure of similarity among faunas sampled over the middle Atlantic continental shelf (Fig. 18-11). The species composition of the benthos shows an apparently transitional change with increasing depth. The overall effect of stability increase with increasing depth is modulated by smaller scale variations in sediment grain size and bottom topography. For example, the continental shelf of the east coast of the United States is marked by a ridge and swale topography. The ridges run parallel to the shoreline. Ridges generally consist of sediments of coarser grain size than those of swales. Species adapted to finer grained sediments occur more abundantly in the swales. Samples from swales can be distinguished faunally from those on ridges.

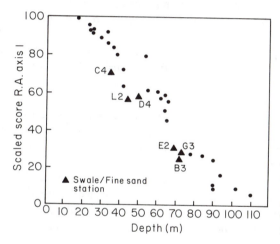

Figure 18-11 Relationship of scores derived from reciprocal averaging ordination analysis (a multivariate statistical technique) for collections of macrobenthos to water depth. Note triangles, which represent samples taken from swales. (Courtesy of D. F. Boesch)

18-5 INTERSPECIES INTERACTIONS, COLONIZATION, AND DYNAMIC EQUILIBRIA

Hard Substrata

Patchy larval recruitment, trophic level interactions, and physical disturbances, such as storms, make a subtidal community a mosaic of areas with differing recent physical and biotic past histories. Colonization by swarms of pelagic larvae provides a source of new interspecies interactions every generation. Although studies of competitive interactions among adults of different species allows a framework for studying competitive exclusion and niche specialization, the dynamic aspects of colonization are an essential part of the study of community structure.

Sutherland (1974) monitored the colonization of ceramic settling plates suspended subtidally below a dock at Beaufort, North Carolina. The fouling epifaunal community contained hydroids, tunicates, bryozoa, and sponges. Development of the community was variable, depending on seasonal patterns of recruitment and yearly variations between seasons (Sutherland and Karlson, 1977). Plates submerged in May 1971 were originally dominated by *Tubularia* and *Bugula*, forming a dense canopy 2 cm in depth. But *Styela* settled beneath the canopy in July and usurped the two former dominants by August. *Styela* persisted as a monoculture for 4 months despite the influxes of other species. Tiles submerged in June were dominated by *Ascidia* and *Pennaria*. So small differences in season greatly altered dominance patterns on different plates. Dominance was partly controlled by first access to the plates, but, in general, no overall species assemblage was eventually dominant on all plates, as might be expected from rocky intertidal studies. Short life span, seasonal changes, lack of strong competitive superiority by one fouling species (or group of species) over the others, and unpredictability of larval recruitment all contribute to between-plate variance in dominance. Despite this variation, diversity, as measured by H', increased to a plateau and remained generally constant (Sutherland, 1974).

Predation in this system was investigated by Karlson (1978), focusing on the urchin *Arbacia punctulata*. In Beaufort Channel, North Carolina, *A. punctulata* is a carnivore, feeding on epifauna that live on wood pilings. Manipulations of *A. punctulata* density showed that this species was primarily responsible for providing open space on pilings. Movement up pilings was halted by the hydroid *Hydractinia enchinata*. Feeding experiments further demonostrated that the sponge *Xestospongia halichondroides* was not a suitable prey species. These two foulers were the only two sessile epifaunal species not exhibiting significant temporal variability.

Subtidal hard-substratum communities need not be unpredictable in competitive outcome. The work of K. Sebens shows that subtidal rockwalls of New England are often dominated by the coelenterate *Alcyonium* and a colonial ascidian. In the absence of disturbance, *Alcyonium* gradually increases in dominance and resists invasion by other species. The tunicate invades open space and is capable of overgrowing small *Alcyonium* colonies. When *Alcyonium* obtains a minimum size, however, it slowly grows and gains space.

Soft Substrata

Successful and sequential recruitment by larvae of different species may also be a significant explanatory factor for spatially variable species compositions of marine benthic habitats. In Long Island Sound subtidal soft-bottom communities are spatially and temporally variable in species composition. Because of tidal currents and storm effects, the muddy bottom of Long Island Sound is continually eroded, presenting newly opened surfaces for colonization by larvae of benthic species. McCall (1977) placed boxes of defaunated mud at the bottom of two sites of central Long Island Sound in July 1972 to stimulate a local disaster. Samples collected 10 days later contained high numbers (ca. 10^5 m^{-2}) of the polychaetes *Streblospio benedicti* and *Capitella capitata* and the amphipod

Ampelisca. These small, sedentary, tube-dwelling deposit feeders were classified as opportunists and were characterized by rapid development, many reproductions per year, high recruitment, and high death rate. Another group, the errant polychaete *Nepthys incisa* and the razor clam *Ensis directus*, is present early in the colonization samples but remained low and in constant abundance throughout the experiment (10^2 m^{-2}). They were classified as conservatives, having slow development, few reproductions per year, low recruitment, and low death rate. Most produce large numbers of planktotrophic larvae. Another relatively conservative set of species composed of *Tellina agilis* and *Nucula annulata* is intermediate in peak abundance, death rate, and life history; *N. annulata* eventually dominates subtidal bottoms in Long Island Sound.

McCall surveyed species distributions in Long Island Sound and found that bottoms dominated by colonists (opportunists) were restricted to shallow water (< 20 m). He inferred that storm waves and tidal turbulence may be especially effective in resuspending the bottom at these depths.

This local depth-dependent variation in storm affects on the bottom mimics the large-scale variation seen on the open continental shelf, as discussed previously. Nearshore shelf habitats are disturbed by storms relatively frequently; biotic assemblages are spatially and temporally variable as a result. McCall's observations suggest two different sets of adaptive strategies with corresponding species assemblages: opportunistic and conservative. When a bottom is disturbed—in this case, by tidal currents or storm events—resident species are eliminated and the first successful colonists belong to the opportunistic species group. More conservative species later colonize the bottom and outcompete opportunists.

The small mactrid bivalve *Mulinia lateralis* is an opportunistic species occurring in both the waters of Long Island Sound and Buzzards Bay (ranging from the Gulf of St. Lawrence to the Yucatan). Dredge, grab, and core samples of bottom muds taken at depths of 0 to 20 m commonly reveal the presence of large numbers of *Mulinia* valves (Levinton and Bambach, 1970). Yet this species is rarely found live in any of these same samples (Levinton and Bambach, 1970). *Mulinia lateralis* occasionally occurs in high densities in Long Island Sound and certain areas of Buzzards Bay. Sanders (1956) shows that *Mulinia lateralis* is a transient opportunistic species fluctuating strongly in numbers from year to year and from place to place in Long Island Sound. *Mulinia* occurs in such great densities and colonizes so strongly that it occurs in thin layers in sediments in cores taken from coastal areas. It usually sets in high densities (ca. 10^4 m^{-2}) and disappears soon after. In August 1966 *M. lateralis* occurred in densities of thousands per square meter but was absent as a living population 2 months later at a subtidal locality off Point Lookout, Milford, Connecticut (Levinton, 1970). A very substantial *Mulinia* invasion occured in subtidal muddy bottoms after the oil spill in West Falmouth, Massachusetts, destroyed local marine faunas in 1969 (Grassle and Grassle, 1974). *Mulinia* invasions do not therefore occur everywhere simultaneously but are spatially and temporally sporadic. Rhoads has examined colonization of benthic fauna on a dump site (see McCall, 1977) and finds that *Mulinia* is an invader of bottoms that have been newly covered with dredged spoils.

We discussed *Mulinia's* short generation time and high fecundity in Chapter 7.

High juvenile mortality, due to clogging of ctenidia by near-bottom turbidity, contributes to the rapid extinction of newly colonized populations. Some evidence exists for a crude deposit-feeding mechanism (Parker, 1975), but *Mulinia* invasions are soon replaced by populations of the deposit-feeding bivalve *Nucula annulata*. Usually these two species do not co-occur in Long Island Sound and are found in adjacent, nonoverlapping patches.

18-6 COMPETITIVE INTERACTIONS

How important is interspecific competition in subtidal soft-bottom benthic communities? Unfortunately, almost all our inferences are supported merely by circumstantial evidence and distributional data of subtidal benthic species. The question of competitive interaction has been the subject of great controversy; and many have favored the hypothesis that competitive interactions do not have a great deal of influence on benthic community structure (e.g., Jones, 1950). Johnson (1971) suggested a hypothesis explaining the structure of subtidal benthic communities. Equilibrium assemblages of benthic species result from interspecific interactions. Community disturbance because of rapid changes in sedimentation, current reworking of the bottom, and disturbances by predators, however, often results in the downgrading of communities to assemblages of colonizing opportunistic species. Thus a benthic realm may be envisaged as a mosaic of assemblages of species at different stages of succession toward an equilibrium state.

In shallow subtidal benthic assemblages of Tomales Bay, California, species found to be low in the successional continuum occurred in many different substratum types. But species occurring in later successional stages were restricted to a more limited range of bottoms. Early colonizing species were therefore more generalized in substratum requirements (Johnson, 1971). This model and these preliminary data are analogous to regarding faunal assemblages in the rocky intertidal as a complex of patches in different stages of succession after a disturbance. Johnson, however, conceives of final competitive dominants as being more specialized to their final habitat with no suggestion as to how that specialization leads to competitive superiority.

In order to examine interspecific competition among soft-bottom species, it is useful to characterize major niche dimensions. They are (a) type of sediment occupied, (b) feeding and living position with respect to the sediment–water interface, and (c) differences in food taken and particle size ingested. An analysis of the life habits of co-occurring species permits an evaluation of the amount of niche overlap and a further investigation of processes that determine niche relations among species. It is useful to consider a guild of species requiring the same spectrum of resources.

Levinton (Levinton, 1977; Levinton and Bambach, 1975) surveyed niche overlap in deposit-feeding benthos of a pair of intergrading communities in Quisset Harbor, Massachusetts. Circular cores of 20 cm diameter by 25 cm deep were taken by divers along transects between the two communities and species abundance was determined. Life position, feeding habit, and interindividual interactions were examined in the laboratory by observing animal behavior in thin aquaria, aided by time-lapse x-radiography of animals and their burrows in sediments. Differences in x-ray opacity between sediments and animals or borrows facilitated a quantitative evaluation of living position. All dominant species were deposit feeders and those feeding below the sediment–water interface had qualitatively similar gut contents. No quantitative evaluation of particle-size specialization was performed, however.

Two communities (Fig. 18-12) were delineated: (a) a shallow-water (1 to 5 m) assemblage living in a sediment of 10 to 80% silt-clay covered by eelgrass and (b) a deeper (5 to 9 m) assemblage living in bare bottoms of 40 to 80% silt-clay. The lateral transition from one community to another is short and usually coincident with the dis-appearance of eelgrass. The two communities were each dominated by a few species living and feeding at different levels below the sediment–water interface (Fig. 18-12). Dominant deposit-feeding species at the same level never co-occurred in this study. In the "eelgrass community" the bivalves *Tellina agilis* and *Cumingia tellinoides* fed with mobile inhalent siphons on surface diatoms. These two species did not co-occur in the community but had contiguous lateral distributions. It appeared, then, that niche separation was complete and that competitive interaction was the probable factor structuring the two communities. The occurrence of coexisting species at different levels below the sedi-ment–water interface has been termed *stratification* (Turpaeva, 1954).

Manipulations of natural populations were not done, but the nature of direct inter-species interactions was experimentally examined in the laboratory. Levinton (1977) described a series of experiments of interindividual interactions of deposit-feeding bi-

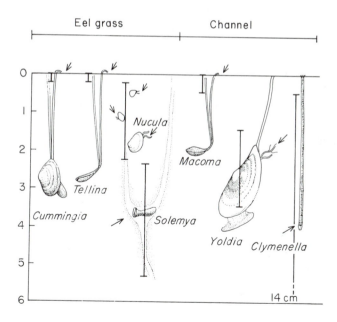

Figure 18-12   The deposit-feeding communities of muddy bottoms of Quisset Harbor, Cape Cod, Massachusetts, showing stratified feeding positions. Range of feeding position is indicated by vertical bar; arrow shows feeding depth for organisms.

valves. Most experiments consisted of placing individuals of different species together, with subsequent x-ray time-lapse photography of sediment trays. Some interactions were neutral, but the extent of direct species interference among the protobranch bivalves (*Nucula, Solemya, Yoldia*) was surprising. A good example is the *Yoldia–Solemya* interaction. The living position of the body of a large *Yoldia* is deeper and overlaps that of *Solemya*. Lateral burrowing by *Yoldia* disrupts *Solemya* burrows (Fig. 18-12). This interaction explains the negative association of these two species in the field. In contrast, individuals of *Nucula proxima,* a codominant with *Solemya* in the eelgrass community, are attracted to burrow openings but do not burrow deeply enough to disrupt directly the living position of *Solemya,* thus permitting their coexistence in the laboratory and the field. These and other observations demonstrate that competition for space is of importance in soft-bottom deposit-feeding bivalvia. Competition for food may be of importance, but evidence is inconclusive.

Similar patterns of deposit-feeding bivalve co-occurrence are found in fossiliferous Silurian strata of Nova Scotia (Levinton and Bambach, 1975). The MacAdam Brook formation shows a transition from high-water-content muds dominated by nonsiphonate *Nucula*-like bivalves to firmer sediments dominated by *Yoldia*-like siphonate forms (Fig. 18-13). The fluid nature of the substratum was determined from examinations of rock-thin sections. Sedimentary rocks with siphonate deposit feeders contained preserved burrows with plastic deformation and well-defined laminae, indicating sediment stability and low water content. Sediments dominated by nonsiphonate forms, however, showed

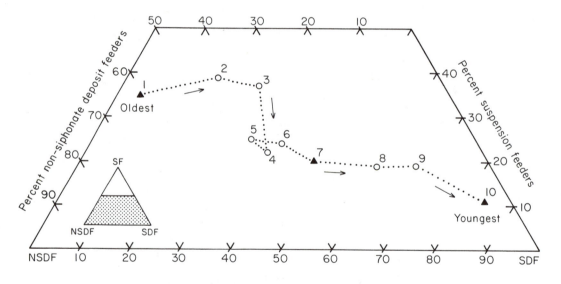

Figure 18-13 Comparative trophic composition of ten faunal subdivisions of the McAdam Brook formation. Note the gradual trend from dominance by nonsiphonate deposit feeders to siphonate deposit feeders. (From Levinton and Bambach, 1975)

no laminae and indistinct burrows, indicating high-water-content-burrowed sediment. In both communities, coexisting deposit-feeding species exhibited a vertical stratification similar to the modern example in Quisset Harbor, Massachusetts (Fig. 18-12).

Patterns of occurrence of mainly suspension-feeding clams living in Mugu Lagoon, southern California, reveal a similar series of stratified living positions. Peterson (1979) demonstrated that experimental removals of a species at a given depth below the sediment–water interface were followed by successful invasions of potentially competing species through larval settlement and improved survival. Transplants within the lagoon demonstrated that somatic growth of the deep-dwelling clam *Sanguinolaria nuttali* was greatly reduced by the deep-dwelling species *Tresus nuttallii* and *Saxidomus nuttalli*. The shallow-dwelling clam *Protothaca staminea,* however, did not affect the growth of *Sanguinolaria.* Because both deep-dwelling and shallow-dwelling clams feed through a siphon at the sediment–water interface, these results suggest that space—and not food—is the principal limiting factor. Clam models produced similar effects to live clams, which also suggests the role of space (Peterson and Andre, 1980). These results further support the contention of Levinton (1972) that niche subdivision among suspension feeders with respect to food type is unlikely. Nevertheless, space competition results in vertical stratification much the same as in infaunal deposit feeders.

18–7 PREDATION

Effects on Diversity

The effects of predation on the structure of subtidal communities are poorly known but are probably of similar importance to those discussed in the intertidal zone (Chapter 16). Many species of bottom-feeding fishes forage on bottom invertebrates and are known to be a major cause of benthic population decrease in summer. Rays feed on smaller bottom invertebrates and rework much of the soft sediment of shallow Caribbean soft bottoms. In the temperate zone bottom-feeding flatfishes forage on small crustacea, polychaetes, and small bivalves (Richards, 1963; Bregnballe, 1961). Bregnballe (1961) calculated that juvenile plaice populations (*Pleuronectes platessa*) could have no measurable influence at all on rapidly reproducing populations of harpacticoid copepods, ostracods, and nematodes. A large decrease was possible for the oligochaete *Paranais litoralis,* however. Benthic predation might thus shift prey species composition to those species with higher intrinsic rates of population increase, which is as predicted by the model discussed in Slobodkin (1961); see Chapter 4. Furthermore, a selective preference for soft forms, such as crustacea and polychaetes (Richards, 1963), might shift species composition toward forms with exoskeletons, such as bivalves. Fish predation may therefore shift competitive superiority toward rapid reproducing forms and toward those species resistant to predation.

Manipulative studies of intertidal communities show that predation influences prey diversity by lowering population size below carrying capacity. Jackson (1972) presents similar evidence for molluscan faunas residing in bottoms covered with turtle grass, *Thalassia testudinum.* Lucinid bivalves occurring in intertidal and very shallow subtidal bottoms of the north coast of Jamaica were tolerant of low oxygen, temperature, and salinity variations (Jackson, 1973). Deep-burrowing intertidal bivalves are only occasionally taken by predators. In going offshore into deeper water, however, the intensity of predation is increased, as shown by frequency of drill holes in shallower-burrowing lucinid bivalve shells (Fig. 18-14). Overall predator diversity was greater in deeper-water sampling sites of Discovery Bay relative to the intertidal flats of Pear Tree Bottom. In Discovery Bay bivalve diversity is higher despite the fact that overall prey biomass is lower. Suppression of competitive interaction may therefore occur in subtidal bottoms with heavy predation.

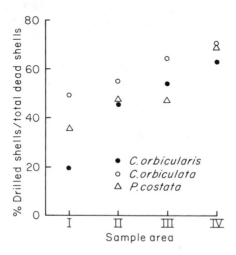

Figure 18-14 Depth distribution of predation by drilling gastropods, preying upon lucinoid bivalves, turtle grass (*Thalassia*) environments, north coast of Jamaica. (After Jackson, 1972, from *Marine Biology*, volume 14)

Although predation should relax competition and permit more species to coexist, intense predation may further diminish species richness as most prey individuals are removed. Species richness should thus have a maximum at intermediate grazing level. Vadas (1977) experimentally manipulated urchin density on an *Agarum* kelp bed and subsequently counted macroalgal species numbers. The pattern of richness matched the expected model (Fig. 18-15). Competition for space presumably reduced diversity at zero grazing. We noted in Chapter 4, however, the limitations of the model. For example, if the urchin preferred macroalgae that are competitively inferior, then modest grazing might enhance competitive dominance. Furthermore, modest grazing often enhances prey productivity (see Cooper, 1973; Fenchel and Kofoed, 1976) and so might increase competitive pressure among the prey species. Consequently, Vadas' model is useful only when predator food preference and effects on prey productivity are well understood.

Summer Migrations

As in the intertidal zone, seasonal influxes of migrating predator populations cause summer declines in subtidal benthic species. The plaice *Pleuronectes platessa* and the flounder *Pleuronectes flesus* migrate into Danish fjords as recently metamorphosed juveniles. Both

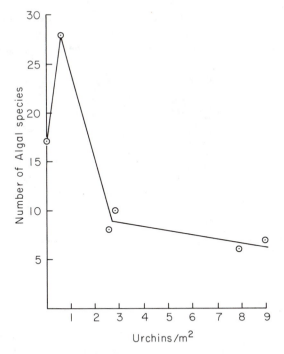

Figure 18-15 The relationship between algal species richness and the population density of sea urchins, shallow-water habitats around San Juan Island, Washington. (After Vadas, 1977)

species cause significant benthic mortality in shallow subtidal bottoms and move into deeper water as they grow (Bregnballe, 1961). In East Coast estuaries, such as Chesapeake Bay, the blue crab *Callinectes Sapidus* decimates populations of benthic organisms. A series of caging experiments showed that populations of the bivalve *Mulinia lateralis* were eliminated by crab predation (Virnstein, 1977). Disturbance of the sediment–water interface by the crab clogged *Mulinia* siphons and also contributed to mortality.

Predator Population Bursts

Although predation is generally an important force in structuring subtidal bottom communities, unusual years of strong recruitment may have major and long-lasting impact on bottom communities. Bursts of predators may result from exceptionally good larval recruitment or from the movement of predators into an area previously inaccessible because of unfavorable conditions, such as low salinity.

Urchins are well known to have bursts of unusually strong larval recruitment. In 1969 a burst of abundance in the urchin *Strongylocentrotus drobachiensis* caused strong declines in both biomass and species richness in seaweed beds in the Strait of Georgia, Canada (Foreman, 1977). Subsequent recovery—mediated by urchin removal—varied, depending on depth. In the first year rapid-growing forms, such as the sea lettuces *Ulva* and *Enteromorpha,* predominated. Forms resistant to grazing, such as articulated coralline algae, were also prominent. In general, community establishment proceeded from shallow to deeper depths. The upper subtidal reached complete recovery within 3 years whereas the laminarian deep-water climax assemblage required at least one more year for recovery. The giant kelp, *Nereocystis luetkeana,* attained maximum biomass at intermediate depths in the third year; it was subsequently eliminated by other seaweeds capable of shading germlings.

The burst of urchins was probably due to a good plankton bloom that enhanced planktotrophic larval success. Besides the strong initial effect on macrophyte biomass, these bursts strongly influenced the composition of the algal community for years. Initial recovery was dominated by fragile, rapid-growing forms that were subsequently eliminated via competition with later colonizing seaweeds. The urchin disturbance also permitted colonization by the giant kelp, *N. leutkeana.* A large buoyed kelp species, *N. luetkeana* is an annual and hence must grow toward the surface every year from the rocky substratum. Eventually colonization by browns, such as *Agarum* and *Laminaria,* forms a low-lying canopy that is, nevertheless, sufficient to inhibit rapid growth of the giant kelp. Therefore urchin disturbance is important in the development of this kelp in the Strait of Georgia and in the general initiation of successional recovery.

Multiple Trophic Level Interactions

Predation in three trophic levels may have complex effects on community structure. The top predator in a three-level system may prevent local extinction of species at the bottom level by reducing the second trophic level. Subtidal kelp forests, off the West Coast of North America, consist of luxuriant stands of many species of macroalgae. Brown algae, such as *Nereocystis* and *Pterygophora,* grow rapidly and form enormous standing crops that are commercially exploited in California. Dense urchin (*Strongylocentrotus franciscanus*) populations, however, often overgraze and eliminate the kelp (North, 1965). Consequently, commercial kelp exploiters have attempted to eliminate urchins. In some areas, populations of sea otters (*Enhydra lutris*) are abundant and prey on urchins, abalone, and other algal grazers.

Estes and Palmisano (1974) compared two Aleutian islands: one with and another without sea otters. Amchitka Island had a dense (20 to 30 animals km^{-2}) otter population for 20 to 30 years, with a dense subtidal vegetation cover and low urchin density. In contrast, Shemya Island lacked sea otters but had a dense urchin population and lacked macrophytic vegetation below the lower intertidal macrophyte zone. Size-frequency distributions of urchins differ in the two islands, with the older age classes lacking at Amchitka. The development of kelp beds at Amchitka shelters the shore from wave action and allows silt to settle out of the water column, killing off species sensitive to turbidity. Competition for space with kelp species further diminishes populations of sessile invertebrates. The presence of sea otters as a keystone species in the third trophic level is therefore regarded as a major determinant of community structure (Estes and Palmisano, 1974).

Subtidal sea pen beds of Puget Sound, Washington, and associated predators further illustrate trophic level interactions at three levels (Birkeland, 1974). A top predator, the sea star *Solaster dawsoni,* consumes a number of predators of the sea pen *Ptilosarcus*

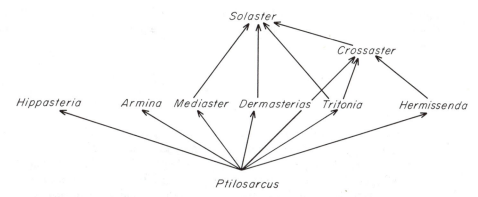

Figure 18-16 The carnivorous food web of sandy bottoms of Puget Sound, Washington, dominated by the top carnivorous starfish, *Solaster*. (From Brikeland, 1974, Ecol. Monogr. vol. 44, pp. 211–232. Copyright 1974, the Ecological Society of America.)

gurneyi (Fig. 18-16). Most sea pen predators are generalists, capable of feeding on other prey when *Ptilosarcus* is rare. But the sea star *Hippasteria* is nearly restricted in diet to sea pens (98.8% of 1800 feeding observations) and so may starve if sea pens are at low densities. This factor is particularly critical at the time of the juvenile sea star's first meal, for failure to find suitable prey can greatly increase the probability of mortality. A more generalized sea star will more easily find a first meal and survive. Other sea pen predators, such as *Mediaster* and *Dermasterias,* are more generalized and consume other prey species when *Ptilosarcus* is rare.

The top predator, *Solaster,* consumes all sea stars but *Hippasteria. Solaster* individuals rapidly crawl off and move away from *Hippasteria,* perhaps repeled by the large sessile pedicellariae of the latter.

Birkeland measured *Solaster* densities and consumption rates of some prey sea star species. In one case, a *Solaster* wrestled with a *Crossaster* for 2 to 3 days before finally ingesting it! Ingestion was completed in another 3 to 4 days. The density (one per 400 m^2), consumption rate (average 4.5 days meal^{-1}), and percentage of the *Solaster* feeding at any one time (19.5%) imply a predation rate of one sea star every 23 days per *Solaster* and four sea stars removed m^{-1}yr^{-1}. Because *Solaster's* diet consists of 23% *Mediaster* and 12.5% *Crossaster,* 9.4% and 38% of their respective populations are removed per year. In both cases, measured larval recruitment is insufficient to yield an increase in the prey populations.

No single factor, therefore, may explain the prevention of overexploitation of sea pens by its major predators. Predation by *Solaster* is sufficient to prevent population increase of *Crossaster* and *Mediaster,* but *Hippasteria* is immune from *Solaster* attack. *Hippasteria* recruitment may be impaired by the spatially patchy and unpredictable recruitment of its prey, *Ptilosarcus.* Community structure is thus influenced by prey patchiness, variation in specialization of predators on prey, and overall level of predation. The immunity of *Hippasteria* from *Solaster* and its specialization on *Ptilosarcus* suggest that an understanding of natural history is essential in judging the outcome of trophic-level interactions.

SUMMARY

1. Remote sampling from ships is a major stumbling block in understanding the subtidal benthos. Although some sampling devices yield quantitative data, different bottom types are often sampled with differing efficiency. Where biomass is extremely low, large nonquantitative samplers have been substituted for smaller quantitative devices. SCUBA diving permits accurate location and sophisticated sampling and field experimentation; it is, however, restricted to depths less than ca. 40 m.

2. Suspension feeders usually dominate sands whereas deposit-feeders dominate fine-grained sediments. Deposit feeders find little suitable material to ingest in sand. Suspension feeders are usually excluded from mud because of the instability of the sediment, the turbidity near the sediment–water interface, and the burrowing activities of deposit feeders, which exascerbate the first two factors.

3. Because remote sampling is generally necessary to explain benthic distributions, much attention has been paid to establishing statistically meaningful associations of benthic species. In some cases, meaningful ecological statements can be made about natural history and biological interactions. More often, a series of correlations is established between biological and physical variables; it is usually difficult to draw firm conclusions because of multiple correlations.

4. Most subtidal benthic assemblages are highly patchy. In some cases, this factor can be related to patchy settlement of certain larval year classes. A mosaic of year classes can be constructed. Some of these patchy settlements can be related to large-scale subtidal disturbances, such as bottom erosion. Dominance can change dramatically over a bottom, which may relate to historical factors, such as what season space is opened for larval recruitment. Thus a subtidal bottom may be a mosaic of patches in various stages since a major disturbance. There is no necessary progression toward a certain dominant; complex competitive interactions might lead to alternative stable dominant species.

5. Competitive interactions seem to organize soft-bottom communities, although predators often alter community structure. Many communities display vertical stratification in which different species occupy differing living positions below the sediment–water interface; this result usually occurs because of interference competition and takes place in both deposit-feeding and suspension-feeding assemblages.

6. Seasonal influxes of predators in temperate and boreal soft bottoms can cause large-scale reductions of benthic standing stock. Subtidal food webs are sometimes complex; geographic differences in predator occurrence can lead to very different community compositions.

19

food supply, trophic structure, and diversity in a depth gradient

19-1 NUTRIENT INPUT TO THE DEEP

Input of Organic Matter

The supply of food to subtidal benthic communities depends on proximity to shore and water depth. Nearshore, shallow-water localities are richly supplied with both benthic and planktonic primary production, much of which enters the food web as organic detritus. Table 19-1 shows the primary productivity of some benthic habitats. Most notable are the high values for macroalgal and seagrass beds, ranging from 200 to 1750 g C m^{-2} y^{-1}, relative to plankton productivity in the same regions (temperate–boreal) of 70 to 200 g C cm^{-2} y^{-1}. In St. Margaret's Bay, Nova Scotia, about 15% of the seaweed production reaches the sediments. The supply of detritus, as measured by sediment traps (at 60- and 65-m depth), coincides with the peak growth period of the seaweeds (Mann, 1976a; Fig. 19-1). Although current transport may carry some organic detritus to great depths—as along submarine canyons to the continental slope and some trenches—the supply of seaweed and seagrass detritus diminishes with depth and distance from shore. Estuaries are also sources of dissolved and particulate nutrients and support rich benthic communities and profitable benthic invertebrate and vertebrate fisheries (Marshall, 1970).

The planktonic supply of organic matter to the benthos similarly decreases with depth and distance from shore. Planktonic production is greater nearshore due to continental nutrient supplies and upwelling systems that recycle lost nutrients from the bottom to the surface. In southern New England temperate estuaries, some 30 to 50% of the annual primary production (Table 19-2) reaches the bottom sediments (Riley, 1956; Marshall, 1970). However, only about 2 to 7% of the surface production reaches the

TABLE 19-1 PRIMARY PRODUCTIVITY OF SOME BENTHIC HABITATS

Habitat	g Cm^{-2}y^{-1}	Source
Macrocystis pyrifera kelp forests, California	550–900	Clendenning, 1960
Laminaria-Agarum kelp beds, Nova Scotia	1750	Mann, 1972
Average seaweed production, shallow bay, Nova Scotia	600	Mann, 1972
Turtle grass, *Thalassia testudinum* bed, Florida	840–1825	Zieman, 1975
Benthic microflora, shoals of southern New England	90	Marshall, 1970
Pennate diatoms, estuarine mud flat, Scotland	31	Leach, 1970
Eelgrass, *Zostera marina* temperate–boreal	15–2700	McRoy and McMillan, 1977

bottom (2000- to 5000-m depth) in the North Atlantic (Riley, 1956). The organic matter reaching the sea floor at great depths has previously been attacked by a variety of decomposers and so is probably far more refractory than organic detritus reaching the bottom on the continental shelf or shallow bottoms adjacent to the shoreline. Amino acids from a core taken at 5454 m (North Atlantic) are in chemically resistant fractions of the sediment and are difficult to extract (Whelan, 1977). Organic material in the deep ocean may recycle on the order of 1000 to 3500 years (Menzel, 1974). Sinking rates of organic particles range from weeks to over a year per 1000 m, although the relatively rapid sinking rate of zooplankton fecal pellets might bring somewhat more labile organic matter to the bottom more rapidly (Smayda, 1970). Consequently, both surface productivity and depth of the traverse of organic matter influence the supply to the bottom (Hargrave, 1973). On the continental slope, turbidity currents and other bottom transport mechanisms may

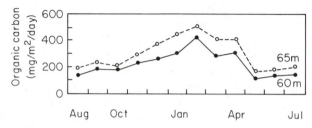

Figure 19-1 Daily rates of deposition of particulate organic carbon in St. Margaret's Bay, Nova Scotia, in traps placed at 60 m and 65 m depth, over a bottom at 70 m depth. (After Mann, 1976, © Blackwell Scientific Publications Limited)

TABLE 19-2 SURFACE PRODUCTION AND SUPPLY OF ORGANIC MATTER TO THE BOTTOM, IN SHALLOW- AND DEEP-WATER HABITATS

Habitat	g $Cm^{-2}y^{-1}$	Source
Long Island Sound	70	Riley, 1956
Long Island Sound, supplied to bottom	25	Riley, 1956
Sargasso Sea	130	Riley, 1970
Sargasso Sea, supplied to bottom	ca. 4	Riley, 1970

deliver organic matter more rapidly to deep waters. Off the coast of southern California, the proximity of the continental slope to shore permits large amounts of kelp detritus to reach the deep-sea bottom.

In shallow waters the supply of organic matter to the benthos takes the form of carcasses of phytoplankton and zooplankton, plus zooplankton fecal pellets. The latter may constitute a major fraction of the organic supply to the bottom during zooplankton blooms (Moore, 1931; Steele and Baird, 1972). In very shallow waters (<5 to 10 m) phytoplankton production may be directly consumed by benthic suspension feeders.

Benthic–Pelagic Coupling

Coupling between the planktonic and benthic systems also diminishes with increasing water depth. The supply of organic matter from the plankton to the bottom is balanced by mineralization of organic detritus by the benthic microbial community and release of dissolved metabolites to the overlying water. Nixon and co-workers (1976) measured the in situ release of nitrogen and oxygen consumption of the seabed in Narragansett Bay, Rhode Island, using opaque chambers placed on the bottom by divers. Oxygen uptake, a useful estimate of metabolic activity in the sediment (Pamatmat, 1971), increased in the spring and early summer and then declined in August. Most of the surface productivity occurs in winter–spring (Pratt, 1965). As temperatures increase in spring, the metabolic potential concomitantly increases and the accumulated organic matter is rapidly oxidized. Later in the summer, the organic matter in the sediment has been reduced to a refractory residue and oxygen consumption decreases.

Ammonia is the principal form of inorganic nitrogen released from the sediment and accounts for the levels found in the overlying water column (6- to 7-m depth). In three different benthic community types, ammonia regeneration from the bottom is 675 to 870 mM m^{-2} y^{-1}. Using measured oxygen consumption, oxygen uptake and ammonia release rates yield an O/N atomic ratio of 27:1 to 33:1 (depending on the lower or upper estimate of ammonia release). But 13.25:1 is an O/N ratio commonly associated with the decomposition in the sea, suggesting a nitrogen deficit in benthic nitrogen release measurements. This deficit is confirmed when phosphorus release is considered. Given measured nitrogen and phosphorus release rates, a N/P ratio of 5.9:1 is calculated instead of the expected 16:1 associated with production and decomposition of organics. The deficit of nitrogen may be balanced by the release of dissolved organic nitrogen (i.e., amino acids, small peptides) or dissolved N_2 produced by denitrifying bacteria. Nixon and colleagues (1976) conclude that the former is more likely because of high measured concentrations of organic nitrogen in the water column and preliminary estimates of organic nitrogen flux from the sediment.

Although nutrient flux from the sediment exerts a major impact on very shallow, tidally mixed embayments, the seasonal breakdown and establishment of a thermocline in deeper basins restrict benthic–pelagic coupling to the fall to late winter-early spring when the thermal structure of the water column permits overturn and transfer of nutrients to surface waters. With increasing depth, however, especially deeper than the continental slope, benthic–pelagic coupling is probably minimal. Over short periods of time the transfer of nutrients from the bottom to the surface is slow enough to conclude that the bottom exerts no major influence on the nutrient regime within the photic zone. Over longer time scales, however, transfer of nutrients upward must occur. So we can conclude that benthic–pelagic coupling has three qualitatively different modes:

1. Release of nutrients to a tidally mixed water column above (very shallow embayments).
2. Release of nutrients to the overlying water, with advective transfer of dissolved nutrients to shallow water in fall and late winter-early spring (shelf depths).
3. Relative decoupling over short time periods (continental rise and deeper).

It is probable that surface currents and deeper circulation result in strong interactions between the continental slope bottoms and the overlying water column; however, the extent of this interaction varies geographically.

Benthic Metabolism

The decrease of organic input with increasing distance from shore results in a concomitant reduction of metabolic activity in the seabed, as measured by oxygen consumption of the benthic community. Oxygen decrease is measured in situ with polarographic electrodes sensing oxygen in chambers on the bottom or in water over cores carefully taken and transported to shipboard. Pamatmat (1971) studied the disposition of oxygen uptake in a benthic community. Over half the oxygen may be lost in chemical oxidation of reduced end products of anaerobic metabolism. The rest is consumed by some chemolithotrophs, aerobic bacteria, meiofauna, and macrofauna. Bacterial biomass and macrofaunal biomass are generally of the same order of magnitude. Because bacteria have a well-known higher oxygen consumption g^{-1}, they must be primarily responsible for the respiratory consumption of oxygen (Mann, 1976). Meiofauna and small macrofauna consume more oxygen than larger macrofauna for the same reason. Organic input, therefore, is met with a pulse of metabolic activity explained principally by anaerobic and aerobic detrital breakdown and respiration by meiofauna and small macrofauna. Organic matter is mostly decomposed and so the organic carbon content decreases with depth into the sediment (Keen and Piper, 1976). Total benthic oxygen consumption decreases greatly with depth from shelf (11 to 180 m, 4 to 40 ml O_2 h^{-1}; Pamatmat and Banse, 1969) to deep sea (1325 to 2900 m, 0.6 to 4.5 ml O_2 h^{-1}; Pamatmat, 1971) depths.

The decrease in organic input with depth also influences the organic content and oxidation state of sediments. At a depth of 70 m in St. Margaret's Bay, Nova Scotia, organic content was 4.5% (Webster et al., 1975). In abyssal regions near the equator or near oceanic peripheries, organic content is 0.5 to 1.5% and sediments are anoxic below the surface. In oceanic regions most distant from shore and beyond the productive equatorial belt, sediment organic content is less than 0.25% and sediments are strongly oxidized (Sokolova, 1972). Sokolova distinguishes between the relatively sterile *oligotrophic* deep oceanic basins distant from shore and the *eutrophic* deep bottoms peripheral to the continental slopes and in equatorial regions. The trend in organic carbon is similar to that found by Sanders and colleagues (1965) except that the latter authors find a great increase in sediment organics at the continental shelf–slope break and low sediment organics in part of the outer shelf, where large numbers of benthic animals are collected. Sanders and his co-workers conclude that standard measures of organic content do not indicate which portion is refractory and unavailable to the benthic community, making small-scale changes difficult to interpret.

Deep-sea microbial dynamics and the potential rate of consumption are poorly understood. An accidental discovery, however, indicates that bacteria–nutrient interactions may be slow indeed. In 1968 an accident at sea (New England continental slope, 1540 m) resulted in the loss overboard of the *Alvin,* a submersible vehicle used by scientists of the Woods Hole Oceanographic Institution. No one was killed, but the crew was unlucky enough to lose their lunches to the deep. The submersible was recovered about a year later with the food (thermos with bouillon, apple, sandwich) showing almost no decomposition (Jannasch et al., 1971). The soup, initially prepared from canned meat extract, was perfectly palatable after 1 year. When kept under refrigeration at 3°C, bacterial attack was immediate and starch and protein fractions spoiled in a few weeks. Therefore the cold temperature of the deep sea (ca. 2 to 4°C) was not an explanatory factor.

This observation was confirmed by incubating deep-sea and shallow-water bacteria on C-14-labeled substrates in experimental in situ chambers at 5300 m and 1830 m (Jannasch et al., 1971; Jannasch and Wirsen, 1973). Both bacterial groups incubated in the deep sea showed 0.4 to 30.9%, but usually less than 2%, of the uptake measured in the laboratory at 4°C over the same period (Table 19-3). Jannasch and colleagues suggest that increased hydrostatic pressure raises the minimal bacterial growth temperature, inactivating the cells when this rise exceeds the environmental temperature. Therefore pressure-adapted bacteria must be adapted to shifting down their optimum growth temperature relative to the hydrostatic pressure in the deep sea (ca. 1 bar 10^{-1} m depth change). In any event, microbial processes in the deep sea are probably very slow, further exacerbating the food shortage for meio- and macro-benthic consumers.

TABLE 19-3 MICROBIAL CONVERSION IN THE DEEP SEA. PERCENT CONVERSION OF C-14-LABELED SUBSTRATES RELATIVE TO LABORATORY CONTROLS (4°C). DATA ARE GIVEN FOR INCORPORATED (I) AND METABOLIZED (M) SUBSTRATE.[a]

C-14-labeled substrate	Taken at 200 m; incubated at 5300 m		Taken at 200 m; incubated at 1830 m		Taken at 1830 m; incubated at 1830 m	
	I	M	I	M	I	M
Acetate	1.14	2.83	0.38	0.37	4.08	2.91
Mannitol	1.14	1.68	7.12	7.02	0.26	0.98
Glutamate	1.62	0.67	0.72	0.81	0.26	0.50
Casamino acids	8.87	7.71	11.74	30.86	1.70	3.08

[a]From Jannasch et al., 1971.

19-2 BIOMASS AND TROPHIC STRUCTURE

Biomass Changes with Depth

The depth-dependent input of organic matter is reflected in an overall decrease of benthic biomass with depth. Many Russian workers have attempted broad classificatory schemes of the oceans based on benthic biomass (e.g., Sokolova, 1972). Belyaev and Ushakov (1957) summarized Antarctic biomass patterns and showed an exponential decline with depth. The meiobenthos (0.5 to 5.0 mm) decreases from 5.0 g m^{-2} in deep-sea eutrophic bottoms (defined earlier) to 0.5 g m^{-2} in oligotrophic bottoms (Sokolova, 1972).

Biomass estimates are highly variable in small anchor-dredge and grab samples, for a patchily distributed large species (e.g., echinoderm) will occasionally be found and

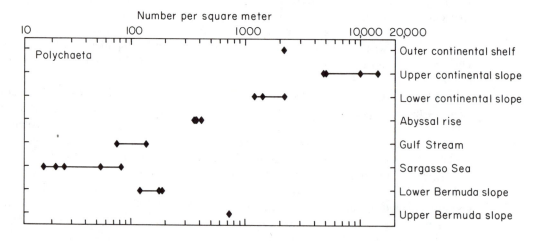

Figure 19-2 The number of polychaetes found as a function of depth, along the Gay Head-Bermuda transect. (After Sanders et al., *Deep-Sea Research,* vol. 12, pp. 845–867, 1965)

swamp out the large numbers of smaller species. Because deep-sea representatives of several groups are the same size or slightly smaller than shelf forms and because of the taxonomic value of the collected material, Sanders and co-workers (1965) report depth changes in numbers rather than biomass. Based on anchor-dredge samples (Fig. 19-2), polychaetes increase from 2000 m^{-2} on the outer shelf to 5 to 15,000 m^{-2} on the upper continental slope. Then density drops with depth to 20 to 100 individuals m^{-2} on the abyssal Sargasso Sea bed. This pattern is repeated in the crustacea.

The influence of available food is further confirmed in comparisons of meiofaunal abundance in the abyssal Pacific with production of the heterotrophic microflora (Sokolova, 1972). The pattern indicates that, as a whole, benthic communities are limited by food supply with increasing depth. In shallow waters where biomass is maximal it is not clear whether food supply is a limiting factor for a given species, for predation, living space, or other limiting factors may regulate population size. In the deep sea even low rates of predation might exert strong regulatory effects on benthic populations (Dayton and Hessler, 1972). So the overall trophic limitation of benthic communities cannot be directly extrapolated to individual species and particular populations.

There are some exceptions to the general pattern of depauperate biomass in the deep sea. In trench bottoms an apparently rapid influx of organic matter along the bottom supports a dense bottom fauna relative to abyssal bottoms. The sloping portions of the trenches are probably sites of rapid downward transport of organic particulates. Because trenches are often near land masses (e.g., Aleutian Trench), a ready source of organic matter is available. Slope habitats often have high biomass as well.

A spectacular and bizarre community of organisms has been discovered in the immediate vicinity of volcanic vents in midoceanic ridges (Ballard, 1977). Vents near the Galapagos Islands spew out water that is anoxic and very hot. Relative to the adjacent ambient water temperature of 2°C, waters in a recently investigated vent area range above 20°C (much hotter vent areas are known). Adjacent to the vent, an incredibly rich biota of invertebrates of large size and great density covers the volcanic rocks. A mussel, *Bathymodiolus,* is unique to the vents and ranges over 20 cm in length. Large limpets, clams, and crabs also abound. Most curious of all is a large tube worm belonging to the Vestimentifera, having no gut and reaching lengths of over 1 m. The worm secretes tubes over 3 m in length (Fig. 19-3).

This community obviously depends on a source of organic matter that differs from that used by typical deep ocean communities at comparable depths (2500 m). It is probable that the sulfur coming from the volcanic vent is the base of a chemoautotrophic bacteria population that, in turn, is the source of food for most of the members of the community. At present, submarine-based direct observations and experimentation are being used in further investigations of this remarkable community.

The vestimentiferan tube worm is clearly the weirdest creature of the lot. Like other members of the phylum Pogonophora, the worms lack a mouth and gut. How does such a large individual obtain sufficient nutrition? Cavanaugh and colleagues (1981) demonstrate that the worms harbor symbiotic chemoautotrophic sulfide-oxidizing bacteria in a vascularized tissue, the trophosome. These bacteria manufacture ATP with the energy generated from sulfide oxidation and reduce CO_2 to organic matter. The bacteria may be

Figure 19-3 Photographs taken from the deep-sea submersible, Alvin, of hard bottoms near a hydrothermal vent of the Galapagos Rift. (a) Dense stand of the vestimentiferan tube-worm. (Photograph by Dr. Kathleen Crane) (b) "Dandelions," siphonophores related to the Portuguese man-of-war, among dead clams and a thermistor probe at right. (Photograph by Dr. R. Hessler) (c) Mats of organisms nicknamed "spaghetti" and tentatively identified as enteropneusts. (Photograph by Dr. J. Childress) (All photographs courtesy Woods Hole Oceanographic Institution)

digested or may excrete dissolved organic compounds that are absorbed by the worm. This symbiosis may extend to other rift macroinvertebrates (e.g., the giant vent clam *Calyptogena magnifica*) and even to shallow-water species living in sulfide-rich environments.

At this writing, much of the recent literature on the biology of the hydrothermal vents can be found cited in Cavanaugh and others (1981).

Trophic Structure and Species Composition

In shallow waters large amounts of suspended detrital particles and phytoplankton provide a food source for suspension feeders. At depth, however, suspended particles with attendant microorganisms occur in a slow drizzle and phytoplankton are absent. This situation is reflected in a shift from the dominance of suspension feeders to deposit-feeding dominance in the deep sea. Figure 19-4 shows a switch of dominance by (mainly) suspension-feeding eulamellibranchiate bivalves to deposit-feeding protobranchiate bivalves on a transect from Cape Cod to Bermuda. At depths of 5500 to 5800 m below the North Pacific gyre, deposit-feeding forms similarly dominate the benthic community. No obligate suspension feeders are found in box cores from the locality (Hessler and Jumars, 1974). Among larger animals and smaller encrusting species obtained by bottom trawls, suspension feeders may be found in comparable localities (Sokolova, 1972), but typical deep-sea benthic animals consume sediment. Carnivores are greatly reduced in number and are always generalized trophically because of the scarcity of prey (Hessler and Jumars, 1974). The rarity of food is reflected in gut development of the bivalve *Abra*

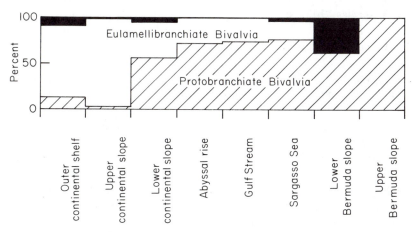

Figure 19-4   The percentage composition of dominant bivalve mollusk taxa in different regions of the Gayhead (Cape Cod, Massachusetts)-Bermuda transect. (After Sanders et al., *Deep-Sea Research*, vol. 12, pp. 845–867, 1965)

profundorum, whose gut is much longer than shallow-water representatives of the same size and taxonomic group (Allen and Sanders, 1966).

Scavengers are apparently common in all deep-sea environments. Large fishes, natantian decapods, and amphipods rapidly approach baited cans lowered to the bottom (Hessler et al., 1972; Isaacs, 1969). Such scavengers are presumably capable of being completely inactive to conserve energy until carrion appears.

19-3 DIVERSITY GRADIENTS

As discussed in Chapter 5, patterns in diversity provide insight into the structure of biotic communities. If more species are present in locality 1 of habitat A relative to locality 2 of habitat A, we may ask which factors allow more species to be supported. Three prominent subtidal gradients in species richness are worth discussing: (a) estuarine diversity gradients, (b) latitudinal diversity gradients, and (c) shelf, deep-sea gradient. General explanations of diversity change are discussed in Chapter 5.

The Change of Species Richness with Depth

Because estuaries are broadly characterized by a decrease of salinity upstream, marine organisms must be able to tolerate increasingly lower salinities. But species richness of marine benthic species steadily decreases upriver. Although there are species of strictly estuarine organisms, benthic species richness is, nevertheless, depauperate relative to open marine bottoms (Sanders, 1968). Many echinoderms and protobranch bivalves cannot penetrate estuaries (see Chapter 17, Section 17-2).

These data ignore the question of why there are fewer species in estuaries in the first place. It is possible that (a) low salinity precludes the successful establishment of most species, (b) other great variability in salinity, temperature, and primary production selects for broad-niched species, permitting few to coexist, and (c) the variable nature of estuaries shifts a dynamic equilibrium of speciation and extinction toward lower diversity.

We discussed latitudinal species richness gradients in Chapter 5. The increase of subtidal benthic diversity toward the tropics is well documented (e.g., Fischer, 1960; Sanders, 1968; Stehli et al., 1967). Sanders (1968) comprehensively examines the latitudinal and deep-sea diversity gradients. Despite a steady decrease of food supply with depth, several studies have documented a dramatic increase of macrofaunal species richness in the deep-sea benthos (Hessler and Sanders, 1967; Sanders and Hessler, 1969; Hessler and Jumars, 1974). Using an epibenthic sled, Hessler and Sanders (1967) were able to get far more animals from the deep sea than previous expeditions. They found that species richness of mud-bottom benthos was greater for slope bottoms than comparable sediments on the New England continental shelf.

Although Sanders and Hessler (1969) showed that species richness increases with increasing depth, subsequent work has demonstrated a decline in richness at depths greater

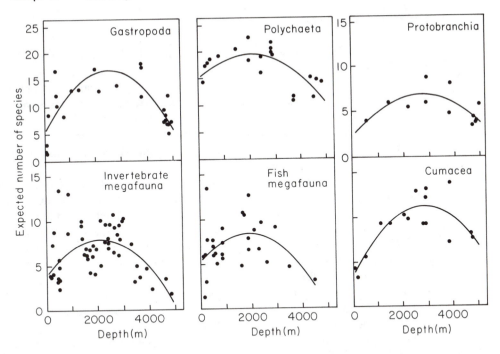

Figure 19-5 Variation in species richness along the depth gradient of the ocean (data compiled by Rex, 1981). Species richness is an estimate for samples of 50 individuals. (See Hurlbert, 1971, for method)

than 2000- to 3000-m depth (Rex, 1973, 1981). For the dominant macrofaunal groups, species richness shows a parabolic relationship with depth (Fig. 19-5); richness increases and then decreases toward abyssal depths. Rex (1973) suggests that the rigorous trophic regime of the abyss may favor a relatively depauperate biota. Interestingly, meiofaunal groups, such as foraminifera, do not follow the parabolic pattern and steadily increase in species richness with increasing depth (e.g., Buzas and Gibson, 1969). Perhaps the trophic limitation of reduced food is not as limiting to species of smaller body size.

The shelf–slope break (100 to 300 m) marks a zone of strong taxonomic change in polychaetes, bivalves, and crustacea; this boundary marks the upper limit of the deep-sea benthos. Below this zone, slope macrobenthic species appear to have much narrower biogeographic ranges than originally supposed (e.g., Sanders and Hessler, 1969). Meiobenthic copepod faunas do not appear to be shared over great distances in the deep sea (Coull, 1972). In contrast, abyssal-plain macrobenthic species seem to have broader geographic distributions (M. Rex, written communication). Macrobenthic diversity in the deep sea is broadly correlated with dispersal mode. The brooding habit of crustacea is correlated with their dominance in species numbers (Sanders and Hessler, 1969). Deep-sea bivalves usually have short-range lecithotrophic larvae and fewer species are found (see Scheltema, 1972).

Explanations for the Depth Gradient

We discuss the correlation of environmental stability with high diversity in Chapter 5, Section 5-3. Jackson (1974) examined bivalve biogeographic range and water depth and found that bivalves able to live in depths less than 1 m have significantly wider geographic distributions than infaunal bivalves restricted to deeper shelf waters. Shallower species must tolerate a wider range of physiological conditions and their broader occurrence probably lowers their probability of speciation or extinction. (The deep sea is enigmatic in this context, for bivalves have broad geographic ranges. This factor may relate to the relative monotony of the deep-sea floor.)

The deep sea can be visualized as a cold (2 to 4°C), physically constant environment of low organic input. Boreal shelfs are seasonal and rich in organic input. Sanders (1968) characterizes the deep-sea benthos as biologically accommodated, with biological inter-actions causing a minimization of competitive interference. Species are visualized as being highly specialized. The very low faunal density, combined with high species diversity, fails to support this assumption. The richness of the biota, together with its rarity, makes the predictability of interacting with the same species twice in a row low indeed. There is some patchiness among deep-sea benthic samples (e.g., Thistle, 1978), but it is apparently in overall abundance of all species as opposed to spatial structuring of different species (Hessler and Jumars, 1974). It seems inconceivable that a population with densities of ca. 115 m^{-2} divided nearly evenly among 20 species has in some subtle way subdivided the resources of a homogeneous mud bottom. Recall that most species are deposit feeders. Some spatial and particle-size differentiation is conceivable, but the depauperate nature of the food supply itself argues for selection favoring trophic generalists. Dayton and Hessler (1972) suggest that biological disturbance might help maintain diversity by altering the course of competitive exclusion. But at this point our knowledge of competition and predation in the deep-sea benthos is insufficient to draw any nonspeculative conclusions.

Huston (1979) proposed that species richness is regulated by the interaction between rates of competitive displacement and the frequency of biological and physical disturbance. The parabolic relation between species richness and depth can be explained as follows. Population growth in shallow water is rapid enough to cause competitive exclusion and depress species richness. In contrast, in the abyss, the frequency of disturbance is very low; competitive exclusion thus reduces species richness. Bottoms of intermediate depths might have intermediate levels of disturbance and competitive displacement; species richness might thus be maximal.

This model is problematic because its premises can never be assessed. It is not clear, on a large scale, that competitive exclusion is a major force in depressing soft-bottom shelf macrobenthic species richness. If anything, many have argued that the relaxation of predation often leads to *increases* of species numbers in shallow soft bottoms. Furthermore, the rate of disturbance would be expected to be sufficient to maintain diversity in shelf habitats. Finally, the progressive increase of meiofaunal diversity with depth weakens this hypothesis.

Stehli and colleagues (1972) document high evolutionary rates of tropical high-diversity fossil biotas relative to high-latitude, low-diversity biotas. Perhaps environmental stability gradients influence the balance of speciation and extinction rates. Presumably speciation is a stochastic process, occurring when a species is subdivided by a sufficient geographic or ecological barrier. But as variable environments select for broad tolerance and hence wide biogeographic range, speciation will be depressed. On the other hand, the extinction of new species isolates will probably be greater in an unpredictable environment, for the new isolate might soon confront a change beyond the range of its adaptability. Net standing diversity is thus low. In contrast, stable environments allow specialization to local conditions simply because unpredictable change is absent, thereby eliminating selection for broad adaptability. A given geographic barrier promotes speciation more easily because there is less migration by wide-dispersing, "catastrophe-adapted" forms. Furthermore, the daughter species have a lower extinction rate because the environment does not suddenly change beyond its newly acquired adaptations. Standing diversity is thus increased. This model assumes that when daughter species are reintroduced, competitive interaction rarely results in exclusion but rather in competitive displacement and specialization (see Schoener, 1974). There may be, of course, an upper limit to the number of species that a habitat can support. But disturbance, predation, and the presence of refuges seem to combine to diminish the probability of competitive extinction except when a novel group of organisms evolves and is far superior to the current dominant. Of course, biological interactions must secondarily alter the balance, but the patterns of interactions and specialization observed may be simply adjustments to standing diversity generated by the balance of speciation and extinction. This balance is probably regulated by environmental stability (Levinton, 1979c).

At abyssal depths the paucity of food might constitute a factor that outweighs the effect of environmental stability. Food availability might be severe enough to keep the population size of many species to levels low enough to increase greatly the probability of extinction. This factor would tip the speciation–extinction balance to low-standing diversity. So in summary, the speciation–extinction balance would favor increasing species richness with depth; however, food scarcity in the abyss would reduce species richness. The parabolic relation of species richness to depth may thus be explained by the influence of the physical and trophic environment on speciation–extinction rates.

Stability may play the dominant role in regulating species diversity; time and biological accommodation, however, may have little to do with the deep-sea diversity gradient. There is no compelling evidence that the deep sea is particularly more ancient than shelf habitats. After all, geologically speaking, the deep Atlantic is a relatively recent habitat. As argued earlier, no compelling reasoning or evidence argues for subtle biological interactions that have led to niche subdivision.

Abele and Walters (1979a,b) have reevaluated the Sanders data on the Gayhead Bermuda transect and found that much of the variance in polychaete species richness may be explained by sediment variation. Species richness increases with depth, however, despite this explained component of the variance. There seems to be a clear distinction only between shelf and deep-sea bottoms. The fact that this difference transcends a great deal of heterogeneity in sampling technique, sediment variation, and geographic variation testifies to its probable reality. As noted, comparable results have been obtained for other groups of organisms (Rex, 1981).

Besides criticizing Sanders' (1968) unequal sampling procedures and failure to account for habitat heterogeneity, Abele and Walters (1979a) propose that area is as good a predictor of polychaete species richness as the stability-time hypothesis. They point out that the greater area of the deep sea might promote lowered extinction of broadly distributed species or might harbor more habitats and hence support more species via habitat heterogeneity.

A study by Bambach (1977) suggests that stability *or* habitat area may be important in regulating species richness. Over long periods of geological time, species richness in nearshore and offshore shelf benthic communities remains constant. Species richness in offshore habitats, however, has increased in a steplike manner several times in the Phanerozoic while remaining constant in nearshore shelf habitats. This fact suggests that some aspect of "offshoreness" promotes increases of species richness. Whether this factor is due to area, habitat heterogeneity, stability, or a combination of the three cannot be discerned from the data.

Data collected on planktonic foraminifera (Stehli et al., 1972) provide further insight into this problem. Stehli and his colleagues evaluated data on the age of origin of living species and genera along a latitudinal gradient in the South Atlantic. This region is of interest because area increases toward the pole, which is the opposite of the confounding correlation between stability and area found in the shelf, deep-sea transect described by Sanders. In the case, species diversity increases with decreasing latitude despite the decrease of area along the same geographic space.

Morphological evolution of foraminifera seems more rapid in low-latitude habitats; generic age increases with increasing latitude. High-latitude habitats favor more generalized forms of relatively great antiquity. This result is in accordance with the stability model discussed earlier, which predicts that unstable environments will select for generalized forms whose descendant species will differ little from their antecedents. In contrast, stable environments will favor the proliferation of new morphologies—hence a reduction in mean generic age. This inference is based on the assumption that generic age is an estimate of morphological change with time. So very probably stability does have an important effect on species richness and the nature of communities. This statement does not imply that area is unimportant. Species spread over wider areas are less likely to become extinct during unfavorable periods. With greater areal coverage, such species would more likely find a refuge during the crisis. If so, it is unlikely that we will ever be able to distinguish effects of stability from area along the shelf, deep-sea gradient. Poor access to experimentation, the lack of a good fossil record, and confounding of correlated variables suggest that Abele and Walters' reinterpretation is interesting but not a unique solution.

Finally, we might note some problems with modes of inference in regard to explanations of species diversity. It is important to remember that short-term ecological explanations for phenomena that may have historical roots instead can be entirely misleading. In Chapter 16, for example, we demonstrated that disturbance is an important force in explaining the number of species in a given microhabitat. Up to a point, increased disturbance will increase the number of coexisting species. There is no reason to believe, however, that this factor can be used to explain differences, say, between the Atlantic and Pacific coasts of the United States. If anything, exposed Pacific coasts are subject to intense disturbance but harbor more species than the Atlantic on exposed rocky shores. Here it is likely that the relative ages of the two oceans, the relative areas, or relative degrees of environmental stability may explain the difference.

SUMMARY

1. The continental shelf, deep-sea depth gradient can be viewed as a gradient of supply of organic matter to the benthos. With increasing depth and distance from shore, the seabed receives progressively less organic matter. Benthic–pelagic decoupling also progressively decreases with depth and distance from shore.

2. The microbial activity within the seabed also decreases with increasing depth. At abyssal depths only refractory organic compounds reach the bottom from the photosynthetically active surface waters. Although there is some locally active organic synthesis, as in the sulfur-based microbial community adjacent to deep-sea volcanic vents, most deep-sea microbial activity is literally in slow motion.

3. The biomass of the benthos decreases with increasing depth; deposit feeders tend to dominate deep-sea benthic communities. Diversity increases with depth, however, and decreases again from the continental rise to the abyssal plain. Intermediate depths seem optimal for maintaining maximal species richness, but the reasons are not clear. Extinction rates may be high on the shelf due to environmental instability. The very low food levels in the abyss may keep populations of species sufficiently small that extinction rates are also very high. Thus the balance of speciation and extinction might tip in favor of maximal standing diversity at intermediate depth where food is not severely limiting and environmental change is not so severe as to cause local extinction very often.

20

coral reefs: limiting factors, morphology, and nutrition of corals

20-1 INTRODUCTION, DEFINITIONS, AND LIMITING FACTORS

Coral reefs are wave-resistant structures notable for their great species richness, topographic complexity, and remarkable beauty. Coral reefs are common in clear open marine water throughout the tropics. Massive reef accretion can be ascribed to the production of calcium carbonate by scleractinian corals and crustose coralline algae; numerous other calcium carbonate-producing algae and invertebrates also contribute to reef growth. The aggregate activities of these organisms has produced the Great Barrier Reef, a 1950-km-long ribbon of coral reefs capped by small tropical islands and stretching along much of the east coast of Australia. This complex of reefs protects the coast of eastern Australia from the wave energy of the Pacific.

The world coral reef biota can be divided into (Fig. 20-1) Atlantic and Indo-Pacific biogeographic provinces, which probably differentiated from a pantropical province in the mid-Miocene. The Indo-Pacific province differs from the Atlantic in

1. Its higher diversity of corals and most associated reef groups.
2. The presence of atolls or rings of islands capping submarine volcanoes, which are rare in the Atlantic province.
3. Extensive development of rich coral populations on intertidal reef flats, with poor intertidal coral development in the Atlantic province.
4. Differences in dominance of some groups.

Details on diversity differences are discussed at the end of Chapter 21.

Diving operations have provided our best source of information on the biotic com-

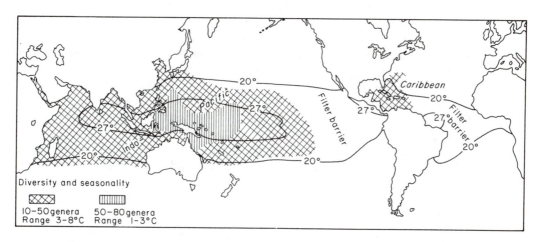

Figure 20-1 Approximate limits of the tropical Indo-Pacific and Caribbean marine coral reef provinces, as compared with minimum average sea temperatures. (After Newell, 1971)

position of reefs and the biological interactions between species. The beauty of coral reefs cannot be appreciated without diving or snorkeling to view the incredible microtopographic complexity and species diversity. The bewildering array of species, competitive interactions, symbioses, predator–prey interactions, and remarkable variety of color have been compared to tropical rain forests (Connell, 1973).

Definitions and General Limiting Factors

Definition.   Coral reefs are compacted and cemented assemblages of skeletons and skeletal sediment of sedentary organisms living in warm marine waters within water depths of strong illumination. They are constructional physiographic features of tropical seas consisting fundamentally of a rigid calcareous framework mainly composed of the interlocked and encrusted skeletons of reef-building *(hermatypic)* corals and crustose coralline algae. The reef framework controls the accumulation of sediment on, in, and around itself (Wells, 1957). The corals belong primarily to the order Scleractinia. Hermatypic corals all have zooxanthellae-endosymbiotic algae that benefit the coral host (see later).

Temperature.   High calcification rates are limited to warm waters. Consequently, coral reefs are restricted to tropical seas (Fig. 20-1), generally between 25° north and 25° south latitude. Well-developed coral reefs are usually not established at temperatures much below 23 to 25°C. Some reefs, however, may develop at temperatures as low as 18°C, as in the Florida Keys of the United States. Reefs at the edge of the above-mentioned latitudinal range may be strongly affected by changes in climate. A 1968 cold event plunged air temperatures below 0°C over most of the Persian Gulf and caused water temperatures to dip to as little as 10°C (Shinn, 1974). Almost all the inshore coral reefs soon completely died off. Some offshore reefs survived and presumably provided propagules for recolonization of the inshore reef zone.

These temperature restrictions apply only to coral reefs and not necessarily to corals per se. McCloskey (1970) studied a community associated with a marine scleractinian coral that lived in shallow waters off South Carolina. Corals with zooxanthellae grew well at temperatures as low as 14.5°C. Some particularly resistant coral species tolerated temperatures as low as 5°C for brief periods. The high rates of calcification ascribed to hermatypic (reef-building) corals thus require both high temperature and the presence of zooxanthellae endosymbionts. Scleractinian corals, such as *Astrangia danae,* have zooxanthellae but live in temperate and boreal waters. Calcium carbonate accretion is correspondingly low.

Light.   After temperature, light is probably the most important limiting factor to well-developed coral reefs because of the symbiosis between hermatypic scleractinian corals and zooxanthellae. Derived from the dinoflagellates, zooxanthellae live within the gastrodermal tissues of scleractinian corals and are apparently essential for rapid calcification. Because light intensity decreases exponentially with increasing depth, active reef building is greatly diminished below depths of 25 m in the Indo-Pacific region (Rosen, 1975). In the Caribbean active hermatypic growth is rare below 75 m, but *Montastrea* and *Agaricia* populations can be found. Normal calcification rates of corals can be cut in half on a cloudy day.

Although active coral growth cannot occur below 15 to 20% of surface light values, some coral species are adapted to diminished light conditions. When brought into shallow waters of high surface illumination, corals typically living in the shade or at depth usually expel zooxanthellae and soon die (Lang, 1971). Corals kept at the same depths in the shade of a boat show no ill effects. Some evidence suggests that zooxanthellae of hermatypic corals acclimatize to the diminished light of an overcast day (Wethey and Porter, 1976).

Wells summarized many factors influencing the growth of corals and concluded that the diversity of hermatypic corals decreases with depth (Wells, 1957; Fig. 20-2). The depth distribution of light intensity and number of coral genera fit well contrasted to the change of temperature and oxygen with depth. So we could believe that the change of coral diversity with depth is controlled by illumination. Other studies show that coral diversity may not simply decrease with depth (Goreau, 1959) and may even increase with depth (Loya and Slobodkin, 1971). Coral species diversity increases with depth in the reefs of Eilat, Israel, because of the greater environmental stability of deeper waters as opposed to the large fluctuations in temperature, desiccation, and salinity in the shallow reef flat. Consequently, light does not exert a simple control on numbers of species of corals with depth.

Salinity.   Hermatypic corals seem to require open-ocean salinity. Well-developed reefs are not generally found in estuarine or excessively hypersaline conditions. Persian

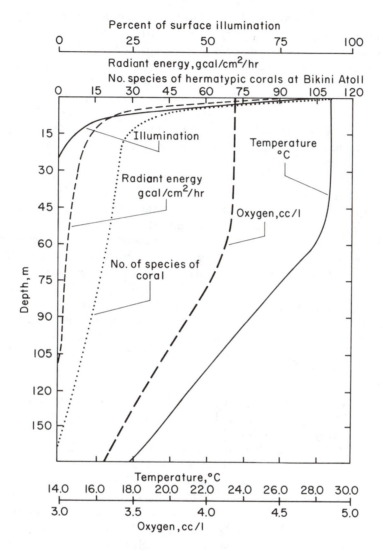

Figure 20-2 Distribution of reef-building corals at Bikini Atoll, as compared to the depth distribution of several physical variables. (After Wells, 1957, courtesy The Geological Society of America)

Gulf reefs, however, develop in salinities of more than 40‰ (Kinsman, 1964). High rains, resulting in excessive runoff, cause extensive damage to corals close to river mouths in such tropical islands as Fiji (Squires, 1962; Goodbody, 1961). Rivers also carry a large suspended sediment load that is also detrimental to corals. Flash floods on the north coast of Jamaica carry large amounts of freshwater and suspended sediment to back-reef lagoons, which typically have lower coral species richness.

Turbidity and sedimentation.   High turbidity and sedimentation rates strongly inhibit reef growth. Turbidity increases light attenuation and thereby decreases photosynthesis by zooxanthellae. Correspondingly, calcification rates are diminished as well. Settling sediment tends to foul the surfaces of coral colonies; horizontal benches often collect sediment and usually support poor reef growth (Loya, 1976a).

The effects of turbidity and sedimentation exert regional differences on reef development. Reefs in the clear waters of the windward sides of Caribbean islands (e.g., north coast of Jamaica) grow more rapidly than on the leeward sides (e.g., south coast of Jamaica) or adjacent to continental coasts (e.g., coast of Venezuela) where sediment accumulates.

Corals show differential adaptations for dealing with turbid environments. Species with large polyps can more easily remove suspended matter than small-polyp forms. Massive and slow-growing hermatypic corals, such as *Platygyra*, produce large amounts of mucus when sedimentation is high. Mucus is transported along the surface of the skeleton of the colony and carries away particulate matter. In contrast, rapidly growing branching forms, such as *Acropora palmata*, produce less mucus because sediment does not tend to settle out on their smaller cross-sectional areas. Richman and others (1975) measured in situ mucus production on the reef at Eilat, Israel, and found production per coral head to be (a) 6.8 ± 1.2 mg head^{-1} d^{-1} for massive forms (e.g., *Platygyra*), (b) 2.1 ± 1.0 for hemispherical species (e.g., *Montipora*), and (c) 1.9 ± 0.7 for branching forms (e.g., *Stylophora*). Corals most resistant to burial have also been found to be resistant to salinity and temperature changes and are thus best adapted for estuarine and low-energy lagoon environments.

Wave energy.   Because coral reefs require clear water and are constructional topographic features, they tend to be located in areas of high wave energy. Erect branching forms, such as the Caribbean *Acropora palmata*, living in reef crest zones must withstand wave shock and are often greater than 2 m across (Fig. 20-3). Hermatypic coral colonies are typically strong and very dense in structure. Such storm events as cyclones and hurricanes, however, often topple coral colonies and exert massive destruction on coral reefs. A cyclone at Heron Island, Australia, obliterated a small area of coral reef, toppling large corals. Within several years many coral colonies were reestablished by larval settlement from the plankton (Woodhead, 1971). In the Caribbean hurricanes exert massive effects on coral reef communities as well. Hurricane Carmen passed over the north coast of Jamaica in the fall of 1974. The ensuing turbulence overturned many large coral heads and tore loose epifaunal organisms from their substrata (Sammarco, 1977). Storms continually renew open substrate for colonization and space occupation by newly settling larval forms. Storms and other major physical disturbances are major sources of change in coral reef communities, suggesting that coral reef communities are not static and constant biomes. Storm damage can, in effect, be a mechanism of coral dispersal, for pieces of living colonies transported to new sites may survive to cement to the bottom and establish a permanent new colony.

Figure 20-3   Colonies of the elkhorn coral, *Acropora palmata,* a dominant of the reef crest of Caribbean coral reefs. Note the preferred orientation of the branches of the colony. (Photograph courtesy James W. Porter)

20-2 REEF TYPES AND DEPTH ZONATION

A complex terminology has been developed to classify coral reefs. For our purposes, we divide reefs into two types: atolls and coastal reefs. Atolls are horseshoe- or ring-shaped arrays of islands consisting of coral reef rock capping an oceanic island of volcanic origin. Coastal reefs border coasts of islands or continents and range in dimension from the enormous Great Barrier Reef of Australia to the small reefs capping the coasts of Eilat,

Israel. They may rest on previous reefs, coastal bedrock, or soft sediment. Even the distinction between atolls and coastal reefs fails to include some reefs whose occurrence is on oceanic volcanic islands but whose morphology resembles coastal reefs. The reader interested in the complexities of reef terminology should consult Stoddart (1969).

Atolls

Atolls mainly occur in the tropical Pacific Ocean. A few are found in the Indian Ocean and the Caribbean (Glynn, 1973). Darwin correctly postulated that atolls could only develop their characteristic array of ring-shaped to horseshoe-shaped emergent coral islands by slow subsidence of volcanic seamounts with continuous and vigorous reef growth toward the sea surface. This hypothesis would imply the presence of great thicknesses of reef rock capping a volcanic basement. This prediction is confirmed by the ca. 1400 m of reef rock capping the volcanic basement of the Enewetak Atoll (Ladd et al., 1953). The reef dates back to the Eocene (40 to 60 million years before present).

    The scheme of atoll development may be summarized as follows. As the volcanic island submerges, coral reefs develop around the fringe of the island. When the volcanic island center plunges below the sea surface, the peripheral reef continues to grow upward, developing a ring-shaped array of islands and leaving a lagoon in the center (Fig. 20-4a).

    Figure 20-4b shows a cross section of a typical atoll. Because the circular array of islands in an atoll usually resides in a unidirectional wind pattern, the windward side of the atoll generally develops different coral reefs than the leeward side of the atoll system. Seaward and windward reefs are the zones of most intense wave energy. At a depth of a few meters, few live corals are to be found and coral rubble accumulates. But large

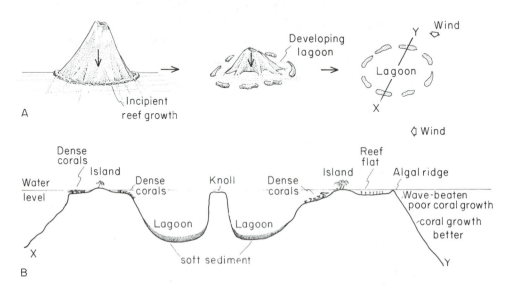

Figure 20-4 (a) The hypothetical origin of a coral atoll; (b) cross section showing major subhabitats.

hermatypic corals grow abundantly below this zone. Below the zone of high wave energy, more delicate and foliose corals are found. On the intertidal margin of windward reefs, a large algal ridge is formed by crustose coralline red algae. It usually rises about a meter above mean low tide level. Behind this algal ridge is a broad intertidal or slightly subtidal reef flat. Changes in salinity and temperature are most extreme on this part of the reef. Corals are most abundant near the lagoonward side of the algal ridge and lagoonward of this zone is a large expanse of *reef flat* in which massive corals as much as 3 m in diameter may occur.

The leeward and seaward sides of the reef experience much less dramatic wave surge; consequently, corals are more abundant in shallower water. The algal ridge is not as well developed. The windward and seaward sides of the reef may be dissected by very large surge channels, sometimes as much as 15 m deep. Sturdy collonies of *Acropora*, *Pocillopora, Millepora,* and *Heliopora* line the walls of the surge channels. Surge and wave energy are important limiting factors in these shallowest portions of the reef zone.

The relatively quiet waters of the lagoon usually contain lagoon reefs and lagoon slope and floor environments (Wells, 1957). Although the lagoon floor is generally between 25 and 50 m deep, occasional pinnacles rise to the surface and are capped by coral reefs similar to the seaward side of the atoll. If passes do not exist from the outside to the lagoon, the atoll floor is normally covered with fine-grained sediments and supports a community typical of tropical calcareous muddy environments. If channels exist for natural reasons or have been blasted for shipping, currents keep the lagoon floor free of fine-grained sediments.

Coastal Reefs

Coastal reefs parallel shorelines and range in size from barrier reefs as much as 2000 km long and fronting a lagoon of 50 km in width to small and discontinuous reefs plastered on the shoreline. Unlike atolls whose morphogeneses are all probably similar in origin, coastal reefs consist of a group of structures of diverse origin. The Great Barrier Reef of Australia, for instance, can consist of geologically ancient reef rock and some areas of the reef tract can be interpreted as being due to continual subsidence (Stoddart, 1969). The Great Barrier Reef tract is an end member in the types of bordering reefs because it fronts a "lagoon" some 40 to 50 km wide protecting the coast of eastern Australia. At the other extreme, fringing reefs cap sand, volcanic rock, or previously dead coral reefs. An example is the fringing reef at Eilat, Israel, studied by Loya and Slobodkin (1971). In the intertidal zone the bottom is capped by a rather continuous table of coral reef rock. At depths of about 30 m, however, patches of coral reef cap sand bodies or preexisting outcrops of rock.

Coastal Reefs of Jamaica

As an example of reef zonation, we shall discuss the windward reef tract on the north coast of Jamaica, West Indies. From the shore, the general morphology of the reef may be described as the following (Fig. 20-5):

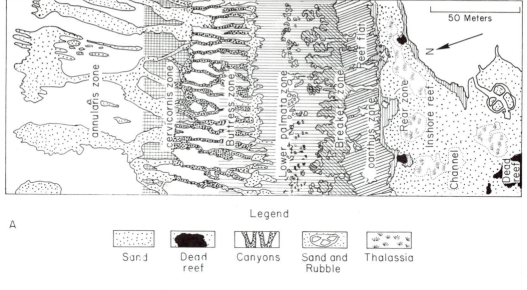

Legend

A

Sand   Dead reef   Canyons   Sand and Rubble   Thalassia

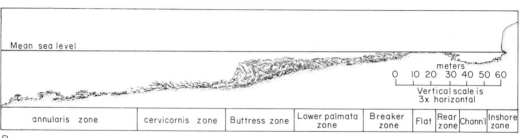

B

Figure 20-5 (a) Map view of coral reef environments, off the north coast of Jamaica. (After Goreau and Land, 1974) (b) Cross-sectional view of depth zonation of the coral reef at Discovery Bay, Jamaica. (After Goreau, 1959. Copyright 1959, the Ecological Society of America)

1. A back-reef zone consisting of shoreline and lagoon (0 to 10 m depth).
2. A reef crest (0 to 20 m).
3. Buttress zone.
4. Staghorn coral zone.
5. A break in slope at depths of 55 to 65 m, leading to a precipitous dropoff at vertical angles of 60 to 90° to water depths of greater than 300 m (Goreau, 1959; Land, 1973; Goreau and Land, 1974).

The shoreward margins of most back-reef lagoons are usually lined with either reef rock or sediment consolidated by mangroves. The nature of the lagoon depends greatly on the hydrography of the region and the openness of the lagoon to the sea. If the lagoon is cut off from strong currents, then the bottom consists of fine-grained sediments. Shallower bottoms support extensive growths of the marine angiosperm *Thallassia testudinum* and other sea grasses. Members of the bivalve family Lucinidae are abundant in soft sediment among the roots of turtle grass and feed mainly on organic detritus falling into an anterior mucous tube (Allen, 1958; Jackson, 1973). Lucinids here live in anoxic sediments and are resistant to physiological stress, such as high temperature and extremes of salinity.

In slightly deeper and less anoxic sediments, species richness increases and a large number of mullusks are capable of surviving, relative to the low-diversity fauna of shallow (less than 1 m) turtle grass flats. In turtle grass meadows greater than 1 m in depth, grazing urchins (e.g., *Tripneustes ventricosus*) may control seagrass biomass. The large sea cucumber *Astichopus multifidus* consumes sediment. Volcano-shaped sediment mounds formed by burrowing callianassid shrimp mottle the sandy bottom.

Also in the lagoon are patch reefs ranging from a meter to 50 to 100 m across. They are often as rich in coral species as the forereef but rarely support the large coral heads found on the outer reef.

As we approach the rearward margin of the reef crest, large numbers of coral species seem to flourish even though wave energy and suspended sediment conditions are less optimal than on the seaward side of the crest. We then go on to a reef flat that is either intertidal or only a few centimeters to 1 m in depth. Here corals are usually rather small. The colonial zoanthid, *Zoanthus* sp., is often abundant. Hermatypic corals in the Caribbean, do not survive intertidal exposure as well as in the Pacific, perhaps because of the smaller tidal excursion in the Caribbean, which may result in long periods of exposure to heat and desiccation when winds move from onshore to offshore.

Just seaward of the reef flat, the reef crest supports large coral heads of the elkhorn coral, *Acropora palmata* (Fig. 20-3). Branches tend to be oriented parallel to major unidirectional currents. At about 6 to 7 m in depth, the colonies are smaller and branches are more flattened. The deeper forereef consists of a broad and relatively flat terracelike area at 10 to 15 m depth, regularly transected at right angles to the shoreline by large and deep surge channels that form a series of large lobes facing seaward (spur and groove topography). These lobes constitute the *buttress zone* and are mainly composed of a framework constructed by the massive hermatype, *Montastrea annularis*. Species richness peaks in the buttress zone.

Below the buttress zone, a broad low-relief bottom is covered with thickets of the staghorn coral, *Acropora cervicornis* (Fig. 20-6). Pieces of coral branches readily break and roll about, making the thicket bottom unstable. Descending the forereef slope, we encounter a steeper slope of high scleractinian coral species richness, the forereef escarpment. From this escarpment a slope descends at vertical angles between 20 to 60° to a second slope break at a depth of 55 to 65 m. This slope is barren of corals, although it contains a rich biota of algae, gorgonians, and burrowing organisms that occupy the predominating soft sediment. A flattened growth form of *Montastrea annularis*

Figure 20-6 The branching coral *Acropora cervicornis,* in thickets, Discovery Bay, Jamaica. (Photograph courtesy James W. Porter)

(Fig. 20-7) can be an important framework builder on coral pinnacles protruding from the forereef slope, but some other species are more important frame builders in the lower half of the forereef zone, such as the foliose hermatype *Agaricia* spp. (Fig. 20-7). An abrupt slope break, the "dropoff," occurs at a depth of about 55 to 65 m, below which the deep forereef descends almost vertically.

Below about 75 m, corals are no longer sufficiently abundant to be principal framework builders. The vertical wall dropping off to greater than 1000 m, however, is actively growing outward. An important framework builder on these walls is the sclerosponge *Ceratoporella nicholsoni,* a potentially large and massive calcifying sponge thought to be related to the extinct Stromatoporoidea (Hartman and Goreau, 1970). They are only present in shallow-reef environments within caves and help define a distinct coral reef cave community, along with articulate brachiopods (Jackson et al., 1971). At these depths, sclerosponges are large and become major framework builders, helping to trap sediment moving downslope on the reef. They are probably of major importance in maintaining the dropoff as a vertical wall. At depths of 200 to 300 m, organisms like stalked crinoids have been found in great abundance on sandy bottoms.

Figure 20-7   The deep-water growth form of the massive reef-building coral, *Montastrea annularis,* in association with several species of *Agaricia*. (Photograph courtesy Philip Dustan)

Indo-Pacific Reef Zonation

Coastal reefs in the Indo-Pacific share many features with those of the Caribbean. Buttresses project seaward and form a spur and groove topography. As in the Caribbean, *Acropora* is dominant in wave-swept areas; coral colony branches tend to align and point into the surf. Behind the reef front, a broad algal rim develops, as on the seaward side of atolls. The rim consists of step pavements of coralline algae, which grow seaward over the spur and groove channels, and bears the brunt of the main attack of waves hitting the coral reef surface. The back reef consists of channels, islands, and relatively quiet lagoons filled with soft sediment. For a detailed account of zonation on the Great Barrier Reef, the reader should consult Manton and Stephenson (1935), and Maxwell (1968).

General features of Indo-Pacific zonation have been summarized by Rosen (1971, 1975), based on exposure to water movement. Three components are important (Fig. 20-8): (a) exposure of the coast, (b) distance from the reef edge exposed to waves, and

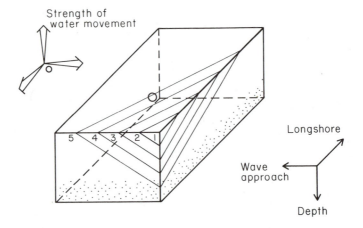

Figure 20-8 Schematic diagram relating major Indo-Pacific shallow-water coral associations to strength of water movement. Water movement is dissected into three major components, arranged perpendicularly. 0 = theoretical point of zero water movement. Sequence of associations is given by numbers: 1 = Calcareous algae; 2 = *Pocillopora;* 3 = *Acropora;* 4 = Faviid; 5 = *Porites.* Stippling suggests increasing effects of loss of illumination with depth. (Modified after Rosen, 1975)

(c) water depth. A given coral assemblage is therefore determined by its position with regard to these three factors. In order of decreasing exposure, the zones are

1. The algal ridge.
2. *Pocillopora* zone of turbulence.
3. *Acropora* exposed zone.
4. *Faviid–Musiid* zone.
5. *Porites* zone.

With the exception of the algal ridge, which is always at the surface, all zones are found in a sequence corresponding to decreasing water movement. Given the three important components, the wave-resistant *Pocillopora* assemblage is to be found in shallow, wave-exposed coasts, nearest the oceanward edge of the reef. *Porites* zone residents will occur in deeper, more protected environments away from the reef edge. In practice, all zones are not found in any one place (Rosen, 1975).

20-3 REEF TOPOGRAPHY: ACCRETION AND EROSION

Genesis of Reef Morphology

The control of reef morphology is a subject in great dispute. We can generally agree that the very thick accumulations of coral reef rock on atolls can only be explained by long-term subsidence and concomitant upward reef growth. However, the complex of terraces and dropoffs found, for example, in Caribbean reefs may be explained by two alternative processes or their combined effects: (a) the constructional growth of reef tracts or (b) erosional processes that occurred during Pleistocene lowerings of sea levels (Purdy, 1974). At low sea level stands an erosional terrain may then have subsequently controlled reef growth when sea level rose as the glaciers retreated. This second process would explain the presence of steep slope dropoffs on oceanic margins of reefs, the presence of barrier reef terraces, and shelf lagoons. The occurrence of blue holes—large and deep depressions in the sea bottom indicating former development of karst or cave topography when sea level was at a lower stand—supports Purdy's hypothesis of extensive erosion at lower Pleistocene sea levels.

Although extensive erosion is likely, massive reef accretion has occurred during postglacial rises in sea level. Because the Pleistocene was a period of several glacial advances and retreats, there were several corresponding rises and falls in sea level. The ability of reef growth to track rises in sea level depends on antecedent topography, the rate of sea-level rise, and wave energy (Adey, 1978). When a rise in sea level inundates a carbonate platform, initial conditions of suspended sediment may be unfavorable for reef growth. The lag time in reef initiation may be too great for sufficient reef growth before water depth has increased to the degree that light is limiting. Windward coasts, however, may favor the growth of species of *Acropora* and coralline red algae; the former are favored in moderate wave energy and explain most rapid reef accretion.

Adey (1978) summarizes the evidence supporting the hypothesis that many reefs are capable of growing with rising sea level during glacial retreats. Reef growth has been vigorous in both the Atlantic and Pacific; rates of 9 to 15 m upward accretion per year have been recorded in the Caribbean.

Curves of reef growth developed from C-14 dating of coral skeletons in both the Atlantic and Pacific often show strong concordance with sea-level rise curves. The role of such organisms as sclerosponges in deep water suggests that significant lateral accretion is also possible and that near-vertical walls can be accretionary features. Coral reef workers have generally thought of Caribbean reefs as relatively depauperate and slow growing. Although the greater species richness found in the Pacific is not in doubt, it is becoming clear that vigorous Caribbean reef growth is the rule and that extensive reef tracts are common (e.g., northern side of Little Bahama Bank on the Nicaraguan rise).

Bioerosion

Although dramatic reef accretion ultimately stems from the contributions of calcium carbonate-producing organisms, numerous species of animals and plants destroy the skeletal output and may even inhibit reef accretion. Urchins and grazing fishes feed on epibionts and concomitantly scrape off bits of calcium carbonate. The majority of sand-sized particles on reefs probably come from grazing activities.

The other major source of bioerosion comes from the activities of endolithic organisms, those creatures that bore into hard substratum. A wide variety of bivalve mollusks, sipunculids, and polychaetes plays a major role in boring and subsequently weakening coral skeletons. Sponges of the family Clionidae are probably the most significant source of bioerosion on coral reefs. They work at reef rock at the base of the intertidal zone and are the cause of overhanging benches of intertidal rock commonly observed in the Caribbean (Neumann, 1966). Clionids are commonly found at the base of hermatypic coral skeletons and may weaken the base to the point of breakage (Goreau and Hartman, 1963). Although chemical dissolution is the mechanism by which the calcium carbonate matrix is weakened, a poorly known type of contractile activity by sponge cells causes a continuous removal of small chips of calcium carbonate. These fragments may be the principal source of silt-sized particles on reefs.

The importance of bioerosion increases with depth, for hermatypic coral growth is adversely affected by diminishing light. The rate of bioerosion is sufficiently important at depth to affect the morphology of hermatypic corals. Although massive and branching forms are common in shallow water, platy forms (e.g., *Agaricia*) predominate in deeper water. If the base of a platy coral is eroded to the point of breakage, the plate will simply fall down to the sediment with the polypoid surface facing upward. The coral colony will therefore survive. Because the deeper reef in Jamaica has a considerable slope, however, breakage of the base of a more massive and hemispherical coral will cause it to roll somewhat; the whole polyp surface may then be smothered in sediment. Furthermore, the platy growth form permits several colonies to grow as a series of adjacent interlocking plates. In some cases, the base may be completely destroyed and yet the colony skeleton will not fall into the sediment (Goreau and Hartman, 1963).

20-4 BIOLOGY OF SCLERACTINIAN CORALS

Morphology and Reproduction

Scleractinian corals are coelenterates closely related to sea anemones that secrete a skeleton of calcium carbonate (aragonite). Some corals are solitary (only one polyp) with polyps as much as 25 to 30 cm in diameter. Most are colonial with hundreds or thousands of polyps averaging about 1 to 15 cm in diameter (Fig. 20-9). A colony of coral polyps is a sheet of live tissue covering a dense massive skeleton of calcium carbonate, formed by gradual accretion by the coral colony at the surface. The periphery of the polyp oral disk region is surrounded by one to several rings of tentacles with varying degrees of adaptation for zooplankton capture and an esophaguslike connection extends from the central mouth to the interior gastrovascular cavity. Tentacles are armed with nematocysts designed to entrap prey.

Many species of corals have been observed to be sequentially hermaphroditic. In some species, individual colonies are of one sex, but sex changes with age. Most commonly, fertilization is internal with extrusion of eggs into the water occurring in a minority of species. Reproductive activity may or may not be seasonal. After fertilization, larvae develop in the gastrovascular cavity of the parent and eventually are ejected through the mouth. The swimming planula larvae are usually elongate or spherical and externally

Figure 20-9 Closeup of the hermatypic coral, *Montastrea cavernosa*, showing expanded polyps. (Photograph courtesy James W. Porter)

ciliated. The larva is the means of dispersal, carrying the species into newly opened environments. Asexual budding allows the parent colony to grow and increase in size (Connell, 1973).

A coral planula spends some time swimming in the plankton; a later stage swims to the bottom, where attachment, settlement, and metamorphosis occur. Connell (1973) studied rates of recruitment of different coral species and found that a wide range of recruitment rates occurred but that there was no obvious explanation for these general variations in space and time. In all areas, however, commoner species had higher recruitment rates and mortality. There were very large differences in recruitment and areas as little as a few hundred meters apart differed strongly in larval recruitment rates. Larval life of planulae in the plankton may be as little as 2 days. This short dispersal phase might result in great variation in settling concentrations over relatively short distances.

Hermatypic corals are those corals that are mainly responsible for the coralline contribution to reef growth. They differ from *ahermatypic* corals in having higher rates of calcification and large numbers of zooxanthallae living within the gastroderm. Ahermatypes have few or no zooxanthallae, calcify at much slower rates, and do not produce large coral heads. The presence of zooxanthallae as symbionts in groups of calcifying organisms seems to be correlated with high calcification rates (as in the giant clams *Hippopus* and *Tridacna*) or inferred to be with such fossil organisms as prorichtofenid brachiopods (Cowan, 1970).

Growth of Corals

Coral-growth analysis allows an understanding of the mechanics of reef growth and reef accretion. The rate of coral reef accretion must be the aggregate of the growth rates of the individual coral heads, with cementation of grains in between the coral colonies. Vaughan's (1915) study of corals of the Dry Tortugas reefs showed that coral heads with a massive growth (Fig. 20-10) form grew more slowly in linear dimensions than corals

Figure 20-10 The shallow-water mound-building form of *Montastrea annularis,* being overtopped by the elkhorn coral, *Acropora palmata*. (Photograph courtesy James W. Porter)

with a ramose (branching) growth form (Fig. 20-3, Fig. 20-6). The staghorn coral, *Acropora cervicornis*, a common coral on reefs in the Caribbean, grows as much as 10 cm yr^{-1}, as measured by branch-tip extension (Shinn, 1966). In contrast, massive hemispherical colonies of coral *Montastrea annularis* accrete at 0.25 to 0.70 cm yr^{-1} in height (Vaughan, 1915). Massive colonies, however, tend to reach a greater total weight of calcium carbonate than the branching forms.

Two types of skeletal addition can be recognized in massive corals (Dustan, 1975; Fig. 20-11): the *axis of upward growth*, representing an increase of skeletal size by the same polyps, and the *axis of polyp addition*, or horizontal axis along which polyps are added through extratentacular budding in conjunction with new skeletal material. Measuring growth along both axes, along with skeletal density (using mercury displacement), a shallow massive growth form of *Montastrea annularis* can be shown to have a higher

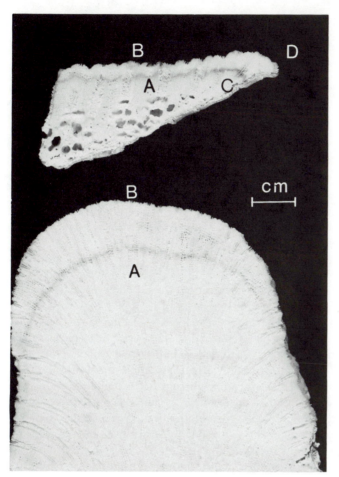

Figure 20-11 *Montastrea annularis.*
Photograph of saw-cut sections of skeletons
after 1-year growth period. Top: deeper
water flat form; bottom: shallow water
mound-shaped form. *A-B* axis of upward
growth used for both forms; *A-B* and *C-D*
are axes of polyp addition used for round
and flat forms, respectively. (From Dustan,
1975, from *Marine Biology,* volume 33)

calcification rate than a deeper platelike form (Dustan, 1975). This difference is mainly
reflected in depth-dependent variation in the axis of upward growth (Fig. 20-12).

The common Caribbean massive hermatypic coral *Montastrea annularis* has dis-
tinctly different growth forms (Fig. 20-7, Fig. 20-10) in shallow and deep-water environments (Goreau and Land, 1974; Dustan, 1975). In shallow water (10 m) the species
grows in massive, hemispherical colonies with the predominant growth vector upward.
In deeper waters (30 m) *M. annularis* assumes a platelike growth form with a predominately horizontal-extending growing edge. Platelike growth form is probably favored to
maximize light capture at low light intensities and to avoid rolling when the base of the
skeleton is bioeroded. The importance of *M. annularis* as a reef framework builder thus
affects the reef profile with depth as the shallow growth morph forms large buttress
structures. Faviid coral skeletons of New Caledonia vary regularly with depth and with
a progressive reduction of septa per corallite and number of corallites per unit area of
colony (Wijsman–Best, 1972). This result is thought to be a response to lowered light,

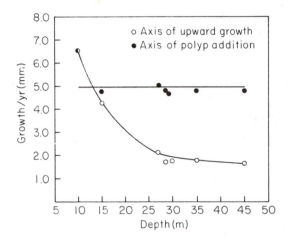

Figure 20-12 Mean colony skeletal extension rates (mm/yr) between 10 and 45 m depth on Dancing Lady Reef, north coast of Jamaica. (After Dustan, 1975, from *Marine Biology,* volume 33)

decreasing the ability of zooxanthellae to aid calcification. Corals in deeper water contain fewer zooxanthellae (Yonge, 1940; Best, 1969).

Several methods have been developed to measure coral skeleton growth. The most simple technique is to measure increments of skeletal addition relative to metal spikes driven into coral heads or to measure distances to branching tips from specified reference points in branching colonies. When introduced in suspension, the dye Alizarin Red-5 is deposited in the coral skeleton and remains as a fixed time plane as future growth occurs (Barnes, 1972). This method can be used in the field by introducing the dye into a polyethylene bag enclosing the coral colony (allowing for some water exchange). The coral skeleton is later sliced with a diamond saw and the distance from the stained time horizon to the new growing surface is measured.

Growth can also be assessed in coral colonies through growth bands, which are visible in diamond-saw sections of coral skeletons and represent variations in skeletal density. Increments correlate with cycles, such as fortnightly tides, and provide time horizons between which growth can be measured. Annual bands corresponding to seasonal temperature change can readily be employed to measure long-term growth rates of coral colonies (Weber et al., 1975). Coral growth was shown by this technique to be inhibited where bottom sediments were resuspended in Bermudian lagoons (Dodge et al., 1974).

Measurement of uptake of the radioisotopes Ca-45 and C-14 permits short-term studies (< 1 h) of calcification and allows controlled laboratory experiments on the effects of temperature and salinity (Goreau, 1959; Goreau and Goreau, 1959). The Ca-45-based estimate of 20 mm yr^{-1} growth of branch tips of the Pacific coral *Porites compressa* compares favorably with more direct measurements (Buddemeier and Kinzie, 1976). In *Pocillopora damicornis*, calcification is six times greater at branch tips than at the lateral regions. Using Ca-45 uptake, Goreau (1959) confirmed the previously discovered greater uptake of calcium in the light. Corals may acclimatize to environmental temperature.

Hermatypic corals participate in a remarkable symbiosis with zooxanthellae. Zooxanthellae cultivated outside their hosts change from a typical spherical shape maintained within the coral endoderm to biflagellate motile dinoflagellates. Zooxanthellae also occur in bivalves, other coelenterates, and in gastropods, but it is not clear as to whether the same dinoflagellate species infects all hosts or a variety of species are host-specific (Taylor, 1971, 1974). Zooxanthellae isolated from different hosts may have differing isozyme patterns (Schoenberg and Trench, 1976). Zooxanthellae taken from one host may or may not be beneficial when introduced into another host (Kinzie and Chee, 1979).

For many years a controversy has raged concerning the nature and cause of this symbiosis. Because zooxanthellae are photosynthetically active, several obvious hypotheses have been proposed

1. They are a possible source of food.
2. They may be a source of oxygen for corals in an oxygen-depleted environment (as reefs are typically oxygen rich, we omit this factor in our discussion).
3. They aid in lipogenesis.
4. They facilitate the excretory processes of hermatypic corals.
5. Through absorption of carbon dioxide they aid or affect calcification rates in hermatypic corals.

Nutrition

The hypothesis that zooxanthellae are a possible source of food for hermatypic corals raises the question of feeding behavior and digestion. In a classic series of experiments, Yonge (1930, 1931) concluded that corals are microcarnivores feeding on zooplankton in the overlying water. The feeding tentacles surrounding the mouth are crowded with nematocysts—eversible structures capable of stinging or entrapping prey. Several amino acids and the peptide glutathione consistently evoke feeding behavior in corals (Lenhoff, 1968). The puncture of prey by nematocysts releases stimulants, which initiates tentacle movement, transfer of food, and opening of the mouth. Ciliary and muscular movement transports the food to the mouth and down the esophagus. Mesenterial filaments extend from the mesentery and secrete digestive enzymes that digest the animal prey tissue.

Reactions of some corals to starvation also suggest their adaptations as microcarnivores. If starved for any length of time, corals extrude their zooxanthellae and subsequently die. This behavior hardly indicates that corals can subsist solely with a zooxanthellae food source. Johannes and co-workers (1970), however, calculated that zooplankton productivity was not adequate to sustain the energy requirements for the reef corals of Bermuda. Although upstream–downstream comparisons of phytoplankton and zooplankton show that reef organisms retain a large amount, the accrual is less than 20%

of net community metabolism (Glynn, 1973). As will be shown, corals differ in their dependence on zooxanthellae.

The suggestion that zooxanthellae provide a food source for reef-dwelling scleractinian corals is supported by other organisms having zooxanthellae, where a clear nutritional involvement has been demonstrated. In the giant clam *Tridacna*, large numbers of zooxanthellae are maintained in ameboid blood cells in the blood sinuses of the mantle tissue. Although the shell's life position is with the umbo of the shell downward, the body is rotated 180° with respect to the shell, a unique adaptation in the Bivalvia. This rotation presents the zooxanthellae-rich mantle tissue to the light so that maximum photosynthesis can occur. Zooxanthellae are concentrated in translucent structures projecting as nodes up from the mantle tissue. Digestive cells carry zooxanthellae from this mantle region and digest the zooxanthellae. Indigestible remains are then carried toward the excretory organs. The entire digestive system of the giant clam has been reduced and highly specialized in order to accommodate this symbiosis (Yonge, 1936b).

Muscatine and Hand (1958) showed that C-14 was fixed by zooxanthellae and demonstrated by autoradiographic studies that it was later found widely dispersed throughout the tissues of coelenterates. Nutrient transport therefore occurs between zooxanthellae and the body of the corals as well. Trench (1974) investigated the tropical coral reef-dwelling coelenterate *Zoanthus sociatus*, a colonial anthozoan with zooxanthellae as intracellular endosymbionts. He demonstrated a direct transfer of photosynthate from zooxanthellae to host; glycerol, glucose, and alanine were released to the anthozoan. *Z. sociatus* was also capable of feeding on zooplankton. Thus the animal maintained the flexibility of autotrophy and heterotrophy. The animal did not digest the zooxanthellae themselves. Zooplankton feeding may provide nitrogen.

Among-species morphological heterogeneity in skeletal morphology and polyp size suggests a spectrum of dependence on zooxanthellae as a food source (Porter, 1976). Polyp diameter P is positively correlated with tentacle length and is a good indicator of zooplankton-capturing ability. The surface area of the coral skeleton covered by live tissue, S, divided by the volume of the skeleton plus tissue, V, is a good index of light-capturing ability. As S/V increases in branching and platy forms, more light can be intercepted because not all incident radiation will be captured by a single intercepting plane. A multilayered morphology, as in the branching Caribbean coral *Acropora palmata*, allows S to be three times the surface area of bottom substrate that they cover (Dahl, 1973). Light can then be intercepted by more surface area of live coral tissue. Zooplankton capture favors a single continuous surface of tissue because feeding structures cannot be saturated with light, as can the photosynthetic zooxanthellae. Branches increase the number of feeding mouths, but a single plane of feeding polyps would effectively remove all the zooplankton.

S/V and polyp diameter are hyperbolically inversely correlated (Fig. 20-13). Thus corals with a shape well adapted to zooplankton capture have large polyps similarly adapted to this function (Porter, 1976). Although this factor suggests a range of dependency of hermatypes on zooxanthellae as food, even species with low values of $(S/V)/P$ show a strong apparent need for such a food source. The large-polyped *Montastrea*

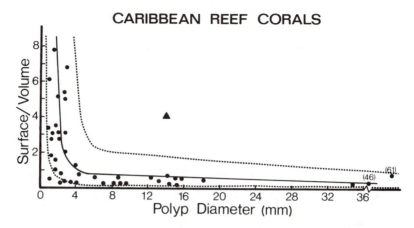

Figure 20-13   The S/V ratio of Caribbean reef-building coral species, as a function of polyp diameter. (After Porter, 1976)

cavernosa can only obtain 10 to 20% of its daily energy requirement during the two most successful hours of its 12-hour night feeding period (Porter, 1974a, 1976).

Dissolved organic matter might be an important source of food for corals as well. Along with many other phyla, several species of coral are capable of absorbing dissolved organic matter (for example, Stephens and Schinske, 1961). Carbon-14-labeled glucose is taken from solution by the coral *Fungia* and uptake rates are independent of light intensity or presence of bacteria. Stephens guessed that the body wall is the site of uptake for organic matter. Although the amounts of glucose necessary for the maintenance of coral metabolism seem too high, given the rate of uptake measured in Stephens' experiments, uptake of organic matter—perhaps by mesentarial filaments—may be an auxiliary method of feeding. The extrusion of mesentarial filaments by a coral colony might provide an extensive and large area for absorption of dissolved organic matter by the colony (Muscatine, 1973).

Lipogenesis

Zooxanthellae apparently are important in lipogenesis for hermatypic corals (Patton et al., 1977). Because lipid constitutes about one-third of the dry weight of anemones and corals, it is probably a primary energy source. The hermatype *Pocillopora capitata* elevates lipid synthesis 300% in the light relative to individuals maintained in darkness. Carbon-14-labeled acetate can be shown to be incorporated by zooxanthellae and used in lipogenesis (Patton et al., 1977); therefore zooxanthellae are responsible for the efficient transfer and conversion of acetate to lipids. It seems likely that a diversity of fatty acids taken up because of the coral's carnivorous habit is quickly oxidized to acetate in digestive cells. The acetate would then be converted to a narrow spectrum of saturated fatty acids with the aid of the energy available from photosynthesis. Acetate produced in lipid breakdown could be recycled to the zooxanthellae for lipid synthesis.

Polyunsaturated fatty acids are less common in corals and may either indicate lipogenesis by the animal instead of by zooxanthellae or may indicate external sources (dietary) for the fatty acids. *Acropora palmata*, a resident of shallow water, contains predominantly saturated fatty acids and so probably depends mainly on zooxanthellae for lipogenesis. Corals in deeper water have greater proportions of unsaturated fatty acids and may thus depend to a lesser extend on algal lipogenesis (Meyers, 1979).

Excretion

Zooxanthellae may facilitate the excretory processes of hermatypic corals. In the giant clam *Tridacna*, Yonge (1936b) demonstrated that clams kept in sealed containers eventually depleted all the phosphorus in the water in the container. Another bivalve mollusk, *Spondylus*, with no zooxanthellae, did not reduce dissolved phosphorus. Phosphorus levels increased because of the excretory activities of the bivalve. Yonge and Nicholls (1931) measured phosphorus excretion by several hermatypic coral genera and the ahermatypic coral *Dendrophyllia*. Although all species excreted phosphorus, the ahermatypic coral, lacking zooxanthellae, excreted more phosphorus than the hermatypic corals. They concluded that zooxanthellae take up phosphate from the overlying water and also remove phosphate as an excretory product from hermatypic corals. Zooxanthellae might therefore by "automatic organs of excretion" (Yonge, 1940). But the strong unidirectional currents common on coral reefs would probably remove any phosphorus excreted in the overlying water, thus preventing poisoning or fouling of the water immediately above the coral colonies (Muscatine, 1973). The major beneficiaries in the removal of excretory products are the zooxanthellae, who gain nutrients not otherwise available in the open water. $CaCO_3$ precipitation can be shown to be inhibited by dissolved phosphate, however. So zooxanthellae might enhance calcification by removal of phosphate excretions (Weber and Woodhead, 1970).

Calcification

Goreau and Goreau (1959) demonstrated that zooxanthellae play a role in the calcification of hermatypic corals by measuring light and dark uptake rates of Ca-45 under both controlled field and laboratory conditions. Their experiments are consistent with field observations that calcification on cloudy days is about 50% of calcification on sunny days. DCMU, an inhibitor of photosynthesis, also lowers rates of calcification in hermatypic corals.

Inhibition of the enzyme carbonic anhydrase also decreases the rate of calcification. Goreau and Goreau (1959) suggested that uptake of carbon dioxide for photosynthesis was the important factor enhancing rates of calcification in the coral. As carbon dioxide is removed, rates of calcium carbonate deposition are increased, perhaps through effects on the carbonate, bicarbonate, carbon dioxide system, leading to deposition of calcium carbonate (see Chapter 1). The skeleton of the coral is secreted by the ectoderm of the basal disk. Calcium carbonate is secreted onto an organic matrix, which provides nucleation sites for crystals of calcium carbonate.

The factor of supersaturation of seawater with respect to calcium carbonate is given by the ion activity product (IAP), divided by the equilibrium constant (K) for the reaction. Calculations by Berner (1971) show that the ratio of IAP to K in seawater is approximately 3.1. This supersaturation suggests that removal of carbon dioxide during photosynthesis by the zooxanthellae might therefore not have an appreciable kinetic effect on the rate of deposition of calcium carbonate. Muscatine (1973) makes this argument from an intuitive point of view by pointing out that the apical polyps of acroporid corals calcify faster than the lateral polyps, although the former have fewer zooxanthellae (Pearse and Muscatine, 1971). Two possible alternative hypotheses are as follows.

1. Zooxanthellae might manufacture some important factor for the organic matrix for the calcium carbonate deposited by corals. It is certainly possible because we know of cases in bivalve mollusks where different types of amino acids influence the mineral phase of $CaCO_3$ precipitated in the molluscan shell. According to this hypothesis, however, zooxanthellae must have evolved the ability to code for proteins that are subsequently secreted and transported to the site of calcium carbonate deposition on the coral skeleton. To date, no such evidence exists.

2. Organic phosphate inhibits calcification and so the removal of phosphate by zooxanthellae during the act of photosynthesis may increase the rate of calcium carbonate deposition. In recent years phosphate has been shown to be an important inhibitor on the surface of crystals, particularly in the case of aragonite, the crystal form of calcium carbonate found in hermatypic corals.

To summarize, zooxanthellae seem to confer several advantages to corals; nutrition, lipogenesis, and calcification are clearly enhanced, It is not clear whether any of these factors are of primary importance—that is, that one particular factor was predominant in favoring selection for the mutualism between zooxanthellae and coelenterates. Because noncalcifying forms—anemones—thrive with zooxanthellae, the roles of nutrition and lipogenesis in maintaining the symbiosis might be emphasized. Yet the advantage to calcifying hermatypes is of obvious importance. Realistically, zooxanthellae probably confer varying advantages, depending on morphology, habitat, and availability of alternative nutritive sources.

SUMMARY

1. Coral reefs are wave-resistant structures dominated by scleractinian corals and calcium carbonate-secreting algae. They are tropical in occurrence and thrive best in clear, open-ocean water at wave-exposed sites. Reef growth is a net result of calcification and upward growth keeping pace with recent global rises in sea level. Some of the topographic features of reefs may be related to erosional processes occurring at Pleistocene low stands of sea level.

2. Scleractinian corals are coelenterates that are closely related to sea anemones and that secrete a skeleton of calcium carbonate. Reef-building (hermatypic) corals harbor large numbers of algal endosymbionts (zooxanthellae) in the gastroderm. The value of zooxanthellae seems to based on a benefit to corals in terms of:

(a) nutrition—zooxanthellae release photosynthate to the coral;

(b) increased calcification rate in the presence of zooxanthellae;

(c) an acceleration of lipogenesis.

21 coral reefs: community structure, diversity patterns, and biogeography

21-1 INTRODUCTION

In many ways coral reefs resemble the rocky intertidal as a natural biotic community. Large numbers of species coexist and depend on a limiting hard substratum. Coral reefs differ, however, in that primary substratum is created by calcifying groups, such as hermatypic corals. Corals, the major space keepers on a subtidal reef, are colonial and grow vegetatively. We might expect the same general processes that control interspecies interactions and species diversity and abundance in the rocky intertidal to be operative on coral reefs as well. Although there is no synthesis quite as elegant as that of Dayton's (1971) study on the rocky intertidal on the northwest coast of the United States, a number of studies done in the 1970s demonstrate that these processes also hold for coral reefs. A more careful examination of this subject must be left for some future monograph by an expert in coral reefs, but a preliminary examination shows that the processes important in controlling rocky intertidal organisms are similarly controlling reef systems, with some interesting differences in resultant community structure.

21-2 MUTUALISM

Mutualistic interactions among species are a major determining force in reef community structure. Many mutualisms are based on protection against predators. Gobiid fishes, for example, often live as commensals in invertebrate burrows in temperate intertidal flats. But in tropical reefs the interaction is more complex because of a fixed relationship between species of the burrowing shrimp *Alpheus* and one or more gobiid fishes. In Eilat reefs each of four shrimp species has a specifically associated fish species that lives in the shrimp-excavated burrow. The shrimp has poor vision and depends on the fish for warning against predators (Karplus et al., 1972, 1974). In one mutualistic pair, the gobiid and the shrimp share in the burrow excavation.

Mutualisms between two mobile species implicitly require interspecies communication or the exchange of information. Such communication has been documented for cleaning symbioses and for shrimp–gobiid fish mutualisms. In the latter case, alpheid shrimp maintain burrows that are used as shelter from predators by both fish and shrimp. The fish "stands guard" by hovering above the burrow entrance or resting at the burrow entrance. When ploughing sand, the shrimp exits from the burrow and communicates with the goby through contact with antennae to fish tail, drops a parcel of sand, and returns into the burrow. When a threatening intruder approaches, the fish flicks its tail and the shrimp either becomes motionless or retreats into the burrow. If the disturbance is severe, the goby flees into the burrow but only after the shrimp (Preston, 1978).

The common Indo-Pacific hermatypic coral *Pocillopora damicornis* harbors an assemblage of crabs, shrimp, and fishes. Sixteen resident species have been found in the coral at Heron Island, Australia (Patton, 1974). Resident crustacea living on coral colonies attack newly introduced fishes. Resident fishes, however, are ignored or recognized after the fishes perform a shivering movement probably derived from courtship behavior (Lassig, 1977). Crustacean symbionts aid *P. damicornis* in protection against the crown-of-thorns sea star, *Acanthaster planci*. When the sea star attempts to mount the coral, the pistol shrimp, *Alpheus luttini,* moves rapidly to the colony's periphery, touches the starfish's arms with its large cheliped, and snaps until the starfish withdraws. Another crab, *Trapezia* sp., grabs tube feet or ambulacral spines and jerks the sea star up and down until the latter retreats. Because of the resident crustacea, *A. planci* avoids the coral and tends to prey on other species (Glynn, 1976).

One of the most remarkable relationships on coral reefs is the cleaning mutualism between cleaner-shrimp or cleaner-fishes and a large number of fish species. Cleaner-fishes and shrimp obtain food by cleaning ectoparasites off fishes. The cleaning fish *Labroides dimidiatus* maintains cleaning stations on Eilat reefs that are visited by about 50 species of fishes each day. Undulating movements of the cleaning-fish are recognized by the visiting species. The cleaning stations are usually localized areas of great species richness (Slobodkin and Fishelson, 1974).

21-3 INTERSPECIFIC COMPETITION

Competition for space between species in the rocky intertidal is an important determinant of how much space a given species occupies. Because corals grow rapidly, we might expect coral species to compete for the limited space on hard substratum. The most rapidly growing corals would be the expected largest space keepers. Goreau and Goreau (1959), however, examined coral growth by using Ca-45 and found no correlation between relative calcification rates and relative abundance of 13 species on Jamaican reefs. Despite its massive form and relatively slow growth, the coral *Montastrea annularis* is a major space keeper on reefs, forming very large buttresses.

Rapidly growing coral species, such as the staghorn coral, *Acropora cervicornis,* and the elkhorn coral, *A. palmata,* are capable of shading more slowly growing coral species that require light for their zooxanthellae symbionts. Rapid growth clearly would be an efficient means of outcompeting neighboring coral species. But the fact that the most rapidly growing corals have not driven all other species to extinction on reefs suggests the importance of other processes. Why is it, for example, that a slowly growing solitary coral maintains space on the reef at all? Might not all such species be driven to extinction merely by the presence of rapidly growing forms?

Catala (1964, cited in Lang, 1971) observed corals in the laboratory in New Caledonia and discovered that a coral would often kill an adjacent neighbor. The observation was confirmed in the field by observing two coral colonies adjacent to each other. When adjacent corals are of different species, a bare zone devoid of living coral tissue often separates the colonies. Lang (1971) placed two supposed ecovariants of the solitary mussid coral *Scolymia lacera* side by side. After several hours one of the two forms extruded mesenterial filaments through openings in the polyp wall and extended them toward the other "ecovariant." Within 12 hours the mesenterial filaments had completely digested part of the second individual, leaving the underlying skeleton exposed. Areas of the victimized coral that were beyond reach of these mesenterial filaments seemed to remain quite healthy. Lang repeated these experiments for many pairs of these same so-called ecovariants and found the same effect; one ecovariant invariably would digest the other. This interaction resulted in the recognition of two species: *Scolymia lacera* and *Scolymia cubensis.*

The study of interspecific aggression was extended to other species of Jamaican corals. The more aggressive *Scolymia* damaged almost every other Jamaican species studied. But many other species of Jamaican corals had the ability to "digest" neighboring corals. A hierarchy exists in which a given coral successfully digests all corals below it in the hierarchy and is digested by all corals above it (Table 21-1). Although this study was done with corals from Jamaica, the hierarchy was preserved in corals from the Caribbean coast of Panama.

It is important to remember that no *Scolymia* interacted with another individual of the same species. When one larva of *Scolymia cubensis* settles beside another of its own species, the adjacent tissues fuse and the coral becomes one large colony (Lang, 1971). Whenever a *S. lacera* is found beside a *S. cubensis,* however, the former succeeds in damaging the latter. *S. lacera* is the largest single-mouthed hermatypic coral found on reefs in the Caribbean. It tends to be solitary and is rarely found with large numbers of other coral species. Lang found that this species was capable of digesting all other species that settled next to it in its relatively deep reef habitat. This gives us an idea as to why rapidly growing species are not the only residents of natural coral reefs: "the race is not only to the swift" (Lang, 1971).

Given the alternatives of interspecific aggression in corals and competition by rapid overgrowth, we might expect a tradeoff between rapid growth and interspecific aggression.

TABLE 21-1 EXTRACOELENTERIC FEEDING INTERACTIONS BY THE HIGHLY AGGRESSIVE JAMAICAN HERMATYPIC SCLERACTINIAN CORALS[a,b]

	M. angulosa	S. lacera	I. sinuosa	M. ferox	M. meandrites	M. reesi	M. aliciae	M. lamarck-iana	M. danaana	S. cubensis	I. rigida
Mussa angulosa	0	0	0	→	⇄	→	→	→	→	→	→
Scolymia lacera		0	0	→	⇄	→	→	→	→	→	→
Isophyllia sinuosa			0	⇄	⇄	→	→	→	→	→	→
Mycetophyllia ferox				0	⇄	⇄	→	→	→	→	→
Meandrina meandrites					0	→	→	→	→	→	→
Mycetophyllia reesi						0	⇆	→	→	→	→
Mycetophyllia aliciae							0	0	0	→	→
Mycetophyllia lamarckiana								0	0	0	⇄
Mycetophyllia danaana									0	0	⇄
Scolymia cubensis										0	0
Isophyllastrea rigida											0

[a]From Lang, 1973.

[b]→ Coral in horizontal column injures the coral in the vertical column.   0 No interaction.   ⇆ Both corals injure each other.   ⇄ Coral in vertical column receives greater damage.

In general, this seems to be true (Lang, 1973; Glynn, 1976). The most aggressive corals tend to be solitary small corals that occupy minor parts of the reef (Lang, 1973). In contrast, the more rapidly growing corals, such as the acroporids, tend to be more weakly aggressive. There is, however, a group of weakly aggressive and relatively slow growing corals, such as *Porites* and *Siderastera,* that are nevertheless reasonably abundant on forereefs and reef crests. It is not clear how this group of species fits in with the general scheme, although *Porites* has a high larval recruitment rate and may be a fugitive colonist (Sammarco, 1977).

These factors indicate the three major mechanisms of interspecific competition among corals:

1. Interspecific digestion.
2. Overgrowth of one species by the other.

3. Shading by foliose and rapidly growing species over less rapidly growing species (e.g., the shading effects by the coral *Agaricia*).

Competitive success in shallow zones of East Pacific reefs is reflected in the dominance by *Pocillopora* in communities of high surface cover and low diversity (Porter, 1974b). The dominance of Pocilloporid corals is principally due to their rapid growth and overtopping characteristics. Diversity (measured by H', see Chapter 5, Section 5-1) increases with depth (Fig. 21-1), away from the zone of *Pocillopora* dominance (Glynn 1976). In the deep reef, higher diversity is accompanied by a low percentage of coral cover. Species dominating the deep reef are slower growing and dominant to *Pocillopora* in extracoelenteric feeding. Furthermore, *Pocillopora* seems to have slower growth in deep reef zones (Glynn, 1976).

The undersides of foliose corals (e.g., *Agaricia*) and coral reef caves harbor hundreds of epibenthic species competing for space (Jackson, 1977). Because well over 90% of the available space is usually occupied, it is paradoxical that so many species can coexist (often greater than 30 species on a single underside of a coral head). The complete cover and the slow vegetative growth of the dominant organisms (sponges, ectoprocts, colonial ascidians) argue against predation as a mechanism in explaining the maintenance of such species richness. If coral heads are turned over, fishes often attack and consume

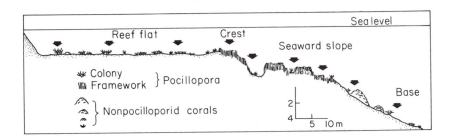

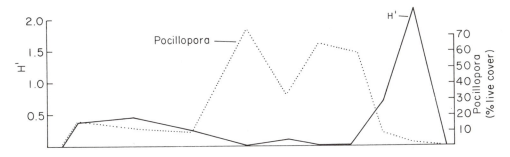

Figure 21-1 Reef coral species diversity (H') and abundance of *Pocillopora* with increasing depth on a coral reef off the Pacific coast of Panama. (After Glynn, 1976, Ecol. Monogr. vol. 46, pp. 431–456. Copyright 1976, the Ecological Society of America)

some of the attached organisms. Therefore the protection afforded by dark crevices may be a deterrent to predation.

   In colonization experiments using asbestos plates, Jackson (1977) demonstrated the eventual successional dominance of colonial organisms (Fig. 21-2) following an early stage of solitary forms (serpulids, bivalves). The competitive superiority of colonial forms is related to (a) their spread on hard substrata by vegetative reproduction (b) their relative insusceptibility to fouling and overgrowth after reaching a critical minimum size, and (c) an apparent resistance to predation in many toxic colonial forms (e.g., sponges). Solitary forms are, in this habitat, essentially fugitive species colonizing newly opened substrata. But temperate and boreal intertidal zones are dominated by solitary forms (e.g., barnacles, mussels) as final stages of competitive dominance. Perhaps higher individual larval recruitment rates counteract the advantage of vegetative growth in the intertidal. Because photopositive larvae of intertidal invertebrates might be concentrated in great densities at the sea surface, there may be a general pattern of high intertidal pelagic larval recruitment relative to the subtidal.

   Which factors permit large numbers of colonial invertebrates to occupy nearly all the space and coexist after a long period of succession? Buss and Jackson (1979) propose that competitive exclusion is delayed through network competitive interactions, where

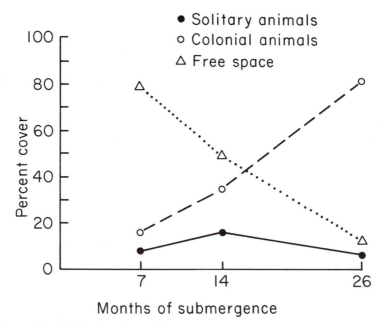

Figure 21-2 Relative abundance of free space, solitary animals, and colonial animals on colonization panels placed at 40 m depth at Discovery Bay, Jamaica. (After Jackson, 1977)

species *A* beats *B*, *B* beats *C*, but *C* beats *A* (see Chapter 4). Networks of overgrowth dominance in asexually growing colonial epibenthic species should result in oscillations (limit cycles) of abundance, alternately favoring different species. Although oscillations will eventually favor the fastest-growing, most-toxic species, the network merry-go-round will delay the rate of eventual exclusion and cause different species to dominate different coral heads due to differing starting complements of species.

Unfortunately, current evidence (e.g., Buss and Jackson, 1979; Jackson, 1979) is insufficient to judge whether networks actually are important. The complexity of cryptic faunas has made it difficult even to document overgrowth. By examining patterns of overgrowth among coexisting epibenthic species, it is clear that overgrowth is common. Sample sizes are too small, however, to be sure of the direction of dominance in all but a few interspecies interactions (e.g., Jackson, 1979). Most competitive interactions, furthermore, are not definitive. In some cases *A* might overgrow *B;* the reverse might be true under different circumstances. Reversal of competitive outcome might thus contribute in the main to coexistence.

Overgrowth patterns in cheilostome ectoprocts have been found to be complex (Jackson, 1979). Variations in competitive outcome occur with variation in encounter angles formed by the intersection of the growth vectors of interacting colonies (Fig. 21-3). Colony surface condition (e.g., fouled versus unfouled) is also an important determining factor in competitive success. So it is likely that coexistence may be mediated by changes in competitive conditions with small-scale changes in growth direction, microtopography, and age of colony. These factors seem just as probable as networks to diminish the rate of competitive exclusion. As a colony ages, for example, its ability to cover space and resist exclusion may be strong at the growing edge of the colony but weak in older portions. Thus another species might be able predictably to take over as the colony in question ages. Exclusion by one species, especially the slow-growing ones studied by Buss and Jackson, would therefore seem unlikely.

The slow growth of the dominating sponges and ectoprocts may account for the indeterminacy of competitive outcome and the lack of dominance by one or by a few species. The long period required for competitive exclusion is based on the very slow growth documented in cryptic sponges and ectoprocts (Jackson, 1977). This time scale would permit the prevention of total competitive overgrowth and dominance by a few species by the mechanism of patch extinction and fugitive colonization. Coral heads periodically topple and smother epibenthic cryptic organisms as a result of storms and bioerosion. An inferior species could theoretically avoid competitive exclusion in a patchy environment if migration among patches is greater than the extinction rate of populations within patches (Horn and MacArthur, 1972). Many if not most of the species on the undersides of foliose corals can avoid extinction by producing swimming larvae before the patch disappears.

It is of theoretical interest that, in the absence of physical or biological disturbance, low-diversity, temperate–boreal rocky intertidal communities show complete dominance by one species in a zone. As discussed in Chapter 16, the competitive ability of the West Coast rocky shore mussel *Mytilus californianus* always results in the usurping of space

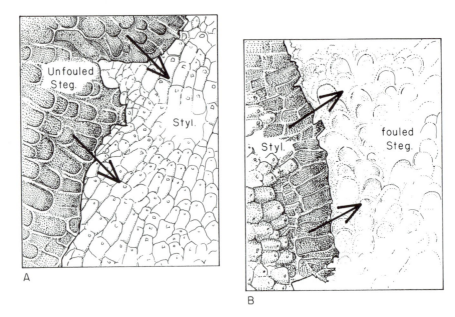

Figure 21-3 Overgrowth of cryptic cheilostome ectoprocts, showing different possible outcomes of overgrowth by one species over another. (a) Unfouled *Steginoporella* sp. overgrows *Stylopoma spongites;* (b) *Stylopoma spongites* overgrows fouled *Steginoporella.* (After Jackson, 1979)

from other competitors (Paine, 1974). In a taxonomically nondiverse system it is possible that the principal competitive contact is usually between distantly related taxa having widely differing competitive abilities to monopolize living space. So if one species encounters another, the probability is great that competitive exclusion will occur despite minor between-patch variation in physical factors. The exclusion is so rapid and the difference in competitive ability so great that the inferior species would always lose. This situation would favor a fugitive strategy of between-patch colonization by the inferior species. If the species are closely related, ecological similarity would slow the process of competitive exclusion, for environmental factors, such as local reversals in resource abundance, and physical factors would alternately favor one or the other species (see discussion in Christiansen and Fenchel, 1976). Under these circumstances small evolutionary changes in one or both species might decrease competitive pressure through niche subdivision (as in *M. edulis* and *M. californianus*). This opportunity would not be as available in competition between distantly related species, for one species would probably be far superior to the other despite minor variations in resource abundance. Thus only disturbance or a strong physiological limitation might prevent the superior species from dominating everywhere (e.g., desiccation controlling the upper limit of mussels).

In a diverse environment the situation is different because large numbers of closely related species are expected to have similar competitive abilities—on a stochastic basis alone. If many species require living space, the similarity will inevitably lead to indeterminacy of competitive outcome. If a major group is competitively inferior to another, evolution of an interference competitive mechanism, such as antibiosis, would further equalize competitive ability. In this case, niche subdivision is also precluded, for all species require the same homogeneous living space. In a situation where specialization is possible (as in *Conus,* described later), closely related species might diverge their resource requirements and thus coexist.

21-4 SPECIES RICHNESS AND THE EXTENT OF NICHE SUBDIVISION

The extent of stereotyped symbioses on reefs has led Grassle (1973) to characterize coral reefs as biologically accommodated communities in which resource limitation has selected for biotic specialization. This concept is somewhat at odds with the dynamic picture afforded by the available data on the guild of corals requiring attachment space and the guild of epibenthic animals living on shaded surfaces. Guilds of reef fishes present a similar picture of lack of specialization (Goldman and Talbot, 1976).

Table 21-2 shows trends of species richness across the Indian and Pacific oceans and the Bahamas. Diversity is highest in the tropical Southeast Asia region, as is the case for hermatypic corals (see later). Although the total number of fish species at One Tree Island (Great Barrier Reef, Australia) is about two and a half that of Tutia Reef in East Africa, single explosive samples collect about the same number of fish species. More species are therefore sampled between East African reef subhabitats than between differing Great Barrier Reef sites, but within-habitat species richness is approximately the same (Goldman and Talbot, 1976). Consequently, the question arises as to how many species

TABLE 21-2   TRENDS IN CORAL REEF FISH SPECIES RICHNESS ACROSS THE INDIAN AND PACIFIC OCEANS AND IN THE BAHAMAS[a]

Region	Coral reef fish species
Bahamas	507
Seychelles	880
Philippines	2177[b]
New Guinea	1700[b]
Australian Great Barrier Reef	1500
Marshall and Mariana	669[b]
Hawaii	448

[a]After Goldman and Talbot, 1976.

[b]Includes freshwater species.

can coexist and whether patterns of specialization are consistent with competitive interaction and niche subdivision.

Part of the fish diversity is between habitat (e.g., Clarke 1977) and specialist species are found restricted to specific microhabitats. The blenny *Aspidontus taenitus*, for instance, is restricted to cleaning-fish stations and mimics the behavior and color of the cleaning wrasse. It strikes fins of fishes awaiting cleaning (Goldman and Talbot, 1976). But within-habitat fish diversity is still great enough to demand explanation. Sale (1977) summarizes records of single rotenone samplings of as many as 200 species of fishes. Coral fish communities contain many coexisting species of herbivorous fishes feeding on algal turf and films of diatoms. Parrot fishes (Scaridae) and surgeon fishes (Acanthuridae) usually dominate but show almost no among-species differences in foods taken (Jones, 1968; Choat, 1969, cited in Sale, 1977). Some differences in microhabitat and feeding behavior were found among surgeon fishes. But similarity of feeding and microhabitat is emphasized by the presence of multispecies feeding schools (Ehrlich and Ehrlich, 1973; Jones, 1968). Bakus (1969) suggests that tropical Labrids are more specialized than temperate species, perhaps indicating competition.

The guild of pomacentrid fish (damselfish) studied by Sale (1974, 1975) shows ecological overlap. On the upper reef slope of Heron Island, three species maintain territories on all available space in coral rubble patches. New fishes rapidly colonize space vacated after the death of a resident. But there is no tendency for a species to be replaced by a conspecific or for a successional sequence of the three species. Coexistence is maintained in the face of near identity of resource requirements and near equality in competitive ability of two of the three species despite a shortage of space. Sale (1977; Sale and Dybdahl, 1975, 1978) suggests that the unpredictability of living space supply and the dispersal of fish larvae by currents create an endless lottery, won by the colonizer who first reaches the newly opened living space. The unpredictable supply of food and shelter (territory) patches favor selection for dispersing larval stages. In this lottery system there is no provision for population growth within a patch, for reproductive products are invariably dispersed away from the patch. This situation is analogous to our discussion of the adaptive value of planktotrophic larvae (Chapter 8). Sale's hypothesis is an important contrast to the thesis that high-diversity biomes, such as the tropics, must support more narrow-niched species. It also ignores the question of why there are more species in the tropics. This dilemma is similar to that of the deep sea and suggests that an evolutionary balance between speciation and extinction is the ultimate basis for explanations of latitudinal diversity gradients and not arguments as to how many species can be "packed" into a given ecological space under certain conditions.

Kohn (1966) found that when several species of the carnivorous snail genus *Conus* co-occurred, their diets were more specialized than when only one species was found. Specialization reduced the diets of species of *Conus* in reef habitats relative to low-diversity nonreef areas. This trend is not continued within the Indo-Pacific Reef province, for species are no more trophically specialized in the highest areas of species richness (fringing reefs of Thailand and Indonesia) than in comparable reef areas elsewhere (Kohn and Nybakken, 1975). In general, species characterized by high overlap in microhabitat differ in preferred prey whereas those with the most similar diets occupy differing microhabitats. Although dietary specialization is about the same in Thailand and Indonesia, relative to other comparable habitats, substratum occupation is not. Ecological overlap in substratum occupation, as measured by Levins' (1968) measure of overlap, is 0.5 to 0.8 in Hawaii and the Maldives but averaged only 0.23 at a station in Indonesia. This analysis indicates that substratum occupation is more restricted in high-diversity environments for *Conus*.

In the context of environmental complexity, *Conus* habitats may be classified into (Kohn, 1967)

1. Sandy substratum of shallow bays, characterized by vast stretches of pure sand with little microtopography.

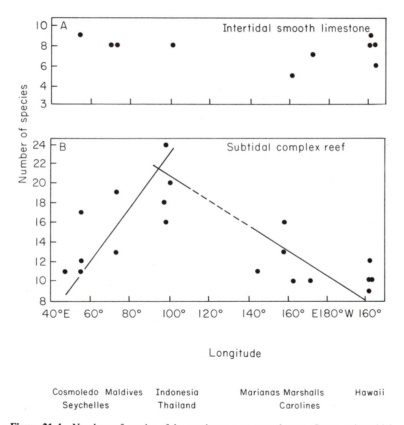

Figure 21-4   Numbers of species of the carnivorous gastropod genus *Conus* on intertidal smooth limestone platforms and subtidal complex reef platforms, plotted against longitude in the Pacific. (After Kohn, 1967)

2. Smooth intertidal limestone platforms, with little topographic relief but the opportunity for gastropods to live in crevices and other shallow surface irregularities, such as a dense algal turf.

3. Subtidal coral reef platforms, topographically more complex habitats with differing microhabitats of varying substratum.

Mean species number in a given Indo-West Pacific reef platform increases from habitat type (1) to (3) (Fig. 21-4). This factor correlates with the greater number of microhabitats in type (3) habitats. A longitudinal species diversity gradient does not exist for species in type (2) habitats, but *Conus* species richness in type (3) habitats follows a trend similar to that of hermatypic corals and reef fishes, with an increase toward southeast Asia. This fact and the relative substratum specialization of Indonesian and Thailand *Conus* assemblages [in type (3) habitats] support the correlation of diversity with microtopographic complexity. Reef fishes have not been as carefully studied, but the co-occurrence of many species in apparently homogeneous microhabitats is in direct contrast to the results of Kohn. It is probable that competitive interaction has influenced the extent of specialization of *Conus* species in subtidal habitats. Species coexisting on intertidal reef rock benches, however, show no significant among-species differences in food preference or microhabitat (Levitan and Kohn, 1980).

21-5 CORAL PREDATORS AND PREDATION ON REEFS

Coral Predators

The large number of coral predators has not been generally appreciated. Robertson (1970) and Glynn (et al., 1972; Glynn, 1976) have listed Pacific coral predators and assessed their importance in regulating coral populations.

Fishes can destroy living corals by feeding directly on corals or by destroying skeletons while feeding on algae or the fauna living within the coral skeleton. Fishes of the genus *Arothron* bite off the branching tips of such corals as *Pocillopora*. *Balistes polylepsis* bites off the protruding surfaces of coral colonies of such coral genera as *Porites* and *Pavona,* searching fragments and the remaining colony for bivalves, gastropods, and other endolithic organisms. The net effect, however, is to destroy the coral colony. Such predation is less intense on Caribbean reefs. Some other species, such as parrot fishes, feed on algae around the periphery of the live coral colony, undercutting the coral head.

The snail *Jenneria pustulata* is an active predator on coral genera, such as *Pocillopora* (Glynn et al., 1972; Glynn, 1976). The species is a nocturnal feeder. Grazed coral surfaces are usually quickly invaded by algae and other organisms (Glynn et al., 1972). Some colid nudibranchs feed on corals and position zooxanthellae in their own bodies for predator defense. Several epitoniid gastropods and a muricid genus feed on corals.

Two species of Pagurid crabs are capable of using their chelae to scrape the branch tips of the coral *Pocillopora.* The species attack the coral and the skeleton and soft parts are removed and ingested. Again, algae tend to invade areas where coral has been removed (Glynn et al., 1972).

Several species of asteroid starfishes feed on corals. Observations seem confined to the Pacific where three genera (*Culcita novaeguineae,* Endean, 1973; *Pharia pyramidata,* Dana and Wolfson, 1970; and *Acanthaster planci,* Glynn et al., 1972; Porter, 1972; Chesher, 1969) are known to feed actively on corals. *A. planci* is a large, multirayed starfish found occasionally on coral reefs throughout the tropical Indo-Pacific region. When feeding, the animal everts its stomach through the mouth and gastric folds are applied to the surface of the coral. Digestive fluids are secreted and digestion takes place extracellularly (Goreau, 1963). *A. planci* mainly consumes scleractinian corals; a few attacks are recorded on hydrocorals and octocorals (Endean, 1969). Sexes are separate and females carry as many as 25 million eggs. At temperatures of 27 to 29°C, larvae in the plankton disperse for approximately 21 days, subsequently settling on coralline algae (Endean, 1973). *A. planci* has a common predator—the giant triton, *Charonia tritonis.* Endean (1973) estimated that 13 out of 52 tritons examined on a reef in the Great Barrier Reef system were mounting or had recently preyed on *A. planci.*

The Acanthaster *Problem*

Widespread explosions of *A. planci* were observed in the Pacific in the 1960s. Starfishes were numerous and significant amounts of coral cover were removed from reefs in Australia, Guam, and many other localities (Chesher, 1969). Devastation of the corals was soon followed by recolonization of hard substratum by algae; vigorous reef accretion was therefore depressed and widespread concern arose as to whether large tracts of reefs would be permanently damaged. Because reefs are often responsible for deflecting wave energy from safe harbors and coasts, the concern was practical as well as aesthetic. The starfish explosion evoked public responses ranging from public hearings (Australia) to expeditions designed to solve the "starfish problem."

A few hypotheses attempted to explain the origin of *Acanthaster* outbursts, but none seems totally satisfactory. Chesher (1969) hypothesized that blasting of channels and passes in various islands in the Pacific created new sites for larval settlement. It is unlikely because much blasting activity occurred during World War II and no starfish explosions were then observed; much open space seems available on many reef areas for starfish settlement. Population explosions may be part of a normal cycle occurring on reefs and should not be tampered with by humans, for they are probably a natural "pruning process" in the normal maintenance of reefs over long periods of time. Endean (1973) points out, however, that it seems improbable to expect such large explosions in such widespread areas all over the Indo-Pacific region. Moreover, observations of natives and the ages of coral heads that seem to have survived previous *Acanthaster* invasions indicate that a supposed cyclicity in population explosions would be well over 100 years. Endean (1973) could not think of any factors that would occur on these time scales. It is tempting to invoke some kind of human involvement in order to explain these population explosions. The question is just what human factor might have suddenly permitted mass explosions of starfishes all over the Indo-Pacific region? Endean excludes the possibility that adult starfishes are food limited in their normal nonexplosive population state because few are typically observed on the Great Barrier Reef (Endean, 1973). In the nonexplosive state, there is little evidence of starfish predation on the reef.

Populations of this species might be limited by a predator (Paine, 1969b). Because the only predator of any importance that has ever been observed is the giant triton, *Charonia tritonis,* human interference becomes possible, for this large snail is an important part of the Indo-Pacific shell trade. Although tritons were important in the culture of the native peoples of the South Pacific, it is likely that large-scale collecting of this species did not actually commence until after World War II, when shell collecting developed into a major worldwide hobby. The sudden increase in shell-collecting pressure might have depleted the populations of tritons to the extent where predator control on the crown-of-thorns starfish was released. It is not clear that shell collecting of tritons is sufficiently widespread to explain the starfish explosions.

Charles Birkeland (in preparation) has discovered that *Acanthaster* explosions are strongly correlated with the occurrence of storms followed by flash floods and influxes of freshwater to coastal habitats. Such influxes might provide large amounts of nutrients to otherwise nutrient-poor coastal waters. The phytoplankton growing under these rich conditions might enhance the survival of *Acanthaster* larvae and permit population bursts. Echinoderms are notable for sporadic success in larval recruitment.

Predator-Prey Interactions

Although the starfish explosions are probably the most interesting aspect of the crown-of-thorns starfish and its natural history, its effects on the coral community under lower densities may also be important in affecting the diversity of coral species. Porter (1972) surveyed coral species richness on a reef that had densities of starfishes of one per 50,000 m^2 compared with reefs having densities of one per 50 m^2, in the eastern Pacific region adjacent to Panama. Larger numbers of coral species occurred on patch reefs with larger numbers of *Acanthaster planci.* Predation may have depressed interspecific competition and subsequent monopolization of the space by fewer species in the community. Glynn (1976), however, showed experimentally that *Acanthaster* prefers nonbranching corals over the dominant *Pocillopora* in Pacific Panamanian reefs. Selective predation lowers diversity (H'), in contrast to Porter's inference. As discussed earlier, symbiotic shrimp and crabs in *Pocillopora* colonies actively defend their home against *Acanthaster* attacks.

Antipredator devices have evolved to a remarkable degree on coral reefs. Most noticeable is the large number of poisonous fishes, invertebrates, and algae. Bakus (1981) found that 73% of the species of sponges, coelenterates, echinoderms, and ascidians he tested were toxic to fishes. Sponges and sea cucumbers are usually deadly and harbor a variety of poisonous organic compounds (Bakus and Green, 1974). The toxic transition element vanadium is found in dry-weight concentrations of 1000 ppm in the tunic of the tunicate *Phallusia nigra* (Stoecker, 1978); similar concentrations of arsenic are found in the sessile "killer" clam *Tridacna* (Benson and Summons, 1981). Thus most easily captured epibenthic or sluggish mobile creatures are well defended against predators.

Crustose corraline and fleshy algae are major components of the epibiota of coral reef environments. A large population of grazing animals of diverse phylogenetic origins exerts strong influence on the biomass and relative abundance of attached algae. Because many grazers are omnivorous to a degree or are capable of significant erosion of calcareous hard substratum, epifauna are affected as well. Some evidence argues for control of reef accretion by grazing organisms that concomitantly erode calcium carbonate (Glynn et al., 1979).

Most grazing is due to the activities of fishes and urchins. Parrot fishes (Scaridae) and surgeon fishes (Acanthuridae) graze fleshy algae with the aid of strengthened and modified teeth. Rasping activities of these fishes account for a significant proportion of the calcareous sediment produced in coral reef environments. In feeding, parrot fishes and surgeon fishes can scrape algae from surfaces or expose algae living in crevices. Grazing algae from the periphery of coral colonies may damage animals as well. Caging experiments show that fish grazing is intense in tropical reef communities and results in dramatic changes in community structure of boh algae and animals (Stephenson and Searles, 1960).

Urchins are probably the most important grazing animals in reef environments. In the Caribbean the urchins *Diadema antillarum* and *Echinometra viridis* dominate shallow-water, hard-substratum environments. The former may also live in turtle grass meadows and is conspicuous for its long delicate spines and usual black color. It may be omnivorous when starved and has extensible tube feet that can transfer drift algae captured on spines to the mouth.

D. antillarium exerts dramatic effects on plant biomass. Many of the population leave patch reefs every night and graze on the surrounding turtle grass flats (*Thalassia testudinum;* Ogden et al., 1973). A grazed halo often surrounds patch reefs as a result. The remainder apparently depend on algae living on the patch reef for their food. Just what effect does this intense grazing have on the organisms living on the surface of the patch reef? An experiment designed to assess the impact of these urchins on their natural habitat was performed by removing approximately 8000 urchins (the whole population) from a patch reef near the West Indies laboratory of Fairleigh Dickinson University, St. Croix, U.S. Virgin Islands (Sammarco et al., 1974). Four other patch reefs served as controls to monitor whether changes occurred in grazed reefs. Algae of each of the five reefs were sampled regularly for the next few weeks. In the ungrazed patch reef the density of algae increased dramatically to as much as 159 g m^{-2} in contrast to an average of 11.9 g m^{-2} in the grazed control patch reefs. The study did not assess whether this large growth of algae was due to settlement of spores or vegetative growth of grazed algae that had been previously present. Observations on patch reefs in Discovery Bay, Jamaica, West Indies, showed that reefs with low densities of *Diadema* had large growths of macroalgae whereas reefs with large numbers of urchins had none.

D. antillarum scrapes hard surfaces effectively enough to diminish the survival of newly settled epibionts severely (Sammarco, 1977, 1980). Surfaces exposed to grazing are soon rendered smooth to the touch; so there is a strong likelihood that a newly settled larva will be scraped away before establishing a colony. Microhabitat strongly influences early survival in that settling on exposed smooth surfaces provides ready vulnerability to a barrage of scraping urchin teeth. In contrast, settlement in crevices or on edges may give the newly settled coral enough time to grow large enough to escape the urchin. Sammarco (1980) demonstrates that microhabitat and size strongly influence the condition of newly established corals subjected to urchin grazing. Some juvenile corals (e.g., *Favia fragum*) survived urchin scraping substantially better than others. Urchin grazing, therefore, may alter the composition of coral assemblages.

Sammarco established a series of cages of varying urchin density on patch reefs in Discovery Bay, Jamaica. At high urchin densities (32 to 64 m^{-2}), most algae and small corals could not survive the grazing intensity. At low urchin densities, coral settlement was maximal. Intermediate urchin densities, however, seemed to be optimal for coral growth. In the absence of grazing, competition for space among algae, corals, and other sessile invertebrates was maximal and coral success was diminished. Therefore intermediate levels of grazing provided enough removal to allow coral success but not too much biological disturbance to result in complete mortality.

Sammarco (1977) also examined the diversity of algae and corals under a range of experimental grazing intensities. Contrary to expectations, diversity decreased steadily with increasing urchin density. Despite obvious crowding, no diminution of algal diversity was observed at zero grazing. Apparently competitive success was not vested in a minority of species and no increase of dominance occurred. In St. Croix, however, a few algal species seem capable of competitive dominance (Sammarco et al., 1974). Under these circumstances we expect a peak of species richness at intermediate urchin density.

21-7 PHYSICAL DISTURBANCE OF CORAL REEFS

Mechanical destruction from storms, hurricanes, and cyclones is a major source of coral mortality. Coral recovery is facilitated by vegetative expansion of surviving colonies and colonization by planula larvae. Stoddart (1963) reported the effect of a particularly strong hurricane in 1961 on the reefs of British Honduras. These reefs were dominated by the elk horn coral *Acropora palmata* and the massive coral *Montastrea annularis*. The hurricane stripped the reef of living corals over an 8-km-long zone and disrupted spur and groove structures over an area 40 km wide. Almost all the branching and unattached corals were destroyed and the reef was covered with rubble (Stoddart, 1969). In the Great Barrier Reef cyclones are frequent and important destroyers of coral formations. In the vicinity of the Heron Island Laboratory at the southern extremity of the Great Barrier Reef, cyclones regularly turn over coral heads and destroy most branching colonies. Stoddart (1969) reports that after the hurricane in 1961 in British Honduras recovery was slow and algae still dominated coral reefs 3 years later. Stephenson and others (1958) suggested that 10 to 20 years is a minimal recovery time for a cyclone hitting Low Isles, Great Barrier Reef.

Unusually low tides, in conjunction with seaward-directed winds, are a principal source of mortality on coral reef flats. On reef flats of the Caribbean side of the Isthmus of Panama, tidal range is minimal and wind patterns primarily control water levels. The combination of low tide and many hours of offshore winds often leaves the reef flat exposed to air for several days and mass mortality ensues (Glynn 1968; Glynn et al., 1972). The net effect is to open space and keep interspecific interactions to a minimum. On East Pacific reefs off Panama the effects of desiccation devastate the reef flat and keep diversity (H') lower than the deep reef (Glynn, 1976).

Another instance of mass mortality has been observed on reef flats of the fringing reef at Eilat, Israel (Fishelson, 1973; Loya, 1976b). The northern part of the Red Sea is surrounded by mountains and is remarkably dry with only about 25 mm yr^{-1} of rainfall. Consequently, evaporation levels are high, ranging from 100 to 200 cm yr^{-1} in the Gulf of Aqaba. Although the tidal range is only 0.5 to 1.0 m, mean sea level fluctuates dramatically when monsoon winds blow toward the southern opening of the Red Sea. This movement of water, plus the high evaporation rate, can cause a lowering of mean sea level in the Gulf of Aqaba. Fishelson (1973) found that in September 1970 coral colonies growing on this reef table were exposed daily for 1.5 to 2.0 hours over 4 to 5 days and extensive mortality occurred. Brainlike corals, such as *Favia* and *Platygyra,* were found more resistant to desiccation than more delicate bushlike forms. Extreme low waters opened space on the reef table, thereby diminishing interspecific competition. The reef table at Eilat has 50% or more open space whereas over 90% of available space is occupied by corals subtidally (see Loya and Slobodkin, 1971).

Physical disturbances reduce the abundance and diversity of the coral reef biota and probably affect coral reefs in much the same way as waves on the rocky intertidal (see Chapter 16). Natural disturbances continually open up space on the reef. These sources of mortality and the effects of grazing suggest that coral reefs are not stable and benign environments. They are dynamic communities experiencing frequent physical disruptions; community structure is regulated by colonization, predation, interspecific competition, and grazing.

21-8 PRODUCTIVITY AND TROPHIC STRUCTURE

Coral reefs are islands of high production in an open sea of very low primary productivity (Odum and Odum, 1955). A list of possible sources of primary production highlights the complexity of the problem:

1. Filamentous algae in live corals.
2. Zooxanthellae.
3. Algal mats on dead coral rock.
4. Fleshy encrusting green algae.
5. Branching algae on dead coral heads.

6. Algae on sand.

7. Zooxanthellae in anemones and *Tridacna*.

8. Phytoplankton.

Sargeant and Austin (1949) show that planktonic algae from the open sea are not a significant source of primary productivity; relatively few phytoplanktivores, such as bivalve mullusks, are present on coral reefs.

    The presence of unidirectional currents flowing across the reef permits simultaneous upstream and downstream measurement of dissolved oxygen. The oxygen increase indicates the net photosynthetic production of the coral reef. At night the oxygen decrease from the upstream to the downstream side estimates total community respiration. Night respiration, plus oxygen increase during the day, gives an idea of total production, uncorrected for atmospheric exchange and losses due to diffusion. This technique, with appropriate attention given to corrections for atmospheric interactions, has yielded primary production estimates of 1500 g C m^{-2} yr^{-1} (Rongelap Atoll, Marshall Islands; Sargeant and Austin, 1949), 3500 g C m^{-2} yr^{-1} (Eniwetok Atoll; Odum and Odum, 1955), and 2900 g C m^{-2} yr^{-1} (Kauai, Hawaii; Kohn and Helfrich, 1957). With the exception of turtle grass flats, which have values as great as ca. 4000 g C m^{-2} yr^{-1}, coral reefs are the most productive marine habitats in the world. All three estimates far exceed 37 g C m^{-2} yr^{-1}, the primary production of the open sea near Hawaii (Kohn and Helfrich, 1957).

    The allocation of primary productivity into different primary producer categories as specified earlier is in dispute. Odum and Odum (1955) examined the biomass of plant and animal tissue in the polyp zone of corals and in a subpolyp zone consisting of algal layers one or more centimeters beneath the coral head surface. They suggested that

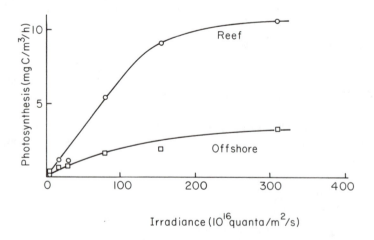

Figure 21-5 Photosynthesis on a coral reef. Comparison is made of photosynthesis in the surface waters of the reef and offshore of the reef. (After Scott and Jitts, 1977, from *Marine Biology*, volume 41)

filamentous algae in colored bands might receive light through a translucent coral skeleton and contribute to primary production of coral reefs. Subsequent work suggests that these filamentous algae probably play a negligible role in primary productivity and may simply be accumulations of pigment layers (Kanwisher and Wainwright, 1967). Kohn and Helfrich (1957) suggested that fleshy green algae might be the major primary producers responsible for the large amounts of primary productivity observed. But not all reef systems examined have large amounts of fleshy green algae. Scott and Jitts (1977) estimated the role of zooxanthellae in Great Barrier Reef primary productivity by determining photosynthesis *(P)* versus irradiance *(I)* for both zooxanthellae and natural reef phytoplankton populations (Fig. 21-5). Phytoplankton productivity was then measured directly with depth and also predicted, using the *P* versus *I* curve (Table 21-3). The daily in situ photosynthesis rates for zooxanthellae in a coral branch tip at a depth of 3 m were calculated from the *P* versus *I* curve. Production per square meter was estimated by assuming 25% coverage of the reef area by actively growing corals. The calculated value of 0.99 g C m^{-2} d^{-1} is three times that of phytoplankton production in waters close to the reef.

TABLE 21-3 DAILY PRIMARY PRODUCTION RATES (CALCULATED FROM PHOTOSYNTHESIS VERSUS IRRADIANCE CURVES) FOR CORAL ZOOXANTHELLAE AND FOR PHYTOPLANKTON IN REEF WATERS OF LIZARD ISLAND, AUSTRALIAN GREAT BARRIER REEF[a]

Sample	Production g C m^{-2} d^{-1}	
	Cloudless day	Cloudy day
Phytoplankton—reef area	0.30	0.29
Phytoplankton—offshore	0.34	0.23
Coral zooxanthellae—reef area	0.99	0.73

[a]From Scott and Jitts, 1977, reprinted from *Marine Biology*, volume 41.

21-9 BIOGEOGRAPHY AND DIVERSITY PATTERNS IN CORAL REEFS

Because coral reefs seem to occupy almost all clear-water tropical areas of the world, they provide a series of biotic islands in which to test evolutionary and ecological theory. The most significant problems in investigating a coral reef are its very complexity and astounding diversity; there is little accurate information on the number of coral species living in more than a few localities. Goreau and Goreau (1959) found over 60 species of corals living in the reef at Ocho Rios, Jamaica. Porter (1974b) found over 40 species of corals living on the Caribbean side of the Isthmus of Panama. Numerous other areas of the Caribbean, however, have many fewer recorded species simply because no careful survey has been made. Wells (1957) estimates that well over 3000 invertebrate species live on such reefs as those in the northern Great Barrier Reef, the Celebes, Palau, or the Marshall Islands. The collection of over 100 polychaete species from a single group of coral rubble (Grassle, 1973) suggests that our work is not nearly over in recording all the species living in coral reefs. Yet a number of conclusions can be drawn about the biogeography of coral reefs and have led to hypotheses concerning the evolution of marine biotas.

Coral reef biotas can be subdivided between Atlantic and Indo-Pacific provinces. In terms of diversity, these two provinces behave as separate systems (Stehli and Wells, 1971). In the Atlantic province a maximum of 24 genera have been observed in Jamaica whereas the maximum number of genera observed in the Pacific is in the Philippines—57 genera. So maximum generic richness in the Indo-Pacific province is about twice that of the Atlantic province. Stehli and Wells (1971) analyzed data from localities all over the world and used spherical harmonic analysis to calculate equal diversity contours for numbers of genera to see if geographic patterns of diversity existed. They found two centers of diversity: one around Jamaica in the Atlantic (probably an artifact of incomplete sampling) and the other located near Indonesia and the Philippines. Numbers of genera decrease in all directions from both centers (Fig. 21-6).

In the Indo-Pacific province generic diversity changes more rapidly with geographic distance on a north–south than on an east–west axis. In the Indo-Pacific province given species drop out in the same sequence in any direction (Wells, 1954; Stehli and Wells, 1971). There is a positive correlation between diversity and temperature, and the Indo-

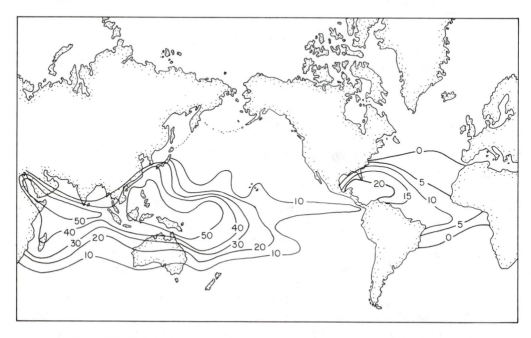

Figure 21-6   Variation in numbers of genera of hermatypic corals in the Indo-Pacific and Caribbean provinces. (After Stehli and Wells, 1971)

Pacific center of diversity coincides with the thermal equator (north of the geographic equator). Although there is a relatively small thermal gradient from east to west, generic diversity tends to decrease in the same manner as it does from north to south. Stehli and Wells suggest that an important factor may be the area of sea floor suitable for coral growth. This type of explanation has also been used to explain why the Atlantic province is depauperate in coral species relative to the Indo-Pacific province (Newell, 1971).

   Coral reefs can serve as a means for understanding ecological and evolutionary processes because of the paleontological record of ancient coral reefs, starting with the origin of the Scleractinia in the Mesozoic. Centers of diversity in the Indonesian region of the Indo-Pacific and the Jamaican region of the Caribbean suggest several alternative hypotheses for the origin of these diversity gradients. One possibility is that genera evolved in centers and subsequently moved out to peripheral areas. Alternatively, coral species once lived all over the Indo-Pacific and Atlantic provinces. A recent period of climatic deterioration, however, resulted in the gradual contraction of areas available for less temperature-tolerant corals.

   If coral reefs have diversity centers that are evolutionary centers as well, we would expect to find more genera of very young geological age in these diversity centers and to find genera of great antiquity in the peripheral areas. To test this hypothesis, Stehli and Wells (1971) used geologic age data for all extant genera of corals and plotted the

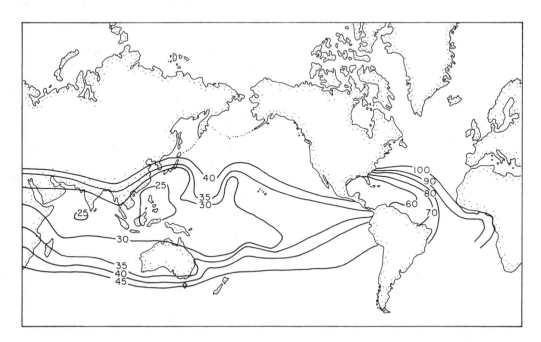

Figure 21-7   The average age (million years) of hermatypic coral genera now living in coral reefs of the Indo-Pacific and Caribbean regions. (After Stehli and Wells, 1971)

average generic age of corals now living at a given coral reef locality. Figure 21-7 shows that the average age of coral genera is less in centers of diversity than in the peripheries and that coral reef generic age is less for Indo-Pacific corals than for Atlantic province corals. The oldest average generic age encountered for a Pacific station was younger than the average generic age encountered for the Atlantic. It might be concluded that evolution is going on twice as fast in the Indo-Pacific as in the Atlantic!

In considering diversity gradients within a given province, a larger proportion of very young genera is found in the Indonesian center of diversity than in the peripheries of the Indo-Pacific province. Stehli and Wells suggest that this kind of age structure implies that genera evolve in the diversity centers and move out toward the peripheries, becoming older with time. It is certainly true that their data are consistent with this model, but other hypotheses suggest the same generic age pattern and hence make the conclusions of Stehli and Wells somewhat more ambiguous.

The biogeographic pattern of species occurrence in the Indo-Pacific region documented by Wells raises questions regarding the center-of-origin hypothesis. Some genera range from the Red Sea to the eastern Pacific; some occur over smaller regions and others only in the region of the Philippines and Indonesia. According to the center-of-origin hypothesis, the genera that are the most biogeographically restricted are the youngest in geologic age and those who have not yet had time to spread out to peripheral regions. One might question, however, the suggestion that resistance to major changes in the environment is independent of coral generic age and survivorship. Suppose, for instance, that we have a coral genus that has existed for some 60 million years. Might we not conclude that the reason for the longevity of this genus has much to do with its ability to survive given climatic events? The great homogeneity of the coral reef province in the Indo-Pacific might simply permit it to live everywhere throughout the province, but its physiological tolerance might allow it to survive major environmental challenges that otherwise contract the other parts of the coral reef biota .

An alternative hypothesis is thus suggested. At one time coral reef genera lived throughout the Indo-Pacific province up to, but perhaps not including, the limits of coral physiological tolerance. Then a major climatic deterioration occurred throughout the tropical world, affecting coral reefs at the periphery of the coral reef province and contracting their biogeographic distribution. Those coral reef genera that had evolved most recently perhaps are the ones that have not yet evolved physiological tolerance capable of resisting large-scale climatic deterioration. They may be specialized derivatives, evolved in a benign climate. Therefore they are the ones that suffer the most and their respective ranges retreat toward the relatively benign climatic center of the Indo-Pacific coral reef province. This process might give us the same diversity and generic age gradients that we observed in coral reefs today. This latter hypothesis might be characterized as the refuge hypothesis (Newell, 1971). "We may ask: Have the more tropical regions of the past been centers of origin and dispersal of taxa? Or, can we consider the present tropical biota a shrunken relic of a more widely distributed biota of, say, Eocene time? I am inclined to give a qualified yes to both questions. . . . " (Newell, 1971) Newell points out that the paleontologic data do not give compelling evidence of systematic displacement of more primitive less successful groups away from the tropics into regions of lower diversity. As mentioned, coral genera and species of the peripheral regions also thrive in the central regions.

Newell proposes the hypothesis that the high generic diversity of the reef biota in the Austral-Asian and the Caribbean areas is mainly an effect of the accumulation of relicts and low extinction rates compared with the peripheral areas. In other words, he suggests that we should find more older forms surviving in the centers of diversity. The data accumulated by Stehli and Wells (1971) contradict this fact, for more young genera are found in the central area than in the peripheral areas.

These problems carry over into our explanations of why the Indo-Pacific province is more diverse than the Caribbean province. Why is it that one area of the world, with an apparently similarly suitable environmental regime for the development of coral reefs, supports 50% more species of corals and probably all other biota than the Caribbean province? Newell (1971) suggests that an area effect may explain the greater diversity of the Indo-Pacific province, being 1,250,000 km² as opposed to 250,000 km² in the Caribbean province. This larger area may lower extinction rates.

Environmental stability may also contribute to the greater species richness in the Indo-Pacific. We discussed the stability hypothesis in Chapters 5 and 19. More stable environments may have reduced extinction rates because of the lower frequency of rapid environmental changes (e.g., the unusual cold event mentioned earlier for the Persian Gulf). Fewer environmental perturbations of relatively small amplitude might select for more specialized species not otherwise able to survive rapid changes in required food or habitat.

Is there any definitive evidence that the Caribbean is less stable than the Indo-Pacific? If temperature is any indication, both open water temperature and water temperatures on reef flats vary more in the Caribbean than in the Indo-Pacific (Wells, 1957). It is not clear, however, that this difference has any importance in local extinction rates. In addition, the uncertain historical component in species richness gradients within the Indo-Pacific province contributes to the uncertainty of the influence of environmental stability.

A brief history of coral reefs and reef communities might help explain biogeographic phenomena (see Newell, 1971). The scleractinians make their first appearance in the middle Triassic, gradually replacing other organisms dominating reefs in the upper Triassic. By Jurassic times the scleractinian community, coexisting with such coralline algae as *Lithothamnion,* was well established. In the Cretaceous period an unusual order of bivalve mollusks, the Rudistids, underwent an adaptive radiation and competed with Scleractinia in reef environments. Rudistids became extinct at the end of the Cretaceous and the Paleocene was without a recognized reef community. During the Eocene period scleractinian corals underwent a second major radiation and all subsequent history has been one of advance and decline of reefs in harmony with oscillations of world climates.

The distribution of coral reefs has reflected changes in world trends in physical environments, such as restriction of epicontinental seas, changes in the size of the tropical belt, and increasing seasonality of climate. From the mid-Miocene to the present there has been a shrinkage in the range of the tropical environment, resulting in a segregation of Pacific and Caribbean provinces. Thus the present centers of high diversity may be regarded as havens rather than principal source areas of dispersal. The uplift of the Isthmus of Panama between the Americas in the latest Miocene segregated a small Panamanian subprovince from the Caribbean area and subsequent isolation has resulted in considerable divergence. The general shrinkage of the tropical climates around the world has produced a shrinkage of the coral reef area in both the Atlantic and Indo-Pacific provinces.

SUMMARY

1. Interspecies interactions and community structure on coral reefs can be visualized as similar to the rocky intertidal. Space is often limiting for epibenthic creatures. Many dominants of coral reefs, however, create their own primary substratum.

2. Mutualistic associations seem common on coral reefs. Many involve protection from predators.

3. Hermatypic corals use shading, overgrowth, and extracoelenteric digestion as mechanisms of space competition. Hierarchies of aggressiveness based on extracoelenteric digestion seem associated with calcification rate; slow-growing species are those with the best-developed means of digesting and holding off neighbors. In cryptic communities colonial invertebrates seem to dominate.

4. Although some groups, such as the carnivorous gastropod genus *Conus,* show niche subdivision, coral reefs are notable for the widespread coexistence of species with strongly overlapping resource requirements. Great controversy exists as to how this coexistence is maintained. Some believe that disturbance and relatively slow rates of competitive exclusion tend to conserve coexisting species. Competitive networks might slow the rate of competitive exclusion.

5. Hermatypic corals are consumed by a variety of predators. Most prominent is the crown-of-thorns starfish, *Acanthaster planci,* which exploded in numbers in the 1960s in the Pacific. One hermatype harbors symbiotic crustacea that actively defend the coral from attack by starfishes.

6. Grazing by sea urchins has a major effect on both algal and animal community composition. In the Caribbean the urchin *Diadema antillarum* scrapes substratum clean of newly settled corals. Algal species richness and coral species composition can be greatly altered with changes in urchin density. Where rapidly recruiting seaweeds are present (e.g., St. Croix), urchin grazing causes a hump in species richness at intermediate urchin densities. Where no rapid recruiter is present, algal density continues to decrease with increasing urchin density.

7. Coral reefs are not benign environments; tropical storms regularly rip through shallow water and destroy large numbers of coral heads. Coral reefs are therefore strongly controlled by disturbance, much as in other shallow-water communities.

8. Coral reefs are islands of high productivity in the middle of barren oceans. Zooxanthellae and fleshy algae probably contribute to this production.

9. Species richness in coral reefs tends to increase toward centers of diversity in both the Atlantic and the Pacific. This factor is especially prominent in the Pacific, where species richness declines with latitude and longitude from a diversity center in the Southwest Pacific. The Pacific is richer in coral reef-associated species than the Atlantic. Generic age tends to decrease with an increase in species richness. This situation may mean that newly evolved genera are derived and specialized and cannot tolerate peripheral, less optimal geographic locales. It is also possible that evolution is progressing more rapidly in the diversity centers.

glossary

Abyssal plain The deep ocean floor, an expanse of low relief at depths of 4000 to 6000 m.

Abyssopelagic The 4000 to 6000-m-depth zone, seaward of the shelf–slope break.

Acclimation Given a change of a single parameter, a readjustment of the physiology of an organism, reaching a new steady state.

Acclimatization An adjustment of the physiological state of an individual in response to a series of intercorrelated parameters changing in concert (as in a seasonal change).

Age structure The relative abundance of different age classes in a population.

Aggregated spatial distribution Case where individuals in a space occur in clusters too dense to be explained by chance.

Ahermatypic Nonreef building (referring to scleractinian corals).

Allele One of several variants that can occupy a locus on a chromosome.

Allopatric speciation The differentiation of geographically isolated populations into distinct species.

Amensal Negatively affecting one or several species.

Amino acids Basic structural unit of proteins.

Anadromous fish Fish that spend most of their life feeding in the open ocean but migrate to spawn in freshwater.

Anoxic Lacking oxygen.

Arrowworms The phylum Chaetognatha, a group of planktonic carnivores.

Asexual reproduction   Reproduction of the individual without the production of gametes and zygotes.

Assimilation efficiency   The fraction of ingested food that is absorbed and used in metabolism.

Assortative mating   When a given genotype mates with another genotype disproportionate to the mating frequency expected from random encounter.

Atoll   A horseshoe or circular array of islands, capping a coral reef system perched around an oceanic volcanic seamount.

Attenuation (light)   Diminution of light intensity; explained in the ocean by absorption and scattering.

Autotrophic algae   Algae capable of photosynthesis and growth with only dissolved inorganic nutrients.

Auxotrophic algae   Algae requiring a few organically derived substances, such as vitamins, along with dissolved inorganic nutrients for photosynthesis.

Bathypelagic   The 2000 to 4000-m-depth zone, seaward of the shelf–slope break.

Benthic–pelagic coupling   The cycling of nutrients between the bottom sediments and overlying water column.

Berm   A broad area of low relief in the upper part of a beach.

Between-habitat comparison   A contrast of diversity in two localities of differing habitat type (e.g., sand versus mud bottoms).

Biogenic graded bedding   A regular change of sediment median grain size with depth below the sediment–water interface caused by the activities of burrowing organisms.

Biogenically reworked zone   The depth zone within a sediment that is actively burrowed by benthic organisms.

Bioluminescence   Light emission, often as flashes, by many marine organisms.

Biomass (= standing crop)   *See* Standing crop.

Blood pigment   A molecule used by an organism to transport oxygen efficiently, usually in a circulatory system (e.g., hemoglobin).

Bloom (phytoplankton)   A population burst of phytoplankton that remains within a defined part of the water column.

Boreal   Pertaining to the Northern Hemisphere, north temperate zone.

Boring   Capable of penetrating a solid substratum by scraping or chemical dissolution.

Brackish sea   Semienclosed water bodies of large extent where tidal stirring and seaward flow of freshwater do not exert enough of a mixing effect to prevent the body of water from having its own internal circulation pattern.

Browsers   Organisms that feed by scraping thin layers of living organisms from the surface of the substratum (e.g., periwinkles feeding on rock surface diatom films; urchins scraping a thin, filmy sponge colony from a rock).

Calcareous   Being made of calcium carbonate.

Capacity adaptation   A physiological adaptation that improves physiological performance but does not make the difference between life and death.

Carrying capacity   The total number of individuals of a population that a given environment can sustain.

Carnivore   An organism that captures and consumes animals.

Catadromous fish   Fishes that spawn in seawater but feed and spend most of their life in estuarine or freshwater.

Chaetognaths   *See* Arrowworms.

Chemosynthesis   Primary production of organic matter, using various substances instead of light as an energy source; confined to a few groups of microorganisms.

Chlorinity   Grams of chloride ions per 1000 grams of seawater.

Chloroplast   The cellular organelle in eukaryotic organisms where photosynthesis takes place.

Coarse-grain exploitation   Exploiting a series of resources disproportionately to their relative abundance in the environment.

Coastal reef   A coral reef occurring near and parallel to a coastline.

Comb jellies   The phylum Ctenophora, a group of gelatinous forms feeding on smaller zooplankton.

Commensal   Having benefit for one but neither positive nor negative effect on another member of a two-species association.

Compensation depth   The depth of the compensation light intensity.

Compensation light intensity   That light intensity at which oxygen evolved from a photosynthesizing organism equals that consumed in its respiration.

Competition   An interaction between or among two or more individuals or species where exploitation of resources by one affects any others negatively.

Competition coefficient   A parameter that quantitatively expresses the negative effect of an individual of one species on an individual of a second species.

Complex life cycle   A life cycle that consists of several distinct stages (e.g., larva and adult).

Conformer   An organism whose physiological state (e.g., body temperature) is identical to, and varies identically with, that of the external environment.

Continental drift   Horizontal movement of continents located in plates moving via sea-floor spreading.

Continental shelf   A broad expanse of ocean bottom sloping gently and seaward from the shoreline to the shelf–slope break at a depth of 100 to 200 m.

Continental slope   *See* Slope.

Convergence   Map expression on the ocean surface of the contact between two water masses converging, one plunging below the other.

Copepod   Order of crustacea found often in the plankton.

Coprophagy   Feeding on fecal material.

Coral reef   A wave-resistant structure resulting from cementation processes and the skeletal construction of hermatypic corals, calcareous algae, and other calcium carbonate-secreting organisms.

Corer   Tubular benthic sampling device that is plunged into the bottom in order to obtain a vertically oriented cylindrical sample.

Coriolis effect   The deflection of air or water bodies, relative to the solid earth beneath, as a result of the earth's eastward rotation.

Critical depth   That depth above which total integrated photosynthetic rate equals total integrated respiration of photosynthesizers.

Critical salinity   A salinity of ca. 5 to 8 ‰ that marks a minimum of species richness in an estuarine system.

Ctenophora   *See* Comb jellies.

Daily estuary   An estuary where tidal movements cause substantial changes in salinity at any one location on a daily basis.

Deep layer   The layer extending from the lowest part of the thermocline to the bottom.

Deep-scattering layer   Well-defined horizons in the ocean that reflect sonar, usually consisting of fishes, squid, or other larger zooplankton.

Demographic   Referring to numerical characteristics of a population (e.g., population size, age structure).

Density (seawater)   Grams of seawater per milliliter of fluid.

Density-dependent factor   Factors, such as resource availability, that vary with population density.

Deposit feeder   An organism that derives its nutrition by consuming some fraction of a soft sediment.

Diatom   Dominant planktonic algal form with siliceous test; occurring as single cells or chains of cells.

Diffusion   The net movement of units of a substance from areas of higher concentration to areas of lower concentration of that substance.

Digestion efficiency   The fraction of living food that does not survive passage through a predator's gut.

Dinoflagellate   Dominant planktonic algal form, occurring as single cells, often biflagellate.

Directional selection   Preferential change in a population, favoring the increase in frequency of one allele over another.

Dissolved organic matter   Dissolved molecules derived from degradation of dead organisms or excretion of molecules synthesized by organisms.

Disturbance   A rapid change in an environment that greatly alters a previously persistent biological community.

Diversity   A parameter describing, in combination, the species richness and evenness of a collection of species. Diversity is often used as a synonym for species richness.

Diversity gradient   A regular change in diversity correlated with a geographic space or gradient of some environmental factor.

Ecological efficiency   Amount of energy extracted from a trophic level, divided by the amount of energy supplied to that level (any quantity could be substituted for energy).

Ekman circulation   Movement of surface water at an angle from the wind, as the result of the Coriolis effect.

Emigration   The departure of individuals from a given area.

Endosymbiotic   Being symbiotic and living within the body of an individual of the associated species.

Environmental stress   Variously defined as (a) an environmental change to which an organism cannot acclimate and (b) an environmental change that increases the probability of death.

Epibenthic (epifaunal or epifloral)   Living on the surface of the bottom.

Epidemic spawning   Simultaneous shedding of gametes by a large number of individuals.

Epipelagic   The 0- to 150-m-depth zone, seaward of the shelf–slope break.

Epiphyte   Microalgal organism living on a surface (e.g., a seaweed frond).

Estuarine flow   Seaward flow of low-salinity surface water, over a deeper and higher salinity layer.

Estuarine realms   Large coastal water regions that have geographic continuity and are bounded landward by a stretch of coastline with freshwater input but bounded seaward by a salinity front.

Estuary   A semienclosed body of water that has a free connection with the open sea and within which seawater is diluted measurably with freshwater that is derived from land drainage.

Euphausiid   Order of holoplanktonic crustacea.

Eutrophic   Water bodies or habitats with high concentrations of nutrients.

Evenness   The component of diversity accounting for the degree to which all species are equal in abundance, as opposed to strong dominance by one or a few species.

Exploitation competition   Competition in which one individual or species exploits a resource more efficiently than a second individual or species.

Fecal pellets   *See* Pellets.

Fecundity   The number of eggs produced per female per unit time (often: per spawning season).

Fine-grain exploitation   Exploiting a series of resources at proportionate rates similar to their relative respective abundances in the environment.

Foliose coral   A coral whose skeletal form approximates a broad flattened plate.

Food chain   An abstraction describing the network of feeding relationships in a community as a series of links of trophic levels, such as primary producers, herbivores, and primary carnivores.

Food chain efficiency   Amount of energy extracted from a trophic level, divided by the amount of energy produced by the next lower trophic level (any quantity could be substituted for energy).

Food web   A network describing the feeding interactions of the species in an area.

Foraminifera   Protozoan group (usually) secreting a calcareous test; both planktonic and benthic representatives.

Forchhammer principle   The constancy of ratios of concentrations of major elements in seawater.

Founder principle   A small colonizing population is genetically unrepresentative of the source population.

Front   A major discontinuity, separating ocean currents and water masses in any combination.

Fugitive species   A species adapted to colonize newly disturbed habitats.

Functional response   The rate of exploitation of a resource as a function of increasing availability (environmental concentration) of the resource.

Fundamental niche   That portion of environmental space in which a species can maintain a continuing population in the absence of interspecific competition or predation.

Gametophyte   Haploid stage in the life cycle of a plant.

Generation time   The time period from birth to average age of reproduction.

Genetic drift   Changes in allele frequencies that can be ascribed to random effects.

Genetic locus   A location on a chromosome (possibly of a diploid organism with variants that segregate according to the rules of Mendelian heredity).

Genetic polymorphism   Presence of several genetically controlled variants in a population.

Genotype   The genetic makeup of an organism. With respect to a given genetic locus: what alleles it carries.

Genus (plural: genera)   The level of the taxonomic hierarchy above the species but below the family level.

Geostrophic flow   Movement of water in the oceans, as a combined response to the coriolis effect and gravitational forces created by an uneven sea surface.

Geotactic   Moving in response to the earth's gravitational field.

Grab   Benthic sampling device with two or more curved metal plates designed to converge when the sampler hits bottom and grab a specified volume of bottom sediment.

Grazer   A predator that consumes organisms far smaller than itself (e.g., copepods graze on diatoms).

Gregarious settling   Settlement of larvae that have been attracted to members of their own species.

Gross primary productivity   The total primary production, not counting the loss in respiration.

Guild   A group of species, possibly unrelated taxonomically, that exploit overlapping resources.

Gyre   Major cyclonic surface current systems in the oceans.

Halocline   Depth zone within which salinity changes maximally.

Hardy–Weinberg law   The frequencies of genotypes in a population at a locus are determined by random mating and allele frequency.

Herbivore   An organism that consumes plants.

Heritable character   A morphological character whose given state can be explained partially by the genotype of the individual.

Hermaphrodite   An individual capable of producing both eggs and sperm during its lifetime.

Hermatypic   Reef building.

Heterotrophic algae   Algae that take up organic molecules as a primary source of nutrition.

Heterozygote   With respect to a given genetic locus, a diploid individual carrying two different alleles.

Highly stratified estuary   An estuary having a distinct surface layer of fresh or very low salinity water, capping a deeper layer of higher salinity, more oceanic water.

Histogram   A multiple-bar diagram representing the frequency distribution of a group as a function of some variable. The frequency of each class is proportional to the length of its associated bar.

Holoplankton   Organisms spending all their life in the plankton.

Homeotherm   An organism that regulates its body temperature despite changes in the external environmental temperature.

Homozygote   With respect to a given genetic locus, a diploid individual carrying two identical alleles.

Hydrographic   Referring to the arrangement and movement of bodies of water, such as currents and water masses.

Hydrothermal vents   Sites in the deep ocean floor where hot, sulfur-rich water is released from geothermally heated rock.

Hypothesis   A refutable statement about one or a series of phenomena.

Infaunal   Living within a soft sediment and being large enough to displace sedimentary grains.

Interference competition   Competition in which one individual actively prevents another from exploiting a resource.

Interspecific competition Condition when one species' exploitation of a limiting resource negatively affects another species.

Interstitial Living in the pore spaces among sedimentary grains in a soft sediment.

Isotonic Having the same overall concentration of dissolved substances as a given reference solution.

K selection Selection for changes in carrying capacity. Implicitly, selection for improved exploitative ability (efficiency).

Keystone species A predator at the top of a food web, or discrete subweb, capable of consuming organisms of more than one trophic level beneath it.

Larva A discrete stage in many species, beginning with zygote formation and ending with metamorphosis.

Larvacea A group of planktonic tunicates that secrete a gelatinous house, used to strain unsuitable particles.

LD_{50} The value of a given experimental variable required to cause 50% mortality.

Leaching The loss of soluble material from decaying organisms.

Lecithotrophic larva A planktonic-dispersing larva that lives off yolk supplied via the egg.

Leeward The side of an island opposite from the one facing a persistent wind.

Leslie matrix A matrix delineating age-specific mortality and reproduction of a set of age classes.

Life table A table summarizing statistics of a population, such as survival and reproduction, all broken down into age classes.

Litter Accumulations of dead leaves in various states of fragmentation and decomposition.

Locus *See* Genetic locus.

Logistic population growth Population growth that is modulated by the population size relative to carrying capacity. Population growth declines as population approaches carrying capacity. Population growth is negative when population size is greater than carrying capacity.

Longshore current A current moving parallel to a shoreline.

Macrobenthos (macrofauna or macroflora) Benthic organisms (animals or plants) whose shortest dimension is greater than 0.5 mm.

Macrofauna Animals whose shortest dimension is greater than or equal to 0.5 mm.

Macrophyte An individual alga large enough to be seen easily with the unaided eye.

Maximum sustainable yield In fisheries biology, the maximum catch obtainable per unit time under the appropriate fishing rate.

Meiobenthos (meiofauna or meioflora) Benthic organisms (animals or plants) whose shortest dimension is less than 0.5 mm but greater than or equal to 0.1 mm.

Meiofauna Animals whose shortest dimension is less than 0.5 mm but greater than or equal to 0.1 mm.

Meroplankton Organisms that spend part of their time in the plankton but also spend time in the benthos (e.g., planktonic larvae of benthic invertebrates).

Mesopelagic The 150- to 2000-m-depth zone, seaward of the shelf–slope break.

Metabolic rate The overall rate of biochemical reactions in an organism. Often estimated by rate of oxygen consumption in aerobes.

Metamorphosis Profound change in morphology experienced as the larva develops into a juvenile adult.

Microbenthos (microfauna or microflora) Benthic organisms (animals or plants) whose shortest dimension is less than 0.1 mm.

Microfauna Animals whose shortest dimension is less than 0.1 mm.

Mixing depth The water depth to which wind energy evenly mixes the water column.

Moderately stratified estuary An estuary where seaward flow of surface low-salinity water and moderate vertical mixing result in a modest vertical salinity gradient.

Mucous-bag suspension feeder Suspension feeder employing a sheet or bag of mucus to trap particles nonselectively.

Mutualism An interaction between two species in which both derive some benefit.

Mutualistic Conferring reciprocal benefit to individuals of two different associated species.

Neap tides Tides occurring when the vertical tidal range is minimal.

Nekton Organisms with swimming abilities that permit them to move actively through the water column and to move against currents.

Neritic Seawater environments landward of the shelf–slope break.

Net primary productivity Total primary production, minus the amount consumed in respiration.

Neuston Planktonic organisms associated with the air–water interface.

Niche A general term referring to the range of environmental space occupied by a species.

Niche overlap An overlap in resource requirements by two species.

Niche width The spread of exploitation ability of a species over a range of alternative resources.

Nitrogen fixation The conversion of gaseous nitrogen to nitrate by specialized bacteria.

Nutrient cycling The pattern of transfer of nutrients between the components of a food web.

Nutrients Those constituents required by plants.

Oceanadromous fish Fishes that spawn and feed as adults in open-ocean seawater.

Oceanic Being associated with seawater environments seaward of the shelf–slope break.

Oceanic ridge A sinuous ridge rising from the deep-sea floor.

Oligotrophic Water bodies or habitats with low concentrations of nutrients.

Omnivory   Being able to feed in more than one distinct way (e.g., an organism capable of carnivory and herbivory).

Optimal foraging theory   A theory designed to predict the foraging behavior that maximizes food intake per unit time.

Organic   Deriving from living organisms.

Organic detritus   Particulate material derived from once-living organisms.

Organic nutrients   Nutrients in the form of molecules synthesized by or originating from the synthesis of other organisms.

Osmoconformer   An organism whose body fluids change directly with a change in the concentration of dissolved ions in the external medium.

Osmoregulator   An organism that regulates the concentration of dissolved ions in its body fluids irrespective of changes in the external medium.

Osmosis   The movement of pure water across a membrane from a compartment with relatively low dissolved ions to a compartment with higher concentrations of dissolved ions.

Outwelling   The outflow of nutrients from an estuary or salt-marsh system to shelf waters.

Overdominance   Selection favoring heterozygotes.

Oxygen dissociation curve   A curve showing the percent saturation of a blood pigment, such as hemoglobin, as a function of oxygen concentration of the fluid.

Oxygen minimum layer   A depth zone, usually below the thermocline, where dissolved oxygen is minimal.

Oxygen technique (primary productivity)   The estimation of primary productivity by the measurement of the rate of oxygen increase.

Parapatric speciation   The differentiation of populations, experiencing some gene flow, into distinct species.

Parasite   An organism living on or in, and negatively affecting, another organism.

Particulate organic matter   Particulate material in the sea derived from the decomposition of the nonmineral constituents of living organisms.

Patchiness   The condition where organisms occur in aggregations.

Pelagic   Living in the water column seaward of the shelf–slope break.

Pellets   Compacted aggregations of particles resulting either from egestion (fecal pellets) or from burrow-constructing activities of marine organisms.

Penetration anchor   In hydraulically burrowing organisms: any device used to penetrate and gain an initial purchase on the sediment so that the body can be thrust in farther.

Peptides   Chains of amino acids; often portions of a protein molecule.

pH   The acidity of water ($-\log_{10}$ of the activity of hydrogen ions in water).

Phi scale   Scale used for measuring the grain size of sediments. $\phi = -\log_2$ (grain diameter).

Photic zone   The depth zone in the ocean extending from the surface to that depth permitting photosynthesis. Defined variably by different authors.

Photorespiration   Enhanced respiration of plants in the light relative to dark respiration.

Photosynthate   The substance synthesized in the process of photosynthesis.

Photosynthetic quotient   In photosynthesis, the moles of oxygen produced, divided by the moles of carbon dioxide assimilated.

Photosynthetic rate   The rate of conversion of dissolved CO_2 or bicarbonate ion to photosynthetic product.

Phototactic   Moving in response to light.

Physiological race   A geographically defined population of a species that is physiologically distinct from other populations.

Phytoplankton   The photosynthesizing organisms residing in the plankton.

Planktivorous   Feeding on planktonic organisms.

Plankton   Organisms living suspended in the water column and incapable of moving against water currents.

Planktotrophic larva   Planktonic-dispersing larva that derives its nourishment by feeding in the plankton.

Planula   The planktonic larval form produced by scleractinian corals and coelenterates.

Plate   Major section of the earth's crust, bounded by such features as midocean ridges.

Pleistocene   Period of time, going back to ca. 2 million years before the present, in which alternating periods of glaciation and deglaciation have dominated the earth's climate.

Poikiloosmotic   Being an osmoconformer.

Poikilotherm   An organism whose body temperature is identical to that of the external environment.

Polyp   An individual of a solitary coelenterate or one member of a coelenterate colony.

Population density   Number of individuals per unit area, or volume, where appropriate.

Predation   The consumption of one organism by another.

Predator   An organism that consumes another living organism (carnivores and herbivores are both predators by this definition).

Primary producer   An organism capable of using the energy derived from light or a chemical substance in order to manufacture energy-rich organic compounds.

Primary production   The production of living matter by photosynthesizing organisms or by chemosynthesizing organisms. Usually expressed as g C m^{-2} y^{-1}.

Province   A geographically defined area with a characteristic set of species or characteristic percentage representation by given species.

Protein polymorphism   Presence of several variants of a protein of a given type (e.g., a certain enzyme, such as carboxylase) in a population.

Pseudofeces   Material rejected by suspension feeders or deposit feeders as potential food before entering the gut.

Pteropods   Group of holoplanktonic gastropods.

Pycnocline   Depth zone within which seawater density changes maximally.

Q_{10}   Increase of metabolic rate with an increase of 10°C.

Quantitative genetics   The study of the genetic basis of traits usually explained by the interaction of a group of genes with the environment.

r_{max}   The intrinsic rate of increase of a population.

r selection   Selection for increased intrinsic rate of increase. Implicitly, selection for the ability to increase rapidly in population at the expense of devoting resources to individual longevity.

Radiocarbon technique (primary productivity)   The estimation of primary productivity by the measurement of radiocarbon uptake.

Radiolaria   Protozoan phylum, planktonic and secreting an often elaborate siliceous test.

Random spatial distribution   Case where individuals are randomly distributed in a space. Probability of an individual being located at any given point is the same irrespective of location in the space.

Realized niche   That part of the fundamental niche to which the species is currently restricted by interactions, such as competition and predation.

Recruitment   The production, successful survival, and colonization by newborn organisms (e.g., recruitment of fish larvae).

Redox potential discontinuity   That depth below the sediment–water interface marking the transition from chemically oxidative to reducing processes.

Red tide   A dense outburst of phytoplankton (usually dinoflagellates) often coloring the water red brown.

Refuge   A device by which an individual can avoid predation.

Regulator   An organism that can maintain some aspect of its physiology (e.g., body temperature) constant despite different and changing properties of the external environment.

Renewable resource   A resource that can be regenerated, as when an exploited population has its own population growth rate.

Reproductive effort   The fraction of assimilated nutrients that are devoted to reproductive behavior and gamete production.

Resistance adaptation   A physiological adaptation that permits the individual to survive an otherwise fatal stress.

Resource   A commodity required by an organism that is potentially in short supply.

Resource switching The switch of an exploiter from one alternate resource to another as the latter increases in frequency.

Respiration Consumption of oxygen in the process of aerobic metabolism.

Respiratory pigment A molecule, polymer, or other complex adapted to bind and transport oxygen efficiently, usually in a circulatory system (e.g., hemoglobin).

Respiratory quotient The ratio of moles of carbon dioxide produced to oxygen consumed in respiration.

Rip current A concentrated rapid current moving offshore from a beach fronting a longshore current.

Rise Bottom of low relief at the base of the continental slope.

Salinity Number of grams of dissolved salts in 1000 g of seawater.

Salps A group of pelagic tunicates, either colonial or solitary, with buccal and atrial siphons on opposite sides of the body.

Salt marsh A coastal habitat consisting of salt-resistant plants residing in an organic-rich sediment accreting toward sea level.

Scavenger An organism that feeds on dead or decomposing animals or macrophytes.

Scleractinia Order of coelenterates, usually producing calcareous skeletons with hexameral symmetry.

Scope for activity The amount of energy available for active movement beyond that being expended at rest.

Scope for growth The surplus of energy available for growth beyond that required for maintenance.

Scyphozoa The true jellyfish.

Sea-floor spreading The horizontal movement of oceanic crust.

Seasonal estuary An estuary where salinity at any one geographic point changes seasonally (e.g., decreases during the spring melt).

Seaward Side of an island that faces the direction of wave action generated by either winds or currents generated by more indirect forces.

Secondary production The production of living material per unit area (or volume) per unit time by herbivores. Usually expressed as g C m^{-2} y^{-1}.

Selection A change in allele frequency over time in a population.

Sequential hermaphrodite An individual that sequentially produces male and then female gametes or vice versa.

Sessile Being immobile due to an attachment to a substratum.

Seston Particulate matter suspended in seawater.

Setules Chitinous projections from copepod maxillipeds that trap food particles.

Shelf–slope break Line demarcating a change from the gentle continental shelf to the much steeper depth gradient of the continental slope.

"Sigma-tee" (σ_T) Parameter expressing the seawater density: $\sigma_t = 1 -$ density of seawater at temperature t and at a pressure of 1 atmosphere.

Silt–clay fraction The percentage (by dry weight) of a soft sediment that can pass through a 0.062-mm sieve.

Siphonophores A group of specialized hydrozoan coelenterates, consisting of large planktonic polymorphic colonies.

Sled A benthic sampling device designed to slide along the sediment surface, digging into the bottom to a depth of only a few millimeters to a couple of centimeters.

Slope A steep-sloping bottom, extending seaward from the edge of the continental shelf and downward toward the rise.

Somatic growth Growth of the body, exclusive of gametes.

Sorting (of a sediment) The range of scatter of particle sizes about the median grain size of a sediment.

Space limited Space is a limiting resource.

Spatial autocorrelation A situation in which some parameter at any location (e.g., population density) can be predicted through a knowledge of the values of the parameter in other locations.

Spatial distribution The arrangement of individuals in a space.

Speciation The process of formation of new species.

Species richness The number of species in an area or biological collection.

Sporophyte Diploid stage in the life cycle of a plant.

Spring diatom increase The major rapid population increase of diatoms occurring in the spring in temperate-boreal latitudes.

Spring tides Fortnightly tides occurring when the vertical tidal range is maximal.

Stability-time hypothesis Higher diversity occurs in habitats that are ancient and stable environmentally.

Standing crop (= biomass) The amount of living material per unit area, or volume, where appropriate. May be expressed as grams of carbon, total dry weight, and so on.

Stock recruitment models Fisheries models that predict the amount of juvenile recruitment as a function of the parent stock.

Stratification In benthos, the presence of different infaunal species at distinct respective horizons below the sediment–water interface.

Subtropical The most equatorward portion of the temperate zone.

Succession A predictable ordering of a dominance of a species or groups of species following the opening of an environment to biological colonization.

Surface layer The layer of the ocean extending from the surface to a depth above which the ocean is homogeneous due to wind mixing.

Survivorship curve The curve describing changes of mortality rate as a function of age.

Suspension feeder An organism that feeds by capturing particles suspended in the water column.

Teleplanic larvae Larvae capable of dispersal over long distances, as across oceans.

Temperate Pertaining to the latitudinal belt between 23°27' and 66°33' north or south latitude.

Tentacle–tube foot suspension feeder Suspension feeder that traps particles on distinct tentacles or tube feet (in echinoderms).

Terminal anchor In hydraulically burrowing organisms: any device used to anchor the leading portion of the burrower, permitting muscular contraction to drag the rest of the body into the sediment.

Territoriality Defense of a specified location against intruders.

Tertiary production The production of living material per unit area (or volume) per unit time by organisms consuming the herbivores. Usually expressed as g C m^{-2} y^{-1}.

Thermocline Depth zone within which temperature changes maximally.

Thermohaline circulation Movement of seawater that is controlled by density differences that are largely explained by temperature and salinity.

Tidal current A water current generated by regularly varying tidal forces.

Tides Periodic movement of water resulting from gravitational attraction between the earth, sun, and moon.

Trade winds Persistent winds at low latitudes in both Northern and Southern hemispheres, blowing toward the west and the equator.

Trench Deep and sinuous depressions in the ocean floor, usually seaward of a continental margin or an arcuate group of volcanic islands.

Trophic level In a food chain, a level containing organisms of identical feeding habits with respect to the chain (e.g., herbivores).

Tropical Being within the latitudinal zone bounded by the two tropics (23°27' north and south latitude).

Turbidity The weight of particulate matter per unit volume of seawater.

Underdominance Selection favoring homozygotes.

Uniform spatial distribution Case in which individuals are more evenly spread in space than would be expected by chance alone.

Upwelling The movement of nutrient-rich water from a specified depth to the surface.

Vents   *See* Hydrothermal vents.

Vertically homogeneous estuary   An estuary where, at any given location, wind or tidal mixing homogenizes salinity throughout the water column.

Vitamin   Chemical substances, required in trace concentrations, acting as cofactors with enzymes in catalyzing biochemical reactions.

Viviparous (development)   Development of an organism, through the juvenile stage, within a parent.

Wash zone   The depth zone in which sediments are disturbed by wave action near the shoreline.

Water mass   A body of water that maintains its identity and can be characterized by such parameters as temperature and salinity.

Westerlies (prevailing westerlies)   Persistent eastward–equatorward winds in midlatitudes in both Northern and Southern hemispheres.

Windward   The side of an island that faces a persistent wind.

Within-habitat comparison   A contrast of diversity between two localities of similar habitat type.

Wrack zone   A bank of accumulated litter at the strandline.

Year-class effect   The common domination of a species population by individuals recruited in one reproductive season.

Zonation   Occurrence of single species or groups of species in recognizable bands that might delineate a range of water depth or a range of height in the intertidal zone.

Zooplankton   Animal members of the plankton.

Zooxanthellae   A group of dinoflagellates living endosymbiotically in association with one of a variety of invertebrate groups (e.g., corals).

references

SOME JOURNALS OF INTEREST TO MARINE ECOLOGISTS

American Association of Petroleum Geologists Bulletin (AAPG Bull.)
American Naturalist (Am. Nat.)
American Zoologist (Am. Zool.)
Applied and Environmental Microbiology
Australian Journal of Marine and Freshwater Research (Aust. J. Mar. Freshw. Res.)
Biological Bulletin (Woods Hole) *(Biol. Bull.)*
Bulletin of Marine Science (Bull. Mar. Sci.)
Cahiers de Biologie Marine (Cahiers Biol. Mar.)
Canadian Journal of Fisheries and Aquatic Science (Can. J. Fish. Aq. Sci.)
Copeia
Crustaceana
Deep Sea Research (Deep Sea Res.)
Ecology
Ecological Monographs (Ecol. Monogr.)
Estuaries
Estuarine and Coastal Marine Science
Evolution
Journal of Animal Ecology (J. Anim. Ecol.)
Journal of Experimental Marine Biology and Ecology (J. Exp. Mar. Biol. Ecol.)
Journal of the Fisheries Research Board of Canada (now *Can. J. Fish. Aq. Sci.)*

Journal of Ichthyology

Journal of the Marine Biological Association of the U.K. (J. Mar. Biol. Ass. U.K.)

Journal of Marine Research (J. Mar. Res.)

Journal of Molluscan Studies

Journal of Plankton Research

Journal of Sedimentary Petrology (J. Sed. Petrol.)

Lethaia

Limnology and Oceanography (Limnol. Oceanogr.)

Marine Biology (Mar. Biol.)

Marine Biology Letters (Mar. Biol. Lett.)

Marine Ecology—Progress Series

Microbial Ecology

Nature

Oecologia

Oikos

Ophelia

Pacific Science

Paleoclimatology Paleogeography Paleoecology (Paleo. Paleo. Paleo.)

Paleobiology

Science

Systematic Zoology (Syst. Zool.)

Veliger

ABBOTT, D. P. 1966. Factors influencing the zoogeographical affinities of Galapagos inshore marine fauna. In *The Galapagos,* R.I. Bowman (Ed.), pp. 108–122. Berkeley: University of California Press.

ABELE, L. G., and K. WALTERS. 1979a. Marine benthic diversity: a critique and alternative explanation. *J. Biogeogr.* 6:115–126.

———. 1979b. The stability-time hypothesis: reevaluation of the data. *Am. Nat.* 114:559–568.

ADEY, W. H. 1978. Coral reef morphogenesis: a multidimensional model. *Science* 202:831–837.

AHMAD, M., and J. L. BEARDMORE. 1976. Genetic evidence that the "Padstow Mussel" is *Mytilus galloprovincialis. Mar. Biol.* 35:139–147.

AHMED, M., and A. K. SPARKS. 1970. Chromosome number, structure and autosomal polymorphism in the marine mussels *Mytilus edulis* and *Mytilus californianus. Biol. Bull.* 138:1–13.

ÅKESSON, B. 1973. Reproduction in the genus *Ophryotrocha* (Polychaeta, Dorvilleidae). *Pubbl. Staz. Zool. Napoli* 39 (Suppl.):377–398.

ALCARAZ, M., G. A. PAFFENHÖFER, and J. R. STRICKLER. 1981. Catching the algae: a first account of visual observations of filter-feeding calanoids. In *Evolution and Ecology of Zooplankton Communities,* W. C. Kerfoot (Ed.). University Press of New England, in press.

ALLAN, J. D. 1976. Life history patterns in zooplankton. *Am. Nat.* 110:165–180.

ALLEN, J. A. 1958. On the basic form and adaptations to habitat in the Lucinacea (Eulamellibranchia). *Phil. Trans. Roy. Soc. London* 684:421–484.

ALLEN, J. A., and H. L. SANDERS. 1966. Adaptations to abyssal life as shown by the bivalve *Abra profundorum* (Smith). *Deep Sea Res.* 13:1175–1184.

ALLER, R. C., and J. Y. YINGST. 1978. Biogeochemistry of tube-dwellings: a study of the sedentary polychaete *Amphitrite ornata* (Leidy). *J. Mar. Res.* 36:201–254.

ALEVIZON, W. S., and M. G. BROOKS. 1975. The comparative structure of two western Atlantic reef-fish assemblages. *Bull. Mar. Sci.* 25:482–490.

ANDERSON, D. M., and D. WALL. 1978. Potential importance of benthic cysts of *Gonyaulax tamarensis* and *G. excavata* in initiating toxic dinoflagellate blooms. *J. Phycol.* 14:224–234.

ANIKOUCHINE, W. A. and R. W. STERNBERG. 1981. *The World Ocean: An Introduction to Oceanography*, Second Edition, 528 pp. Englewood Cliffs, N.J.: Prentice-Hall, Inc.

ANSELL, A. D. 1961. The functional morphology of the British species of Veneracea (Eulamellibranchia). *J. Mar. Biol. Assoc. U.K.* 41:489–515.

ANSELL, A. D., P. SIVADAS, B. NARAYANAN, and A. TREVALLION. 1972. The ecology of two sandy beaches in south west India. III. Observations on the population of *Donax incarnatus* and *D. spiculum*. *Mar. Biol.* 17:318–332.

ANTIA, N. J., T. BISALPUTRA, J. Y. CHENG, and J. P. KALLEY. 1975. Pigment and cytological evidence for reclassification of *Nannochloris oculata* and *Monallantus salina* in the Eustimatophyceae. *J. Phycol.* 11:339–343.

ARMSTRONG, R. A., and R. McGEHEE. 1980. Competitive exclusion. *Am. Nat.* 115:151–170.

ATKINS, D. 1936. On the ciliary mechanisms and interrelationships of lamellibranchs. I. Some new observations on sorting mechanisms in certain lamellibranchs. *Quart. J. Micro. Sci.* 79:181–308.

AUSMUS, B. S. 1973. The use of the ATP assay in terrestrial decomposition studies. In *Modern Methods in the Study of Marine Ecology*, T. Rosswall (Ed.), *Bull. Ecol. Res. Comm. Stockholm*, no. 17, Stockholm, pp. 223–234.

AZAM, F., and R. E. HODSON. 1977. Size distribution and activity of marine microheterotrophs. *Limnol. Oceanogr.* 22:492–501.

BAER, J. G. 1952. *Ecology of Animal Parasites*. Urbana: University of Illinois Press.

BAILEY, N. J. T. 1964. *The Elements of Stochastic Processes with Applications to the Natural Sciences*, New York: John Wiley and Sons.

BAKUS, G. J. 1969. Energetics and feeding in shallow marine waters. *Int. Rev. Gen. Exp. Zool.* 4:275–369.

———. 1981. Chemical defense mechanisms on the Great Barrier Reef, Australia. *Science* 211:497–499.

BAKUS, G. J., and G. GREEN. 1974. Toxicity in sponges and holothurians: a geographic pattern. *Science* 185:951–953.

BALAGOT, B. P. 1971. Microgeographic variation at two biochemical loci in the blue mussel, *Mytilus edulis*. M.S. Dissertation. Stony Brook: State University of New York.

BALLARD, R. D. 1977. Notes on a major oceanographic find. *Oceanus* 20:35–44.

BAMBACH, R. K. 1977. Species richness in marine benthic habitats throughout the Phanerozoic. *Paleobiology* 3:152–157.

BARKLEY, E. 1940. Nahrung und Filter Apparat des Walkrebses *Euphausia superba* Dana. *Z. Fisch. Hilfswissensch.* Beib. 1, Walforsch. 1:65–156.

BARNES, D. J. 1972. The structure and formation of growth ridges in scleractinian coral skeletons. *Proc. Roy. Soc. London* 182:331–350.

BARNES, H. 1956. *Balanus balanoides* (L.) in the Firth of Clyde: the development and annual variation of the larval population, and the causative factors. *J. Anim. Ecol.* 25:72–84.

——. 1959. Stomach content and micro-feeding of some common cirripedes. *Can. J. Zool.* 37:231–236.

——. 1962. Note on variations in the release of nauplii of *Balanus balanoides* with special reference to the spring diatom outburst. *Crustaceana* 4:118–122.

BARNES, H., and M. BARNES. 1968. Egg numbers, metabolic efficiency of egg production and fecundity; local and regional variations in a number of common cirripedes. *J. Exp. Mar. Biol. Ecol.* 2:135–153.

BARRETT, I., and F. J. HESTER. 1964. Body temperature of yellowfish and skipjack tunas in relation to sea surface temperature. *Nature* 203:96–97.

BARSDATE, R. J., R. T. PRENTKI, and T. FENCHEL. 1974. Phosphorous cycle of model ecosystems: significance for decomposer food chains and effect of bacterial grazers. *Oikos* 25:239–251.

BARSOTTI, G. and C. MELUZZI. 1968. Osservazioni su *Mytilus edulis* L. e *M. galloprovincialis*. *Conchiglie* 4:50–58.

BATTAGLIA, B. 1958. Balanced polymorphism in *Tisbe reticulata*, a marine copepod. *Evolution* 12:358–364.

——. 1959. Ecological differentiation and incipient intraspecific isolation in marine copepods. *Ann. Biol.* 33:259–268.

BAYLOR, E. R., and W. H. SUTCLIFFE. 1962. Adsorption of phosphates onto bubbles. *Deep Sea Res.* 9:120–124.

——1963. Dissolved organic matter in sea water as a source of particulate food. *Limnol. Oceanogr.* 8:369–371.

BAYNE, B. L. 1964. Primary and secondary settlement in *Mytilus edulis* L. (Mollusca). *J. Anim. Ecol.* 33:513–523.

——. (Ed.) 1976. *Marine Mussels: Their Ecology and Physiology*, 506 pp. Cambridge, U.K: Cambridge University Press.

BAYNE, B. L., J. WIDDOWS, and R. J. THOMPSON. 1976a. Physiology:I. In *Marine Mussels: Their Ecology and Physiology*, B. L. Bayne (Ed.), pp. 121–206. Cambridge: Cambridge University Press.

BAYNE, B. L., R. J. THOMPSON, and J. WIDDOWS. 1976b. Physiology:II. In *Marine Mussels: Their Ecology and Physiology*, B. L. Bayne (Ed.), pp. 207–260. Cambridge: Cambridge University Press.

BAYNE, B. L., J. WIDDOWS, and R. J. THOMPSON. 1976c. Physiological integrations. In *Marine Mussels: Their Ecology and Physiology*, B. L. Bayne (Ed.), pp. 261–291. Cambridge: Cambridge University Press.

BELYAEV, G. M., and P. V. USHAKOV. 1957. Some regularities in the quantitative distribution of the benthic fauna in Antarctic waters. *Dokl. Akad. Nauk SSSR Biol.* 112:116–119.

BENSON, A. A., and R. E. SUMMONS. 1981. Arsenic accumulation in Great Barrier Reef invertebrates. *Science* 211:482–483.

BERNER, R. A. 1971. *Principles of Chemical Sedimentology*, 240 pp. New York: McGraw-Hill.

BEST, M. B. 1969. Etude systématique et écologique des Madréporaires de la région de Banyuls-sur-Mer (Pyrénées-Orientales). *Vie et Milieu* 20A:293–326.

BIGELOW, H. B. 1926. Plankton of the offshore waters of the Gulf of Maine. *Bull. U.S. Bur. Fish.* 40:1–507.

BIRD, E. G. 1972. Some aspects of the biology of polychaete larvae in Southampton water with notes on *Sagitta setosa*. M.Sc. Dissertation. U.K.: University of Southampton.

BIRKELAND, C. 1974. Interactions between a sea pen and seven of its predators. *Ecol. Monogr.* 44:211–232.

BLOOM, S. A. 1975. The motile escape response of a sessile prey: a sponge-scallop mutualism. *J. Exp. Mar. Biol. Ecol.* 17:311–321.

BOADEN, P. J. 1962. Colonization of graded sand by an interstitial fauna. *Cahiers Biol. Mar.* 3:245–248.

BOESCH, D. F. 1979. Benthic ecological studies: macrobenthos. In *Middle Atlantic Outer Continental Shelf Environmental Studies, Volume IIb. Chemical and Biological Benchmark Studies.* Unpublished report. Gloucester Point, Virginia: Virginia Institute of Marine Sciences.

BØHLE, B. 1972. Effects of adaptation to reduced salinity on filtration activity and growth of mussels *(Mytilus edulis)*. *J. Exp. Mar. Biol. Ecol.* 10:41–49.

BOLD, H. C., and M. J. WYNNE. 1978. *Introduction to the Algae,* 706 pp. Englewood Cliffs, N.J.: Prentice-Hall, Inc.

BOOTHE, P. N., and G. A. KNAUER. 1972. The possible importance of fecal material in the biological amplification of trace and heavy metals. *Limnol. Oceanogr.* 17:270–274.

BOUCOT, A. J. 1975. *Evolution and Extinction Rate Controls,* 427 pp. Amsterdam: Elsevier.

BOYD, C. M. 1976. Selection of particles sizes by filter-feeding copepods: a plea for reason. *Limnol. Oceanogr.* 21:175–179.

BOYNTON, W. R., W. M. KEMP, and C. G. OSBORNE. 1980. Nutrient fluxes across the sediment–water interface in the turbid zone of a coastal plain estuary. In *Estuarine Perspectives,* V. S. Kennedy (Ed.), pp. 93–109. New York: Academic Press.

BRADLEY, B. P. 1978. Genetic and physiological adaptation of the copepod *Eurytemora affinis* to seasonal temperatures. *Genetics* 90:193–205.

BRAFIELD, A. E., and G. E. NEWELL. 1961. The behaviour of *Macoma balthica* (L.). *J. Mar. Biol. Assoc. U.K.* 41:81–87.

BRAND, L. E., L. S. MURPHY, R. R. L. GUILLARD, and H. -t. LEE. 1981. Genetic variability and differentiation in the temperature niche component of the diatom *Thalassiosira pseudonana*. *Mar. Biol.,* In Press.

BREGNBALLE, F. 1961. Plaice and flounder as consumers of the microscopic bottom fauna. *Meddel. Danmarks Fisk. Hav.* 3:133–182.

BRETSKY, P. W. 1969. Evolution of Paleozoic benthic marine invertebrate communities. *Paleo. Paleo. Paleo.* 6:45–59.

———. 1973. Evolutionary patterns in the Paleozoic bivalvia: documentation and some theoretical considerations. *Geol. Soc. Am. Bull.* 84:2079–2096.

BRETSKY, P. W. and D. M. LORENZ. 1970. An essay on genetic-adaptive strategies and mass extinctions. *Geol. Soc. Am. Bull.* 81:2449–2456.

BRINKHUIS, B. H. 1976. The ecology of salt-marsh fucoids. I. Occurrence and distribution of *Ascophyllum nodosum* ecads. *Mar. Biol.* 34:325–338.

BRINKHUIS, B. H., and R. F. JONES. 1976. The ecology of temperate salt-marsh fucoids. II. *In Situ* growth of transplanted *Ascophyllum nodosum*. *Mar. Biol.* 34:339–348.

BRINKHUIS, B. H., N. R. TEMPEL, and R. F. JONES. 1976. Photosynthesis and respiration of exposed salt-marsh fucoids. *Mar. Biol.* 34:349–359.

BROOKS, J. L. 1950. Speciation in ancient lakes. *Quart. Rev. Biol.* 25:30–60; 131–176.

BROOKS, J. L., and S. I. DODSON. 1965. Predation, body size and composition of plankton. *Science* 150:28–35.

BROWN, J. H., and C. R. FELDMETH. 1971. Evolution in constant and fluctuating environments: thermal tolerances of desert pupfish *(Cyprinodon). Evolution* 25:390–398.

BROWN, S. C. 1969. The structure and function of the digestive system of the mud snail *Nassarius obsoletus* (Say). *Malacologia* 9:447–500.

BROWN, T. E., and F. L. RICHARDSON. 1966. The effect of growth environment on the physiology of algae: light intensity. *J. Phycol.* 4:38–54.

BROWN, W. L., and E. O. WILSON. 1956. Character displacement. *Syst. Zool.* 5:49–64.

BUCHANAN, J. B. 1967. Dispersion and demography of some infaunal echinoderm populations. *Symp. Zool. Soc. London* 20:1–11.

BUDDEMEIER, R. W., and R. A. KINZIE. 1976. Coral growth. *Oceanogr. Mar. Biol. Ann. Rev.* 14:183–225.

BUNT, J. S. 1963. Diatoms of Antarctic sea ice as agents of primary production. *Nature* 199:1255–1257.

BURKENROAD, M. N. 1943. A possible function of bioluminescence. *J. Mar. Sci.* 5:161–164.

BURKHOLDER, P. 1956. Studies on the nutritive value of *Spartina* grass growing in the marsh areas of central Georgia. *Bull. Torrey Cany. Bot. Club* 83:327–334.

BUSS, L. W., and J. B. C. JACKSON. 1979. Competitive networks: nontransitive competitive relationships in cryptic coral reef environments. *Am. Nat.* 113:223–234.

BUZAS, M. A. 1969. Foraminiferal species densities and environmental variables in an estuary. *Limnol. Oceanogr.* 14:411–422.

BUZAS, M. A., and T. G. GIBSON. 1969. Species diversity: benthonic foraminifera in western north Atlantic. *Science* 163:72–75.

CALABRESE, A. 1969. *Mulinia lateralis:* molluscan fruit fly? *Proc. Natl. Shellfish Assoc.* 59:65–66.

———. 1970. Reproductive cycle of the coot clam, *Mulinia lateralis* (Say), in Long Island Sound. *Veliger* 12:265–269.

CALOW, P. 1975. Defecation strategies of two freshwater gastropods, *Ancylus fluviatilis* Mull, and *Planorbis contortus* Linn. (Pulmonata) with a comparison of field and laboratory estimates of food absorption rate. *Oecologia* (Berlin) 20:51–63.

CAMMEN, L., P. RUBLEE, and J. HOBBIE. 1978. The significance of microbial carbon in the nutrition of the polychaete *Nereis succinea* and other aquatic deposit feeders. University of North Carolina, *Sea Grant Publ. UNC-SG-78-12.*

CAMPBELL, R. D. 1968. Host specificity, settling and metamorphosis of the two-tentacled hydroid *Proboscidactyla falvicirrata. Pacific Sci.* 22:336–339.

CAPERON, J. 1967. Population growth in micro-organisms limited by food supply. *Ecology* 48:715–722.

CARLUCCI, A. F., and P. M. BOWES. 1970. Vitamin production and utilization by phytoplankton in mixed culture. *J. Phycol.* 6:393–400.

CARPENTER, E. J. 1972. Nitrogen fixation by *Oscillatoria (Trichodesmium) thiebautii* in the southwestern Sargasso Sea. *Deep Sea Res.* 20:285–288.

CARPENTER, E. J., and R. R. L. GUILLARD. 1971. Intraspecific differences in nitrate half-saturation constants for three species of marine phytoplankton. *Ecology* 52:183–185.

CARPENTER, E. J., C. C. REMSEN, and S. W. WATSON. 1972. Utilization of urea by some marine phytoplankters. *Limnol. Oceanogr.* 17:265–269.

CARRICKER, M. R. 1961. Comparative functional morphology of boring mechanisms in gastropods. *Am. Zool.* 1:263–266.

CAVALLI–SFORZA, I. L., and W. F. BODMER. 1971. *The Genetics of Human Populations.* San Francisco: Freeman.

CAVANAUGH, C. M., S. L. GARDINER, M. L. JONES, H. W. JANNASCH, and J. B. WATERBURY. 1981. Procaryotic cells in the hydrothermal vent tubeworm *Riftia pachyptila* Jones: possible chemoautotrophic symbionts. *Science* 213:340–342.

CHAMBERLAIN, J. A. 1978. Mechanical properties of coral skeleton: compressive strength and its adaptive significance. *Paleobiology* 4:419–435.

CHAPMAN, A. R. O. 1979. *Biology of Seaweeds,* 134 pp. Baltimore: University Park Press.

CHAPMAN, V. J. 1940. Succession on the New England salt marshes. *Ecology* 21:279–282.

––––––. 1960. *Salt Marshes and Salt Deserts of the World.* London: Leonard Hill.

CHARLES, G. H. 1961. The orientation of *Littorina* species to polarised light. *J. Exp. Biol.* 38:189–202.

CHARNOV, E. L. 1976. Optimal foraging: attack strategy of a mantid. *Am. Nat.* 110:141–151.

CHESHER, R. H. 1969. Destruction of Pacific corals by the sea star *Acanthaster planci. Science* 165:280–283.

CHIPPERFIELD, P. N. J. 1953. Observations on the breeding and settlement of *Mytilus edulis* (L.) in British waters. *J. Mar. Biol. Assoc. U.K.* 32:449–476.

CHOAT, J. H. 1969. Studies on labroid fishes. Ph.D. Dissertation. Australia: University of Queensland.

CHRISTENSEN, A. M. 1970. Feeding biology of the sea-star *Astropecten irregularis* Pennant. *Ophelia* 8:1–134.

CHRISTIANSEN, F. B., and T. FENCHEL. 1976. *Theories of Populations in Biological Communities,* 144 pp. Berlin: Springer-Verlag.

CHRISTIANSEN, F. B., and O. FRYDENBERG. 1973. Selection component analysis of natural polymorphisms using population samples including mother-offspring combinations. *Theor. Pop. Biol.* 4:425–445.

––––––. 1974. Geographical patterns in four polymorphisms in *Zoarces viviparus* as evidence of selection. *Genetics* 77:765–770.

CIFELLI, R. 1969. Radiation of Cenozoic planktonic foraminifera. *Syst. Zool.* 18:154–168.

CLARK, P. J., and F. C. EVANS. 1954. Distance to nearest neighbor as a measure of spatial relationships in populations. *Ecology* 35:445–453.

CLARK, R. B. 1962. Observations of the food of *Nephtys. Limnol. Oceanogr.* 7:380–385.

––––––. 1964. *The Dynamics of Metazoan Evolution.* Oxford, U.K.: Clarendon Press.

CLARKE, G. L. 1939. The utilization of solar energy by aquatic organisms. Problems in lake biology. *Publ. Amer. Assoc. Advance. Sci.* 10:27–38.

CLARKE, R. D. 1977. Habitat distribution and species diversity of Chaetodontid and Pomacentrid fishes near Bimini, Bahamas. *Mar. Biol.* 40:277–289.

CLENDENNING, K. A. 1960. Organic productivity of giant kelp areas. *Qu. Progr. Rept* (1 July–30 Sept 1959) *Kelp Invest. Proj. Univ. Calif. Inst. Mar. Res. IMR Ref. 60-61:*1–11.

CODY, M. L. 1974. *Competition and the Structure of Bird Communities*, 318 pp. Princeton: Princeton University Press.

COLE, H. A., and E. W. KNIGHT–JONES. 1949. The setting behaviour of larvae of the European flat oyster *Ostrea edulis* L., and its influence on methods of cultivation and spat collection. *Fish. Invest. (London)* II 17:1–39.

CONNELL, J. H. 1961a. The influence of interspecific competition and other factors on the distribution of the barnacle *Chthamalus stellatus*. *Ecology* 42:710–723.

———. 1961b. Effects of competition, predation by *Thais lapillus* and other factors on natural populations of the barnacle *Balanus balanoides*. *Ecol. Monogr.* 31:61–104.

———. 1970. A predator–prey system in the marine intertidal region. I. *Balanus glandula* and several predatory species of *Thais*. *Ecol. Monogr.* 40:49–78.

———. 1973. Population ecology of reef-building corals. In *Biology and Geology of Coral Reefs*, O. A. Jones, and R. Endean (Eds.), Vol. 2, pp. 205–245. New York: Academic Press.

———. 1975. Some mechanisms producing structure in natural communities. In *Ecology and Evolution of Communities*, M. L. Cody and J. M. Diamond (Eds.), pp. 460–490. Cambridge, Mass.: Belknap Press.

CONNELL, J. H., and R. O. SLATYER. 1977. Mechanisms of succession in natural communities and their role in community stability and organization. *Am. Nat.* 111:1119–1144.

CONOVER, R. J. 1956. Oceanography of Long Island Sound, 1952–1954. VI. Biology of *Acartia clausi* and *Acartia tonsa*. *Bull. Bingham Oceanogr. Coll.* 15:156–233.

———. 1960. The feeding behavior and respiration of some marine planktonic crustacea. *Biol. Bull.* 119:399–415.

———. 1966. Assimilation of organic matter by zooplankton. *Limnol. Oceanogr.* 11:339–345.

———. 1968. Zooplankton—life in a nutritionally dilute environment. *Am. Zool.* 8:107–118.

CONOVER, S. A. M. 1956. Oceanography of Long Island Sound. 1952–1954. IV. Phytoplankton. *Bull. Bingham Oceanogr. Coll.* 15:62–112.

COOPER, D. C. 1973. Enhancement of net primary productivity by herbivore grazing in aquatic laboratory microcosms. *Limnol. Oceanogr.* 18:31–37.

COPELAND, M., and H. L. WIEMAN. 1924. The chemical sense and feeding behavior of *Nereis virens* Sars. *Biol. Bull. Woods Hole* 47:231–238.

CORLISS, J. B., J. DYMOND, L. I. GORDON, J. M. EDMOND, R. P. VON HERZEN, R. D. BALLARD, K. GREEN, D. WILLIAMS, A. BAINBRIDGE, K. CRANE, and T. H. VAN ANDEL. 1979. Submarine thermal springs on the Galapagos Rift. *Science* 203:1073–1083.

COSTLOW, J. D., C. G. BOOKHOUT, and R. MONROE. 1960. The effect of salinity and temperature on larval development of *Sesarma cinereum* (Bosc) reared in the laboratory. *Biol. Bull. Woods Hole* 118:183–202.

COTT, H. B. 1940. *Adaptive Coloration in Animals*. London: Methuen.

COULL, B. C. 1972. Species diversity and faunal affinities of meiobenthic copepoda in the deep-sea. *Mar. Biol.* 14:48–51.

COULL, B. C. and W. B. VERNBERG. 1970. Harpacticoid copepod respiration: *Enhydrosoma propinquum* (Brady) and *Longipedia helgolandica* (Klie). *Mar. Biol.* 5:341–344.

COULL, B. C. and others. 1977. Quantitative estimates of the meiofauna from the deep sea off North Carolina USA. *Mar. Biol.* 39:233–240.

COWAN, R. 1970. Analogies between the recent bivalve Tridacna and the fossil brachipods Lyttoniacea and Richthofeniacea. *Paleo. Paleo. Paleo.* 8:329–344.

CRISP, D. J. 1961. Territorial behaviour in barnacle settlement. *J. Exp. Biol.* 38:429–446.

———. 1964. The effects of the severe winter of 1962–1963 on marine life in Britain. *J. Anim. Ecol.* 33:165–210.

CRISP, D. J., and H. BARNES. 1954. The orientation and distribution of barnacles at settlement, with particular reference to surface contour. *J. Anim. Ecol.* 23:142–162.

CRISP, D. J., and A. J. SOUTHWARD. 1961. Different types of cirral activity of barnacles. *Phil. Trans. Roy. Soc. London* 243B:273–308.

CROKER, R. A. 1967. Niche diversity in five sympatric species of intertidal amphipods (Crustacea:Haustoriidae). *Ecol. Monogr.* 27:173–200.

CUSHING, D. H. 1959. On the nature of production in the sea. *Fish. Invest. (London) Ser. II* 22:1–40.

———. 1971. Upwelling and the production of fish. *Adv. Mar. Biol.* 9:255–334.

———. 1975. *Marine Ecology and Fisheries*, 278 pp. Cambridge, U.K.: Cambridge University Press.

CUSHING, D. H., and J. P. BRIDGER. 1966. The stock of herring in the North Sea and changes due to fishing. *Fish. Invest. London, Ser. 2*, 25:1–123.

CUSHING, D. H., and J. G. K. HARRIS. 1973. Stock and recruitment and the problem of density dependence. *Rapp. Proces.-Verb. Cons. Int. Expl. Mer.* 164:142–155.

DAHL, A. L. 1973. Surface area in ecological analysis: Quantification of benthic coral-reef algae. *Mar. Biol.* 23:239–249.

DALE, N. G. 1974. Bacteria in intertidal sediments: factors related to their distribution. *Limnol. Oceanogr.* 19:509–518.

DALES, R. P. 1963. *Annelids*, 200 pp. London: Hutchinson and Co.

DANA, T., and A. WOLFSON. 1970. Eastern Pacific crown-of-thorns starfish in the lower Gulf of California. *Trans. San Diego Soc. Nat. Hist.* 16:83–90.

DAVID, P. M. 1961. The influence of vertical migration on speciation in the oceanic plankton. *Syst. Zool.* 10:10–16.

DAVIS, C. O., R. J. HARRISON, and R. C. DUGDALE. 1973. Continuous culture of marine diatoms under silicate limitation. I. Synchronized life cycle of *Skeletonema costatum*. *J. Phycol.* 9:175–180.

DAYTON, P. K. 1971. Competition, disturbance and community organization: the provision and subsequent utilization of space in a rocky intertidal community. *Ecol. Monogr.* 41:351–389.

———. 1973a. Dispersion, dispersal and persistence of the annual intertidal alga, *Postelsia palmaeformis*. *Ecology* 54:433–438.

———. 1973b. Two cases of resource partitioning in an intertidal community: making the right prediction for the wrong reason. *Am. Nat.* 107:662–670.

DAYTON, P. K., and R. R. HESSLER. 1972. Role of biological disturbance in maintaining diversity in the deep sea. *Deep Sea Res.* 19:199–208.

DEEVEY, E. S. 1947. Life tables for natural populations of animals. *Quart. Rev. Biol.* 22:283–314.

DEEVEY, G. B. 1948. Zooplankton of Tisbury Great Pond. *Bull. Bingham Oceanogr. Coll.* 12:1–44.

————. 1956. Oceanography of Long Island Sound. 1952–1954. V. Zooplankton. *Bull. Bingham Oceanogr. Coll.* 15:113–155.

DEFANT, A. 1961. *Physical Oceanography, Volumes I and II.* London: Pergamon Press.

DE LIGNY, W. and E. M. PANTELOURIS. 1973. Origin of the European eel. *Nature* 246:518–519.

DE ZWAAN, A., and T. C. M. WIJSMAN. 1976. Anaerobic metabolism in bivalvia (Mollusca): characteristics of anaerobic metabolism. *Comp. Biochem. Phys.* B43:47–54.

DIAMOND, J. 1974. Colonization of exploded volcanic islands by birds: the supertramp strategy. *Science* 184:803–806.

DOBZHANSKY, T., F. J. AYALA, G. L. STEBBINS, and J. W. VALENTINE. 1977. *Evolution,* pp. 1–572. San Francisco: Freeman.

DODGE, R. E., R. C. ALLER, and J. THOMSON. 1974. Coral growth related to resuspension of bottom sediments. *Nature 247:574–577.*

DOTY, M. S. 1957. Rocky intertidal surfaces. In *Treatise on Marine Ecology and Paleoecology,* I, *Marine Ecology,* J. W. Hedgpeth (Ed.), pp. 535–585. Geol. Soc. Am. Mem. 67, I.

DREBES, G. 1977. Sexuality. In *The Biology of Diatoms,* D. Werner (Ed.), pp. 250–283. Oxford, U.K.: Blackwell Scientific Publications.

DROOP, M. R. 1968. Vitamin B_{12} and marine ecology. IV. The kinetics of uptake, growth and inhibition in *Monochrysis lutheri. J. Mar. Biol. Ass. U.K.* 48:689–733.

————. 1973. Some thoughts on nutrient limitation in algae. *J. Phycol.* 9:264–272.

DUGDALE, R. C. 1967. Nutrient limitation in the sea: dynamics, identification and significance. *Limnol. Oceanogr.* 12:685–695.

————. 1976. Nutrient cycles. In *The Ecology of the Seas,* D. H. Cushing and J. J. Walsh (Eds.), pp. 141–172. Oxford, U.K.: Blackwell Scientific Publications.

DUGDALE, R. C., and J. J. GOERING. 1967. Uptake of new and regenerated forms of nitrogen in primary productivity. *Limnol. Oceanogr.* 12:196–206.

DUNNILL, R. M., and D. V. ELLIS. 1969. The distribution and ecology of sub-littoral species of *Macoma* (Bivalvia) off Moresby Island and in Satellite Channel. *Veliger* 12:201–206.

DUSTAN, P. 1975. Growth and form in the reef-building coral *Montastrea annularis. Mar. Biol.* 33:101–107.

DUURSMA, E. K. 1961. Dissolved organic carbon, nitrogen and phosphorous in the sea. *Netherland J. Sea Res.* 1:1–148.

ECKMAN, J. E. 1979. Small-scale patterns and processes in a soft-substratum, intertidal community. *J. Mar. Res.* 37:437–457.

EDWARDS, G. A., and L. IRVING. 1943. The influence of temperature and season upon the oxygen consumption of the sand crab, *Emerita talpoida* (Say). *J. Cell. Comp. Physiol.* 21:169–182.

EHRENFELD, D. W., and A. L. KOCH. 1967. Visual accommodation in the green turtle. *Science* 155:827–828.

EHRLICH, P. R., and A. H. EHRLICH. 1973. Coevolution: heterotypic schooling in Caribbean reef fishes. *Am. Nat.* 107:157–160.

ELNER, R. W., and R. N. HUGHES. 1978. Energy maximization in the diet of the Shore Crab, *Carcinus maenus* (L.) *J. Anim. Ecol.* 47:103–116.

ENDEAN, R. 1973. Population explosions of *Acanthaster planci* and associated destruction of hermatypic corals in the Indo-west Pacific region. In *Biology and Geology of Coral Reefs,* O. A. Jones and R. Endean (Eds.), vol. 2, pp. 389–438. New York: Academic Press.

————. 1976. Destruction and recovery of coral reef communities. In *Biology and Geology of Coral Reefs*, O. A. Jones and R. Endean (Eds.), vol. 3, pp. 215–254. New York: Academic Press.

ENRIGHT, J. T., and W. M. HAMNER. 1967. Vertical diurnal migration and endogenous rhythmicity. *Science* 157:937–941.

EPPLEY, R. W. 1977. The growth and culture of diatoms. In *The Biology of Diatoms*, D. Werner (Ed.), pp. 24–64. Oxford, U.K.: Blackwell Scientific Publications.

EPPLEY, R. W., E. M. RENGER, E. L. VENRICK, and M. M. MULLIN. 1973. A study of plankton dynamics and nutrient cycling in the central gyre of the North Pacific Ocean. *Limnol. Oceanogr.* 18:534–551.

EPPLEY, R. W., J. N. ROGERS, and J. J. McCARTHY. 1969. Half saturation constants for uptake of nitrate and ammonium by marine phytoplankton. *Limnol. Oceanogr.* 14:912–920.

EPPLEY, R. W., and W. H. THOMAS. 1969. Comparison of half-saturation constants for growth and nitrate uptake of marine phytoplankton. *J. Phycol.* 5:375–379.

ESAIAS, W. E., and H. C. CURL. 1972. Effect of dinoflagellate bioluminescence on copepod ingestion rates. *Limnol. Oceanogr.* 17:901–906.

ESTERLY, C. O. 1917. Occurrence of a rhythm in the geotropism of two species of plankton copepods when certain recurring external conditions are absent. *Univ. California Publ. Zool.* 16:393–400.

————. 1919. Reactions of various plankton animals with reference to their diurnal migrations. *Univ. California Publ. Zool.* 19:1–83.

ESTES, J. A., and J. F. PALMISANO. 1974. Sea Otters: their role in structuring nearshore communities. *Science* 185:1058–1060.

EVANS, R. G. 1947. The intertidal ecology of selected localities in the Plymouth neighbourhood. *J. Mar. Biol. Assoc. U.K.* 27:173–218.

FAGER, E. W. 1957. Determination and analysis of recurrent groups. *Ecology* 38:586–595.

FALCONER, D. S. 1960. *Introduction to Quantitative Genetics*, 365 pp. New York: Ronald Press.

FANKBONER, P. V. 1971. Intracellular digestion of symbiotic zooxanthellae by host amoebocytes in giant clams (Bivalvia: Tridacnidae), with a note on the nutritional role of the hypertrophied siphonal epidermis. *Biol. Bull. Woods Hole* 141:222–234.

FELLER, R. J., G. L. TAGHON, E. D. GALLAGHER, G. E. KENNY, and P. A. JUMARS. 1979. Immunological methods for food web analysis in a soft-bottom benthic community. *Mar. Biol.* 54:61–74.

FENCHEL, T. 1965. Feeding biology of the sea-star *Luidia sarsi* Düben and Koren. *Ophelia* 2:223–236.

————. 1968. The ecology of marine microbenthos II. The food of marine benthic ciliates. *Ophelia* 5:73–121.

————. 1969. The ecology of marine microbenthos. IV. Structure and function of the benthic ecosystem, its chemical and physical factors and the microfauna communities with special reference to the ciliated protozoa. *Ophelia* 6:1–182.

————. 1970. Studies on the decomposition of organic detritus derived from the turtle grass *Thalassia testudinum*. *Limnol. Oceanogr.* 15:14–20.

————. 1972. Aspects of decomposer food chains in marine benthos. *Verh. Dtsch. Zool. Ges.* 65:14–22.

———. 1974. Intrinsic rate of natural increase: the relationship with body size. *Oecologia (Berlin)* 14:317–326.

———. 1975a. Factors determining the distribution patterns of mud snails (Hydrobiidae). *Oecologia (Berlin)* 20:1–17.

———. 1975b. Character displacement and coexistence in mud snails (Hydrobiidae). *Oecologia (Berlin)* 20:19–32.

———. 1977. Aspects of the decomposition of seagrasses. In *Seagrass Ecosystems*, C. P. McRoy and C. Helfferich (Eds.), pp. 123–145. New York: Marcel Dekker.

FENCHEL, T., and H. BLACKBURN. 1979. *Bacteria and Mineral Cycling*. Berlin: Springer-Verlag.

FENCHEL, T., J. P. FRIER, and S. KOLDING. 1978. The evolution of competing species. In *Marine Organisms: Genetics, Ecology, Evolution*, B. Battaglia and J. L. Beardmore (Eds.), pp. 289–301. New York: Plenum.

FENCHEL, T., and P. HARRISON. 1976. The significance of bacterial grazing and mineral cycling for the decomposition of particulate detritus. In *The Role of Terrestrial and Aquatic Organisms in Decomposition Processes*, J. M. Anderson and A. MacFadyen (Eds.), pp. 285–299. Oxford, U.K.: Blackwell Scientific.

FENCHEL, T., and B. B. JØRGENSEN. 1977. Detritus food chains of aquatic ecosystems: the role of bacteria. In *Adv. Microb. Ecol.*, M. Alexander (Ed.), vol. 1, pp. 1–58. New York: Plenum.

FENCHEL, T., and L. H. KOFOED. 1976. Evidence for exploitative interspecific competition in mud snails (Hydrobiidae). *Oikos* 27:367–376.

FENCHEL, T., L. H. KOFOED, and A. LAPPALAINEN. 1975. Particle size-selection of two deposit feeders: the amphipod *Corophium volutator* and the prosobranch *Hydrobia ulvae*. *Mar. Biol.* 30:119–128.

FENCHEL, T. M., C. P. McROY, J. C. OGDEN, P. PARKER, and W. E. RAINEY. 1979. Symbiotic cellulose degradation in green turtles. *Appl. Environ. Microbiol.* 37:348–350.

FENCHEL, T., and R. J. RIEDL. 1970. The sulfide system: a new biotic community underneath the oxidized layer of marine sand bottoms. *Mar. Biol.* 7:255–268.

FENICAL, W. 1975. Halogenation in the Rhodophyta: a review. *J. Phycol.* 11:245–259.

FERGUSON, R. L., and M. B. MURDOCH. 1975. Microbial ATP and organic carbon in sediments of the Newport River estuary. In *Estuarine Research*, L. E. Cronin (Ed.), vol. 1, pp. 229–250. New York: Academic Press.

FIELD, I. A. 1922. Biology and economic value of the sea mussel *Mytilus edulis*. *Bull. U.S. Bur. Fish.* 38:127–259.

FISCHER, A. G. 1960. Latitudinal variations in organic diversity. *Evolution* 14:64–81.

FISH, J. D., and S. FISH. 1974. The breeding cycle and growth of *Hydrobia ulvae*. *J. Mar. Biol. Assoc. U.K.* 54:685–697.

FISHELSON, L. 1973. Ecological and biological phenomena influencing coral-species composition on the reef tables at Eilat (Gulf of Aqaba, Red Sea). *Mar. Biol.* 19:183–196.

FISHER, N. S., L. B. GRAHAM, E. J. CARPENTER, and C. F. WURSTER. 1973. Geographic differences in phytoplankton sensitivity to PCB's. *Nature* 241:548–549.

FISHER, W. K., and G. E. MacGINITIE. 1928. The natural history of an echiuroid worm. *Ann. Mag. Nat. Hist. Ser. 10, 1*:204–213.

FISHER-PIETTE, E. 1934. Sur l'équilibre des faunas: interactions des moules, des poupres et des cirripedes. *C. R. Soc. Biogeogr.* 92:47–48.

FLEMING, R. H. 1957. General features of the ocean. In *Treatise on Marine Ecology and Paleo-ecology*, I, Marine Ecology, J. W. Hedgpeth (Ed.), pp. 87–107. Geol. Soc. Am. Mem. 67, I.

FLESSA, K. W. 1975. Area, continental drift and mammalial diversity. *Paleobiology* 1:189–194.

FLESSA, K. W., and J. S. LEVINTON. 1975. Phanerozoic diversity patterns: tests for randomness. *J. Geol.* 83:239–248.

FOGG, G. E. 1975. Primary productivity. In *Chemical Oceanography*, J. P. Riley and G. Skirrow (Eds.), vol. 2, pp. 385–453. London: Academic Press.

FOREMAN, R. E. 1977. Benthic community modification and recovery following intensive grazing by *Strongylocentrotus droebachiensis*. *Helgol. wiss. Meeresunters.* 30:468–484.

FOX, D. L., C. H. OPPENHEIMER, and J. S. KITTREDGE. 1953. Microfiltration in oceanographic research. II. Retention of colloidal micelles by adsorptive filters and by filter feeding invertebrates; proportions of dispersed organic to dispersed inorganic matter and to organic solutes. *J. Mar. Res.* 12:233–243.

FRAENKEL, G. 1927. Beiträge zur Geotaxis und Phototaxis von *Littorina*. *Z. Vergl. Physiol.* 5:585–597.

———. 1966. The heat resistance of intertidal snails at Shirahama, Wakayamaken, Japan. *Publ. Seto Mar. Biol. Lab.* 14:185–195.

FRANCIS, L. 1973a. Clone specific segregation in the sea anemone *Anthopleura elegantissima*. *Biol. Bull.* 144:64–72.

———. 1973b. Intraspecific aggression and its effect on the distribution of *Anthopleura elegantissima*. *Biol. Bull.* 144:73–92.

FRANK, P. W. 1965. The biodemography of an intertidal snail population. *Ecology* 46:831–844.

FRANKENBURG, D., S. L. COLES, and R. E. JOHANNES. 1967. The potential trophic significance of *Callinassa major* fecal pellets. *Limnol. Oceanogr.* 12:113–120.

FRETTER, V., and A. GRAHAM. 1962. *British Prosobranch Molluscs*, pp. 1–755. London: Ray Society.

FROST, B. W. 1972. Effects of size and concentration of food particles on the feeding behavior of the marine planktonic copepod *Calanus pacificus*. *Limnol. Oceanogr.* 17:805–815.

FUTUYMA, D. J. 1973. Community structure and stability in constant environments. *Am. Nat.* 107:443–446.

GADGIL, M., and W. H. BOSSERT. 1970. Life historical consequences of natural selection. *Am. Nat.* 104:1–24.

GADGIL, M., and O. T. SOLBRIG. 1972. The concept of "R" and "K" selection: evidence from wild flowers and some theoretical considerations. *Am. Nat.* 106:14–31.

GAGE, J., and A. D. GEEKIE. 1973. Community structure of the benthos in Scottish Sea-lochs. II. Spatial pattern. *Mar. Biol.* 19:41–53.

GAINEY, L. F., and M. J. GREENBERG. 1977. Physiological basis of the species abundance-salinity relationship in molluscs: a speculation. *Mar. Biol.* 40:41–49.

GALLAGHER, J. C. 1980. Population genetics of *Skeletonema costatum* (Bacillariophyceae) in Narragansett Bay. *J. Phycol.* 16:464–474.

GALTSOFF, P. S. 1964. The American oyster *Crassostrea virginica* Gmelin. *Bull. U.S. Bur. Fish.* 64:1–480.

GANNING, B. 1967. Laboratory experiments in the ecological work on rockpool animals with special notes on the ostracod *Heterocypris salinus*. *Helgol. wiss. Meeresunters.* 15:27–40.

GESSNER, F., and W. SCHRAM. 1971. Salinity. Plants. In *Marine Ecology,* O. Kinne (Ed.), vol. 1, part 2, pp. 705–820. London: Wiley–Interscience.

GHISELIN, M. T. 1969. The evolution of hermaphroditism among animals. *Quart. Rev. Biol.* 44:189–208.

GIESEL, J. T. 1970. On the maintenance of a shell pattern and behavior polymorphism in *Acmaea digitalis,* a limpet. *Evolution* 24:98–119.

GILBERT, W. H. 1968. Distribution and dispersion patterns of the dwarf tellin clam, *Tellina agilis.* *Biol. Bull.* 135:419–420.

———. 1970. Territoriality observed in a population of *Tellina agilis* (Bivalvia: Mollusca). *Biol. Bull.* 139:423–424.

GLEMAREC, M. 1973. The benthic communities of the European North Atlantic continental shelf. *Oceanogr. Mar. Biol. Ann. Rev.* 11:263–289.

GLUDE, J. B. 1954. Survival of soft-shell clams, *Mya arenaria,* buried at various depths. *Maine, Dep. Sea Shore Fish. Res. Bull.* 22:1–26.

GLYNN, P. W. 1968. Mass mortalities of echinoids and other reef flat organisms coincident with midday, low water exposures in Puerto Rico. *Mar. Biol.* 1:226–243.

———1973a. Ecology of a Caribbean coral reef. The *Porites* reef-flat biotope: Part II. Plankton community with evidence for depletion. *Mar. Biol.* 22:1–21.

———. 1973b. Aspects of the ecology of coral reefs in the Western Atlantic region. In *The Biology and Geology of Coral Reefs,* O. A. Jones and R. Endean (Eds.), vol. II, pp. 271–324. New York: Academic Press.

———. 1976. Some physical and biological determinants of coral community structure in the eastern Pacific. *Ecol. Monogr.* 46:431–456.

GLYNN, P. W., R. H. STEWART, and J. E. McCOSKER. 1972. Pacific coral reefs of Panama: structure, distribution and predators. *Geol. Rundschau* 61:483–519.

GLYNN, P. W., G. W. WELLINGTON, and C. BIRKELAND. 1979. Coral reef growth in the Galapagos: limitation by sea urchins. *Science* 203:47–49.

GOLDBERG, E. D. 1965. Minor elements in sea water. In *Chemical Oceanography,* J. P. Riley and G. Skirrow (Eds.), pp. 163–196. Volume I. London: Academic Press.

GOLDMAN, B., and F. H. TALBOT. 1976. Aspects of the ecology of coral reef fishes. In *Biology and Geology of Coral Reefs,* O. A. Jones and R. Endean (Eds.), vol. 3, pp. 125–153. New York: Academic Press.

GOOCH, J. L. 1975. Mechanisms of evolution and population genetics. In *Marine Ecology,* O. Kinne (Ed.), part 2, pp. 349–409. London: Wiley-Interscience.

GOOCH, J. L., and T. J. M. SCHOPF. 1970. Population genetics of marine species of the phylum Ectoprocta. *Biol. Bull.* 138:138–156.

GOODBODY, I. 1961. Mass mortality of a marine fauna following a tropical rain. *Ecology* 42:150–155.

GORDON, D. C. 1966. The effects of the deposit feeding polychaete *Pectinaria gouldii* on the intertidal sediments of Barnstable Harbor. *Limnol. Oceanogr.* 11:327–332.

GORDON, J., and M. R. CARRIKER. 1978. Growth lines in a bivalve mollusk: subdaily patterns and dissolution of the shell. *Science* 202:519–521.

GOREAU, T. F. 1959. The ecology of Jamaican coral reefs. I. Species composition and zonation. *Ecology* 40:67–90.

GOREAU, T. F., and N. I. GOREAU. 1959. The physiology of skeleton formation in corals. II. Calcium deposition by hermatypic corals under various conditions in the reef. *Biol. Bull.* 117:239–250.

GOREAU, T. F., and W. D. HARTMAN. 1953. Boring sponges as controlling factors in the formation and maintenance of coral reefs. In *Mechanisms of Hard Tissue Destruction,* pp. 25–54. Washington, D.C.: American Association for the Advancement of Science Publ. No. 75.

GOREAU, T. F., and L. S. LAND. 1974. Fore-reef morphology and depositional processes, north Jamaica. In *Reefs in Time and Space,* L. F. Laporte (Ed.), pp. 77–89. Soc. Econ. Pal. Min. Spec. Publ. No. 18.

GOSS–CUSTARD, J. D. 1977. Predator responses and prey mortality in the red shank *Tringa totanus* (L.) and a preferred prey *Corophium volutator* (Pallas). *J. Anim. Ecol.* 46:21–36.

GRAHAM, A. 1938. The structure and function of the alimentary canal of aeolid molluscs with a discussion on their nematocysts. *Trans. Roy. Soc. Edinburgh* 59:267–307.

GRAHAME, J. 1977. Reproductive effort and r- and k-selection in two species of *Lacuna* (Gastropoda:Prosobranchia). *Mar. Biol.* 40:217–224.

GRANT, P. R. 1972. Convergent and divergent character displacement. *Biol. J. Linn. Soc.* 4:39–68.

GRASSLE, J. F. 1973. Variety in coral reef communities. In *Biology and Geology of Coral Reefs,* O. A. Jones and R. Endean (Eds.), vol. 1, pp. 247–270. New York: Academic Press.

GRASSLE, J. F., and J. P. GRASSLE. 1974. Opportunistic life histories and genetic systems in marine benthic polychaetes. *J. Mar. Res.* 32:253–284.

GRASSLE, J. P., and J. F. GRASSLE. 1976. Sibling species in the marine pollution indicator, *Capitella capitata* (Polychaeta). *Science* 192:567–569.

GRAY, I. E. 1960. The seasonal occurrence of *Mytilus edulis* on the Carolina coast as a result of transport around Cape Hatteras. *Biol. Bull.* 119:550–559.

GRAY, J. S. 1966. The attractive factor of intertidal sands to *Protodrilus symbioticus. J. Mar. Biol. Assoc. U.K.* 46:627–645.

————. 1974. Animal-sediment relationships. *Oceanogr. Mar. Biol. Ann. Rev.* 12:223–261.

GREEN, G. 1977. Ecology of toxicity in marine sponges. *Mar. Biol.* 40:207–215.

GREEN, R. H. 1971. A multivariate statistical approach to the Hutchinsonian niche: bivalve molluscs of central Canada. *Ecology* 52:543–556.

GUILLARD, R. R. L. 1968. B_{12} specificity of marine centric diatoms. *J. Phycol.* 4:59–64.

GUILLARD, R. R. L., and J. A. HELLEBUST. 1971. Growth and the production of extracellular substances by two strains of *Phaeocystis poucheti. J. Phycol.* 7:330–338.

GUILLARD, R. R. L., and P. KILHAM. 1977. The ecology of marine planktonic diatoms. In *The Biology of Diatoms,* D. Werner (Ed.), pp. 372–469. Oxford, U.K.: Blackwell Scientific Publications.

GULLAND, J. A. 1972. *The Fish Resources of the Ocean.* London: Fishing News (Books).

————. 1974. *The Management of Marine Fisheries,* 198 pp. Seattle: University of Washington.

GUNNERSON, C. G., and K. O. EMERY. 1962. Suspended sediment and plankton over San Pedro Basin, California. *Limnol. Oceanogr.* 7:14–20.

GUNTHER, E. R. 1936. A report on oceanographical investigations in the Peru coastal current. *Discovery Repts.* 13:109–276.

HAINES, E. B. 1975. Nutrient inputs to the coastal zone: the Georgia and South Carolina shelf. In *Estuarine Research*, L. E. Cronin (Ed.), vol. 1, pp. 303–324. New York: Academic Press.

———. 1979. Interactions between Georgia salt marshes and coastal waters: a changing paradigm. In *Ecological Processes in Coastal and Marine Systems*, R. J. Livingston (Ed.), pp. 35–46. New York: Plenum.

HALLAM, A. 1967. The interpretation of size-frequency distributions in molluscan death assemblages. *Palaeontology* 10:25–42.

HANCOCK, D. A. 1965. Adductor muscle size in Danish and British mussels in relation to starfish predation. *Ophelia* 2:253–267.

HARBISON, G. R., and V. L. McALISTER. 1980. Fact and artifact in copepod feeding experiments. *Limnol. Oceanogr.* 25:971–981.

HARDEN JONES, F. R. 1968. *Fish Migration*, 325 pp. London: Edw. Arnold.

HARDY, A. C. 1954. *The Open Sea, Its Natural History: The World of Plankton*, 355 pp. London: Collins.

HARDY, A. C., and R. BAINBRIDGE. 1954. Experimental observations on vertical migrations of plankton animals. *J. Mar. Biol. Assoc. U.K.* 33:409–448.

HARGER, J. R. 1968. The role of behavioral traits in influencing the distribution of two species of sea mussel, *Mytilus edulis* and *Mytilus californianus*. *Veliger* 11:45–49.

———. 1970. Comparisons among growth characteristics of the two species of sea mussel, *Mytilus edulis* and *Mytilus californianus*. *Veliger* 13:44–56.

———. 1972. Competitive coexistence: maintenance of interacting associations of the sea mussels *Mytilus edulis* and *Mytilus californianus*. *Veliger* 14:387–410.

HARGRAVE, B. T. 1970a. The utilization of benthic microflora by *Hyalella azteca*. *J. Anim. Ecol.* 39:427–437.

———. 1970b. The effect of a deposit-feeding amphipod on the metabolism of the benthic microflora. *Limnol. Oceanogr.* 15:21–30.

———. 1972. Prediction of egestion by the desposit-feeding amphipod *Hyalella azteca*. *Oikos* 23:116–124.

———. 1973. Coupling carbon flow through some pelagic and benthic communities. *J. Fish. Res. Bd. Canada* 30:1317–1326.

———. 1976. The central role of invertebrate faeces in sediment decomposition. In *The Role of Terrestrial and Aquatic Organisms in Decomposition Processes*, J. M. Anderson and A. MacFadyen (Eds.), pp. 301–321. Oxford, U.K.: Blackwell Scientific Publications.

HARRISON, P. G., and K. H. MANN. 1975. Detritus formation from eelgrass (*Zostera marina* L.): the relative effects of fragmentation leaching and decay. *Limnol. Oceanogr.* 20:924–934.

HART, T. J. 1942. Phytoplankton periodicity in Antarctic surface waters. *Discovery Repts.* 21:263–348.

HARTMAN, W. D., and T. F. GOREAU. 1970. Jamaican coralline sponges: their morphology, ecology and fossil relatives. *Symp. Zool. Soc. Lond.* 25:205–243.

HARVEY, H. W. 1926. Nitrate in the sea. *J. Mar. Biol. Assoc. U.K.* 14:71–88.

HARVEY, N. W., L. H. N. COOPER, M. V. LeBOUR, and F. S. RUSSELL. 1935. Plankton production and its control. *J. Mar. Biol. Assoc. U.K.* 20:407–441.

HATTON, H. 1938. Essais de bionomie explicative sur quelques espéces intercotidales d'algues et d'animaux. *Ann. Inst. Océanogr.* 17:241–338.

HAVEN, D. S., and R. MORALES-ALAMO. 1966. Aspects of biodeposition by oysters and other invertebrate filter feeders. *Limnol. Oceanogr.* 11:487–498.

HEINLE, D. R., R. P. HARRIS, J. F. USTACH, and D. A. FLEMER. 1977. Detritus as food for estuarine copepods. *Mar. Biol.* 40:341–353.

HELLEBUST, J. A. 1971. Glucose uptake by *Cyclotella cryptica:* dark induction and light inactivation of transport. *J. Phycol.* 7:345–349.

HELLEBUST, J. A., and J. LEWIN. 1977. Heterotrophic nutrition. In *The Biology of Diatoms*, D. Werner (Ed.), pp. 169–197. Oxford, U.K.: Blackwell Scientific Publications.

HEPPLESTON, P. B. 1971. The feeding ecology of oystercatchers (*Haematopus ostralegus* L.) in winter in northern Scotland. *J. Anim. Ecol.* 4:651–672.

HESSLER, R. R., J. D. ISAACS, and E. L. MILLS. 1972. Giant amphipod from the abyssal Pacific Ocean. *Science* 175:636–637.

HESSLER, R. R., and P. A. JUMARS. 1974. Abyssal community analysis from replicate box cores in the central north Pacific. *Deep Sea Res.* 21:185–209.

HESSLER, R. R., and H. L. SANDERS. 1967. Faunal diversity in the deep-sea. *Deep Sea Res.* 14:65–78.

HOBBIE, J. E., B. J. COPELAND, and W. G. HARRISON. 1975. Sources and fates of nutrients of the Pamlico River Estuary, North Carolina. In *Estuarine Research*, L. E. Cronin (Ed.), vol. I., pp. 287–302. New York: Academic Press.

HOLLING, C. S. 1965. The functional response of predators to prey density and its role in mimicry and population regulation. *Mem. Entomol. Soc. Canada* no. 45, pp. 1–60.

HOLME, N. A. 1950. Population dispersion in *Tellina tenuis* da Costa. *J. Mar. Biol. Assoc. U.K.* 29:267–280.

HOLME, N. A., and A. D. MCINTYRE. 1971. *Methods for the Study of Marine Benthos*, 334 pp. Oxford, U.K.: Blackwell Scientific Publications.

HONJO, S. 1978. Sedimentation of materials in the Sargasso Sea at a 5,367 m deep station. *J. Mar. Res.* 36:469.

HORN, H. S., and R. H. MACARTHUR. 1972. Competition among fugitive species in a harlequin environment. *Ecology* 53:749–752.

HUGHES, T. G. 1973. Deposit feeding in *Abra tenuis* (Bivalvia: Tellinacea). *J. Zool. London* 171:499–512.

HULBERT, E. M. 1970. Competition for nutrients by marine phytoplankton in oceanic, coastal and estuarine regions. *Ecology* 51:475–484.

HURLBERT, S. H. 1971. The nonconcept of species diversity: a critique and alternative parameters. *Ecology* 52:577–586.

HURLBERT, S. H. 1978. The measurement of niche overlap and some relatives. *Ecology* 59:67–77.

HUSTON, M. 1979. A general hypothesis of species diversity. *Am. Nat.* 113:81–101.

HUTCHINSON, G. E. 1957. Concluding remarks. *Cold Spring Harbor Symp. Quant. Biol.* 22:415–427.

———. 1961. The paradox of the plankton. *Am. Nat.* 95:137–145.

———. 1967. *A Treatise on Limnology*, vol. 2. pp. 1–1115. New York: Wiley.

HUTNER, S. H., and J. J. A. MCLAUGHLIN. 1958. Poisonous tides. *Sci. Amer.* 199:92–98.

HYLLEBERG, J. 1975a. The effect of salinity and temperature on egestion in mud snails (Gastropoda:Hydrobiidae): a study on niche overlap. *Oecologia* 21:279–289.

—————. 1975b. Selective feeding by *Abarenicola pacifica* with notes on *Abarenicola vagabunda* and a concept of gardening in lugworms. *Ophelia* 14:113–137.

—————. 1977. *Økoligiske Problemstillinger*. Arhus, Denmark: Inst. fur Genetik & Økologi.

HYMAN, L. H. 1940. *The Invertebrates: Protozoa through Ctenophora*, vol. 1, 726 pp. New York: McGraw-Hill.

—————. 1951. *The Invertebrates: Platyhelminthes and Rhynchocoela*, vol. 2, pp. 459–531. New York: McGraw-Hill.

—————. 1955. *The Invertebrates: Echinodermata*, vol. 4, 761 pp. New York: McGraw-Hill.

—————. 1959. *The Invertebrates: Smaller Coelomate Groups*, vol. 5, 783 pp. New York: McGraw-Hill.

—————. 1967. *The Invertebrates: Mollusca I*, vol. 6, 792 pp. New York: McGraw-Hill.

IMBRIE, J., and N. G. KIPP. 1968. A new micropaleontological method for quantitative paleoclimatology: application to a late Pleistocene Caribbean core. In *Late Cenozoic Glacial Ages*, K. K. Turekian, (Ed.), pp. 71–81. New Haven: Yale University Press.

INNES, D. J., and L. E. HALEY. 1977. Inheritance of a shell-color polymorphism in the mussel. *J. Hered.* 68:203–204.

IRVINE, G. V. 1973. The effect of selective feeding by two species of sea urchins on the structuring of algal communities. M.S. thesis. Seattle: University of Washington.

ISAACS, J. D. 1969. The nature of oceanic life. *Sci. Amer.* 221:146–162.

IVERSON, R. L., L. K. COACHMAN, R. T. COONEY, T. S. ENGLISH, J. J. GOERING, G. L. HUNT, M. C. MACAULEY, C. P. MCROY, W. S. REEBURG, and T. E. WHITLEDGE. 1979. Ecological significance of fronts in the southeastern Bering Sea. In *Ecological Processes in Coastal and Marine Systems*, R. J. Livingstone (Ed.), pp. 437–466. New York: Plenum.

IVLEV, V. S. 1961. *Experimental Ecology of the Feeding of Fishes*, 302 pp. New Haven: Yale University Press.

JACKSON, G. A. 1977. Nutrients and production of giant kelp, *Macrocystis pyrifera*, off southern California. *Limnol. Oceanogr.* 22:979–995.

JACKSON, J. B. C. 1972. The ecology of the molluscs of *Thalassia* communities, Jamaica, West Indies. II. Molluscan population variability along an environmental stress gradient. *Mar. Biol.* 14:304–337.

—————. 1973. The ecology of molluscs of *Thalassia* communities, Jamaica, West Indies. I. Distribution, environmental physiology, and ecology of common shallow-water species. *Bull. Mar. Sci.* 23:313–350.

—————. 1974. Biogeographic consequences of eurytopy and stenotopy among marine bivalves and their evolutionary significance. *Am. Nat.* 108:541–560.

—————. 1977. Habitat area, colonization, and development of epibenthic community structure. In *Proc. 11th Europ. Mar. Biol. Symp.*, B. F. Keegan, P. O. Ceidigh, and P. J. S. Boaden (Eds.), pp. 349–358. New York: Pergamon Press.

—————. 1979. Overgrowth competition between encrusting Cheilostome ectoprocts in a Jamaican cryptic reef environment. *J. Anim. Ecol.* 48:805–823.

JACKSON, J. B. C., T. F. GOREAU, and W. D. HARTMAN. 1971. Recent brachiopod-coralline sponge communities and their paleoecological significance. *Science* 173:623–625.

JANNASCH, H. W., K. EIMHJELLEN, C. WIRSEN, and A. FARMANFARMAIAN. 1971. Microbial degradation of organic matter in the deep sea. *Science* 171:672–675.

JANNASCH, H. W., and G. E. JONES. 1959. Bacterial populations in seawater as determined by different methods of enumeration. *Limnol. Oceanogr.* 4:128–139.

JANNASCH, H. W., and P. H. PRITCHARD. 1972. The role of inert particulate matter in the activity of aquatic microorganisms. *Memorie Ist. Ital. Idrobiol.* 29 (Suppl.):289–308.

JANNASCH, H. W., and C. O. WIRSEN. 1973. Deep-sea microorganisms: In situ response to nutrient enrichment. *Science* 180:641–643.

JERLOV, N. G. 1951. Optical studies of ocean waters. *Repts. Swedish Deep Sea Exped.* 3:1–59.

JOHANNES, R. E., S. L. COLES, and N. T. KUENZEL. 1970. The role of zooplankton in the nutrition of some scleractinian corals. *Limnol. Oceanogr.* 15:579–586.

JOHANNES, R. E., S. J. COWARD, and K. L. WEBB. 1969. Are dissolved amino acids an energy source for marine invertebrates? *Comp. Biochem. Physiol.* 29:283–288.

JOHANNES, R. E., and M. SATOMI. 1966. Composition and nutritive value of fecal pellets of a marine crustacean. *Limnol. Oceanogr.* 11:191–197.

JOHNSON, M. S. 1971. Adaptive lactate dehydrogenase variation in the crested blenny *Anoplarchus*. *Heredity* 27:205–226.

JOHNSON, M. W. 1939. The correlation of water movements and dispersal of pelagic larval stages of certain littoral animals, especially the sand crab, *Emerita*. *J. Mar. Res.*, pp. 236–245.

JOHNSON, R. G. 1971. Animal-sediment relations in shallow water benthic communities. *Mar. Geol.* 11:93–104.

JOHNSON, W. S., A. GIGNON, S. L. GULMAN, and H. A. MOONEY. 1974. Comparative photosynthetic capacities of intertidal algae under exposed and submerged conditions. *Ecology* 55:450–453.

JONES, N. S. 1950. Marine bottom communities. *Biol. Rev.* 25:283–313.

JONES, R. S. 1968. Ecological relationships in Hawaiian and Johnston Island Acanthuridae (Surgeon fishes). *Micronesica* 4:309–361.

JONES, W. E., and A. DEMETROPOULOS. 1968. Exposure to wave action: measurements of an important ecological parameter on rocky shores on Anglesey. *J. Exp. Mar. Biol. Ecol.* 2:46–63.

JØRGENSEN, B. B. 1977. Bacterial sulfate reduction within reduced microniches of oxidized marine sediments. *Mar. Biol.* 41:7–17.

JØRGENSEN, B. B., and T. FENCHEL. 1974. The sulfur cycle of a marine sediment model system. *Mar. Biol.* 24:189–201.

JØRGENSEN, C. B. 1959. Quantitative aspects of filter feeding in invertebrates. *Biol. Rev.* 30:391–454.

———. 1960. Efficiency of particle retention and rate of water transport in undisturbed lamellibranchs. *Jour. du Cons. Exp. Mer* 26:94–116.

———. 1966. *Biology of Suspension Feeding*, pp. 1–357. Oxford, U.K.: Pergamon Press.

———. 1976. August Pütter, August Krogh, and modern ideas on the use of dissolved organic matter in aquatic environments. *Biol. Rev.* 51:291–328.

JØRGENSEN, C. B., and E. D. GOLDBERG. 1953. Particle filtration in some ascidians and lamellibranchs. *Biol. Bull. Woods Hole* 105:477–489.

JØRGENSEN, E. G. 1977. Photosynthesis. In *The Biology of Diatoms*, D. Werner (Ed.), pp. 150–168. Oxford, U.K.: Blackwell Scientific Publications.

JUMARS, P. A. 1974. Two pitfalls in comparing communities of differing diversities. *Am. Nat.* 108:389–391.

———. 1975. Environmental grain and polychaete species diversity in a bathyal benthic community. *Mar. Biol.* 30:253–266.

KANWISHER, J. W. 1957. Freezing and drying in intertidal algae. *Biol. Bull. Woods Hole* 109:56–63.

KANWISHER, J. W., and S. A. WAINWRIGHT. 1967. Oxygen balance in some reef corals. *Biol. Bull.* 133:378–390.

KARLSON, R. 1978. Predation and space utilization patterns in a marine epifaunal community. *J. Exp. Mar. Biol. Ecol.* 31:225–239.

KARPLUS, I., R. SZLEP, and M. TSURNAMAL. 1972. Associative behavior of the fish *Cryptocentrus cryptocentrus* (Gobiidae) and the pistol shrimp *Alpheus djiboutensis* (Alpheidae) in artificial burrows. *Mar. Biol.* 15:95–104.

———. 1974. The burrows of Alpheid shrimp with gobiid fish in the northern Red Sea. *Mar. Biol.* 24:259–268.

KEATING, K. I. 1977. Allelopathic influence on Blue-Green bloom sequence in a eutrophic lake. *Science* 196:885–886.

———. 1978. Blue-green algal inhibition of diatom growth: transition from mesotrophic to eutrophic community structure. *Science* 199:971–973.

KEEN, M. J., and D. J. PIPER. 1976. Kelp, methane and an impenetrable reflector in a temperate bay. *Can. J. Earth Sci.* 13:312–318.

KHAILOV, K. M., and Z. P. BURLAKOVA. 1969. Release of dissolved organic matter by marine seaweeds and distribution of their total organic production to inshore communities. *Limnol. Oceanogr.* 14:521–527.

KHLEBOVICH, V. V. 1968. Some peculiar features of the hydrochemical regime and the fauna of mesohaline waters. *Mar. Biol.* 2:47–49.

KIERSTEAD, H., and L. B. SLOBODKIN. 1953. The size of water masses containing plankton blooms. *J. Mar. Res.* 12:141–147.

KIM, Y. S. 1969. An observation on the opening of bivalve molluscs by starfish *Asterias amurensis*. *Bull. Fac. Fish. Hokkaido Univ.* 20:60–64.

KING, R. J., and W. SCHRAMM. 1976. Photosynthetic rates of benthic marine algae in relation to light intensity and seasonal variations. *Mar. Biol.* 37:215–222.

KINNE, O. 1970. Temperature. Animals–Invertebrates. pp. 407–514. In *Marine Ecology*, O. Kinne (Ed.), vol. 1, part 1. London: Wiley-Interscience.

———. 1971. Salinity. Animals–Invertebrates. In *Marine Ecology*, O. Kinne (Ed.), vol. 1, pp. 821–995. Environmental Factors, part 2. London: Wiley-Interscience.

KINSMAN, D. J. J. 1964. Reef coral tolerance of high temperatures and salinities. *Nature* 202:1280–1282.

KINZIE, R. A. 1968. The ecology of the replacement of Pseudosquilla ciliata (Fabricius) by Gonodactylus falcatus (Forskal) (Crustacea: Stomatopoda) recently introduced into the Hawaiian Islands. *Pacific Sci.* 22:465–475.

KINZIE, R. A., and G. S. CHEE. 1979. The effect of different zooxanthellae on the growth of experimentally reinfected hosts. *Biol. Bull.* 156:315–327.

KNIGHT–JONES, E. W. 1953. Laboratory experiments on gregariousness during setting in *Balanus balanoides* and other barnacles. *J. Exp. Biol.* 30:584–598.

KNUDSEN, J. 1967. The deep sea Bivalvia. *John Murray Exped. 1933–1934. Sci. Rept.* 11:237–343.

KOBLENTZ–MISHKE, I. J., V. V. VOLKOVINSKY, and J. B. KABANOVA. 1970. Plankton primary production of the world ocean. In *Scientific Exploration of the South Pacific*, W. S. Wooster (Ed.), pp. 183–193. Washington, D.C.: National Academy of Sciences.

KOCH, A. L., A. CARR, AND D. W. EHRENFELD. 1969. The problem of open-sea navigation: the migration of the Green Turtle to Ascension Island. *J. Theoret. Biol.* 22:163–179.

KOEHL, M. A. R. 1976. Mechanical design in sea anemones. In *Coelenterate Ecology and Behavior*, G. O. Mackie (Ed.), pp. 23–31. New York: Plenum.

KOEHL, M. A. R., and S. A. WAINWRIGHT. 1977. Mechanical adaptations of a giant kelp. *Limnol. Oceanogr.* 22:1067–1071.

KOEHN, R. K., R. MILKMAN, and J. B. MITTON. 1976. Population genetics of marine pelecypods. IV. Selection, migration and genetic differentiation in the blue mussel *Mytilus edulis*. *Evolution* 30:2–32.

KOFOED, L. H. 1975. The feeding biology of *Hydrobia ventrosa* (Montagu). I. The assimilation of different components of the food. *J. Exp. Mar. Biol. Ecol.* 19:233–241.

KOHLER, A. C., and D. N. FITZGERALD. 1969. Comparisons of food of cod and haddock in the Gulf of St. Lawrence and on the Nova Scotia Banks. *J. Fish. Res. Bd. Canada* 26:1273–1287.

KOHN, A. J. 1956. Piscivorous gastropods of the genus *Conus*. *Proc. Nat. Acad. Sci.* 42:168–171.

———. 1959. The ecology of *Conus* in Hawaii. *Ecol. Monogr.* 29:47–90.

———. 1966. Food specialization in *Conus* in Hawaii and California. *Ecology* 47:1041–1043.

———. 1967. Environmental complexity and species diversity in the gastropod genus *Conus* on Indo-west Pacific reef platforms. *Am. Nat.* 101:251–259.

———. 1971. Diversity, utilization of resources, and adaptive radiation in shallow-water marine invertebrates of tropical oceanic islands. *Limnol. Oceanogr.* 16:332–348.

KOHN, A. J., and P. HELFRICH. 1957. Primary organic productivity of a Hawaiian coral reef. *Limnol. Oceanogr.* 2:241–251.

KOHN, A. J., and J. W. NYBAKKEN. 1975. Ecology of *Conus* on eastern Indian Ocean fringing reefs: diversity of species and resource utilization. *Mar. Biol.* 29:211–234.

KOLLER, G. 1930. Versuche an marinen Wirbellosen über die Aufnahme gelöster Nährstoffe. *Zeits. Vergl. Phys.* 2:437–447.

KREBS, J. R. 1978. Optimal foraging: decision rules for predators. In *Behavioural Ecology*, J. R. Krebs and N. B. Davies (Eds.), pp. 23–63. Sunderland, Mass.: Sinauer Associates.

KREBS, J. R., J. T. ERICHSEN, M. I. WEBBER, and E. L. CHARNOV. 1977. Optimal prey selection in the Great Tit *(Parus major)*. *Anim. Behav.* 25:30–38.

KREMER, J. N., and S. W. NIXON. 1977. *A Coastal Marine Ecosystem*, 217 pp. *Simulation and Analysis*. Berlin: Springer-Verlag.

KRÜGER, F. 1969. Zur Ernährungsphysiologie von *Arenicola marina* L. *Helgl. Wiss. Meeresunters.* 22:149–200.

LA BARBERA, M. 1978. Particle capture by a Pacific brittle star: experimental test of the aerosol suspension feeding model. *Science* 201:1147–1149.

LACK, D. 1947. *Darwin's Finches*, 204 pp. Cambridge: Cambridge University Press.

LADD, H. S., E. INGERSON, R. C. TOWNSEND, M. RUSSELL, and H. K. STEPHENSON. 1953. Drilling on Eniwetok Atoll, Marshall Islands. *Am. Ass. Petr. Geol. Bull.* 37:2257–2280.

LAM, R. K., and B. W. FROST. 1976. Model of copepod filtering response to changes in size and concentration of food. *Limnol. Oceanogr.* 21:490–500.

LAND, L. S. 1973. Contemporaneous dolomitization of middle Pleistocene reefs by meteoric water, north Jamaica. *Bull. Mar. Sci.* 23:64–92.

LANDENBERGER, D. E. 1968. Studies on selective feeding in the Pacific starfish *Pisaster* in southern California. *Ecology* 49:1062–1075.

LANDRY, M. R. 1976. The structure of marine ecosystems: an alternative. *Mar. Biol.* 35:1–7.

———. 1977. A review of important concepts in the trophic organization of pelagic ecosystems. *Helgol. wiss. Meeresunters* 30:8–17.

LANG, J. C. 1971. Interspecific aggression by scleractinian corals. I. The rediscovery of *Scolymia cubensis* (Milne Edwards and Haime). *Bull. Mar. Sci.* 21:952–959.

———. 1973. Interspecific aggression by scleractinian corals. 2. Why the race is not only to the swift. *Bull. Mar. Sci.* 23:260–279.

LANGE, R. 1968. The relation between the oxygen consumption of isolated gill tissue of the common mussel, *Mytilus edulis* L., and salinity. *J. Exp. Mar. Biol. Ecol.* 2:37–45.

LASSEN, H. H. 1975. The diversity of freshwater snails in view of the equilibrium theory of biogeography. *Oecologia* 19:1–8.

LASSEN, H. H. and M. E. CLARK. 1979. Comparative fecundity in three Danish mudsnails (Hydrobiidae). *Ophelia* 18:171–178.

LASSIG, B. R. 1977. Communication and coexistence in a coral reef community. *Mar. Biol.* 42:85–92.

LEACH, J. H. 1970. Epibenthic algal production in an intertidal mudflat. *Limnol. Oceanogr.* 15:514–521.

LE DANOIS, Y. 1959. Adaptations morphologiques et biologiques des poissons des massifs corralien. *Bul. de l'I. Franc. A.N.* 21:1304–1325.

LEE, J. J., J. H. TIETJEN, N. M. SAKS, G. G. ROSS, H. RUBIN, and W. A. MULLER. 1975. Educing and modeling the functional relationships within sublittoral salt-marsh aufwuchs communities inside one of the black boxes. In *Estuarine Research*, L. E. Cronin, (Ed.), vol. 1, pp. 710–738. New York: Academic Press.

LENT, C. M. 1969. Adaptations of the ribbed mussel, *Modiolus demissus* (Dillwyn). Effects and adaptive significance. *Am. Zool.* 9:283–292.

LEVINS, R. 1968. *Evolution in Changing Environments.* Princeton, N.J.: Princeton University Press.

LEVINTON, J. S. 1970. The paleoecological significance of opportunistic species. *Lethaia* 3:69–78.

———. 1971. Control of tellinacean (Mollusca:Bivalvia) feeding behavior by predation. *Limnol. Oceanogr.* 6:660–662.

———. 1972a. Stability and trophic structure in deposit-feeding and suspension-feeding communities. *Am. Nat.* 106:472–486.

———. 1972b. Spatial distribution of *Nucula proxima* (Protobranchia): an experimental approach. *Biol. Bull.* 143:175–183.

———. 1973. Genetic extinction hypothesis and its critics. *Geology* 1:157–158.

———. 1974. Trophic group and evolution in bivalve molluscs. *Palaeontology* 17:579–000.

———. 1975. Levels of genetic polymorphism at two enzyme encoding loci in eight species of the genus *Macoma* (Mollusca: Bivalvia). *Mar. Biol.* 33:41–47.

———. 1977. The ecology of deposit-feeding communities: Quisset Harbor, Massachusetts. In *Ecology of Marine Benthos*, B. C. Coull (Ed.), pp. 191–228. Columbia: University of South Carolina Press.

————. 1979a. Deposit-feeders, their resources, and the study of resource limitation. In *Ecological Processes in Coastal and Marine Ecosystems,* R. J. Livingston (Ed.), pp. 117–141. New York: Plenum.

————. 1979b. The effect of density on deposit-feeding populations: movement, feeding and floating of *Hydrobia ventrosa.* Montagu (Gastropoda, Prosobranchia). *Oecologia* 43:27–39.

————. 1979c. A theory of diversity equilibrium and morphological evolution. *Science* 204:335–336.

————. 1980a. Particle feeding by deposit-feeders: models, data and a prospectus. In *Marine Benthic Dynamics,* B. C. Coull and K. L. Tenore (Eds.), pp. 423–438. Columbia: University of South Carolina Press.

————. 1980b. Genetic divergence in estuaries. In *Estuarine Perspectives,* V. S. Kennedy (Ed.), pp. 509–520. New York: Academic Press.

LEVINTON, J. S., and R. K. BAMBACH. 1970. Some ecological aspects of bivalve mortality patterns. *Amer. J. Sci.* 268:97–112.

————. 1975. A comparative study of Silurian and recent deposit-feeding bivalve communities. *Paleobiology* 1:97–124.

LEVINTON, J. S., and T. BIANCHI. 1981. The role of microbial organisms in the growth of mud snails (Hydrobiidae). *J. Mar. Res.* (in press).

LEVINTON, J. S., and D. L. FUNDILLER. 1975. An ecological and physiological approach to the study of biochemical polymorphisms, pp. 165–178. *Proc. 9th Europ. Symp. Mar. Biol.* Aberdeen, Scotland: Aberdeen University Press.

LEVINTON, J. S., and H. H. LASSEN. 1978. Experimental mortality studies and adaptation at the LAP locus in *Mytilus edulis.* In *Marine Organisms: Genetics, Ecology and Evolution,* B. Battaglia and J. L. Beardmore (Eds.), pp. 229–254. New York: Plenum.

LEVINTON, J. S., and G. R. LOPEZ. 1977. A model of renewable resources and limitation of deposit-feeding benthic populations. *Oecologia (Berlin)* 31:177–190.

LEVINTON, J. S., G. R. LOPEZ, H. H. LASSEN, and U. RAHN. 1977. Feedback and structure in deposit-feeding marine benthic communities. In *Biology of Benthic Organisms,* B. F. Keegan, P. O.Ceidigh, P. J. S. Boaden (Eds.), pp. 409–416. New York: Pergamon Press.

LEVINTON, J. S. and S. STEWART. 1981. Marine succession: The effect of deposit-feeding snails on population growth of *Paranais litoralis. J. Exp. Mar. Biol. Ecol.* (in press).

LEVINTON, J. S., and T. H. SUCHANEK. 1978. Geographic variation, niche breadth, and genetic differentiation at different geographic scales in the mussels, *Mytilus californianus* and *M. edulis. Mar. Biol.* 49:363–375.

LEVITAN, P. J., and A. J. KOHN. 1980. Microhabitat resource use, activity patterns, and episodic catastrophe: *Conus* on tropical intertidal reef rock benches. *Ecol. Monogr.* 50:55–75.

LEWIN, J. C., and R. R. L. GUILLARD. 1963. Diatoms. *Ann. Rev. Microbiol.* 17:373–414.

LEWIN, J. C., T. HRUBY, and D. MACKAS. 1975. Blooms of surf-zone diatoms along the coast of the Olympic peninsula, Washington: V. Environmental conditions associated with blooms (1971 and 1972). *Estuar. Coast. Mar. Sci.* 3:229–242.

LEWIN, J. C., and R. A. LEWIN. 1960. Auxotrophy and heterotrophy in marine littoral diatoms. *Can. J. Microbiol.* 6:127–134.

LEWIS, J. B. 1963. Environmental and tissue temperatures of some tropical intertidal marine animals. *Biol. Bull. Woods Hole* 124:277–284.

LEWIS, J. R. 1964. *The Ecology of Rocky Shores,* 323 pp. London: English University Press.

LI, C. C. 1955. *Population Genetics.* Chicago: University of Chicago Press.

LILLY, S. J., J. F. SLOANE, R. BASSINGDALE, F. J. EBLING, and J. A. KITCHING. 1953. The ecology of Lough Ine with special reference to water currents. IV. The sedentary fauna of sublittoral boulders. *J. Anim. Ecol.* 22:187–122.

LIMBAUGH, C. 1961. Cleaning symbiosis. *Sci. Amer.* 1961 (August):42–49.

LIPPS, J. H. 1970. Plankton evolution. *Evolution* 24:1–22.

LITTLER, M. M., and D. S. LITTLER. 1980. The evolution of thallus form and survival strategies in benthic marine macroalgae: field and laboratory tests of a functional form model. *Am. Nat.* 116:25–44.

LOCKWOOD, A. P. M. 1976. Physiological adaptation to life in estuaries. In *Adaptation to Environment,* R. C. Newell (Ed.), pp. 315–392. London: Butterworths.

LOOSANOFF, V. L., and J. B. ENGLE. 1947. Effect of different concentrations of micro-organisms on the feeding of oysters *(O. virginica). Fish. Bull. U.S. Fish Wildl. Serv.* 51:31–57.

LOOSANOFF, V. L., and C. A. NOMEJKO. 1951. Existence of physiologically-different races of oysters, *Crassostrea virginica. Biol. Bull. Woods Hole* 101:151–156.

LOPEZ, G. R., and J. S. LEVINTON. 1978. The availability of microorganisms attached to sediment particles as food for *Hydrobia ventrosa. Oecologia* 32:263–275.

LOPEZ, G. R., J. S. LEVINTON, and L. B. SLOBODKIN. 1977. The effect of grazing by the detritivore *Orchestia grillus* on *Spartina* litter and its associated microbial community. *Oecologia (Berlin)* 30:111–127.

LORENZEN, C. J. 1971. Continuity in the distribution of surface chlorophyll. *J. Con. Int. Explor. Mer.* 34:18–23.

LOYA, Y. 1976a. Effects of water turbidity and sedimentation on the community structure of Puerto Rican atolls. *Bull. Mar. Sci.* 26:450–466.

———. 1976b. Skeletal regeneration in a Red Sea scleractinian coral population. *Nature* 261:490–491.

LOYA, Y., and L. B. SLOBODKIN. 1971. The coral reefs of Eilat (Gulf of Eilat, Red Sea). *Symp. Zool. Soc. London* 28:117–139.

LUBCHENCO, J. 1978. Plant species diversity in a marine intertidal community: importance of herbivore food preference and algal competitive abilities. *Am. Nat.* 112:23–39.

———. 1980. Algal zonation in the New England rocky intertidal community: an experimental analysis. *Ecology* 61:333–344.

LUBCHENCO, J., and J. CUBIT. 1980. Heteromorphic life histories of certain marine algae as adaptations to variations in herbivory. *Ecology* 61:676–687.

LUBCHENCO, J., and B. A. MENGE. 1978. Community development and persistence in a low rocky intertidal zone. *Ecol. Monogr.* 59:67–94.

LUTZ, R. A., and D. C. RHOADS. 1977. Anaerobiosis and a theory of growth line formation. *Science* 198:1222–1227.

MACARTHUR, R. H. 1965. Patterns of species diversity. *Biol. Rev.* 40:510–533.

———. 1972. *Geographical Ecology.* New York: Harper and Row.

MACARTHUR, R. H., and R. LEVINS. 1964. Competition, habitat selection and character displacement in a patchy environment. *Proc. Natl. Acad. Sci. U.S.* 51:1207–1210.

MacArthur, R. H., and E. R. Pianka. 1966. On the optimal use of a patchy environment. *Am. Nat.* 100:603–609.

MacArthur, R. H., and E. O. Wilson. 1967. *The Theory of Island Biogeography*, 203 pp. Princeton: Princeton University.

MacGinitie, G. E. 1939a. The method of feeding of *Chaetopterus*. *Biol. Bull. Woods Hole* 77:115–118.

———. 1939b. The method of feeding of tunicates. *Biol. Bull. Woods Hole* 77:443–447.

MacGinitie, G. E., and N. MacGinitie. 1949. *Natural History of Marine Animals*. New York: McGraw-Hill.

MacIsaac, J. J., and R. C. Dugdale. 1969. The kinetics of nitrate and ammonia uptake by natural populations of marine phytoplankton. *Deep Sea Res.* 16:45–57.

MacIsaac, J. J., and R. C. Dugdale. 1972. Interactions of light and inorganic nitrogen in controlling nitrogen uptake in the sea. *Deep Sea Res.* 19:209–232.

Mackas, D. L. and C. M. Boyd. 1979. Spectral analysis of zooplankton spatial heterogeneity. *Science* 204:62–64.

Madin, L. P. 1974. Field observations on the feeding behavior of salps (Tunicata:Thaliacea). *Mar. Biol.* 25:143–147.

Magnus, D. 1964. Gezeitenströmung und Nahrungsfiltration bei Ophiuren und Crinoiden. *Helgol. Wiss. Meeresunters.* 10:104–117.

Malone, T. C. 1977. Environmental regulation of phytoplankton productivity in the Lower Hudson Estuary. *East. Coast. Mar. Sci.* 5:157–171.

Malone, T. C., and M. Chervin. 1979. The production and fate of phytoplankton size fractions in the plume of the Hudson River, New York Bight. *Limnol. Oceanogr.* 24:683–696.

Mangum, C. P. 1964. Studies on speciation in maldanid polychaetes of the North American Atlantic coast. II. Distribution and competitive interaction of five sympatric species. *Limnol. Oceanogr.* 9:12–26.

Mann, K. H. 1972. Ecological energetics of the seaweed zone in a marine bay on the Atlantic coast of Canada. *Mar. Biol.* 14:199–209.

———. 1975. Relationship between morphometry and biological functioning in three coastal inlets of Nova Scotia. In: *Estuarine Research*, L. E. Cronin, (Ed.), vol. 1, pp. 634–644. New York: Academic Press.

———. 1976. Decomposition of marine macrophytes. In *The Role of Terrestrial and Aquatic Organisms in Decomposition Processes*, J. M. Anderson and A. MacFadyen (Eds.), pp. 247–267. Oxford, U.K.: Blackwell Scientific Publications.

Manton, S. M., and T. A. Stephenson. 1935. Ecological surveys of coral reefs. *Sci. Rept. Great Barrier Reef Expedition* 3:273–312.

Margalef, R. 1958. Temporal succession and spatial heterogeneity in phytoplankton. In *Perspectives in Marine Biology*, A. A. Buzzati–Traverso (Ed.), pp. 323–349. Berkeley: University of California Press.

———. 1962. Succession in marine populations. *Advg. Front. Pl. Sci.* 2:137–188.

Margolin, A. S. 1976. Swimming of the sea cucumber *Parastichopus californicus* (Stimpson) in response to sea stars. *Ophelia* 15:105–114.

Marshall, N. 1970. Food transfer through the lower trophic levels of the benthic environment. In *Marine Food Chains*, J. H. Steele (Ed.), pp. 52–66. Berkeley: University of California Press.

MARSHALL, S. M., and A. P. ORR. 1955. *The Biology of a Marine Copepod Calanus finmarchicus* (Gunnerus). Edinburgh: Oliver and Boyd.

————. 1958. On the biology of *Calanus finmarchicus*. X. Seasonal changes in oxygen consumption. *J. Mar. Biol. Assoc. U.K.* 37:459–472.

MARTIN, R. 1966. On the swimming behaviour and biology of *Notarchus punctatus* Phillip (Gastropoda, Opisthobranchia). *Pubbl. Staz. Zool. Napoli* 35:61.

MATHERS, N. F. 1975. Environmental variability at the phosphoglucose isomerase locus in the genus *Chlamys. Biochem. System. Ecol.* 3:123–127.

MAUZEY, K. P., C. BIRKELAND, and P. K. DAYTON. 1968. Feeding behavior of asteroids and escape responses of their prey in the Puget Sound region. *Ecology* 49:603–619.

MAXWELL, W. G. H. 1968. *Atlas of the Great Barrier Reef*, 258 pp. Amsterdam: Elsevier.

MAY, R. M. 1976. Models for single populations. In *Theoretical Ecology: Principles and Applications*, R. M. May (Ed.), pp. 4–35. Philadelphia: W. B. Saunders.

MAYNARD, N. G. 1974. The distribution of diatoms in the surface sediments of the Atlantic Ocean and their relationship to the biological and physical oceanography of the overlying waters. Ph.D. Dissertation. University of Miami.

MAYR, E. 1963. *Animal Species and Evolution*, 797 pp. Cambridge, Mass.: Harvard University Press.

McALICE, B. J. 1970. Observations on the small scale distribution of estuarine phytoplankton. *Mar. Biol.* 7:100–111.

McALLISTER, C. D., T. R. PARSONS, K. STEPHENS, and J. D. STRICKLAND. 1961. Measurements of primary production in coastal sea water using a large-volume plastic sphere. *Limnol. Oceanogr.* 6:237–258.

McALLISTER, C. D., N. SHAH, and J. D. H. STRICKLAND. 1964. Marine phytoplankton photosynthesis as a function of light intensity: a comparison of methods. *J. Fish. Res. Bd. Canada* 21:159–181.

McALLISTER, D. E. 1961. A collection of oceanic fishes from off British Columbia with a discussion of the evolution of black peritoneum. *Bull. Natl. Mus. Canada* 172:39–43.

McCALL, P. L. 1977. Community patterns and adaptive strategies of the infaunal benthos of Long Island Sound. *J. Mar. Res.* 35:221–266.

McCAMMON, H. M. 1969. The food of articulate brachiopods. *J. Paleontol.* 43:976–985.

McCLOSKEY, L. R. 1970. The dynamics of a community associated with a marine scleractinian coral. *Int. Rev. Ges. Hydrobiol.* 55:13–81.

McHUGH, J. L. 1967. Estuarine nekton. In *Estuaries*, G. H. Lauff (Ed.), pp. 581–620. Washington, D.C.: American Association for the Advancement of Science.

————. 1974. The role and history of the International Whaling Commission. In *The Whale Problem, a Status Report*, W. E. Schevill (Ed.), pp. 305–335. Cambridge, Mass.: Harvard University Press.

McINTYRE, A. D. 1969. Ecology of marine meiobenthos. *Biol. Rev.* 44:245–290.

————. 1971. Efficiency of benthos sampling gear. In *Methods for the Study of Marine Benthos*, N. A. Holme and A. D. McIntyre (Eds.), 334 pp. Oxford, U.K.: Blackwell Scientific Publications.

McINTYRE, A. D., and A. ELEFTHERIOU. 1968. The fauna of a flatfish nursery ground. *J. Mar. Biol. Assoc. U.K.* 48:113–142.

McIntyre, A. D., A. L. S. Munro, and J. H. Steele. 1970. Energy flow in a sand ecosystem. In *Marine Food Chains,* J. H. Steele (Ed.), pp. 19–31. Berkeley: University of California Press.

McLaren, I. A. 1963. Effects of temperature on growth of zooplankton and the adaptive value of vertical migration. *J. Fish. Res. Bd. Canada* 20:685–727.

————. 1974. Demographic strategy of vertical migration by a marine copepod. *Am. Nat.* 108:91–102.

McRoy, C. P., and C. McMillan. 1977. Production ecology and physiology of seagrasses. In *Seagrass Ecosystems,* C. P. McRoy and C. Helfferich (Eds.), pp. 53–87. New York: Marcel Dekker.

Mead, A. D. 1900. On the correlation between growth and food supply in starfish. *Am. Nat.* 34:17–23.

Meadows, P. S., and A. Reid. 1966. The behaviour of *Corophium volutator* (Crustacea:Amphipoda). *J. Zool. London* 150:387–399.

Meeuse, B. J. D. 1956. Free sulfuric acid in the brown alga *Desmarestia. Biochim. Biophys. Acta* 19:372–374.

Menge, B. A. 1972. Foraging strategy of a starfish in relation to actual prey availability and environmental predictability. *Ecol. Monogr.* 42:25–50.

————. 1974. Effect of wave action and competition on brooding and reproductive effort in the sea-star *Leptasterias hexactis. Ecology* 55:84–93.

Menge, B. A., and J. P. Sutherland. 1976. Species diversity gradients: synthesis of the roles of predation, competition, and temporal heterogeneity. *Am. Nat.* 110:351–369.

Menge, J. L., and B. A. Menge. 1974. Role of resource allocation, aggression and spatial heterogeneity in coexistence of two competing starfish. *Ecol. Monogr.* 44:189–209.

Menzel, D. W. 1974. Primary productivity, dissolved and particulate organic matter, and sites of oxidation of organic matter. In *The Sea,* E. D. Goldberg (Ed.), vol. 5, pp. 659–678. New York: Wiley.

Menzel, D. W., E. M. Hulburt, and J. H. Ryther. 1963. The effects of enriching Sargasso Sea water on the production and species composition of the phytoplankton. *Deep Sea Res.* 10:209–219.

Menzel, D. W., and J. H. Ryther. 1960. The annual cycle of primary production in the sea off Bermuda. *Deep Sea Res.* 6:351–357.

————. 1961a. Nutrients limiting the production of phytoplankton in the Sargasso Sea, with special reference to iron. *Deep Sea Res.* 7:276–281.

————. 1961b. Zooplankton in the Sargasso Sea off Bermuda and its relation to organic production. *J. Con. Int. Explor. Mer.* 26:250–258.

Menzies, R. J., and J. Imbrie. 1958. On the antiquity of the deep sea bottom fauna. *Oikos* 9:192–210.

Meyer, D. L. 1973. Feeding behavior and ecology of shallow-water unstalked crinoids (Echinodermata) in the Caribbean Sea. *Mar. Biol.* 22:105–129.

Meyers, P. A. 1979. Polyunsaturated fatty acids in coral: indicators of nutritional sources. *Mar. Biol. Lett.* 1:69–75.

Mihursky, J. A., and V. S. Kennedy. 1967. Water temperature criteria to protect aquatic life. *Am. Fish. Soc. Spec. Publ.* 4:20–32.

MILLER, J. M., and M. L. DUNN. 1980. Feeding strategies and patterns of movement in juvenile estuarine fishes. In *Estuarine Perspectives,* V. S. Kennedy (Ed.), pp. 449–464. New York: Academic Press.

MILLER, R. L. 1979. Sperm chemotaxis in the Hydromedusae. I. Species-specificity and sperm behavior. *Mar. Biol.* 53:99–114.

MILLER, R. S. 1967. Pattern and process in competition. *Adv. Ecol. Res.* 4:1–74.

MITTON, J. B. 1977. Shell color and pattern variation in *Mytilus edulis* and its adaptive significance. *Chesapeake Sci.* 18:387–389.

MOLL, R. A. 1977. Phytoplankton in a temperate-zone salt marsh: net production and exchanges with coastal waters. *Mar. Biol.* 42:109–118.

MØLLER, D. 1969. The relationship between Arctic and coastal cod in their immature stages illustrated by frequencies of genetic characters. *Fisk Dir. Ser. Havunders.* 15:220–233.

MOORE, H. B. 1931. The muds of the Clyde Sea area. III. Chemical and physical conditions; rate and nature of sedimentation; and fauna. *J. Mar. Biol. Assoc. U.K.* 17:325–358.

MORGAN, P. R. 1972. *Nucella lapillus* (L.) as a predator of edible cockles. *J. Exp. Mar. Biol. Ecol.* 8:45–52.

MORRIS, A. W., and P. FOSTER. 1971. The seasonal variation of dissolved organic carbon in the inshore waters of the Menai Strait in relation to primary production. *Limnol. Oceanogr.* 16:987–989.

MORRISON, S. J., J. D. KING, R. J. BOBBIE, R. E. BECHTOLD, and D. C. WHITE. 1977. Evidence for microfloral succession on allochthonous plant litter in Apalachicola Bay, Florida, USA. *Mar. Biol.* 41:229–240.

MULLIN, M. M. 1963. Some factors affecting the feeding of marine copepods of the genus *Calanus. Limnol. Oceanogr.* 8:239–250.

MULLIN, M. M. and E. R. BROOKS. 1970. Growth and metabolism of two planktonic, marine copepods as influenced by temperature and type of food. In *Marine Food Chains,* J. H. Steele, (Ed.), pp. 74–95. Berkeley, Ca.: Univ. of California Press.

MULLIN, M. M., and E. R. BROOKS. 1976. Some consequences of distributional heterogeneity of phytoplankton and zooplankton. *Limnol. Oceanogr.* 21:784–796.

MULLIN, M. M., E. F. STEWART, and F. J. FUGLISTER. 1975. Ingestion by planktonic grazers as a function of concentration of food. *Limnol. Oceanogr.* 20:259–262.

MUNK, W. 1955. The circulation of the oceans. *Sci. Am.* 1955:96–102.

MURPHY, G. I. 1968. Pattern in life history and the environment. *Am. Nat.* 102:391–403.

MUSCATINE, L. 1967. Glycerol excretion by symbiotic algae from corals and *Tridacna* and its control by the host. *Science* 156:519.

———. 1973. Nutrition of corals. In *The Biology and Geology of Coral Reefs,* vol. 2, O. A. Jones and R. Endean (Eds.), pp. 77–115. New York: Academic Press.

MUSCATINE, L., and C. HAND. 1958. Direct evidence for transfer of materials from symbiotic algae to the tissues of a coelenterate. *Proc. Nat. Acad. Sci. U.S.* 44:1259–1263.

MUSCATINE, L., and H. M. LENHOFF. 1963. Symbiosis: on the role of algae symbiotic with *Hydra. Science* 142:956–958.

MUSCATINE, L., R. R. POOL, and E. CERNICHIARI. 1972. Some factors influencing selective release of soluble organic material by zooxanthellae from reef corals. *Mar. Biol.* 13:298–308.

NAIR, N. B., and A. D. ANSELL. 1968a. Characteristics of penetration of the substratum by some marine bivalve molluscs. *Proc. Mal. Soc. London* 38:179–197.

————. 1968b. The mechanism of boring in *Zirphaea crispata* (L.) (Bivalvia:Pholadidae). *Proc. Roy. Soc. London* 170B:155–173.

NEUMANN, A. C. 1966. Observations on coastal erosion in Bermuda and measurements of the boring rate of the sponge, *Cliona lampa*. *Limnol. Oceanogr.* 11:92–108.

NEUMANN, G. and W. J. PIERSON. 1966. *Principles of Physical Oceanography.* Englewood Cliffs, N.J.: Prentice-Hall, Inc.

NEWELL, G. E. 1958a. The behaviour of *Littorina littorea* (L.) under natural conditions and its relation to position on the shore. *J. Mar. Biol. Assoc. U.K.* 37:229–239.

————. 1958b. An experimental analysis of the behaviour of *Littorina littorea* (L.) under natural conditions and in the laboratory. *J. Mar. Biol. Assoc. U.K.* 37:241–266.

NEWELL, N. D. 1971. An outline history of tropical organic reefs. *Novitates* 2465:1–37.

NEWELL, R. C. 1962. Behavioural aspects of the ecology of *Peringia (= Hydrobia) ulvae* (Pennant) (Gasteropoda, Prosobranchia). *Proc. Zool. Soc. London* 140:49–75.

————. 1964. Some factors controlling the upstream distribution of *Hydrobia ulvae* (Pennant)(Gastropoda, Prosobranchia). *Proc. Zool. Soc. London* 142:85–106.

NEWELL, R. C. 1965. The role of detritus in the nutrition of two marine deposit feeders, the prosobranch *Hydrobia ulvae* and the bivalve *Macoma balthica*. *Proc. Zool. Soc. London* 143:25–45.

————. 1970. *The Biology of Intertidal Animals,* 555 pp. New York: American Elsevier.

NEWELL, R. C., and L. H. KOFOED. 1977. Adjustment of the components of energy balance in the gastropod *Crepidula fornicata* in response to thermal acclimation. *Mar. Biol.* 44:275–286.

NEWELL, R. C., V. I. PYE, and M. AHSANULLAH. 1971. Factors affecting the feeding rate of the winkle *Littorina littorea*. *Mar. Biol.* 9:138–144.

NICHOLS, F. H. 1970. Benthic polychaete assemblages and their relationship to the sediment in Port Madison, Washington. *Mar. Biol.* 6:48–57.

NICOL, E. A. 1930. The feeding mechanism, formation of the tube, and physiology of digestion in *Sabella pavonina*. *Trans. Roy. Soc. Edinburgh* 56:537–596.

NICOLAISEN, W., and E. KANNEWORFF. 1969. On the burrowing and feeding habits of the amphipods *Bathyporeia pilosa* Lindstrom and *Bathyporeia sarsi* Watkin. *Ophelia* 6:231–250.

NIKOLSKY, G. V. 1963. *The Ecology of Fishes,* 352 pp. New York: Academic Press.

NINIVAGGI, D. V. 1979. Particle retention efficiencies in *Temora longicornis* (Muller). M. S. Thesis. Stony Brook: State University of New York.

NIXON, S. W., and C. A. OVIATT. 1973. Ecology of a New England Salt Marsh. *Ecol. Monogr.* 43:363–498.

NIXON, S. W., C. W. OVIATT, and S. S. HALE. 1976. Nitrogen regeneration and the metabolism of coastal marine bottom communities. In *The Role of Terrestrial and Aquatic Organisms in Decomposition Processes,* J. M. Anderson and A. MacFadyen (Eds.), pp. 269–283. Oxford, U.K.: Blackwell Scientific Publications.

NORGÅRD, S. W., A. SVEC, S. LIAAEN-JENSEN, A. JENSEN and R. R. L. GUILLARD. 1974. Chloroplast pigments and algal systematics. *Biochem. Syst. Ecol.* 2:3–6.

NORTH, W. J. 1965. *Kelp Habitat Improvement Project, Ann. Rept. 1964–1965.* Pasadena: California Institute of Technology.

NORTON–GRIFFITHS, M. 1967. Some ecological aspects of the feeding behaviour of the oystercatcher *Haematopus ostralegus* on the edible mussel *Mytilus edulis*. *Ibis* 109:412–424.

OCKELMANN, K. W. 1965. Developmental types in marine bivalves and their distribution along the Atlantic coast of Europe. *Proc. First Europ. Malacol. Congr., 1962,* pp. 25–35.

ODUM, E. P. 1969. The strategy of ecosystem development. *Science* 164:262–270.

———. 1980. The status of three ecosystem-level hypotheses regarding salt marsh estuaries: tidal subsidy, outwelling, and detritus-based food chains. In *Estuarine Perspectives,* V. S. Kennedy, (Ed.), pp. 485–495. New York: Academic Press.

ODUM, E. P., and A. A. DE LA CRUZ. 1967. Particulate organic detritus in a Georgia salt marsh-estuarine ecosystem. In *Estuaries,* G. H. Lauff (Ed.), pp. 383–388. Washington, D.C.: American Association for the Advancement of Science.

ODUM, E. P., and A. E. SMALLEY. 1959. Comparison of population energy flow of a herbivorous and deposit-feeding invertebrate in a salt marsh ecosystem. *Proc. Natl. Acad. Sci. U.S.A.* 45:617–622.

ODUM, H. T. 1957. Primary production in eleven Florida springs and a marine turtle-grass community. *Limnol. Oceanogr.* 2:85–97.

ODUM, H. T., and E. P. ODUM. 1955. Trophic structure and productivity of a windward coral reef community on Eniwetok Atoll. *Ecol. Monogr.* 25:295–320.

ODUM, W. E., J. S. FISHER, and J. C. PICKRAL. 1979. Factors controlling the flux of particulate matter from estuarine wetlands. In *Ecological Processes in Coastal and Marine Systems,* R. J. Livingston (Ed.), pp. 69–80. New York: Plenum.

ODUM, W. E., and E. J. HEALD. 1975. The detritus-based food web of an estuarine mangrove community. In *Estuarine Research,* L. E. Cronin (Ed.), vol. 1, pp. 265–286. New York: Academic Press.

OGDEN, J. C., R. A. BROWN, and N. SALESKY. 1973. Grazing by the echinoid *Diadema antillarum* Philippi: formation of halos around West Indian Patch reefs. *Science* 182:715–717.

O'GOWER, A. K., and P. I. NICOL. 1968. A latitudinal cline of hemoglobins in a bivalve mollusc. *Heredity* 23:485–491.

OHNO, S., L. CHRISTIAN, M. ROMERO, R. DOFUKU, and C. IVEY. 1973. On the question of the American eel, *Anguilla rostrata,* versus European eel, *Anguilla anguilla. Experientia* 29:891.

OKUBO, A. 1971. Oceanic diffusion diagrams. *Deep Sea Res.* 18:789–802.

———. 1978. Horizontal dispersion and critical scales for phytoplankton patchiness. In *Spatial Pattern in Plankton Communities,* J. H. Steele (Ed.), pp. 21–42. New York: Plenum.

OTSUKI, A. and T. HANYA. 1972. Production of dissolved organic matter from dead green algal cells. I. Aerobic microbial decomposition. *Limnol. Oceanogr.* 17:248–257.

PAINE, R. T. 1966. Food web complexity and species diversity. *Am. Nat.* 100:65–75.

———. 1969a. The *Pisaster-Tegula* interaction: prey patches, predator food preference and intertidal community structure. *Ecology* 50:950–961.

———. 1969b. A note on trophic complexity and community stability. *Am. Nat.* 103:91–93.

———. 1971a. Energy flow in a natural population of the herbivorous gastropod *Tegula funebralis. Limnol. Oceanogr.* 16:86–98.

———. 1971b. A short-term experimental investigation of resource partitioning in a New Zealand rocky intertidal habitat. *Ecology* 52:1096–1106.

———. 1974. Intertidal community structure. Experimental studies on the relationship between a dominant competitor and its principal predator. *Oecologia (Berlin)* 15:93–120.

———. 1976. Size-limited predation: an observational and experimental approach with the *Mytilus–Pisaster* interaction. *Ecology* 57:858–873.

PAINE, R. T., C. J. SLOCUM, and D. O. DUGGINS. 1979. Growth and longevity in the crustose red alga *Petrocelis middendorfi. Mar. Biol.* 51:185–192.

PAINE, R. T., and R. L. VADAS. 1969a. Calorific values of benthic marine algae and their postulated relation to invertebrate food preference. *Mar. Biol.* 4:79–86.

———. 1969b. The effects of grazing by sea urchins, *Strongylocentrotus* spp., on benthic algal populations. *Limnol. Oceanogr.* 14:710–719.

PALMER, A. R. 1965. Biomere—a new kind of biostratigraphic unit. *J. Paleontol.* 39:151–153.

PALMER, A. R. 1979. Fish predation and the evolution of gastropod shell sculpture: experimental and geographic evidence. *Evolution* 33:697–713.

PAMATMAT, M. M. 1971. Oxygen consumption by the seabed. IV. Shipboard and laboratory experiments. *Limnol. Oceanogr.* 16: 536–550.

PAMATMAT, M. M., and K. BANSE. 1969. Oxygen consumption by the seabed. II. *In situ* measurements to a depth of 180 m. *Limnol. Oceanogr.* 14:250–259.

PANDIAN, T. J. 1975. Mechanisms of heterotrophy. In *Marine Ecology,* O. Kinne (Ed.), vol. 2, part 1, pp. 61–249. New York: Wiley-Interscience.

PARKER, R. H. 1975. *The Study of Benthic Communities, a Model and a Review,* 279 pp. Amsterdam: Elsevier.

PARSONS, T. R., and J. D. H. STRICKLAND. 1962. On the production of particulate organic carbon by heterotrophic processes in seawater. *Deep Sea Res.* 8:211–222.

PARSONS, T. R., and M. TAKAHASHI. 1973. *Biological Oceanographic Processes,* 186 pp. Oxford, U.K.: Pergamon Press.

PATTERSON, M. J., J. BLAND, and E. W. LINDGREN. 1978. Physiological response of symbiotic polychaetes to host saponins. *J. Exp. Mar. Biol. Ecol.* 33:51–56.

PATTON, J. S., S. ABRAHAM, and A. A. BENSON. 1977. Lipogenesis in the intact coral *Pocillopora capitata* and its isolated zooxanthellae: evidence for a light-driven carbon cycle between symbiont and host. *Mar. Biol.* 44:235–247.

PATTON, W. K. 1974. Community structure among the animals inhabiting the coral *Pocillopora damicornis* at Heron Island, Australia. In *Symbiosis in the Sea,* W. B. Vernberg (Ed.), pp. 219–244. Columbia: University of South Carolina Press.

PEARSE, V. B., and L. MUSCATINE. 1971. Role of symbiotic algae (zooxanthellae) in coral calcification. *Biol. Bull.* 141:350–363.

PEDDICORD, R. K. 1977. Salinity and substratum effects on condition index of the bivalve *Rangia cuneata. Mar. Biol.* 39:351–360.

PETERSEN, C. G. J. 1913. Valuation of the sea. II. The animal communities of the sea bottom and their importance for marine zoogeography. *Rept. Dan. Biol. Stn.* 21:1–44.

———. 1918. The sea bottom and its production of fish food. *Rept. Dan. Biol. Stn.* 25:1–62.

PETERSON, C. H. 1977. Competitive organization of the soft-bottom macrobenthic communities of southern California lagoons. *Mar. Biol.* 43:343–359.

———. 1979. Predation, competitive exclusion, and diversity in the soft-sediment benthic communities of estuaries and lagoons. In *Ecological Processes in Coastal and Marine Ecosystems,* R. L. Livingston (Ed.), pp. 233–264. New York: Plenum.

PETERSON, C. H., and S. V. ANDRE. 1980. An experimental analysis of interspecific competition among marine filter feeders in a soft-sediment environment. *Ecology* 61:129–139.

PETRAITIS, P. S. 1979. Behavior of *Littorina littorea* and its role in maintaining refuges for sessile organisms. Ph.D. Dissertation. Stony Brook: State University of New York.

PIANKA, E. R. 1966. Latitudinal gradients in species diversity: a review of concepts. *Am. Nat.* 100:33–46.

PIANKA, E. R., and W. S. PARKER. 1975. Age-specific reproductive tactics. *Am. Nat.* 109:453–464.

PIELOU, E. C. 1969. *An Introduction to Mathematical Ecology,* pp. 1–286. New York: Wiley-Interscience.

PIMM, S. L., and J. H. LAWTON. 1977. Number of trophic levels in ecological communities. *Nature* 268:329–331.

PLATT, T. 1971. The annual production by phytoplankton in St. Margaret's Bay, Nova Scotia. *J. Cons. Int. Explor. Mer.* 33:324–333.

———. 1972. Local phytoplankton abundance and turbulence. *Deep Sea Res.* 19:183–187.

———. 1975. Analysis of the importance of spatial and temporal heterogeneity in the estimation of annual production by phytoplankton in a small, enriched marine basin. *J. Exp. Mar. Biol. Ecol.* 18:99–110.

PLATT, T., L. M. DICKIE, and R. W. TRITES. 1970. Spatial heterogeneity of phytoplankton in a near-shore environment. *J. Fish. Res. Bd. Canada* 27:1453–1473.

POHLO, R. 1969. Confusion concerning deposit-feeding in the Tellinacea. *Proc. Mal. Soc. London* 38:361–364.

POMEROY, L. R. 1959. Algal productivity in salt marshes of Georgia. *Limnol. Oceanogr.* 4:386–397.

POMEROY, L. R., H. M. MATHEWS, and H. S. MIN. 1963. Excretion of phosphate and soluble organic phosphorous compounds by zooplankton. *Limnol. Oceanogr.* 8:50–55.

PORTER, J. W. 1972. Predation by *Acanthaster* and its effect on coral species diversity. *Am. Nat.* 106:487–492.

———. 1974a. Zooplankton feeding by the Caribbean reef-building coral *Montastrea cavernosa*. *Proc. 2nd Intl. Coral Reef Symp.* 1:111–125.

———. 1974b. Community structure of coral reefs on opposite sides of the Isthmus of Panama. *Science* 186:543–545.

———. 1976. Autotrophy, heterotrophy, and resource partitioning in Caribbean reef-building corals. *Am. Nat.* 110:731–742.

PORTER, K. G. 1973. Selective grazing and differential digestion of algae by zooplankton. *Nature* 244:179–180.

———. 1976. Enhancement of algal growth and productivity by grazing zooplankton. *Science* 192:1332–1334.

PORTER, K. G., and J. W. PORTER. 1979. Bioluminescence in marine plankton: a coevolved antipredation system. *Am. Nat.* 114:458–461.

POTTS, W. T. W. and G. PARRY. 1964. *Osmotic and Ionic Regulation in Animals.* Oxford, U.K.: Pergamon Press.

POULET, S. A. 1972. Grazing of *Pseudocalanus minutus* on naturally occurring particulate matter. *Limnol. Oceanogr.* 18:564–573.

———. 1974. Seasonal grazing of *Pseudocalanus minutus* on particles. *Mar. Biol.* 25:109–123.

PRATT, D. M. 1965. The winter–spring diatom flowering in Narragansett Bay. *Limnol. Oceanogr.* 10:173–184.

PRATT, D. M., and D. A. CAMPBELL. 1956. Environmental factors affecting growth in *Venus mercenaria. Limnol. Oceanogr.* 1:2–17.

PRESTON, J. L. 1978. Communication systems and social interactions in a goby-shrimp symbiosis. *Anim. Behav.* 26:791–802.

PRÉZELIN, B. B., and R. S. ALBERTE. 1978. Photosynthetic characteristics and organization of chlorophyll in marine dinoflagellates. *Proc. Nat. Acad. Sci. U.S.A.* 75:1801–1804.

PRÉZELIN, B. B., and B. M. SWEENEY. 1978. Photoadaptation of photosynthesis in *Gonyaulax polyedra. Mar. Biol.* 48:27–35.

PRITCHARD, D. W. 1967. Observations of circulation in coastal plain estuaries. In *Estuaries,* G. H. Lauff (Ed.), pp. 37–44. Washington, D.C.: American Association for the Advancement of Science.

PROVASOLI, L. 1958. Nutrition and ecology of protozoa and algae. *Ann. Rev. Microbiol.* 12:279–308.

PURCELL, E. M. 1977. Life at low Reynolds number. *Am. J. Phys.* 45:3–11.

PURCHON, R. D. 1955. The structure and function of the British Pholadidae (rock-boring Lamel-libranchia). *Proc. Zool. Soc. London* 124:859–911.

PURDY, E. G. 1974. Reef configurations: cause and effect. In *Reefs in Time and Space,* L. F. Laporte (Ed.), pp. 9–76. Soc. Econ. Pal. Min. Spec. Publ. 18.

RAGOTZKIE, R. A. 1959. Plankton productivity in estuarine waters of Georgia. *Publ. Inst. Mar. Sci. Univ. Texas* 6:146–158.

RAMUS, J., S. I. BEALE, D. MAUZERALI, and K. L. HOWARD. 1976. Changes in photosynthetic pigment concentration in seaweeds as a function of water depth. *Mar. Biol.* 37:223–229.

RANDALL, J. E. 1958. A review of the labrid fish genus *Labroides* with description of two new species and notes on ecology. *Pacific Sci.* 12:327–347.

RANWELL, D. S. 1972. *Ecology of Salt Marshes and Sand Dunes,* 258 pp. London: Chapman and Hall.

RASMUSSEN, E. 1973. Systematics and ecology of the Isefjord marine fauna (Denmark). With a survey of the eelgrass *(Zostera)* vegetation and its communities. *Ophelia* 11:1–495.

———. 1977. The wasting disease of eelgrass *(Zostera marina)* and its effects on environmental factors and fauna. In *Seagrass Ecosystems,* C. P. McRoy and C. Helfferich (Eds.), pp. 1–51. New York: Marcel Dekker.

RAUP, D. M. 1972. Taxonomic diversity during the Phanerozoic. *Science* 177:1065–1071.

RAUP, D. M., S. J. GOULD, T. J. M. SCHOPF, and D. S. SIMBERLOFF. 1973. Stochastic models of phylogeny and the evolution of diversity. *J. Geol.* 81:525–542.

RAY, D. L. 1959. Nutritional physiology of *Limnoria.* In *Marine Boring and Fouling Organisms,* D. L. Ray (Ed.), pp. 46–61. Seattle: University of Washington Press.

RAYMONT, J. E. G. 1963. *Plankton and Productivity in the Oceans,* 660 pp. Oxford, U.K.: Pergamon Press.

RAYMONT, J. E. G. and M. N. E. ADAMS. 1958. Studies on the mass culture of *Phaeodactylum. Limnol. Oceanogr.* 3:119–136.

READ, K. R. H. 1962. The hemoglobin of the bivalve mollusc, *Phacoides pectinatus* Gmelin. *Biol. Bull.* 123:605–617.

REDFIELD, A. C. 1956. The biological control of chemical factors in the environment. *Amer. Scientist* 46:205–221.

REDFIELD, A. C., B. H. KETCHUM, and F. A. RICHARDS. 1963. The influence of organisms on the composition of seawater. In *The Sea,* M. N. Hill (Ed.), vol. 2, pp. 26–77. New York: Wiley.

REDFIELD, A. C., and A. B. KEYS. 1938. The distribution of ammonia in the waters of the Gulf of Maine. *Biol. Bull.* 74:83–92.

REID, R. G. B., and A. REID. 1969. Reeding processes of the genus *Macoma* (Mollusca: Bivalvia). *Can. J. Zool.* 47:649–657.

REID, R. G. B., and A. M. REID. 1974. The carnivorous habit of members of the septibranch genus *Cuspidaria* (Mollusca:Bivalvia). *Sarsia* 56:47–56.

REISH, D. J. 1970. The effects of varying concentrations of nutrients, chlorinity, and dissolved oxygen on polychaetous annelids. *Water Res.* 4:721–735.

REMANE, A., and C. SCHLIEPER. 1971. *Biology of Brackish Water,* pp. 1–372. New York: Wiley-Interscience.

REX, M. A. 1973. Deep-sea species diversity: decreased gastropod diversity at abyssal depths. Science 181:1051–1053.

———. 1981. Community structure in the deep-sea benthos. *Ann. Rev. Ecol. Syst.* (in press).

RHOADS, D. C. 1963. Rates of sediment reworking by *Yoldia limatula* in Buzzards Bay, Massachusetts, and in Long Island Sound. *J. Sed. Petrol.* 33:723–727.

———. 1967. Biogenic reworking of intertidal and subtidal sediments in Barnstable Harbor and Buzzards Bay, Massachusetts. *J. Geol.* 75:461–474.

———. 1970. Mass properties, stability and ecology of marine muds related to burrowing activity. In *Trace Fossils,* T. P. Crimes and J. C. Harper (Eds.), pp. 391–406. Liverpool: Seel House Press.

———. 1973. The influence of deposit-feeding benthos on water turbidity and nutrient recycling. *Amer. J. Sci.* 273: 1–22.

———. 1974. Organism-sediment relations on the muddy sea floor. *Oceanogr. Mar. Biol. Ann. Rev.* 12:263–300.

RHOADS, D. C., and G. PANNELLA. 1970. The use of molluscan shell growth patterns in ecology and paleoecology. *Lethaia* 3:143–161.

RHOADS, D. C., and D. J. STANLEY. 1965. Biogenic graded bedding. *J. Sed. Petrol.* 35:956–963.

RHOADS, D. C., K. L. TENORE, and M. BROWNE. 1975. The role of resuspended bottom mud in nutrient cycles of shallow embayments. In *Estuarine Research,* L. E. Cronin (Ed.), vol. 1, pp. 563–579. New York: Academic Press.

RHOADS, D. C., J. Y. YINGST, and W. J. ULLMAN. 1978. Seafloor stability in central Long Island Sound: Part I. Temporal changes in erodibility of fine-grained sediment. pp. 221–244. In *Estuarine Interactions,* M. L. Wiley (Ed.), New York: Academic Press.

RHOADS, D. C. and D. K. YOUNG. 1970. The influence of deposit-feeding organisms on sediment stability and community trophic structure. *J. Mar. Res.* 28:150–178.

———. 1971. Animal-sediment relationships in Cape Cod Bay. II. Reworking by *Molpadia oolitica* (Holothuroidea). *Mar. Biol.* 11:255–261.

RICHARDS, S. W. 1963. The demersal fish population of Long Island Sound. *Bull. Bingham Oceanogr. Coll.* 18:1–101.

RICHMAN, S., D. R. HEINLE, and R. HUFF. 1977. Grazing by adult estuarine calanoid copepods of the Chesapeake Bay. *Mar. Biol.* 42:69–84.

RICHMAN, S., Y. LOYA, and L. B. SLOBODKIN. 1975. The rate of mucus production by corals and its assimilation by the reef copepod *Acartia negligens. Limnol. Oceanogr.* 20:918–923.

RICHMAN, S., and J. N. ROGERS. 1969. The feeding of *Calanus helgolandicus* on synchronously growing populations of the diatom *Ditylum brightwelli. Limnol. Oceanogr.* 14:701–709.

RICKER, W. E. 1954. Stock and recruitment. *J. Fish. Res. Bd. Can.* 11:559–623.

RICKETTS, E. F., J. CALVIN, and J. W. HEDGPETH. 1968. *Between Pacific Tides,* 614 pp. Stanford, Ca.: Stanford University Press.

RIEDL, R. 1970. Water movement. Animals. In *Marine Ecology,* O. Kinne (Ed.), pp. 1123–1156. Vol. 1, part 1. London: Wiley-Interscience.

RILEY, G. A. 1946. Factors controlling phytoplankton abundance on George's Bank. *J. Mar. Res.* 6:54–73.

———. 1956. Oceanography of Long Island Sound, 1952–1954. IX. Production and utilization of organic matter. *Bull. Bingham Oceanogr. Coll.* 15:324–344.

———. 1970. Particulate organic matter in sea water. *Adv. Mar. Biol.* 8:1–118.

RILEY, G. A., and D. F. BUMPUS. 1946. Phytoplankton-zooplankton relationships on Georgia's Bank. *J. Mar. Res.* 6:33–47.

RILEY, G. A., H. STOMMEL, and D. F. BUMPUS. 1949. Quantitative ecology of the plankton of the Western North Atlantic. *Bull. Bingham, Oceanogr. Coll.* 12:1–169.

ROBERTSON, D. R. 1972. Social control of sex reversal in a coral-reef fish. *Science* 177:1007–1009.

ROBERTSON, R. 1963. Wentletraps (Epitoniidae) feeding on sea anemones and corals. *Proc. Mal. Soc. London* 35:51–63.

———. 1970. Review of the predators and parasites of stony corals, with special reference to symbiotic prosobranch gastropods. *Pacific Sci.* 24:43–54.

RODINO, E., and A. COMPARINI. 1978. Genetic variability in the European eel, *Anguilla anguilla* L. In *Marine Organisms: Genetics, Ecology, Evolution,* B. Battaglia and V. L. Beardmore (Eds.), pp. 389–424. New York: Plenum.

ROMAN, M. R. 1977. Feeding of copepod *Acartia tonsa* on the diatom *Nitzschia closterium* and brown algae *(Fucus vesiculosus). Mar. Biol.* 42:149–155.

ROOT, R. B. 1967. The niche exploitation pattern of the blue-grey gnatcatcher. *Ecol. Monogr.* 37:317–350.

ROPES, J. W. 1968. The feeding habits of the green crab, *Carcinus maenas* (L.). *Fish. Bull. U.S.* 67:183–203.

ROSEN, B. R. 1971. Principal features of reef coral ecology in shallow water environments of Mahé, Seychelles. *Symp. Zool. Soc. London,* 8:163–183.

———. 1975. The distribution of reef corals. *Rep. Underwater Ass.* 1:1–16.

ROTHSCHILD, M. 1936. Gigantism and variation in *Peringia ulvae* Pennant 1777, caused by infection with larval trematodes. *J. Mar. Biol. Assoc. U.K.* 20:537–546.

ROUGHGARDEN, J. 1971. Density dependent natural selection. *Ecology* 52:453–468.

———. 1972. Evolution of niche width. *Am. Nat.* 106:683–718.

RUBENSTEIN, D. I., and M. A. R. KOEHL. 1977. The mechanisms of filter feeding: some theoretical considerations. *Am. Nat.* 111:981–994.

RUSSELL, E. 1966. An investigation of the palatability of some marine invertebrates to four species of fish. *Pacific Sci.* 20:452–460.

RUSSELL HUNTER, W. D. 1970. *Aquatic Productivity,* 306 pp. London: Macmillan.

RUSSELL HUNTER, W. D., M. L. APLEY, and R. D. HUNTER. 1972. Early life-history of *Melampus* and the significance of semilunar synchrony. *Biol. Bull.* 143:623–656.

RYLAND, J. S. 1959. Experiments on the selection of algal substrates by Polyzoa larvae. *J. Exp. Biol.* 36:613–631.

———. 1975. Parameters of the lophophore in relation to population structure in a bryozoan community. *Proc. 9th Europ. Mar. Biol. Symp.* 1975:363–393.

RYTHER, J. H. 1956. Photosynthesis in the ocean as a function of light intensity. *Limnol. Oceanogr.* 1:61–70.

———. 1969. Photosynthesis and fish production in the sea. *Science* 166:72–76.

RYTHER, J. H., and W. M. DUNSTAN. 1971. Nitrogen, phosphorous, and eutrophication in the coastal marine environment. *Science* 171:1008–1013.

RYTHER, J. H., and R. R. L. GUILLARD. 1959. Enrichment experiments as a means of studying nutrients limiting phytoplankton production. *Deep Sea Res.* 6:65–69.

RYTHER, J. H., and C. S. YENTSCH. 1958. Primary production of continental shelf waters off New York. *Limnol. Oceanogr.* 3:327–335.

SALE, P. F. 1974. Mechanisms of co-existence in a guild of territorial fishes at Heron Island. In *Second Intl. Symp. Coral Reefs,* A. Cameron, et al. (Ed.), vol. 1, pp. 195–206. Brisbane, Australia: Great Barrier Reef Committee.

———. 1975. Patterns of use of space in a guild of territorial reef fishes. *Mar. Biol.* 29:89–97.

———. 1977. Maintenance of high diversity in coral reef fish communities. *Am. Nat.* 111:337–359.

SALE, P. F., and R. DYBDAHL. 1975. Determinants of community structure for coral reef fishes in an experimental habitat. *Ecology* 56:1343–1355.

———. 1978. Determinants of community structure for coral reef fishes in isolated coral heads at lagoonal and reef slope sites. *Oecologia* 34:57–74.

SALTON, M. R. 1960. *Microbial Cell Walls,* pp. 1–94. New York: Wiley.

SAMMARCO, P. W. 1977. The effects of grazing by *Diadema antillarum* Philippi on a shallow-water coral reef community. Ph.D. Dissertation. Stony Brook: State University of New York.

———. 1980. *Diadema* and its relationship to coral spat mortality: grazing, competition and biological disturbance. *J. Exp. Mar. Biol. Ecol.* 45:245–272.

SAMMARCO, P. W., J. S. LEVINTON, and J. C. OGDEN. 1974. Grazing and control of coral reef community structure by *Diadema antillarum* Philippi (Echinodermata: Echinoidea): a preliminary study. *J. Mar. Res.* 32:47–53.

SANDERS, H. L. 1956. Oceanography of Long Island Sound, 1952–1954. X. The biology of marine bottom communities. *Bull. Bingham Oceanogr. Coll.* 15:345–414.

———. 1958. Benthic studies in Buzzards Bay. I. Animal-sediment relationships. *Limnol. Oceanogr.* 3:245–258.

———. 1960. Benthic studies in Buzzards Bay. III. The structure of the soft-bottom community. *Limnol. Oceanogr.* 5:138–153.

———. 1968. Marine benthic diversity: a comparative study. *Am. Nat.* 102:243–282.

SANDERS, H. L., E. M. GOUDSMIT, E. L. MILLS, and G. E. HAMPSON. 1962. A study of the intertidal fauna of Barnstable Harbor, Massachusetts. *Limnol. Oceanogr.* 7:63–79.

SANDERS, H. L., and R. R. HESSLER. 1969. Ecology of the deep-sea benthos. *Science* 163:1419–1424.

SANDERS, H. L., R. R. HESSLER, and G. R. HAMPSON. 1965. An introduction to the study of deep-sea benthic faunal assemblages along the Gay Head–Bermuda transect. *Deep Sea Res.* 12:845–867.

SANDERS, H. L., P. C. MANGELSDORF, and G. R. HAMPSON. 1965. Salinity and faunal distribution in the Pocasset River, Massachusetts. *Limnol. Oceanogr.* 10:R216–R229.

SARGENT, M. C. and T. S. AUSTIN. 1949. Organic productivity of an atoll. *Trans. Am. Geophys. Un.* 30:245–249.

SCHAEFER, R. D., and C. E. LANE. 1957. Some preliminary observations bearing on the nutrition of *Limnoria*. *Bull. Mar. Sci.* 7:289–296.

SCHANTZ, E. J. 1971. The dinoflagellate poisons. In *Microbial Toxins*, S. Kadis, A. Ciegler, and S. Ajl (Eds.), vol. 7, pp. 3–25. New York: Academic Press.

SCHELTEMA, R. S. 1961. Metamorphosis of the veliger larvae of *Nassarius obsoletus* (Gastropoda) in response to bottom sediment. *Biol. Bull.* 120:92–109.

———. 1971. Larval dispersal as a means of genetic exchange between geographically separated populations of shallow-water benthic marine gastropods. *Biol. Bull.* 140:284–322.

———. 1972. Reproduction and dispersal of bottom dwelling deep-sea invertebrates: a speculative summary. In *Barobiology and the Experimental Biology of the Deep Sea*, R. W. Brauer (Ed.), pp. 58–66. Chapel Hill: University of North Carolina.

SCHEVILL, W. E. 1974. *The Whale Problem, a Status Report*, 419 pp. Cambridge, Mass.: Harvard University Press.

SCHMIDT, J. 1923. Breeding places and migrations of the eel. *Nature* 111:51–54.

———. 1925. The breeding places of the eel. *Smithsonian Rept.* 1924:279–316.

SCHMIDT–NIELSEN, K. 1975. *Animal Physiology*, pp. 1–699. Cambridge, U.K.: Cambridge University Press.

SCHNEIDER, D. C. 1977. Selective predation and the structure of marine benthic communities. Ph.D. Dissertation. Stony Brook: State University of New York.

———. 1978. Equalisation of prey numbers by migratory shorebirds. *Nature* 271:353–354.

SCHOENBERG, D. A., and R. K. TRENCH. 1976. Specificity of symbioses between marine cnidarians and zooxanthellae. In *Coelenterate Ecology and Behavior*, G. O. Mackie (Ed.), pp. 423–432. New York: Plenum.

SCHOENER, A. 1967. Post larval development of five deep-sea ophiuroids. *Deep Sea Res.* 14:645–660.

———. 1968. Evidence for reproductive periodicity in the deep sea. *Ecology* 49:81–87.

———. 1974. Experimental zoogeography: colonization of marine mini-islands. *Am. Nat.* 108:715–739.

SCHOENER, T. W. 1965. The evolution of bill size differences among sympatric congenetic species of birds. *Evolution* 19:189–213.

———. 1968. The *Anolis* lizards of Bimini: resource partitioning in a complex fauna. *Ecology* 49:704–726.

———. 1974a. Resource partitioning in ecological communities. *Science* 185:27–39.

———. 1974b. Some methods for calculating competition coefficients from resource-utilization spectra. *Am. Nat.* 108:332–340.

SCHOLANDER, P. F., L. VAN DAM, J. W. KANWISHER, H. T. HAMMEL, and M. S. GORDON. 1957. Supercooling and osmoregulation in arctic fish. *J. Cell. Comp. Physiol.* 49:5–24.

SCHOPF, T. J. M. 1964. Survey of genetic differentiation in a coastal zone invertebrate: the ectoproct *Schizoporella errata. Biol. Bull.* 146:78–87.

———. 1974. Permo-Triassic extinctions: relation to sea-floor spreading. *J. Geol.* 82:129–143.

SCHOPF, T. J. M., J. B. FISHER, and C. A. F. SMITH. 1978. Is the marine latitudinal diversity gradient merely another example of the species area curve? In *Marine Organisms: Genetics, Ecology, Evolution,* B. Battaglia and J. Beardmore (Eds.), pp. 365–386. New York: Plenum.

SCHOPF, T. J. M., and J. L. GOOCH. 1971. A natural experiment using deep-sea invertebrates to test the hypothesis that genetic homozygosity is proportional to environmental stability. *Biol. Bull.* 141:401.

SCHRAMM, W. 1968. Okologisch-physiologische Untersuchungen zur Aurstrocknungs- und Temperaturresistenz an *Fucus vesiculosus* L. der westlichen Ostsee. *Int. Rev. Ges. Hydrobiol.* 53:469–510.

SCHUBEL, J. R. 1972. Distribution and transportation of suspended sediment in Upper Chesapeake Bay. In *Environmental Framework of Coastal Plain Estuaries,* B. W. Nelson (Ed.), pp. 151–168. Boulder, Col.: Geol. Soc. America, Mem. 133.

———. 1974. Effects of Tropical Storm Agnes on the suspended solids of the Northern Chesapeake Bay. In *Suspended Solids in Water,* R. J. Gibbs (Ed.), pp. 113–132. New York: Plenum.

SCHUBEL, J. R., and D. J. HIRSCHBERG. 1978. Estuarine graveyards, climatic change, and the importance of the estuarine environment. In *Estuarine Interactions,* M. L. Wiley (Ed.), pp. 285–303. New York: Academic Press.

SCHWENKE, H. 1970. *Water movement. Plants.* In *Marine Ecology,* O. Kinne (Ed.), vol. 1, part 7, pp. 1091–1121. London: Wiley-Interscience.

SCOTT, B. D., and H. R. JITTS. 1977. Photosynthesis of phytoplankton and zooxanthellae on a coral reef. *Mar. Biol.* 41:307–315.

SEBENS, K. P. 1977. Habitat suitability, reproductive ecology and the plasticity of body size in two sea anemone populations *(Anthopleura elegantissima* and *A. xanthogrammica).* Ph.D. Dissertation. Seattle: University of Washington.

———. 1979. The energetics of asexual reproduction and colony formation in benthic marine invertebrates. *Am. Zool.* 19(3):683–697.

SEED, R. 1969. The ecology of *Mytilus edulis* L. (Lamellibranchiata) on exposed rocky shores. II. Growth and mortality. *Oecologia* (Berlin) 3:317–350.

———. 1971. A physiological and biochemical approach to the taxonomy of *Mytilus edulis* L. and *M. galloprovincialis* LmK. Cahiers Biol. Mar. 12:291–322.

———. 1976. Ecology. In *Marine Mussels: Their Ecology and Physiology,* B. L. Bayne (Ed.), pp. 13–65. Cambridge, U.K.: Cambridge University Press.

———. 1978. Systematics and evolution of *Mytilus galloprovincialis* LmK. In *Marine Organisms: Genetics, Ecology and Evolution,* B. Battaglia and J. L. Beardmore (Eds.), pp. 447–468. New York: Plenum.

SEGESTRÅLE, S. G. 1962. Investigations on Baltic populations of the bivalve *Macoma balthica* (L.). Part II. What are the reasons for the periodic failure of recruitment and the scarcity of *Macoma* in the deeper waters of the inner Baltic? *Soc. Sci. Fenn., Comm. Biol.* 24:1–26.

SEIGLER, D. and P. W. PRICE. 1976. Secondary compounds in plants: primary functions. *Am. Nat.* 110:101–105.

SELF, R. F., and P. A. JUMARS. 1978. New resource axes for deposit feeders? *J. Mar. Res.* 36:627–641.

SELLMER, G. P. 1967. Functional morphology and ecological life history of the gem clam, *Gemma gemma* (Eulamellibranchiata, Veneridae). *Malacologia* 5:137–223.

SEPKOSKI, J. J. 1978. A kinetic model of Phanerozoic taxonomic diversity. I. Analysis of marine orders. *Paleobiology* 4:223–251.

SEPKOSKI, J. J., and M. A. REX. 1974. Distribution of freshwater mussels: coastal rivers as biogeographic islands. *Syst. Zool.* 23:165–188.

SHELBOURNE, J. E. 1957. The feeding and condition of plaice larvae in good and bad plankton catches. *J. Mar. Biol. Assoc. U.K.* 36:539–552.

SHEPARD, F. P. 1963. *Submarine Geology*, 2nd Ed. New York: Harper and Row.

SHICK, J. M., R. J. HOFFMAN, and A. N. LAMB. 1979. Asexual reproduction, population structure, and genotype-environment interactions in sea anemones. *Am. Zool.* 19:699–713.

SHINN, E. A. 1966. Coral growth rate, an environmental indicator. *J. Paleontol.* 40:233–240.

———. 1976. Coral reef recovery in Florida and the Persian Gulf. *Environ. Geol.* 1:241–254.

SICK, K. 1961. Haemoglobin polymorphism in fishes. *Nature* 192:894–896.

———. 1965. Haemoglobin polymorphism of cod in the North Sea and the North Atlantic Ocean. *Hereditas* 54:49–69.

SIEBURTH, J. M. 1960. Acrylic acid, an "antibiotic" principle in *Phaeocystis* in Antarctic waters. *Science* 132:676–677.

———. 1961. Antibiotic properties of acrylic acid, a factor in the gastrointestinal antibiosis of polar marine animals. *J. Bacteriol.* 82:72–79.

SILVER, M. W., A. L. SHANKS, and J. D. TRENT. 1978. Marine snow: microplankton habitat and source of small-scale patchiness in pelagic populations. *Science* 201:371–373.

SIMBERLOFF, D. S. and E. O. WILSON. 1969. Experimental zoogeography of islands. The colonization of empty islands. *Ecology* 50:278–296.

SLOBODKIN, L. B. 1954. Population dynamics in *Daphnia obtusa* Kurz. *Ecol. Monogr.* 24:69–88.

———. 1961. *Growth and Regulation of Animal Populations,* 184 pp. New York: Holt, Rinehart and Winston.

———. 1964. Ecological populations of hydrida. *J. Anim. Ecol.* 33(Suppl.):131–148.

———. 1970. Summary. In *Marine Food Chains,* J. H. Steele (Ed.), pp. 537–540. Berkeley: University of California Press.

SLOBODKIN, L. B., and L. FISHELSON. 1974. The effect of the cleaner-fish *Labroides dimidatus* on the point diversity of fishes on the reef front at Eilat. *Am. Nat.* 108:369–376.

SMAYDA, T. J. 1970. The suspension and sinking of phytoplankton in the sea. *Oceanogr. Mar. Biol. Ann. Rev.* 8:353–414.

———. 1973. The growth of *Skeletonema costatum* during a winter–spring bloom in Narragansett Bay. *Norw. J. Bot.* 20:219–247.

SMITH, D. L., L. MUSCATINE, and D. LEWIS. 1969. Carbohydrate movement from autotrophs to heterotrophs in parasitic and mutualistic symbioses. *Biol. Rev.* 44:17–90.

SMITH, F. E. 1969. Effects of enrichments in mathematical models. In *Eutrophication: Causes, Consequences, and Correctives,* pp. 631–645. Washington, D. C.: National Academy of Sciences.

SMITH, O. L., H. H. SHUGART, R. V. O'NEILL, R. S. BOOTH, and D. C. MCNAUGHT. 1971. Resource competition and an analytical model of zooplankton feeding on phytoplankton. *Am. Nat.* 109:571–591.

SNEATH, P. H. A., and R. SOKAL. 1974. *Numerical Taxonomy: The Principles and Practice of Numerical Classification.* San Francisco: Freeman.

SNYDER, T. P., and J. L. GOOCH. 1973. Genetic differentiation in *Littorina saxatilis* (Gastropoda). *Mar. Biol.* 22:177–182.

SOKOLOVA, M. N. 1970. Weight characteristics of meiobenthos in different regions of the deep sea trophic areas of the Pacific Ocean. *Okeanologia* (in Russian) 10:348–356.

———. 1972. Trophic structure of deep-sea macrobenthos. *Mar. Biol.* 16:1–12.

SOMERO, G. N., and A. L. DE VRIES. 1967. Temperature tolerance of some Antarctic fishes. *Science* 156:257–258.

SOUTHWARD, A. J. 1964. The relationship between temperature and rhythmic cirral activity in some Cirripedia considered in connection with their geographical distribution. *Helgol. wiss. Meeresuntersuch.* 10:391–403.

SPIGHT, T. M. 1973. Ontogeny, environment, and shape of a marine snail *Thais lamellosa* Gmelin. *J. Exp. Mar. Biol. Ecol.* 13:215–228.

———. 1974. Sizes of populations of a marine snail. *Ecology* 55:712–729.

———. 1977. Diversity of shallow-water gastropod communities on temperate and tropical beaches. *Am. Nat.* 111:1077–1097.

SPILLER, J. 1977. Evolution of Turritellid gastropods from the Miocene and Pliocene of the Atlantic coastal plain. Ph.D. Dissertation. Stony Brook: State University of New York.

SQUIRES, D. F. 1962. Corals at the mouth of the Rewa River, Viti Levu, Fiji, *Nature* 195:361–362.

STAIGER, H. 1957. Genetical and morphological variation in *Purpura lapillus* with respect to local and regional differentiation of population groups. *Ann. Biol.* 33:251–258.

STALLARD, M. O., and D. J. FAULKNER. 1974. Chemical constituents of the digestive gland of the sea hare *Aplysia californica.* I. Importance of diet. *Comp. Biochem. Physiol.* 49B:25–36.

STANLEY, S. M. 1969. Bivalve mollusk burrowing aided by discordant shell ornamentation. *Science* 166:634–635.

———. 1970. *Relation of Shell Form to Life Habits in the Bivalvia.* Geol. Soc. Amer. Mem. 125, pp. 1–296.

———. 1975. A theory of evolution above the species level. *Proc. Natl. Acad. Sci. U.S.A.* 72:646–650.

STASEK, C. R. 1961. The ciliation and function of the labial palps of *Acila castrensis* (Protobranchia, Nuculidae), with an evaluation of the role of the protobranch organs of feeding in the evolution of the Bivalvia. *Proc. Zool. Soc. London* 137:511–538.

STAUBER, L. A. 1950. The problem of physiological species with special reference to oysters and oyster drills. *Ecology* 31:107–118.

STAVN, R. H. 1971. The horizontal-vertical distribution hypothesis: Langmuir circulation and *Daphnia* distributions. *Limnol. Oceanogr.* 16:453–466.

STEELE, J. H. 1974. *The Structure of Marine Ecosystems,* pp. 1–128. Cambridge, Mass.: Harvard University Press.

———. 1976. Patchiness. In *Ecology of the Seas,* D. H. Cushing and J. J. Walsh (Eds.), pp. 98–115. Oxford, U.K.: Blackwell Scientific Publications.

STEELE, J. H., and I. E. BAIRD. 1968. Production ecology of a sandy beach. *Limnol. Oceanogr.* 13:14–25.

———. 1972. Sedimentation of organic matter in a Scottish sea loch. *Mem. Ist. Ital. Idrobiol.* 29 (Suppl.):74–88.

STEELE, V. J., and D. H. STEELE. 1972. The biology of *Gammarus* (Crustacea, Amphipoda) in the northwestern Atlantic. V. *Gammarus oceanicus* Segerstråle. *Can. J. Zool.* 50:801–813.

STEEMAN NIELSEN, E. 1952. The use of radioactive carbon (C_{14}) for measuring organic production in the sea. *J. Con. Int. Explor. Mer.* 18:117–140.

———. 1958. The balance between phytoplankton and zooplankton in the sea. *J. Cons. Int. Explor. Mer.* 23:178–198.

———. 1975. *Marine Photosynthesis,* 141 pp. Amsterdam: Elsevier.

STEHLI, F. G., R. DOUGLAS, and I. KAFESCEGLIOU. 1972. Models for the evolution of planktonic foraminifera. In *Models in Paleobiology,* T. J. M. Schopf (Ed.), pp. 116–128. San Francisco: Freeman, Cooper and Co.

STEHLI, F. G., A. L. MCALESTER, and C. E. HELSLEY. 1967. Taxonomic diversity of recent bivalves and some implications for geology. *Geol. Soc. Amer. Bull.* 78:455–466.

STEHLI, F. G., and J. W. WELLS. 1971. Diversity and age patterns in hermatypic corals. *Syst. Zool.* 20:115–126.

STEIDINGER, K. A. 1973. Phytoplankton ecology: a conceptual review based on eastern Gulf of Mexico research. *CRC Critical Reviews in Microbiology* 3:49–68.

STEIDINGER, K. A., and R. M. INGLE. 1972. Observations of the 1971 summer red tide in Tampa Bay, Florida. *Environ. Lett.* 3:271–277.

STEPHENS, G. C. 1964. Uptake of organic material by aquatic invertebrates. III. Uptake of glycine by brackish-water annelids. *Biol. Bull. Woods Hole* 126:150–162.

———. 1975. Uptake of naturally occurring primary amines by marine annelids. *Biol. Bull. Woods Hole* 149:397–407.

STEPHENS, G. C., and R. A. SCHINSKE. 1957. Uptake of amino acids from sea water by ciliary-mucoid filter feeding animals. *Biol. Bull.* 113:356–357.

———. 1961. Uptake of amino acids by marine invertebrates. *Limnol. Oceanogr.* 6:175–181.

STEPHENSON, W., R. ENDEAN, and I. BENNETT. 1958. An ecological survey of the marine fauna of Low Isles, Queensland. *Aust. J. Mar. Freshw. Res.* 9:261–318.

STEPHENSON, W., and R. B. SEARLES. 1960. Experimental studies on the ecology of intertidal environments at Heron Island. I. Exclusion of fish from beach rock. *Aust. J. Mar. Freshw. Res.* 11:241–267.

STIMSON, J. 1973. The role of the territory in the ecology of the intertidal limpet *Lottia gigantea* (Gray). *Ecology* 54:1020–1030.

STODDART, D. R. 1963. Effects of hurricane Hattie on the British Honduras reefs and cays, October 30–31, 1961. Atoll. *Res. Bull.* 95:1–142.

———. 1969. Ecology and morphology of recent coral reefs. *Biol. Rev.* 44:433–498.

———. 1973. Coral reefs: the last two million years. *Geography* 58:313–323.

STOECKER, D. 1978. Resistance of a tunicate to fouling. *Biol. Bull.* 155:615–626.

STOECKER, D., R. R. L. GUILLARD, and R. M. KAVEE. 1981. Selective predation by *Favella ehrenbergii* (Tintinnia) on and among Dinoflagellates. *Biol. Bull.* In Press.

STAARUP, B. J. 1970. On the ecology of turbellarians in a sheltered brackish shallow-water bay. *Ophelia* 7:185–216.

STRATHMAN, R. 1974. The spread of sibling larvae of sedentary marine invertebrates. *Am. Nat.* 108:29–44.

STRICKLAND, J. D. H. 1965. Production of organic matter in the primary stages of the marine food chain. In *Chemical Oceanography*, J. P. Riley and G. Skirrow (Eds.), vol. 1, pp. 477–610. London: Academic Press.

STRUHSAKER, J. W. 1968. Selection mechanisms associated with intraspecific variation in *Littorina picta* (Prosobranchia: Mesogastropoda). *Evolution* 22:459–480.

SUCHANEK, T. H. 1978. The ecology of *Mytilus edulis* L. in exposed rocky intertidal communities. *J. Exp. Mar. Biol. Ecol.* 31:105–120.

SUCHANEK, T. H., and J. S. LEVINTON. 1974. Articulate brachiopod food. *J. Paleontol.* 48:1–5.

SUTHERLAND, J. P. 1970. Dynamics of high and low populations of the limpet *Acmaea scabra* (Gould). *Ecol. Monogr.* 40:169–188.

―――. 1974. Multiple stable points in natural communities. *Am. Nat.* 108:859–873.

SUTHERLAND, J. P., and R. H. KARLSON. 1977. Development and stability of the fouling community at Beaufort, North Carolina. *Ecol. Monogr.* 47:425–446.

SVERDRUP, H. U., M. W. JOHNSON, and R. H. FLEMING. 1942. *The Oceans*, Englewood Cliffs, N.J.: Prentice-Hall, Inc.

SWEDMARK, B. 1964. The interstitial fauna of marine sand. *Biol. Rev.* 39:1–42.

TAGHON, G. L., A. R. M. NOWELL, and P. A. JUMARS. 1980. Induction of suspension feeding in spionid polychaetes by high particulate fluxes. *Science* 21:562–564.

TAGHON, G. L., R. F. L. SELF, and P. A. JUMARS. 1978. Predicting particle selection by deposit-feeders: a model and its implications. *Limnol. Oceanogr.* 23:752–759.

TANADA, T. 1951. The photosynthetic efficiency of carotenoid pigments in *Navicula minima*. *Am. J. Bot.* 38:276–283.

TAYLOR, D. L. 1971. Ultrastructure of the "zooxanthella" *Endodinium chattonii* in situ. *J. Mar. Biol. Assoc. U.K.* 51:227–234.

―――. 1974. Symbiotic marine algae: taxonomy and biological fitness. In *Symbiosis in the Sea*, W. Vernberg (Ed.), pp. 245–262. Columbia: University of South Carolina Press.

TEAL, J. M. 1962. Energy flow in the salt marsh ecosystem of Georgia. *Ecology* 43:614–624.

TEAL, J. M., and J. KANWISHER. 1961. Gas exchange in a Georgia salt marsh. *Limnol. Oceanogr.* 6:388–399.

TENORE, K. R., R. B. HANSON, B. E. DORNSEIF, and C. N. WEIDERHOLD. 1979. The effect of organic nitrogen supplement on the utilization of different sources of detritus. *Limnol. Oceanogr.* 24:350–355.

TERUMOTO, I. 1964. Frost resistance in some marine algae from the winter intertidal zone. *Low Temp. Sci. (Japan) 22B:19–28.*

THEEDE, H. 1963. Experimentelle Untersuchungen über die Filtrationsleistung der Miesmuschel *Mytilus edulis* L. *Kiel. Meeresfors.* 19:20–41.

THISTLE, D. 1978. Harpacticoid dispersion patterns: implications for deep-sea diversity maintenance. *J. Mar. Res.* 36:377–397.

THOMPSON, T. E. 1960. Defensive adaptations in opisthobranchs. *J. Mar. Biol. Ass. U. K.* 39:123–134.

————. 1969. Acid secretion in Pacific Ocean gastropods. *Austr. J. Zool.* 17:755–764.

THORSON, G. 1946. Reproduction and larval development of Danish marine bottom invertebrates with special reference to the planktonic larvae in the sound (Øresund). Medd. Komm. Danm. Fiskeri-og Havunders. *Kobh. Ser. Plankt.* 4:1–523.

————. 1950. Reproductive and larval ecology of marine bottom invertebrates. *Biol. Rev.* 25:1–45.

————. 1966. Some factors influencing the recruitment and establishment of marine benthic communities. *Netherl. J. Sea Res.* 3:267–293.

THRESHER, R. E. 1978. Polymorphism, mimicry, and the evolution of the Hamlets (*Hypoplectrus*, Serranidae). *Bull. Mar. Sci.* 28(2):345–353.

TITMAN, D. 1976. Ecological competition between algae: experimental confirmation of resource-based competition theory. *Science* 192:463–464.

TREVALLION, A. 1971. Studies on *Tellina tenuis* Da costa. III. Aspects of general biology and energy flow. *J. Exp. Mar. Biol. Ecol.* 7:95–122.

TREVALLION, A., R. R. C. EDWARDS, and J. H. STEELE. 1970. Dynamics of a benthic bivalve. In *Marine Food Chains*, J. H. Steele (Ed.), pp. 285–295. Berkeley: University of California Press.

TRUEMAN, E. R. 1971. The control of burrowing and the migratory behavior of *Donax denticulatus* (Bivalvia: Tellinacea). *J. Zool. London* 165:453–469.

————. 1975. *The Locomotion of Soft-Bodied Animals*, pp. 1–200. Bristol, England: Arnold.

TRUEMAN, E. R., and A. D. ANSELL. 1969. The mechanism of burrowing into soft substrates by marine animals. *Oceanogr. Mar. Biol. Ann. Rev.* 7:315–366.

TUCKER, D. W. 1959. A new solution to the Atlantic eel problem. *Nature* 183:495–501.

TUNNICLIFFE, V. J., and M. J. RISK. 1977. Relationships between *Macoma balthica* and bacteria in intertidal sediments: Minas Basin, Bay of Fundy. *J. Mar. Res.* 35:499–507.

TURNER, J. T. 1977. Sinking rates of fecal pellets from the marine copepod *Pontella meadii*. *Mar. Biol.* 40:249–259.

TURPAEVA, E. P. 1954. Feeding and trophic classification of benthic invertebrates. *Trud. Inst. Oceanol.* (in Russian). 7:259–299.

UNDERWOOD, A. J. 1974. On models for reproductive strategy in marine benthic invertebrates. *Am. Nat.* 108:874–878.

USHAKOV, B. P. 1968. Cellular resistance adaptation to temperature and thermostability of somatic cells with special reference to marine animals. *Mar. Biol.* 1:153–160.

VADAS, R. L. 1977. Preferential feeding: an optimization strategy in sea urchins. *Ecol. Monogr.* 47:337–371.

VAHL, O. 1972. Efficiency of particle retention in *Mytilus edulis* L. *Ophelia* 10:17–25.

VALENTINE, J. W. 1966. Numerical analysis of marine molluscan ranges on the extratropical northeastern Pacific shelf. *Limnol. Oceanogr.* 11:198–211.

————. 1971. Resource supply and species diversity patterns. *Lethaia* 4:51–61.

————. 1973. *Evolutionary Ecology of the Marine Biosphere*, 511 pp. Englewood Cliffs, N.J.: Prentice-Hall, Inc.

————. 1976. Genetic strategies of adaptation. In *Molecular Evolution*, R. J. Ayala, (Ed.), pp. 78–94. Sunderland, Mass.: Sinauer Associates.

VALENTINE, J. W., and E. M. MOORES. 1972. Global tactonics and the fossil record. *J. Geol.* 80:167–184.

VALIELA, I., and J. TEAL. 1979. The nitrogen budget of a salt marsh ecosystem. *Nature* 280:652–656.

VALIELA, I., J. M. TEAL, S. VOLKMANN, D. SCHAFER, and E. J. CARPENTER. 1978. Nutrient and particulate fluxes in a salt marsh ecosystem: tidal exchanges and inputs by precipitation and groundwater. *Limnol. Oceanogr.* 23:798–812.

VALIELA, I., J. E. WRIGHT, J. M. TEAL, and S. B. VOLKMANN. 1977. Growth production and energy transformations in the salt-marsh killifish *Fundulus heteroclitus*. *Mar. Biol.* 40:135–144.

VANCE, R. R. 1973a. On reproductive strategies in marine benthic invertebrates. *Am. Nat.* 107:339–352.

———. 1973b. More on reproductive strategies in marine benthic invertebrates. *Am. Nat.* 107:353–361.

VAN VALEN, L. 1974. Predation and species diversity. *J. Theoret. Biol.* 44:19–21.

VASSALLO, M. T. 1969. The ecology of *Macoma inconspicua* (Broderip and Sowerby, 1829) in central San Francisco Bay. Part I. The vertical distribution of the Macoma community. *Veliger* 11:223–234.

VAUGHAN, V. W. 1915. The geological significance of the growth rate of the Floridian and Bahamian shoal-water corals. *J. Nat. Acad. Sci. U.S.A.* 5:591–600.

VENRICK, E. L. 1972. Small-scale distributions of oceanic diatoms. *Fish. Bull. U.S. Nat. Mar. Fish. Serv.* 70:363–372.

VERMEIJ, G. J. 1971a. Temperature relationships of some tropical Pacific intertidal gastropods. *Mar. Biol.* 10:301–314.

———. 1971b. Substratum relationships of some tropical Pacific intertidal gastropods. *Mar. Biol.* 10:315–320.

———. 1972a. Endemism and environment: some shore molluscs of the tropical Atlantic. *Am. Nat.* 106:89–101.

———. 1972b. Intraspecific shore-level size gradients in intertidal molluscs. *Ecology* 53:693–700.

———. 1973. Morphological patterns in high intertidal gastropods: adaptive strategies and their limitations. *Mar. Biol.* 20:319–346.

———. 1977. Patterns in crab claw size: the geography of crushing. *Syst. Zool.* 26:138–152.

VERWEY, J. 1952. On the ecology of distribution of cockle and mussel in the Dutch Waddensea, their role in sedimentation and the source of their food supply, with a short review of the feeding behavior of bivalve mollusks. *Arch. Neerl. Zool.* 10:172–239.

VINCE, S., I. VALIELLA, N. BACKUS, and J. M. TEAL. 1976. Predation by the salt marsh killifish *Fundulus heteroclitus* (L.) in relation to prey and habitat structure: consequences for prey distribution and abundance. *J. Exp. Mar. Biol. Ecol.* 23:255–256.

VINOGRADOV, M. E. 1968. *Vertical Distribution of the Oceanic Zooplankton,* 339 pp. Washington, D.C.: U.S. Department of Commerce. (Translated from Russian.)

VIRNSTEIN, R. W. 1977. The importance of predation by crabs and fishes on benthic infauna in Chesapeake Bay. *Ecology* 58:1199–1217.

VLYMEN, W. J. 1970. Energy expenditure of swimming copepods. *Limnol. Oceanogr.* 15:348–356.

VOGEL, S. and W. L. BRETZ. 1972. Interfacial organisms: passive ventilation in the velocity gradients near surfaces. *Science* 175:210–211.

WAINWRIGHT, S. A., and J. R. DILLON. 1969. On the orientation of seafans (genus Gorgonia). *Biol. Bull.* 136:130–139.

WAINWRIGHT, S. A., and M. A. R. KOEHL. 1976. The nature of flow and the reaction of benthic Cnidarians to it. In *Coelenterate Ecology and Behavior*, G. O. Mackie (Ed.), pp. 5–21. New York: Plenum.

WALKER, F. T., and W. D. RICHARDSON. 1955. An ecological investigation of *Laminaria cloustoni* Edm. (*L. Hyperborea* Fosl.). *J. Ecol.* 43:26–38.

WALNE, P. R. 1963. Observations of the food value of seven species of algae to the larvae of *Ostrea edulis*. *J. Mar. Biol. Ass. U. K.* 43:767–784.

WALSH, J. J. 1971. Relative importance of habitat variables in predicting the distribution of phytoplankton at the ecotone of the Antarctic upwelling ecosystem. *Ecol. Monogr.* 41:291–309.

———. 1975. A spatial simulation model of the Peru upwelling ecosystem. *Deep Sea Res.* 22:201–236.

———. 1976. Models of the sea. In *The Ecology of the Seas*, D. H. Cushing and J. J. Walsh (Eds.), pp. 388–407. Oxford, U. K.: Blackwell Scientific Publications.

WARNER, G. F., and J. D. WOODLEY. 1975. Suspension-feeding in the brittle-star *Ophiothrix fragilis*. *J. Mar. Biol. Ass. U. K.* 55:199–210.

WARNER, R. R. 1975. The adaptive significance of sequential hermaphroditism in animals. *Am. Nat.* 109:61–82.

WATERBURY, J. B., S. W. WATSON, R. R. L. GUILLARD, and L. E. BRAND. 1979. Widespread occurrence of a unicellular, marine, planktonic cyanobacterium. *Nature* 277:293–294.

WATERMAN, T. H., R. F. NUNNEMACHER, F. A. CHACE, and G. L. CLARKE. 1939. Diurnal migrations of deep-water plankton. *Biol. Bull.* 76:256–279.

WEBB, K. L., R. E. JOHANNES, and S. J. COWARD. 1971. Effects of salinity and starvation on release of dissolved free amino acids by *Dugesia dorotocephala* and *Bdelloura candida* (Platyhelminthes, Turbellaria). *Biol. Bull. Woods Hole* 141:364–371.

WEBER, J. N., P. DEINES, E. W. WHITE, and P. H. WEBER. 1975. Seasonal high and low density bands in reef coral skeletons. *Nature* 255:697.

WEBER, J. N., and P. M. WOODHEAD. 1970. Carbon and oxygen isotope fractionation in the skeleton carbonate of reef-building corals. *Chem. Geol.* 6:93–117.

WEBSTER, T. J., M. A. PARANJAPE, and K. H. MANN. 1975. Sedimentation of organic matter in St. Margaret's Bay, N. S. *J. Fish. Res. Bd. Canada* 32:1399–1407.

WEISER, W., and J. KANWISHER. 1961. Ecological and physiological studies on marine nematodes from a small salt marsh near Woods Hole, Massachusetts. *Limnol. Oceanogr.* 6:262–270.

WELLS, J. W. 1957. Coral reefs. In *Treatise on Marine Ecology and Paleoecology*, I, Ecology, J. W. Hedgpeth (Ed.), pp. 609–631. Geol. Soc. America Mem. 67, vol. 7.

WERNER, D. 1977. Silicate metabolism. In *The Biology of Diatoms*, D. Werner (Ed.), pp. 110–149. Oxford, U. K.: Blackwell Scientific Publications.

WERNER, E. and B. WERNER. 1954. Uber den Mechanismus des Nahrungserwerbs der Tunicaten speciell der Ascidien. *Helgol. Wiss. Meeresunters.* 5:57–92.

WETHEY, D. S., and J. W. PORTER. 1976. Sun and shade differences in productivity of reef corals. *Nature* 262:281–282.

WEYMOUTH, F. W., and H. C. MCMILLIN. 1931. The relative growth and mortality of the Pacific razor clam (*Siliqua patula* Dixon) and their bearing on the commercial fishery. *Bull. U.S. Bur. Fish.* 46:543–567.

WHEELER, P. A., B. B. NORTH, and G. S. STEPHENS. 1974. Amino acid uptake by marine phytoplankters. *Limnol. Oceanogr.* 19:249–259.

WHELAN, J. K. 1977. Amino acids in a surface sediment core of the Atlantic abyssal plain. *Geochim. Cosmochim. Acta* 41:803–810.

WHITLATCH, R. B. 1974. Food resource partitioning in the deposit-feeding polychaete *Pectinaria gouldi. Biol. Bull.* 147:227–235.

———. 1981. Patterns of resource utilization and coexistence in marine intertidal deposit-feeding communities. *J. Mar. Res.* 38:743–765.

WHITTAKER, R. H. and P. P. FEENEY. 1971. Allelochemics: chemical interactions between species. *Science* 171:757–770.

WIJSMAN–BEST, M. 1972. Systematics and ecology of New Caledonian Faviinae (Coelenterata-Scleractinia). *Bijdragentot de Dierkunde* 42:1–71.

WILKINS, N. P., and N. F. MATHERS. 1975. Phenotypes of phosphoglucose isomerase in some bivalve molluscs. *Comp. Biochem. Phys.* 48B:599–611.

WILLIAMS, G. C. 1966. Natural selection, the costs of reproduction and a refinement of Lack's principle. *Am. Nat.* 100:687–690.

WILLIAMS, G. C., R. K. KOEHN, and J. B. MITTON. 1973. Genetic differentiation without isolation in the American eel, *Anguilla rostrata. Evolution* 27:192–204.

WILLIAMS, J. G. 1979. The influence of adults on the settlement of spat of the clam, *Tapes japonica. J. Mar. Res.* 38:729–741.

WILSON, D. P. 1937. The influence of the substratum on the metamorphosis of *Notomastus* larvae. *J. Mar. Biol. Assoc. U. K.* 22:227–243.

———. 1948. The relationship of the substratum to the metamorphosis of *Ophelia* larvae. *J. Mar. Biol. Assoc. U. K.* 27:723–760.

———. 1954. The attractive factor in the settlement of *Ophelia bicornis* Savigny. *J. Mar. Biol. Assoc. U. K.* 33:361–380.

———. 1955. The role of micro-organisms in the settlement of *Ophelia bicornis* Savigny. *J. Mar. Biol. Assoc. U. K.* 34:513–543.

WINN, H. E., M. SALMON, and N. ROBERTS. 1964. Sun-compass orientation by parrot fishes. *Z. Tierpsychol.* 21:798–812.

WISELY, B. 1960. Observations on the settling behaviour of larvae of the tubeworm *Spirorbis borealis* Daudin (Polychaete). *Austrl. J. Mar. Freshw. Res.* 11:55–72.

WOOD, E. J. F. 1953. Heterotrophic bacteria in marine environments of eastern Australia. *Aust. J. Mar. Freshw. Res.* 4:160–200.

WOOD, L. 1968. Physiological and ecological aspects of prey selection by the marine gastropod *Urosalpinx cinerea* (Prosobranchia:Muricidae). *Malacologia* 6:267–320.

WOOD, L., and W. J. HARGIS. 1971. Transport of bivalve larvae in a tidal estuary. In *Fourth European Marine Biology Symposium,* D. J. Crisp (Ed.), pp. 29–44. London: Cambridge University Press.

WOODHEAD, A. D. 1977. Reproductive ecology of spiny dogfish, *Squalus acanthias. Bull. Mount Desert Is. Biol. Lab.* 16:103–106.

WOODHEAD, P. M. J. 1971. Surveys of coral recolonization on reefs damaged by starfish and by a cyclone. In *Appendix E. Rept. of Commission on the Crown of Thorns Starfish,* pp. 34–40. Canberra, Australia: Prime Ministers Office.

WOODIN, S. A. 1974. Polychaete abundance patterns in a marine soft-sediment environment: the importance of biological interactions. *Ecol. Monogr.* 44:171–187.

———. 1976. Adult-larval interactions in dense infaunal assemblages: patterns of abundance. *J. Mar. Res.* 34:25–41.

———. 1978. Refuges, disturbance, and community structure: a marine soft-bottom example. *Ecology* 59:274–284.

WOODWELL, G. M., C. A. S. HALL, D. E. WHITNEY, and R. A. HOUGHTON. 1979. The Flax Pond ecosystem study: exchanges of inorganic nitrogen between an estuarine marsh and Long Island Sound. *Ecology* 60:695–702.

WOODWELL, G. M., D. E. WHITNEY, C. A. S. HALL, and R. A. HOUGHTON. 1977. The Flax Pond ecosystem study: exchanges of carbon between a salt marsh and Long Island Sound. *Limnol. Oceanogr.* 22:833–838.

WUST, G., W. BROGMUS, and E. N. NOODT. 1954. Die zonale Verteilung von Salzgehalt, Niederschlag, Verdungstung, Temperatur und Dichte an der Oberflache der Ozeane. *Kieler Meeresforsch* 10:137–161.

WYATT, T. 1972. Some effects of food density on the growth and behavior of plaice larvae. *Mar. Biol.* 14:210–216.

YINGST, J. Y., and D. C. RHOADS. 1978. Seafloor stability in central Long Island Sound: Part II. Biological interactions and their potential importance for seafloor erodibility. In *Estuarine Interactions*, M. L. Wiley (Ed.), pp. 245–260. New York: Academic Press.

YONGE, C. M. 1928. Structure and function of the organs of feeding and digestion in the septibranchs, *Cuspidaria* and *Poromya*. *Phil. Trans. Roy. Soc. London* 26B:221–263.

———. 1930. Studies on the physiology of corals. I. Feeding mechanisms and food. *Sci. Repts. Great Barrier Reef Exped.* 1:13–57.

———. 1931. The significance of the relationship between corals and zooxanthellae. *Nature* 128:309–310.

———. 1936a. The protobranchiate mollusca: a functional interpretation of their structure and evolution. *Phil. Trans. Roy. Soc. London* 230:79–147.

———. 1936b. Mode of life, feeding, digestion, and symbiosis with zooxanthellae in the Tridacnidae. *Sci. Rept. Great Barrier Reef Exped.* 1:283–321.

———. 1940. The biology of reef-building corals. *Sci. Repts. Great Barrier Reef Exped., 1928–1929* 1:353–361.

———. 1950. On the structure and adaptations of the Tellinacea, deposit-feeding eulamellibranchia. *Phil. Trans. Roy. Soc. London* 234B:29–76.

———. 1963. Rock-boring organisms. In *Mechanisms of Hard Tissue Destruction*. R. F. Sognnaes (Ed.), Washington, D.C.: American Association for the Advancement of Science.

YONGE, C. M., and A. G. NICOLLS. 1931. Studies on the physiology of corals. V. The effect of starvation in light and in darkness on the relationship between corals and zooxanthellae. *Sci. Rept. Great Barrier Reef Exped.* 1:177–211.

YOUNG, D. K. 1971. Effects of infauna on the sediment and seston of a subtidal environment. *Vie et Milieu* 22(Suppl.):557–571.

ZARET, T. M. 1972. Predator–prey interaction in a tropical lacustrine system. *Ecology* 53:248–257.

ZARET, T. M., and J. S. SUFFERN. 1976. Vertical migration in zooplankton as a predator avoidance mechanism. *Limnol. Oceanogr.* 21:804–813.

ZEUTHEN, E. 1953. Oxygen uptake as related to body size in organisms. *Quart. Rev. Biol.* 28:1–12.

ZIEMAN, J. C. 1975. Quantitative and dynamic aspects of the ecology of Turtle Grass, *Thalassia testudinum*. In *Eustuarine Research,* L. E. Cronin (Ed.), Vol. 1, pp. 541–579. New York: Academic Press.

ZIPSER, E. and G. J. VERMEIJ. 1978. Crushing behavior of tropical and temperate crabs. *J. Exp. Mar. Biol. Ecol.* 31:155–172.

ZOBELL, C. E. 1938. Studies on the bacterial flora of marine bottom deposits. *J. Sed. Petrol.* 8:10–18.

———. 1946. Studies in redox potential of marine sediments. *AAPG Bull.* 30:477–513.

ZOBELL, C. E., and C. B. FELTHAM. 1938. Bacteria as food for certain marine invertebrates. *J. Mar. Res.* 1:312–327.

index

DATE DUE

DATE DUE		
MAR 0 1 1990		
OCT 1 8 1990		
APR 1 7 1994		
NOV 0 2 1996		
GAYLORD		PRINTED IN U.S.A